Colorimetric Determination
of Nonmetals

CHEMICAL ANALYSIS

A SERIES OF MONOGRAPHS ON
ANALYTICAL CHEMISTRY AND ITS APPLICATIONS

Editors

P. J. ELVING · J. D. WINEFORDNER

Editor Emeritus: **I. M. KOLTHOFF**

VOLUME 8

Second Edition

A WILEY-INTERSCIENCE PUBLICATION

JOHN WILEY & SONS

New York / Chichester / Brisbane / Toronto

Colorimetric Determination of Nonmetals

Editors

DAVID F. BOLTZ
(deceased)

JAMES A. HOWELL
Western Michigan University

Authors

David F. Boltz
K. L. Cheng
H. K. L. Gupta
Larry G. Hargis
W. J. Holland
James A. Howell
Robert J. Jakubiec
Ralph A. Johnson
Charles H. Lueck
Stephen Megregian

B. G. Messick
James M. Pappenhagen
Gordon A. Parker
Gordon D. Patterson, Jr.
G. Victor Potter
R. A. Smith
Michael J. Taras
Louis A. Trudell
E. R. Wright
Bennie Zak

A WILEY-INTERSCIENCE PUBLICATION

JOHN WILEY & SONS

New York / Chichester / Brisbane / Toronto

Library of Congress Cataloging in Publication Data

Boltz, David Ferdinand, 1918–1976, ed.

    Colorimetric determination of nonmetals.

    (Chemical analysis; v. 8)
    "A Wiley-Interscience publication."
    Includes indexes.
    1. Colorimetric analysis. I. Howell, James A.,
joint ed. II. Title. III. Series.

QD113.B6   1978   543'.085   77-12398
ISBN 0-471-08750-5

Printed in the United States of America

10 9 8 7 6 5 4 3 2 1

PREFACE

Professor David F. Boltz, editor of the first edition of this book, passed away on January 11, 1976. Prior to his death he had done much of the work toward this revision. As a former student and close friend of Professor Boltz, I accepted the responsibility for completing the final editing. The many contributions to analytical chemistry made by Professor Boltz over the years are not likely to be soon forgotten. This revision of the first edition of this book was one of his last major contributions reflecting his dedication to the advancement of analytical chemistry and his gratitude to all those practicing analytical chemists, whose contributions have made this book possible.

This second edition, devoted exclusively to the colorimetric determination of traces of nonmetals, is the result of widespread acceptance and success of the first edition, the development of new and improved methods for the determination of nonmetals in the past twenty years, and the continued interest and use made of these methods in solving many practical analytical problems. New chapters on carbon and oxygen have been included, in this edition, and the introductory chapter, "Principles and Practices of Colorimetric Analysis," has been omitted.

Most of the procedures presented in this book have been compiled from the literature by the contributing authors. It was not practical for them to thoroughly explore al the potential hazards that might be involved in the implementation of the procedures described. This responsibility must necessarily rest with the readers, who is urged to seriously consider the potential hazards of all procedures and chemicals before implementing any of the methods given in this book.

Professor Boltz was appreciative of the authors' dedication and cooperation in preparing this revised edition. I am equally grateful, and would also like to express my appreciation to Professor George Schenk of Wayne State University for his invaluable assistance in helping complete the final steps in the preparation of this edition. Completion of the manuscript would not have been possible without the cooperation, encouragement, and patience of Mrs. Dorothy Boltz as well as David and Betsy Boltz.

JAMES A. HOWELL

Kalamazoo, Michigan
September 1977

v

CONTENTS

Colorimetric Determination
of Nonmetals

BORON

D. F. BOLTZ and H. K. L. GUPTA[1]

Department of Chemistry, Wayne State University, Detroit, Michigan

Boron, which constitutes only about 0.001% of the earth's crust, is an element of considerable commercial importance, often making its determination at microgram levels highly advantageous. The incorporation of 0.005% to 0.0005% of boron in steel produces hardened steels. However, traces of boron at the ppm level in the fabricating materials used in nuclear reactors is of very critical concern because of the high cross-sectional area of boron for thermal neutrons.

[1] Present address: Texasgulf Canada Ltd., Timmins, Ontario, Canada.

Aluminum wire having high electrical conductivity often contains boron. Boron, although considered an essential element for plant growth, is considered harmful at higher concentrations. Thus, a concentration of boron in irrigation water exceeding 2 ppm is undesirable. The use of borax in cleansing formulations and boric acid as a preservative often results in the necessity of analyzing waters and foods for traces of boron.

In recent years, there has been considerable research on the development of new and improved spectrophotometric methods for the determination of traces of boron. A classical approach has involved the formation of colored borate esters with derivatives of anthraquinone in concentrated sulfuric acid solution. Quinalizarin, carminic acid, and 1,1'-dianthrimide are typical reagents belonging to this category. A second type of color-forming reaction is characterized by the formation of a colored complex in a virtually nonaqueous solution when an unsaturated hydroxy ketone, for example, curcumin, is the chromogenic agent. A new development has been the use of cationic dyes and complex ions to form ion pair complexes which are extractable from aqueous solution with immiscible organic solvents. Progress has also been made in the separation of boron by ion exchange and extraction methods. As in all trace determinations, special care is necessary to avoid contamination from boron in reagents or origination from borosilicate glassware. Often, compensation techniques can be utilized to minimize such errors. Quartz, platinum, Teflon, and polyethylene are satisfactory materials for many items of apparatus to be used in the determination of boron.

I. SEPARATIONS

Inasmuch as colorimetric methods are used primarily to determine trace amounts of boron, the specific constituents comprising the matrix of the sample will determine whether or not a preconcentration or an analytic separation is necessary.

A. PRECIPITATION METHOD

Aluminum can be separated from boron by the precipitation of aluminum with 8-quinolinol from ammoniacal solution (88). Other cations such as iron(III), nickel, zinc, and titanium(IV) may also be removed by precipitation with this reagent (125). The removal of interfering ions by precipitation often leads to loss of boron through coprecipitation. The precipitation of barium borotartrate from ammoniacal solution has been used to isolate boron (13,15,50) from silicate, iron, aluminum, molybdenum, and vanadium (147). The tartrate obtained upon dissolution of the precipitate has been used to determinine boron colorimetrically (14). The removal of large amounts of fluoride by precipitation of thorium fluoride has been recommended (11).

B. DISTILLATION METHOD

Whenever it is necessary to isolate traces of boron from relatively large concentrations of diverse ions derived from the matrix of the sample, one of the most suitable methods involves the volatilization of methyl borate, $B(OCH_3)_3$. This distillation method, first utilized by Gooch (51,52) and by Rosenbladt (119) in 1887, has been studied and used very extensively (41,85,86,105).

Methyl borate is distilled as an azeotrope with methanol at $54.6°C$. The composition of the azeotrope is 75% by weight methyl borate, essentially an equimolar mixture of methyl borate and methanol:

$$H_3 BO_3 + 4CH_3 OH \longrightarrow [B(OCH_3)_3 + CH_3 OH] + 3H_2 O$$

The boiling points for methyl borate and methanol are $64°$ and $68°C$, respectively (126). The methyl borate is collected in a basic solution of either calcium hydroxide or sodium hydroxide so the hydrolysis of the ester generates the borate ion:

$$B(OCH_3)_3 + 3OH^- \longrightarrow 3CH_3 OH + BO_3^{3-}$$

The addition of glycerol to the sodium hydroxide solution has been suggested in order to reduce the loss of boron when this basic solution is evaporated to dryness prior to the development of a colored system (132). The removal of the glycerol by ignition is necessary, especially, if the curcumin colorimetric method is to be used. The addition of ammonia to the saponifying solution has also been recommended (28,29).

The percentage recovery of boron at the microgram level in the distillation process is probably less than 100%, although complete recovery has been reported (75). The conflicting reports in the literature can be attributed to the following possible sources of error. A low recovery results from incomplete volatilization of methyl borate. Large amounts of gelatinous silica presumably are a cause of incomplete recovery (85). The possible loss of boric acid or boron trifluoride may exist if a condenser is not used in certain dissolution steps. Traces of boric acid may be lost in the evaporation of the distillate to remove water prior to color development. Often, an apparent incomplete recovery of boron may be due to a diminution of sensitivity arising from a difference in ionic strength or composition of the sample solution relative to the standard solutions. Sometimes, the extent of incomplete recovery may not be apparent as a result of preparing the calibration graph from samples to which boron was added prior to distillation. Onishi, Ishiwatari, and Nagai (105) obtained an 87%

BORON

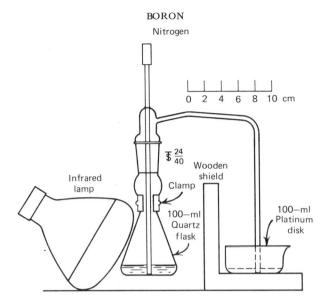

Fig. 1. Quartz distillation apparatus as described by Luke and Flaschen (87). (Reproduced through permission of *Analytical Chemistry*.)

recovery in isolating 0.2 to 1 μg boron from 1 g uranium as uranyl sulfate. Thus, a 5 to 15% correction factor may be necessary for a specific material, but this factor should be rather constant. The deleterious effect of fluoride ion in the distillation of methyl borate has been eliminated by Gaestel and Hure (48) by the addition of aluminum chloride.

Two general types of distillation apparatus are used. The apparatus shown in Fig. 1 is designed for the distillation of boron in one installment, while the cyclic apparatus depicted in Fig. 2 permits the use of small amounts of methanol which are recycled. In the cyclic distillation apparatus, methyl borate in the distillate is hydrolyzed and the liberated methanol is returned to the distilling flask. Trentelman and Welthuijsen (141) employed a novel technique of distilling the methyl borate directly into an anion exchange column of Dowex 2 (OH form). The ester is hydrolyzed, the borate is retained on ion exchange resin, and the methanol is returned to distillation flask.

The following general procedure is given as a guide and is applicable to the apparatus shown in Fig. 1. Transfer 1–2 ml of sample solution containing 1 μg or less boron to the clean and dry 100-ml conical quartz flask and then add 1.0 ml low-boron sulfuric acid, 30 ml redistilled "acetone-free" methanol, and several silicon carbide chips. Attach distillation head and, by means of a polyethylene graduated cylinder or pipet, transfer 10 ml of an approximately 0.05 M calcium hydroxide suspension to a 100-ml platinum dish. (If a platinum dish is not available, a Teflon evaporating dish or a Vycor glass evaporating dish

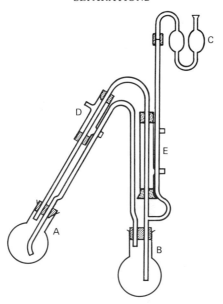

Fig. 2. Cyclic distillation apparatus. (This apparatus is commercially available from the Scientific Glass Apparatus Co., Inc., and constructed according to ASTM specifications.)

may be used.) Adjust the sidearm or tip of the condenser to dip to within 1 mm of the bottom of the dish. While passing nitrogen through the flask at the rate of 50 ml per min, employ a 500-W infrared lamp or heating mantle to heat the distillation flask to about 100°C. Continue the distillation until the methanol has been distilled.

C. EXTRACTION METHOD

Boron may be separated from many cations by conversion to the tetrafluoroborate ion, followed by its extraction as an ion pair complex with the tetraphenylarsonium ion with chloroform (38). Treatment with sodium hydroxide and evaporation of the chloroform extract in a platinum dish with an infrared lamp gives a residue that can be dissolved in trichloroacetic acid and used for the colorimetric determination of boron by the curcumin method (30). The tetrafluoroborate ion also forms extractable complexes with many cationic dyes, such as methylene blue (37,112), monomethylthionin (110), rhodamine B (107), methyl violet (113), and malachite green (77). 1,2-Dichloroethane is a satisfactory extractant for many of these dye—tetrafluoroborate complexes. A ferroin—tetrafluoroborate complex can be extracted with *n*-butyronitrile (7).

Materova and Rozhanskaya (92) quantitatively extracted quinoline tetrafluoroborate from an $8M$ hydrochloric acid solution with methyl isobutyl ketone and from $3M$ hydrochloric acid using cyclohexone. Tantalum, tin, molybdenum, arsenic, cadmium, titanium, niobium, and zinc are coextracted. Fukushi and Kakita (47) extracted interfering ions with methyl isobutyl ketone as a preliminary step to determining the boron in iron and steel by the methylene blue method. Boric acid forms a complex with 2-ethyl-1,3-hexanediol which is extractable with chloroform (1) or methyl isobutyl ketone (94). Methyl isobutyl ketone has also been used to extract submilligram quantities of boron as the tetrabutylammonium–tetrafluoroborate complex (90). Ethanol has been found to extract borate from a sodium chloride residue (122).

D. ION EXCHANGE METHOD

The ion exchange method of effecting analytic separations is particularly attractive in the determination of boron because boric acid is a weak acid and, therefore, should not be retained by either a strong acid cation exchange resin or by a weak acid anion exchange resin. Hence, a strong acid cation exchange resin can be used to remove cations (78,91,103), and a weak anion exchange resin can be used to remove anions of strong acids. The ion exchange method is inherently much faster and simpler than the distillation method.

Calkins and Stenger (21) used Dowex 50-X8 (20–50 mesh) to remove peroxytitanium(IV) cations from solution prior to using the carminic acid coloremetric method to determine boron in effluents. Capelle (25) also used this same resin (100–200 mesh) to remove iron(III), aluminum, titanium(IV), uranyl, and zirconium(IV) ions before determining boron in steel by the colorimetric azomethine method (128). Zeo Carb-225 was used by Towndrow and Webb (140) to separate boron from aluminum and other elements in silicon–aluminum alloys. The 1,1′-dianthrimide method was then used to determine the boron. In determining traces of boron in rocks and minerals, Fleet (45) added mannitol to the acidified solution of the fusion mixture to inhibit loss of boric acid due to adsorption by the precipitated silica and then employed the batch (noncolumn) ion exchange technique to remove cations with Dowex 50. After removal of the resin by filtration, the boron in an aliquot of the filtrate was determined by the carminic acid method. Vasilevskaya and Lenskaya (144) removed interfering cations with a strong acid cation exchange resin, KU-2, in the colorimetric determination of boron by the salicylic acid–crystal violet method. In using either Dowex 50 or Amberlite IR-100, Martin and Hayes (91) found that boric acid could be separated from iron(III), aluminum, beryllium, magnesium, zinc, nickel, tin(IV), titanium, uranium, and zirconium.

The following general procedure is applicable to the use of a strong acid cation exchange resin for separating boron as boric acid. Adjust the solution to pH 2 using $2N$ sulfuric acid or $2N$ sodium hydroxide and then transfer the solution to a 50-cm polyethylene tube (I.D. = 0.5 in.) filled with a strong acid cation exchange resin in the hydrogen ion form (approximately 50 mesh). This column should be prewashed with 50 ml 1:4 sulfuric acid and then with 200 ml distilled water. Adjust the stopcock so that the flow rate is about 1 ml per minute while using about 50 ml $0.1N$ sulfuric acid to rinse the column. Make the effluent basic with a $5N$ sodium hydroxide solution and evaporate carefully to dryness in a platinum dish.

Strong base anion exchange resins, for example, Dowex 1 or Amerblite IRA-400, remove borates as well as other anions of weak and strong acids. Wolszon, Hayes, and Hill (148) reported that the order of elution from an Amberlite IRA-400 column, initially in OH form, is borate, chloride, nitrate, arsenite, phosphate and arsenate. Schutz (127) reported in the analysis of fertilizers that phosphate can be separated from borate by eluting through a strong base anion exchange resin column, Amberlite IRA-400, in the formate form. However, this separation involved high concentrations of boron (>10 mg) and phosphate. A strong base anion exchange resin, Dowex 1-X10, saturated with gluconic acid may be washed with $0.1M$ sodium chloride to remove borate prior to the colorimetric determination of boron (26). Maximum exchange of borate or boric acid with a strong base anion exchange resin takes place in the pH range of 7.5–9.5 according to Everest and Popiel (44), with no exchange taking place below pH 5. By conversion of boric acid to tetrafluoroboric acid, the exchange with a strong base exchange resin in OH form is very high in the pH range of 2–6 (92). Muzzarelli (98) separated boron from uranium by retaining the borate on a strong base anion exchanger, Dowex 1-X8, (100–200 mesh) in Cl^- form, while the uranyl nitrate passes through the column. The borate was eluted with $5M$ hydrochloric acid.

A mixed-ion exchange bed consisting of a strong acid cation exchange resin and a weak acid anion exchange resin as a means of separating cations and anions of strong acids from borate has been investigated (148). Because of the slow elution of boron from the resin itself, the use of a specific flow rate and the application of a correction for boron coming from the resin were necessary in determining boron by the carminic acid method (23). Cerrai and Testa (26) employed a mixed resin bed with boron being separated from silicate prior to photometric determination by passing the solution successively through a strong base anion exchanger and strong acid cation exchanger in a procedure developed by Ryabchikov and Kuril'chikov (121). Lang (80) removed interfering ions in fertilizers by successive elutions of borate through cation and anion exchange resins and then determining the boron by the carminic acid method. Muto (97) used successively a strong acid cation resin, Amberlite IR-120, in the hydrogen ion form, and a strong base anion resin, Amberlite IRA-400, in the chloride ion form, in a procedure for the spectrophotometric determination of boron.

Ion exchange membranes coupled with electrolysis serve to separate cations and the anions from borates. Thus, Morrison and Rupp (95) removed sodium ions, resulting from the dissolution of silicon in sodium hydroxide, from borate and silicate by employing a cation-permeable membrane, Amberplex C-1, to separate the analyte and catholyte compartments of a special polyethylene electrolysis cell. Logie (83) used this same technique in the colorimetric determination of boron in sodium metal by the curcumin method. When manganese, aluminum, and chromium are also present, their anions in basic solution remain in the analyte compartment. Therefore, acidification of the analyte and the use of a strong acid cation ion exchange column is recommended.

E. ELECTRODEPOSITION METHOD

The mercury cathode may be used to separate bismuth, copper, cadmium, cobalt, chromium, gallium, germanium, gold, iron, molybdenum, nickel, silver, tin, and zinc from a dilute sulfuric acid solution containing boric acid. Aluminum, titanium, uranium, vanadium, and zirconium are not deposited at the mercury cathode. Bush and Higgs (20) used a mercury cathode to remove iron, nickel, cobalt, and copper in order to isolate boric acid prior to titration of the mannitol–boric acid complex. Microgram quantities of boron in nickel sheet were determined colorimetrically following the removal of nickel by electro-deposition at the mercury cathode. By using the nickel sample as the anode, dissolution of the sample and separation were performed simultaneously, with the total salt concentration of resultant solution being kept to a minimum (27). Green (55) removed interfering metals by electrolysis with a mercury cathode prior to determining boron in cast iron by the chromotrope method. The use of a water cooled electrolysis cell is recommended.

II. METHODS OF DETERMINATION

A. CURCUMIN METHOD

Curcumin is an extensively used and very sensitive organic reagent for the detection and determination of boron. This chromogenic agent derives its name from curcuma, the rhizome of a tropical species (*C. longa*), or turmeric. Curcumin is isolated from the alcoholic extract of the roots of this plant. Hence, the alternate name "turmeric yellow" is sometimes used. Boric acid forms two different red dyes, rosocyanin and rubrocurcumin in its reaction with curcumin, depending primarily on whether oxalic acid is absent or present.

Boric acid reacts with curcumin, when protonated by a mineral acid, to form a red dye, rosocyanin. This reaction is rather slow and is inhibited by water (123). Because of the 2:1 ratio for this curcumin–boron complex (39), rosocyanin has inherently a higher molar absorptivity than rubrocurcumin.

Silverman and Trego (129) proposed a 2:1 curcumin:boric acid complex. Hayes and Metcalfe (68) found evidence for a 3:1 curcumin:boric acid complex. The following formula was proposed by Spicer and Strickland to illustrate the "structure" for rosocyanin (133).

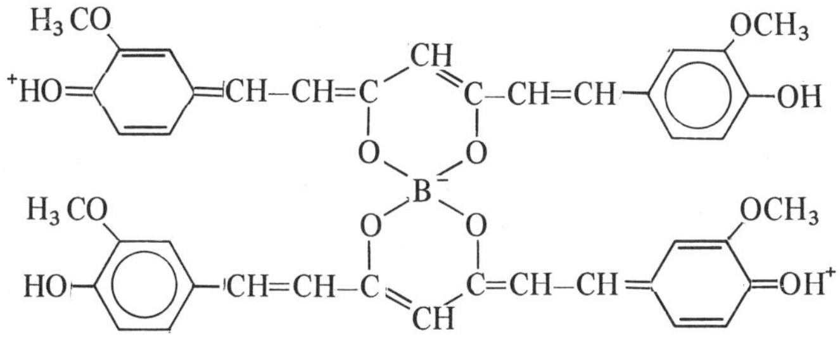

In the presence of oxalic acid, a red 2:2:2 complex of curcumin, boric acid, and oxalic acid is formed, called rubrocurcumin. Evaporation of the reaction mixture to dryness is necessary for maximum color development. Water is especially deleterious to the formation of rubrocurcumin; and if a mineral acid is also present, the concomitant formation of rosocyanin in the solution could be expected. The following structure has been proposed by Spicer and Strickland (134) to account for the cationic nature of rubrocurcumin in an ion exchange study and its slight hypsochromic shift as compared to the more ionic rosocyanin.

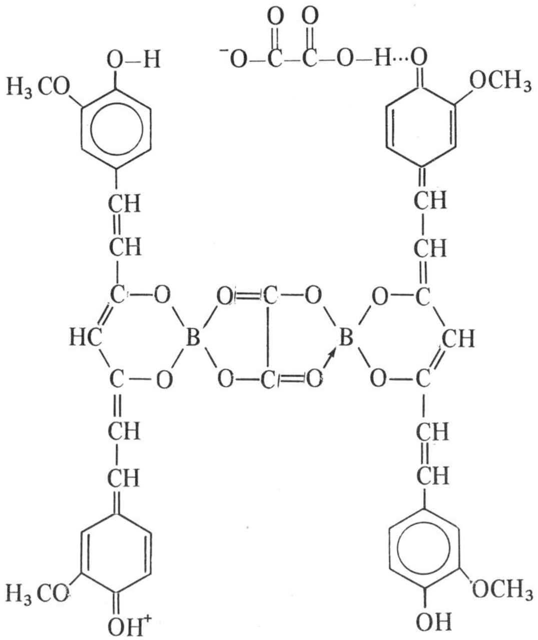

1. Curcumin (Rosocyanin) Method

The sensitivity of this method is highly dependent upon the presence of water and the amount of excess curcumin remaining in protonated state. The elimination of water and the minimization of the absorptivity due to the excess reagent are of critical importance. Careful evaporation of the sample solution (87,132), use of a homogeneous phenol—acetic acid melt (56), use of acetic anhydride with hydrochloric acid as catalyst (32), use of propionic anhydride with oxalyl chloride as catalyst (143), or solvent extraction with a methyl isobutyl ketone—chloroform—phenol mixture (139) have been proposed as approaches to circumvent the erratic results obtained with appreciable and variable concentrations of water in the reaction mixture. Hayes and Metcalfe (68) diluted the reaction mixture with ethanol to decrease the acidity and reduce the color due to the excess curcumin, while Grinstead and Snider (58) added ammonium acetate to decrease the acidity and thereby increase the sensitivity by avoiding dilution. Germanium(IV) nitrate, nitrite, and fluoride ions are the main interferences. However, if over 20 μg chromate, vanadate, chlorate, iodate, iron(III), or aluminum is present, a preliminary separation or pretreatment procedure should be considered (123). The following procedure is essentially that developed by Uppstrom which eliminates water through the use of propionic anhydride, destroys excess protonated curcumin by the addition of acetate ion, and permits the option of solvent extraction for sample solutions containing less than 0.01 ppm boron.

Special Solutions for Procedures A1

Curcumin Solution. This solution should be freshly prepared weekly by dissolving 0.125 g curcumin in 100 ml glacial acetic acid, and should be stored in a polyethylene bottle.

Sulfuric—Acetic Acid Solution. Mix equal volumes of concentrated sulfuric acid (sp. gr. 1.84; 98% H_2SO_4) and glacial acetic acid.

Buffer Solution "C". Dissolve 250 g ammonium acetate and 300 ml glacial acetic acid in distilled water and dilute to 1 liter. Store this solution in a polyethylene bottle.

Standard Buffer Solution. Mix 90 ml 95% ethanol, 180 g ammonium acetate, and 135 ml glacial acetic acid. Dilute to 1 liter and store in a polyethylene bottle.

Extractant. Mix 100 ml methyl isobutyl ketone, 150 ml chloroform, and 1 g phenol and store in a dark bottle in a refrigerator.

Propionic Anhydride

Oxalyl Chloride

Procedure A 1a (143)

Transfer a 1.00-ml sample of aqueous solution containing 0.2–1 µg boron to a small polyethylene beaker. Add 2.0 ml glacial acetic acid and 5.0 ml propionic anhydride and mix by swirling. Add dropwise 0.50 ml oxyalyl chloride and allow to stand for 30 min. Cool to room temperature and add 4.0 ml of the sulfuric–acetic acid solution and 40 ml of the curcumin solution, mix thoroughly, and allow to stand for 45 min. Now add 20 ml of the standard buffer solution, mix thoroughly, and cool to room temperature before measuring the absorbance at 545 nm in 1.000-cm cells against a reagent blank solution.

Procedure A1b

Transfer 5.00 ml aqueous sample solution containing 0.015–0.075 µg boron to a polyethylene beaker, add 10.0 ml glacial acetic acid and 25.0 ml propionic anhydride, and mix by swirling. Add 2.0 ml oxalyl chloride and allow to stand for 30 min. Cool to room temperature, add 15.0 ml of the sulfuric–acetic acid solution and 15.0 ml of the curcumin solution, mix thoroughly, and allow to stand for 4 hr.

Transfer the reaction mixture to a 250-ml separatory funnel using 100 ml distilled water to effect a quantitative transfer. Add 5.0–10.0 ml of the extractant, stopper, and shake carefully. Centrifuge if an opalescence develops and separate the two layers. Transfer by pipet a portion of the colored layer to a 5.000-cm absorption cell and measure the absorbance at 545 nm against a reagent blank solution.

Procedure A1c (58)

Transfer by microburet 0.50 ml aqueous sample containing 0.1–1.5 µg boron to a 60-ml polyethylene bottle, add 3.00 ml of the curcumin solution and 3.00 ml of the sulfuric–acetic acid solution, and mix thoroughly by swirling. Allow the solution to stand in the dark for 2.5 hr. With a buret, add 15.0 ml buffer solution "C" mix thoroughly, and measure the absorbance at 555 nm against a reagent blank solution in 1.000-cm cells.

2. Curcumin (Rubrocurcumin) Method

Although this method is less sensitive then the curcumin (rosocyanin) method, the rapid formation of the rubrocurcumin, the use of a nonsulfuric acid medium, and its applicability to methyl borate distillates has resulted in widespread use of the curcumin (rubrocurcumin) method. The main interferences are germanium(IV), nitrate, nitrite, fluoride, and perchlorate ions (123). A preliminary treatment to eliminate nitrates or perchlorates by heating the dried residue of the distillate in a platinum dish at dull red heat for 1 min or longer until all evolution of gas from the melt ceases has been suggested (132). The fluoride interference can be minimized by addition of aluminum chloride to the sample solution prior to distillation (48). Germanium can be precipitated as the sulfide, although about 48 hr are required for the colloidal germanium sulfide to coagulate and special care is necessary in washing the precipitate (71).

Procedure A2a is essentially that developed by Spicer and Strickland (132) based on the original procedure of Naftel (99). Glycerol is used to minimize loss of boron as methyl borate during initial evaporation. About 4% loss in boron can be expected in this procedure because of the preliminary treatment. Rigorous control of each experimental step is recommended for reproducible results (87).

Special Solutions for Procedure A2a

Curcumin–Oxalic Acid Solution. Prepare 1 liter of a water–ethanol solution having a specific gravity of 0.813 at 20°C (approx. 35 ml water per liter absolute ethanol). Add 15.0 g oxalic acid dihydrate, $H_2C_2O_4 \cdot 2H_2O$, 25.0 ml concentrated hydrochloric acid, and 75.0 ml water to about 700 ml of the water–ethanol solution in a 1-liter volumetric flask. Stir until dissolution of the oxalic acid is complete and then add 0.35 g finely powdered pure curcumin with stirring until completely dissolved. Cool to 20°C, dilute to volume with the water–ethanol mixture, and store this reagent in the dark. This solution should be prepared fresh each week.

Glycerol–Sodium Hydroxide Solution. Dissolve 1.0 g sodium hydroxide pellets, 0.1 g sodium chloride, and 3 ml glycerol in 100 ml distilled water. Store in a polyethylene bottle.

Standard Boron Solution. Dissolve 0.1430 g boric acid, H_3BO_3, in distilled water and dilute to 1 liter. Dilute 10.0 ml of this solution to 100 ml to obtain a standard solution containing 2.5 μg boron per ml.

Procedure A2a

Collect the methyl borate from distillation in 2 ml of the glycerol–sodium hydroxide solution in a platinum dish (about 8 cm in diameter). The distillate should contain 0.5–5 μg boron as borate. Evaporate the distillate to dryness on a water bath adjusted to 85–90°C. Heat the platinum dish on a hot plate at approximately 200°C for 10 min in order to remove most of the glycerol. Finally, heat the platinum dish for about 30 sec at dull red heat over a burner flame. Cool and add 5.0 ml of the curcumin–oxalic acid solution by means of pipet, rinsing the inside of the dish as the reagent is being added. Float the platinum dish upon the surface of the water bath which is thermostatically controlled to maintain a temperature of $55° \pm 2°C$. Evaporate the solution to dryness and continue heating for 30 min to ensure complete removal of all hydrochloric acid. Cool the dish and contents to room temperature in a desiccator.

Add 10.0 ml 95% ethanol to the residue in the platinum dish and stir with a glass rod having a rubber policeman or a polyethylene rod. Transfer the solution to a 50-ml volumetric flask and rinse the dish with a water—ethanol solution. Dilute to volume with the water—ethanol solution, mix well, and then filter the solution through a dry filter paper of coarse porosity. Measure the absorbance at 555 nm using a 1.000- or 5.000-cm cell, depending on the amount of boron, within 2 hr of color development against a reagent blank solution.

Preparation of Calibration Graph. Add 0, 0.5, 1.0, 2.0, and 3.0 ml standard boron solution to 2.0 ml of the glycerol—sodium hydroxide solution in platinum dishes and follow the same procedure as used for the sample solution.

Special Solutions for Procedure A2b

Curcumin Solution. Dissolve 0.125 g curcumin in 500 ml 95% ethanol, filter, and store in an amber flint glass or polyethylene bottle. Prepare a fresh solution every two weeks.

Oxalic Acid Solution. Dissolve 50 g oxalic acid dihydrate, $H_2C_2O_4 \cdot 2H_2O$, in 450 ml acetone and filter. Dilute the filtrate to 500 ml with acetone and store in a flint glass or polyethylene bottle. Prepare a fresh solution every two weeks.

Calcium Hydroxide Suspension. Heat 2.5 g calcium carbonate in a platinum crucible at a temperature of about $500°C$, then gradually heat to about $1000°C$, and ignite for 30 min at this temperature. Cool the calcium oxide melt, grind in an agate mortar, dissolve the powder in 500 ml redistilled water, and store in polyethylene bottle.

Procedure A2b

Transfer a 5-ml aliquot of the sample solution containing $1-10$ μg boron to a porcelain casserole. Add 5 ml of the calcium hydroxide suspension if the sample does not contain calcium hydroxide from the previous methyl borate distillation procedure, or 2 ml of the calcium hydroxide suspension if the sample solution already contains calcium hydroxide. The solution must be basic. Evaporate to dryness on a steam bath, cool, add 1 ml 1:4 hydrochloric acid and 5 ml of the oxalic acid solution, and mix with a quartz or polyethylene stirring rod. Add 2.00 ml of the curcumin solution by means of a microburet and mix thoroughly.

Evaporate to dryness in a water bath maintained at $55° \pm 2°C$ and continue heating at this temperature for 30 min. Cool to room temperature, dissolve residue in 25 ml acetone, filter through a sintered glass funnel using slight suction, and collect the filtrate in a large test tube. Wash the sintered glass funnel with 25 ml acetone in 5-ml increments. Transfer filtrate and washings to a 100-ml volumetric flask, dilute to the mark with distilled water, and mix well. Measure the absorbance at 540 nm in 1.000-cm cells against a reagent blank solution similarly prepared.

Preparation of Calibration Graph. Transfer 0, 1, 3, 5, 7, and 10 ml standard boron solution containing 1 μg per ml to six evaporating dishes and add 5 ml calcium hydroxide suspension to each. Evaporate to dryness on a steam bath and cool. Follow the procedure as outlined for the analysis of the sample solution.

B. CARMINIC ACID METHOD

Carminic acid, a glucoside of carmine, gives the same blue color with boric acid in concentrated sulfuric acid solution as does carmine because it is only the conjugated anthraquinone which is responsible for the color reaction. Since

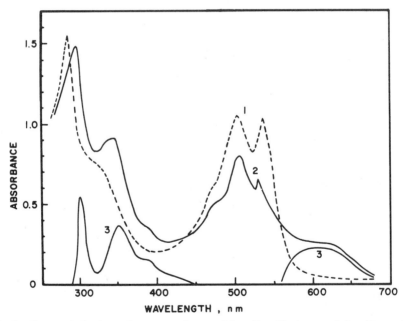

Fig. 3. Characteristic absorption spectra for carminic acid and boron-carminic acid complex in sulfuric acid: (1) reagent blank (1 mg reagent per 25 ml sulfuric acid) against sulfuric acid; (2) boron-carminic acid complex (15.5 μg boron per 25 ml sulfuric acid solution) against 95% sulfuric acid; (3) boron-carminic acid complex (15.5 μg boron per 25 ml sulfuric acid solution) against reagent blank solution.

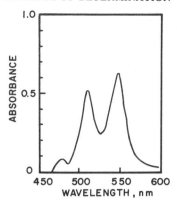

Fig. 4. Characteristic absorption spectrum of boron-carminic acid complex in sulfuric acid-acetic acid solution containing 10 μg boron and 1 mg carminic acid in 25 ml of 20:80 sulfuric acid-acetic acid mixture against reagent blank solution.

Hatcher and Wilcox (67) reported the use of carmine as a chromogenic agent for boron, both carmine and carminic acid have been used because of the rather wide optimum concentration ranges, relatively few interferences, and moderately low reagent blanks (18,22). A critical spectrophotometric investigation of this method has shown that increased sensitivity results (Figs. 3 and 4) when either absorbance measurements are made at 300 nm or a sulfuric acid–acetic acid reaction medium is used (64). Germanium(IV), titranium(IV), and fluoride ions are the main interfering ions although the tolerance for germanium and titanium is ten times larger (about 0.4 ppm) in sulfuric acid–acetic acid mixture than in the sulfuric acid solution. Sodium, potassium, calcium, magnesium, manganese(II), iron(II), nickel, zinc, cerium(III), molybdenum(VI), chloride, perchlorate, sulfate, arsenate, and phosphate do not interfere (64).

The carminic acid method has been used to determine boron in rocks (45), in titanium (21), in molybdenum (137), in waters, in soil extracts, in plant materials (67), in aluminum–uranium alloys (115), and in biologic material (131).

Special Solutions for Procedure B

Standard Boron Solution I. Dissolve 1.430 g boric acid in water and dilute to 1 liter with 95% sulfuric acid. Dilute 10.00 ml of this stock solution to 1 liter with 95% sulfuric acid to obtain a solution containing 2.50 μg boron per ml.

Standard Boron Solution II. Dissolve 1.430 g boric acid in water and dilute to 1 liter. Dilute 100 ml of this solution to 1 liter with glacial acetic acid and transfer 40.0 ml of this latter solution to a 1-liter volumetric flask. Add 21.8 ml acetic anhydride (97% reagent grade) and dilute to volume with glacial acetic acid. Each milliliter of this standard solution contains 1.00 μg boron.

Carminic Acid Solution I. Dissolve 1.00 g carminic acid in 1 liter 95% sulfuric acid and keep refrigerated.

Carminic Acid Solution II. Dilute 18.0 ml carminic acid solution I to 100 ml with 95% sulfuric acid solution.

Carminic Acid Solution III. Dissolve 1.00 g carminic acid in 1 liter glacial acetic acid, filter, and dilute 400 ml of the filtrate to 1 liter with glacial acetic acid.

Procedure B1

Transfer 1–10 ml of a concentrated sulfuric acid solution containing 2–35 μg boron to a 25-ml volumetric flask. If necessary, dilute to 10 ml with 95% sulfuric acid and add 10 ml carminic acid solution I (0.1%). Prepare a reagent blank solution using 10 ml 95% sulfuric acid. Place flasks in water bath maintained at $50 \pm 2°C$, heat for 15–20 min, cool to room temperature, and dilute to volume with 95% sulfuric acid. Measure the absorbance at 615 nm in 1.000-cm cells against the reagent blank solution.

Preparation of Calibration Graph. By means of a microburet, transfer 0, 1.00, 2.00, 4.00, 7.00, and 10.00 ml standard boron solution I to six 25-ml volumetric flasks, dilute each to 10 ml with 95% sulfuric acid, and continue with Procedure B1.

Procedure B2

Transfer 1–10 ml of a concentrated sulfuric acid solution containing 1–10 μg boron to a 25-ml volumetric flask. If necessary, dilute to 10 ml with 95% sulfuric acid and add 10 ml carminic acid solution II (0.018%). Prepare a reagent blank solution using 10 ml of the 95% sulfuric acid instead of the sample solution. Place the flasks in a water bath maintained at $50 \pm 2°C$, heat for 15–20 min, remove flasks, and allow to cool to room temperature. Dilute to volume with 95% sulfuric acid, mix thoroughly, and measure the absorbance at 300 nm against a reagent blank solution.

Preparation of Calibration Graph. With a microburet, transfer 0, 0.50, 1.00, 2.00, 3.00, and 4.00 ml standard boron solution I to six 25-ml volumetric flasks, dilute to 10 ml with 95% sulfuric acid, and continue using Procedure B2.

Procedure B3

Transfer a 1- to 5-ml aliquot of an acetic acid solution of boric acid containing 0.3–8 μg boron to a 25-ml volumetric flask. Dilute to about 8 ml with acetic acid, add 10 ml carminic acid solution III, cool in ice, and add 5 ml 98% sulfuric acid. Prepare a reagent blank solution using 8 ml acetic acid instead of sample solution. Heat the flasks in a water bath maintained at 50 ±2°C for 10–15 min, cool to room temperature, dilute to volume with acetic acid, and mix thoroughly. After 1 hr, measure the absorbance at 548 nm using a reagent blank solution in the reference cell.

Preparation of Calibration Graph. Using a microburet, transfer 0, 0.50, 1.50, 3.00, 5.00, and 8.00 ml standard boron solution II to six 25-ml volumetric flasks and continue with Procedure B3.

C. QUINALIZARIN METHOD

Quinalizarin (1,2,5,8-tetrahydroxyanthraquinone) reacts with boric acid in concentrated sulfuric acid to form a blue 1:1 complex. Unfortunately, the reagent is also highly colored, with an absorbance maximum which overlaps that of the boron–quinalizarin complex. The maximum difference in absorbance for the two colored systems in sulfuric acid occurs at approximately 610 nm (Fig. 5). Discrepant sensitivities cited for this reagent in various publications (11,54,74,89,108,118) can be attributed to the different concentrations of reagent used and other variations in procedural parameters. The use of an acetic acid–sulfuric acid medium results in about a tenfold increase in sensitivity (63).

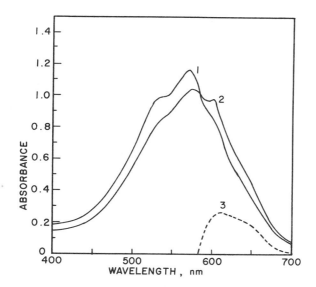

Fig. 5. Characteristic absorption spectra for boron–quinalizarin complex in sulfuric acid: (1) reagent against sulfuric acid; (2) boron–quinalizarin complex against sulfuric acid; (3) boron–quinalizarin complex against reagent blank (1 ppm B).

1. Quinalizarin Method Using Sulfuric Acid Medium

The sulfuric acid concentration should be maintained between 91% and 93% by weight for maximum color development (130). Germanium(IV), fluoride, and oxidants are the main interferences.

Special Solutions for Procedure C1

Quinalizarin Solution. Dissolve 0.0900 g quinalizarin in 1 liter 93% by weight sulfuric acid and store in an amber flint glass bottle.

Standard Boron Solution. Prepare 0.5, 1.0, 2.0, 3.0, and 5.0 μg per ml boron solutions in 93% sulfuric acid by diluting the stock solution of procedure A2a.

Sulfuric Acid 93% by Weight. Transfer 50 ml distilled water to a large casserole and slowly add about 100 ml concentrated sulfuric acid (sp. gr. 1.84; 98% by weight sulfuric acid). Cool, transfer to a 1-liter volumetric flask, dilute to the mark with 98% sulfuric acid, mix throughly, and store in a flint glass bottle.

Procedure C1

Transfer 2.00 ml of a sulfuric acid solution containing 1–12 μg boron to a 25-ml volumetric flask and add 10 ml 98% sulfuric acid. Cool the solution, add

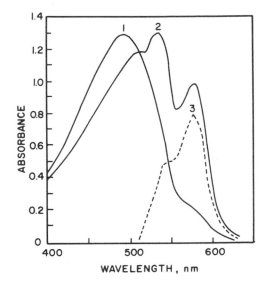

Fig. 6. Characteristic absorption spectra for boron–quinalizarin complex in sulfuric–acetic acid medium: (1) reagent blank solution against acetic acid; (2) boron–quinalizarin complex against acetic acid; (3) boron–quinalizarin complex in sulfuric–acetic acid medium against reagent blank solution (0.32 ppm B).

10.0 ml quinalizarin solution, dilute to volume with 93% sulfuric acid, and mix thoroughly. After 45 min, measure the absorbance at 610 nm using a reagent blank solution in the reference cell.

2. Quinalizarin Method Using Sulfuric Acid–Acetic Medium

The following modified quinalizarin method (63) using an acetic acid-sulfuric acid medium is more sensitive than the conventional quinalizarin ⌐method using a sulfuric acid medium (Fig. 6). Germanium(IV), selenium(IV), titanium(IV), fluorode, and nitrate ions interfere.

Special Solutions for Procedure C2

Standard Boron Solution. This solution preparation is exactly the same as that for standard boron solution II under the carminic acid method (Procedure B1) previously described.

Quinalizarin Solution. Dissolve 0.160 g quinalizarin in 1 liter glacial acetic acid and filter.

Procedure C2

Transfer 5 ml acetic acid solution containing 0.5–8 μg boron to a 25-ml volumetric flask, dilute to about 8 ml with acetic acid, and add 10.0 ml of the quinalizarin solution. Slowly add 5.0 ml concentrated sulfuric acid (sp. gr. 1.84, 98% by weight sulfuric acid) and mix thoroughly. Cool the solution to room temperature, dilute to volume with glacial acetic acid, and mix well. After 1 hr, measure the absorbance at 577 nm in 1.00-cm cells against a reagent blank solution prepared by this same procedure. Prepare a calibration graph by measuring 0, 0.50, 1.50, 3.00, 5.00, and 7.00 ml of the standard boron solution into six 25-ml volumetric flasks and continue with Procedure C2.

D. 1,1′-DIANTHRIMIDE METHOD

1,1′-Dianthrimide, an aminoanthraquinone (1,1′-iminodianthraquinone), reacts with boric acid in concentrated sulfuric acid to give a blue 1:1 complex. However, heating is required for full color development. The absorbance of the reagent is relatively low at the wavelength of measurement in the 620- to 640-nm region (Fig. 7). This reagent was first used by Rudolph and Flickinger (120). Although there are numerous recommendations regarding the optimum temperature and heating time required for development of the color (9,17,28,35,42,53), the 80°C temperature and 4.5 hr of heating as first suggested in the factorial experiment by Danielsson (33) has been confirmed (62). In determining 5 μg boron per 25 ml solution, it was found that

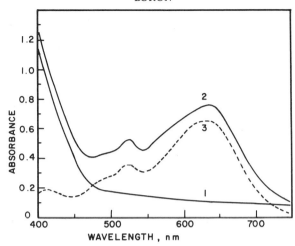

Fig. 7. Characteristic absorption spectra for boron–1,1'-dianthrimide complex in sulfuric acid: (1) reagent blank solution against sulfuric acid; (2) boron–1,1'-dianthrimide complex against sulfuric acid; (3) boron–1,1'-dianthrimide complex against reagent blank solution (0.4 ppm B).

less than 100 μg tellurium, 30 μg germanium, 200 μg bromide, 200 μg iodide, 0.9 μg fluoride, and 1 mg phosphorus as orthophosphate were permissible (62). The optimum range for the general procedure is 1–10 μg boron. This method has been used for the determination of boron in solids (53,93), in plants (42), in low-alloy steels (34,35), in titanium alloys (28), and in aluminum (17).

Special Solutions

Standard Boron Solution. The preparation of this solution is exactly the same as that for the standard boron solution I previously described in the carminic acid method (Method IIB).

1,1'-Dianthrimide Solution. Dissolve 0.1000 g 1,1'-dianthrimide in 500 ml 95% by weight sulfuric acid.

Procedure

Transfer a 5-ml or 10-ml aliquot containing 0.5–10 μg boron in approximately 95% by weight sulfuric acid to a 25-ml volumetric flask. (The flask should be given a pretreatment consisting of being filled with 95% sulfuric acid and heated for 2–3 hr at about 100°C.) A reagent blank should be prepared simultaneously using an equivalent volume of the 95% by weight sulfuric acid. Add 10 ml of the 1,1'-dianthrimide solution to each flask, loosely insert the stoppers in the flasks, and heat in a thermostatically controlled water bath at $80 \pm 2°C$ for 4.5 hr. Cool the flasks to room temperature, dilute to volume with 95% by weight sulfuric acid, and mix thoroughly. Measure the absorbance at 635 nm using 1.000-cm cells against a reagent blank solution.

Preparation of Calibration Graph. Transfer 0, 0.50, 1.00, 2.00, 3.00, and 4.00 ml of the standard boron solution to six pretreated 25-ml volumetric flasks, and dilute to about 5 or 10 ml with 95% sulfuric acid. Add 10.0 ml 1,1'-dianthrimide solution to each flask, loosely insert the stoppers in the flasks, and heat on a water bath at $80 \pm 2°C$ for 4.5 hr. Cool to room temperature, dilute to volume with 95% by weight sulfuric acid, and mix throughly. Measure the absorbance at 635 nm using 1.000-cm cells against a reagent blank.

E. METHYLENE BLUE–TETRAFLUOROBORATE METHOD

Ducret (37) first proposed the separation and determination of traces of boron by extraction of the methylene blue–tetrafluoroborate complex with dichloroethane. Further studies of this method were made by Pasztor and Bode (110,112). The methylene blue-tetrafluoroborate method has been applied successfully to the determination of boron in steel (12,57,84,96,112,121), in tin (81), in zirconium (136), and in rocks (73). The suitability of polyethylene and Teflon ware is an advantage of this method.

Perhaps the most critical step in this method involves the complete conversion of boric acid to the tetrafluoroborate. In very acidic solution (about $1N$), the conversion is essentially quantitative but is a slow reaction, with perhaps a 2-hr reaction time necessary to assure complete conversion. In less acidic solution (pH 1), the reaction is much slower, and the tendency for the hydrolytic formation of HBF_3OH^- is enhanced. Inasmuch as the HBF_3OH^- species is not extractable, the use of an excess of hydrofluoric acid and a long reaction time is recommended. Use of an extremely large excess of hydrofluoric acid is to be avoided, inasmuch as the reagent blank becomes very large as a result of methylene blue–fluoride being extracted, especially in the presence of a high concentration of sulfuric acid (145). The acidity and the time of extraction are important because the methylene blue should exist predominantly as a cation. The maximum difference in the efficiency of extraction of the methylene blue–tetrafluoroborate and methylene blue with dichloroethane was found to be at pH 1 or slightly less (112). Other dyes that have been used in place of methylene blue, such as methyl violet (113), malachite green (77), rhodamine B (106), and thionine derivatives (111), have their respective acidities for optimum extraction. Instead of using a dye, a cationic complex ion, for example, tris(1,10-phenanthroline)–iron(II), can be used to form an ion-associated complex with the tetrafluoroborate which is extractable (7,102).

Special Solutions

Methylene Blue Solution. Dissolve 3.739 g methylene blue (C. I. Basic Blue 9 or methythionine chloride) in distilled water and dilute to 1 liter. Dilute 100 ml of this solution to 1 liter with distilled water to obtain a $0.001M$ solution.

Standard Boron Solution. Dissolve 0.5716 g boric acid, H_3BO_3, in distilled water, transfer to a 1-liter volumetric flask, and dilute to volume. Transfer 25.0 ml of this stock solution to a 250-ml volumetric flask and dilute to volume. Each ml of this dilute standard solution contains 10.0 μg boron. Store both solutions in polyethylene bottles.

Standard Tetrafluoroborate Solution. Transfer 50.0 ml of the dilute standard boron solution (1 ml = 10.0 μg boron) to a polyethylene bottle, add 5 ml 5% hydrofluoric acid, and allow to stand for 24 hr. Transfer this solution to a 500-ml volumetric flask, dilute to volume, and mix. Transfer this solution to a dry polyethylene bottle. Each ml of this solution contains 1.0 μg boron as BF_4^-. A reagent blank solution should be prepared using 20.0 ml distilled water in place of the sample solution.

Transfer 20.0 ml of a neutral sample solution containing 2–20 μg boron to a calibrated 100-ml polyethylene cylinder. Add 10.0 ml of a $2.5N$ sulfuric acid solution and 5 ml of a 5% sodium fluoride solution, mix thoroughly, and allow to stand at room temperature for 2 hr. Dilute the solution to 50 ml with distilled water, mix thoroughly, and add 10.0 ml of the methylene blue reagent and 25.0 ml 1,2-dichloroethane. Stopper the cylinder, shake for 1 min, and, after allowing the layers to separate, pipet 5.0 ml of the organic layer to a 25-ml volumetric flask and dilute to volume with 1,2-dichlorethane. Measure the absorbance at 660 nm in 1.000-cm cells against a reagent blank solution.

Preparation of Calibration Graph. By means of a microburet, transfer 0, 1.0, 2.0, 4.0, 6.0, 8.0, and 10.0 ml of the standard tetrafluoroborate solution to seven calibrated 100-ml polyethylene cylinders, dilute each to 50 ml with distilled water, and mix. Add 10.0 ml of the methylene blue solution and 25.0 ml 1,2-dichloroethane, stopper the cylinder, and shake for 1 min. Allow the layers to separate, pipet 5.0 ml of the organic layer to a 25-ml volumetric flask, and dilute to volume with 1,2-dichloroethane. Measure the absorbance of each solution at 660 nm using the solvent in the reference cell. Subtract the absorbance of reagent blank solution from all absorbance readings and plot corrected absorbance against micrograms of boron.

F. OTHER REAGENTS

In addition to the five chromogenic reagents discussed in this section, numerous other reagents have been suggested for the determination of traces of

TABLE I. Other Chromogenic Reagents for Boron

Reagent	Reaction medium	General comments (Wavelength λ of meas., Å; optimum concr. range, ppm; Sandell sensitivity s (124))	Reference
Azomethine	aqueous	λ = 4.15; 0.02–0.1	(128)
Diaminochrysazin	95.4% $H_2 SO_4$	λ = 525; s = 0.020	(29)
Diaminoanthrarufin	95.4% $H_2 SO_4$	λ = 605; s = 0.0025	(29)
Tribromoanthrarufin	95.4% $H_2 SO_4$	λ = 625; s = 0.0009	(29)
Tetrabromochrysazin	96% $H_2 SO_4$	λ = 540; s = 0.020	(149)
Chromotrope 2B	conc. $H_2 SO_4$	λ = 620; s = 0.1	(8, 76)
1-Hydroxy-4-p-toluidinoanthraquinone	83–84% $H_2 SO_4$	λ = 525–550	(10, 116, 142)
H-Resorcinol	0.1M HOAc	λ = 530; s = 0.0036	(59)
Victoria violet	aqueous pH 8.75	λ = 540; 2–20	(117)
Chromotropic acid	aqueous, pH 7	λ = 316; 0–2.4	(79)
1,1'-Bis(6-chloroanthraquinonyl)amine	96% $H_2 SO_4$	λ = 641; s = 0.0005	(60)
5-Benzamido-6'-chloro-1,1'-bis(anthraquinonyl)amine	96% $H_2 SO_4$	λ = 635; 0.1–5; s = 0.0005	(61)
5-p-Toluidino-1,1'-bis(anthraquinonyl)amine	96% $H_2 SO_4$	λ = 720; 0.1–4; s = 0.0004	(61)
Salicylate, iron(II), crystal violet	aqueous; $C_6 H_6$ extn.	λ = 570; 0.01–0.1	(146)
Barium chloroanilate	aqueous; pH 8, tartrate	λ = 350; 0.3–10	(135)
Molybdate, fluoride, tin(II)	aqueous	λ = 750; 2–50	(24)
Nile blue	aqueous, fluoroborate	λ = 647; 0.05–0.5	(49)

boron (26,54,60,61,100,147). Several of these reagents, for example, diamino-chrysazin and tetrabromochrysazin, have been found to offer advantages in specific applications (41,42). Additional reagents which may have potential applications are listed in Table I.

In addition to the spectrophotometric methods based on the presence of boric acid in the sample, there are several methods based on the presence of boron hydrides. Offner (104) developed a spectrophotometric method for determining traces of pentaborane, $B_5 H_9$, based on the formation of a yellow pentaboranepyridine adduct in toluene. Conformity to Beer's law was observed in the 2- to 12-μg per ml concentration range. The method was applied to the determination of trace amounts of pentaborane in air. Decaborane, $B_{10} H_{14}$, dissolves in an aqueous triethanolamine solution, presumably forming hypo-borate, which exhibits a strong absorbance at 270 nm (70). Colorimetric methods based on the use of N,N-diethyl nicotinamide (69), quinoline, which forms a red adduct in xylene (70), and β-naphthoquinoline (16) have also been reported. The sensitivity of the β-naphthoquinoline method is 0.1 μg decaborane per ml.

III. APPLICATIONS

A. BORON IN WATER (2,82)

Most natural waters contain traces of boron which may be augmented by industrial wastes and cleaning products. The following procedure developed by Bunton and Tain (19) utilizes a preliminary removal of cations by means of an ion exchange column before determination of the boron by the curcumin method. The use of this preliminary treatment is recommended whenever the hardness exceeds 100 ppm (calcium carbonate).

Special Solutions

Standard Boron Solution. Dissolve 0.5716 g anhydrous boric acid, H_3BO_3 in redistilled water and dilute to volume in a 1-liter volumetric flask. Transfer 10.0 ml of this stock solution to a 1-liter volumetric flask and dilute to the mark with redistilled water. Each milliliter of this standard boron solution contains 1.00 μg boron.

Curcumin Solution. Dissolve 0.100 g finely ground curcumin and 12.5 g oxalic acid in about 150 ml 95% ethanol in a 250-ml volumetric flask. Add 10.5 ml concentrated hydrochloric acid and dilute to volume with 95% ethanol.

Procedure

Transfer by pipet a 25.0-ml aliquot of the water sample to the top of the ion exchange column and adjust stopcock so that the flow rate is about 0.1 ml per sec. (The column should be 12–15 mm in diameter and 15–20 cm in length. The strong cation exchange resin, for example, Dowex 5-X-8, should be in the acid form and washed free of excess acid.) Collect the effluent from column in a 50-ml volumetric flask.[2] Elute with 3- to 5-ml portions of distilled water until the effluent reaches the graduation mark.

Mix the flask thoroughly before transferring by pipet 2.0 ml to a Vycor evaporating dish. Add 4.0 ml of the curcumin reagent and mix thoroughly by swirling the dish. Partially immerse the quartz dish in a water bath maintained at $55 \pm 2°C$, evaporate to dryness, and allow the residue to remain at this temperature for an additional 15 min. Cool the dish, add 10 ml 95% ethanol, stir

[2] If the carminic acid or the quinalizarin method is preferred, collect effluent in a platinum dish containing a sodium hydroxide solution. Evaporate to dryness at low temperature and continue to heat for 10 min. Cool, dissolve, and transfer with about 8 ml acetic acid to a 25-ml volumetric flask. Use either Procedure II B3 or II C2.

TABLE II. Summary: Determination of Traces of Boron in Metals and Selected Materials

Metal or materials	Method	Reference
Nickel	mercury cathode, curcumin	(27)
	distillation, curcumin	(86)
	curcumin	(6)
Tin	methylene blue—BF_4	(81)
Silicon	Hydrothermal refining distillation, curcumin	(87)
Beryllium, zirconium, thorium, uranium	distillation, diaminochysazin	(41)
Uranium	distillation, curcumin	(105)
Uranium	solvent extn., curcumin	(31)
Titanium	distillation, 1,1'-dianthrimide	(28)
Titanium	ion exchange, carminic acid	(21)
Molybdenum	distillation, carminic acid	(137)
Germanium, germanium oxide	distillation, curcumin	(85)
Magnesium, beryllium	distillation, curcumin	(68)
Zirconium, hafnium, titanium	curcumin	(43)
Aluminum alloys	curcumin	(109)
Iron, aluminum, magnesium	curcumin	(68)
Zirconium alloys	curcumin	(68)
Zircaloy	distillation, curcumin	(46)
Aluminum	Ion exchange, 1,1'-dianthrimide	(140)
Steel	methylene blue—fluoroborate	(12, 112)
Steel	1-hydroxy-4-p-toluidinoanthra-quinone	(10)
Steel	1,1'-dianthrimide	(34)
Steel	curcumin	(65, 130)
Stainless steel	curcumin	(68)
Organic	curcumin	(68)
Fertilizers	ion exchange; carminic acid	(87)
Uranium dioxide	curcumin	(68)
Uranium oxides	methylene blue—fluoroborate	(10)
Rocks	ion exchange; carminic acid	(45)
Seawater	curcumin	(56)
Seawater	automated; curcumin	(72)
Water	carminic acid	(4)
Water	Nile blue—fluoroborate	(49, 101)

with a polyethylene rod, and swirl to effect complete dissolution. Transfer this red solution to a 25-ml volumetric flask, dilute to the mark with 95% ethanol, and mix. Measure the absorbance at 560 nm, using 1.0-cm cells (longer cells are recommended for very low concentrations), against 95% ethanol in the reference cell. Prepare a reagent blank by taking 25 ml of redistilled water instead of the sample and follow this same procedure. Subtract this absorbance from that of sample.

Preparation of Calibration Graph. Transfer 0, 0.25, 0.50, 0.75, 1.0, and 1.25 ml of the standard boron solution to six Vycor evaporating dishes and add from a buret sufficient redistilled water to give a total volume of 2.0 ml. Add 4.0 ml of the curcumin solution to each solution and mix by swirling carefully. Partially immerse dishes in water bath maintained at $55 \pm 2°C$ evaporate to dryness, and continue heating at this temperature for an additional 15 min. Cool, add 10 ml 95% ethanol to each residue, and swirl until dissolution is complete. Transfer each solution to a 25-ml volumetric flask using 95% ethanol to rinse. Dilute to volume with 95% ethanol, mix thoroughly, and measure the absorbance at 560 nm against 95% ethanol.

An automated module system, Autolab, employing the curcumin method of Uppstrom (143) has been used to determine 0.1–6 ppm boron in seawater (72).

B. BORON IN CARBON STEELS

The following method based on the separation of boron by the distillation of methyl borate followed by the use of curcumin as the chromogenic agent is applicable to low concentrations of boron (less than 0.01% in plain carbon steels (3)). Danielsson (34) has also reported satisfactory results in determining boron in iron and low-alloy steels using the dianthrimide method.

Special Apparatus

Cyclic Distillation Apparatus. See Section IB and Fig. 2.

Special Solutions

Standard Boron Solution. Transfer 0.2288 g boric acid, H_3BO_3, to a 500-ml volumetric flask, dilute to mark with redistilled water, and mix. Transfer 25.0 ml of this stock solution to a 1-liter volumetric flask, dilute to volume, mix, and store in a polyethylene bottle. Each ml of this solution contains 2.0 μg boron.

Calcium Hydroxide Suspension. Ignite 7 g pure calcium carbonate in a platinum dish at 500–600°C, gradually increase the temperature to 1000°C, and continue ignition for 30 min. Cool the residue and then grind to a fine powder in an agate mortar. Transfer 2.8 g of the calcium oxide powder to 500 ml cold freshly distilled water in a polyethylene bottle. Mix thoroughly before using this reagent.

Curcumin Reagent Solution. Dissolve 0.10 g curcumin in 400 ml 95% ethanol and filter through previously alcohol-washed filter paper into a polyethylene bottle. Store in a dark place. This solution should be prepared fresh every two to three weeks.

Oxalic Acid Solution. Dissolve 50 g oxalic acid, $H_2C_2O_4 \cdot 2H_2O$, in 450 ml acetone and filter.

Absolute Methanol

Low Boron Steel (NBS Sample No. 55, Ingot Iron)

Procedure

Transfer a 0.5-g steel sample containing less than 0.002% boron, or a 0.25-g sample if the steel contains 0.002% to 0.005% boron, to the 100-ml quartz flask A (Fig. 2). Transfer a "low boron" steel sample, the same weight as the sample, to another 100-ml quartz flask. Treat this "low boron" sample and test sample in an identical manner in following procedure.

Acid-Soluble Boron. Add 50 ml absolute methanol and 5 ml of the calcium hydroxide suspension to the 100-ml quartz flask B. Add sufficient calcium hydroxide (0.5 ml) to trap C to form a liquid seal. Add 10 ml analytical reagent-grade phosphoric acid to flask A containing the sample. Assemble the apparatus (Fig. 2) and adjust the flow of water through both the auxiliary condenser D and the condenser connected with flask B. Use a heating mantle or small flame to heat flask A gently until the reaction ceases. Remove the heat from flask A and disconnect water supply to condenser D.

Place flask B in a hot water bath and heat until about 25 ml methanol has distilled over into flask A. Now place both flasks A and B in hot water baths and heat so that the methanol will cycle evenly between flask A and flask B for 1 hr.

Remove from the hot water baths and transfer the solution from flask B and from trap C to a 300-ml unscratched porcelain casserole. Rinse the flask and trap with water, then with 0.1 ml 1:9 hydrochloric acid, and again with water, adding the washings to the casserole. Evaporate the solution to dryness on a steam bath and cool to room temperature. Reserve for use in the color development for the determination of acid-soluble boron.

Acid-Insoluble Boron. Dilute the solution remaining in flask A to 90 ml with 1:8 hydrochloric acid. Filter through a 9-cm fine-porosity ashless filter paper, for example, Whatman No. 42, containing a small amount of ashless filter paper pulp. Wash the flask thoroughly with hot 2:98 hydrochloric acid and police the flask to remove insoluble matter. Wash the residue on the filter paper about 20 times with hot 2:98 hydrochloric acid to remove the iron and then about 15 times with cold water to remove hydrochloric acid. Transfer the paper and residue to a 20-ml platinum crucible. Add 5 ml of the calcium hydroxide suspension and evaporate to dryness. Ignite at 600–700°C until the carbon is removed. Add 1 g sodium carbonate and fuse the residue, finally tilting and heating the crucible so that the fusion is collected in a ball. Cool, remove the bulk of the melt by exerting slight pressure on the crucible wall, transfer the melt to a clean, dry flask A, and cool to 10–15°C. Add 4 ml phosphoric acid to the crucible and warm to dissolve the adhering fusion mixture. Cool and add to flask A containing the bulk of the melt, keeping the solution cold. Rinse the crucible twice with 3-ml portions of phosphoric acid and add rinsings to flask A. Assemble the cyclic distillation apparatus and repeat procedure as previously described for acid-soluble boron. Reserve the residue in a 300-ml casserole for color development and determination of acid-insoluble boron.

Color Development. To the residue in each casserole, add 1 ml 1:4 hydrochloric acid and 5 ml of the oxalic acid solution. Mix thoroughly and add 2.0 ml of the curcumin reagent. When dissolution of the residue is complete, evaporate to dryness in a water bath maintained at $55 \pm 2°C$ and bake for 30 min at this same temperature. Cool the casserole and residue to room temperature and add 25 ml acetone. When the residue is dissolved completely, filter through a sintered glass crucible of fine porosity into a 100-ml volumetric flask. Rinse the crucible and contents with 25 ml acetone, using about 5 ml for each washing. Dilute to a 100-ml volume with cold water and mix. Measure the absorbance in 1.0- or 2.0-cm absorption cells using distilled water in the reference cell. Subtract the absorbance of the reagent blank solution from the absorbance of the blank. The total boron is obtained from the sum of the acid-insoluble boron and acid-soluble boron determinations.

Preparation of Calibration Graph. Use a microburet to transfer 0, 1.0, 2.0, 3.0, 4.0, and 5.0 ml of the standard boron solution (1 ml = 2.0 μg boron) to a 100-ml quartz flask A. Add 5 ml of the calcium hydroxide suspension to each flask and evaporate the contents of each flask to dryness. Add 0.5 g of the "low boron" steel to each flask and continue using the procedure given under "acid-soluble boron" and "color development" in preceding paragraphs. Construct the calibration graph by plotting corrected absorbance of the standard samples against μg boron in the final 100 ml of the colored solution.

C. BORON IN ALLOY STEELS

1. Methylene Blue–Tetrafluoroborate Method (12,112,145)

This method is based on the conversion of the boron in the steel to tetrafluoroborate, BF_4^-, the formation of a dye–tetrafluoroborate complex, the extraction of this complex with 1,3-dichloroethane, and the measurement of the absorbance of the extract. This procedure, based on Ducret's method (37), is essentially that developed by Pasztor, Bode, and Fernando (112) and is applicable to the determination of 0.1 to 20 μg boron.

Special Solutions for Procedure C1

See methylene blue–tetrafluoroborate method IIE.

Procedure C1

Dissolve 0.1 g of an iron or steel sample in 5.0 ml 2.5N sulfuric acid and 5.0 ml 1.25M phosphoric acid in a 100-ml boron-free flask. Use 10.0 ml 2.5N sulfuric acid (no phosphoric acid) in dissolution of stainless steel. Attach a boron-free glass or polyethylene reflux condensing tube to the flask and place on a hot plate at 95–98°C and reflux until dissolution is complete. Cool and add 1 ml distilled water through the condenser. Two reagent blank solutions should be prepared using the same amounts of acid and weight of boron-free steel.

Acid-Soluble Boron. Transfer the solution through a polyethylene funnel fitted with a fine-porosity filter paper to a polyethylene graduated cylinder. Wash the filter paper with a minimum of water (3–4 ml). Remove the filter paper and residue for acid-insoluble boron. Dilute the solution in the polyethylene graduated cylinder to 20.0 ml.

Color Development. Add 5.0 ml 5% sodium fluoride solution, stopper the cylinder, shake to mix thoroughly, and allow to stand at room temperature for 2 hr. Add 0.1M potassium permanganate solution dropwise until a pink hue is detected and then add 2.0 ml of a 4% ammonium iron(II) sulfate solution. From the volume of solution in the graduated cylinder, calculate the volume of rinse water required to give a total volume of 50 ml upon transfer to a 100-ml calibrated polyethylene flask or bottle. After the transfer, add 10.0 ml of the 0.001M methylene blue reagent solution and 25.0 ml dichloroethane, stopper the flask or bottle, and shake for 1 min. After the layers separate completely, pipet 5.0 ml of the organic layer and dilute to 25.0 ml with dichloroethane in a 25-ml volumetric flask. Measure the absorbance at 660 nm using 1.0-cm cells and 1,2-dichloroethane in the reference cell. Subtract the absorbance reading of the reagent blank solution from the measured absorbance. From the calibration graph, determine the amount of acid-soluble boron.

Acid-Insoluble Boron. Transfer the filter paper and residue to a platinum crucible and ignite in a muffle furnace at 250°C and finally at 550°C. Add 1.00 g anhydrous sodium carbonate to the ignited residue and fuse. Cool the crucible and its contents to room temperature and dissolve the melt in 5.0 ml 2.5N sulfuric acid and 5.0 ml 1.25M phosphoric acid.

Transfer the solution to a polyethylene graduated cylinder, using sufficient rinsings to give a volume of 20.0 ml. Continue with the color development procedure as previously given after the "acid-soluble boron" section. The amount of boron corresponding to the corrected absorbance is the acid-insoluble boron. The total boron is equal to the sum of acid-soluble and acid-insoluble boron.

Preparation of Calibration Graph. Transfer 0, 2.0, 5.0, 10.0, 15.0, and 20.0 ml of the standard tetrafluoroborate solution to 100-ml graduated polyethylene cylinders. Add sufficient distilled water to give a total volume of 20.0 ml and continue with the procedure given under "color development" beginning with "Add 0.1M potassium permanganate solution dropwise . . . ."

2. Carminic Acid Method

This procedure is modified slightly from that presented in the first edition by Porter and Shubert (114).

Special Solutions for Procedure C2

Carminic Acid Solution. Dissolve 0.25 g carminic acid in 1 liter concentrated sulfuric acid and shake the solution 10 times at intervals over a 2-hr period.

Procedure C2

Transfer 1.00 g of the sample to a 100-ml boron-free glass or quartz flask and attach an air condenser. Add 10 ml hydrochloric acid and 10 drops nitric acid and warm, but do not boil. If metal remains after 20 or 30 min, add five more drops of nitric acid, boil off nitrous oxide fumes, cool the flask, rinse, and remove the condenser.

Dilute to approximately 30 ml with water and filter through a 9-cm No. 40 Whatman paper (or equivalent) into a flask and reserve. Wash the paper and residue about five times with 2:98 hydrochloric acid and then two or three times with water. Sprinkle the paper while still in the funnel with 0.5 g sodium carbonate, transfer to a platinum crucible, and add 1 or 2 ml water. Dry the crucible and its contents and then place it in a muffle furnace at approximately 700°C to burn off the paper. Add 2.0 g sodium carbonate and fuse over a blast furnace, cool, take up in water, acidify with hydrochloric acid, and add to the reserved filtrate.

Dilute the combined sample so that 1 ml contains 0.5—5 µg boron and pipet 2 ml to a dry flask. Add two drops hydrochloric acid, stopper with the air condenser, and, while holding at a 60-degree angle, slowly pipet through the condenser 20 ml concentrated sulfuric acid and allow to drain while cooling the flask. Remove the condenser and add with a pipet 20 ml of the prepared carminic acid solution, mix well, cover, and allow to stand for 30 min.

Determine the absorbance of the solution in a spectrophotometer at 615 nm against a reagent blank solution. Prepare a concentration-versus-absorbance graph by treating varying amounts of a standard boron solution in a similar manner.

D. BORON IN ALUMINUM ALLOYS

The following procedure is based on the use of carminic acid and is suitable for the determination of 0.005 to 0.05% boron in aluminum and its alloys (4).

Special Solutions

Carminic Acid Solution. Transfer 0.46 g carminic acid to a large polyethylene beaker and add 500 ml concentrated sulfuric acid. Use a magnetic stirrer with a Teflon-covered stirring bar to stir until dissolution is complete and store in a dry polyethylene bottle.

Standard Boron Solution. Dissolve 0.2857 g boric acid, H_3BO_3, in warm water (35°C), cool, transfer to a 1-liter volumetric flask, dilute to volume, and mix thoroughly. Each milliliter of this solution contains 50.0 μg boron.

Procedure

Prepare a reagent blank solution by following this procedure using the same amounts of each reagent.

Transfer a 1.000-g sample to a 250-ml conical flask and add 15 ml bromine water and then 10 ml concentrated hydrochloric acid. If the reaction appears to be too vigorous, cool the reaction mixture. As soon as the sample is completely dissolved, heat the solution to expel the excess bromine, filter the solution through a filter paper of coarse porosity, for example, Whatman 41, and collect the filtrate in a 50-ml volumetric flask. Rinse the conical flask and filter paper several times with hot water but do not use more than 15 ml and set aside.

Dry the filter paper and add 0.25 g anhydrous sodium carbonate to any residue on it. Place the filter paper and its contents in a platinum crucible, char the filter paper, and increase heat gradually until the paper is burned and a melt forms. When the melt is clear, cool to room temperature, dissolve in 5 ml hot water, and add dropwise a 1:1 sulfuric acid solution until the effervescence stops. Transfer to the 50-ml volumetric flask containing the initial filtrate, dilute to the mark with distilled water, and mix well.

Color Development Procedure. Transfer a 2.00-ml aliquot of the solution to a dry 50-ml ground-glass stoppered conical flask, add 10 ml concentrated sulfuric acid, mix, and cool to room temperature. Add 10.0 ml of the carminic acid solution, stopper, and mix. After 45 min, measure the absorbance at 585 nm in a 1.000-cm cell against the reagent blank solution.

Preparation of Calibration Graph. With a microburet, transfer 0, 1.0, 2.0, 4.0, 6.0, 8.0, and 10.0 ml of the standard boron solution to seven 50-ml volumetric flasks. Add 10 ml 1:1 sulfuric acid to each flask, dilute to 40 ml with distilled water, cool to room temperature, dilute to volume, and mix thoroughly. Follow the color development procedure. Plot absorbance versus μg boron per 22 ml of the final solution.

E. BORON IN SOILS AND PLANTS

The following three procedures for the determination of available boron in soils, total boron in soils, and total boron in plant tissue are those developed by Dible, Truog, and Berger (36) using curcumin as the chromogenic reagent.

Special Solutions

Curcumin—Oxalic Acid Solution. Dissolve 0.04 g of finely ground curcumin and 5 g of oxalic acid in 100 ml of 95% ethanol and store in an amber flint glass bottle in a refrigerator. Prepare fresh weekly.

Standard Boron Solution. Two micrograms (2.0 μg) boron per ml.

Procedure

Color Development Procedure. Transfer a 1-ml aliquot of sample solution containing 0.1–2 μg boron to a 250-ml polyethylene or Teflon beaker, add 4 ml of the curcumin-oxalic acid solution, and mix thoroughly by rotating the beaker. Evaporate on a water bath maintained at $55 \pm 3^\circ$C and continue to heat the residue at this same temperature for 30 min to ensure complete dryness. Cool and add 25 ml 95% ethanol, filter through fine-porosity filter paper, and measure the absorbance at 540 nm in 1.000-cm cells against the reagent blank solution.

Determination of Available Soil Boron. Transfer a 10.0-g sample of soil, air dried and 20-meshed, to a 125-ml boron-free flask and add 20 ml distilled water. Attach a reflux condenser and boil for 5 min. Disconnect the condenser and filter or centrifuge the suspension to obtain a clear extract. Take a 1.00-ml aliquot of the extract and continue with the color development procedure given in the preceding paragraph.

Determination of Total Boron in Soil. Fuse 0.5 g soil with 3 g anhydrous sodium carbonate in a platinum crucible, cool the melt, and place the crucible in a 250-ml polyethylene beaker containing about 50 ml distilled water. Place a cover glass on the beaker and add 4N sulfuric acid intermittently until the melt has disintegrated and the solution is at pH 6–6.8. Transfer the solution to a 500-ml volumetric flask, rinse the beaker and crucible several times with distilled water, and add the washings to the flask. The volume of solution in the flask should not exceed 150 ml. Add 95% ethanol to the flask until the volume is almost 500 ml, mix thoroughly, add sufficient sodium carbonate to make the solution slightly basic, and then adjust to the mark with ethanol. Filter or centrifuge the solution to obtain a clear supernatant. Transfer a 400-ml aliquot of the clear solution to 500-ml polyethylene beaker and add 100 ml distilled water. Evaporate to a small volume, transfer to a platinum dish, evaporate to dryness, and ignite just to destroy any organic matter. Cool, add 5.00 ml 0.10N hydrochloric acid, and triturate thoroughly with a polyethylene rod. Take a 1.00-ml aliquot of this solution and continue with the color development procedure.

Determination of Total Boron in Plant Tissue (36,99). Transfer a 0.25- to 0.5-g sample of plant material, oven dried and ground, in a quartz crucible and ash in muffle furnace at 550°C. Dissolve the ash in 5 ml 0.10N hydrochloric acid and dilute to 10.0 ml. Take a 1.00-ml aliquot of this clear solution and continue with the color development procedure.

Preparation of Calibration Graph. Use a microburet to transfer 0, 0.20, 0.40, 0.60, 0.80, and 1.00 ml of the standard boron solution to six 250-ml beakers. Add 1.00, 0.80, 0.60, 0.40, and 0.20 ml, respectively, of distilled water to first five beakers and continue with the color development procedure. Plot absorbance versus micrograms of boron per ml.

REFERENCES

1. Agazzi, E. J., *Anal. Chem.*, **39**, 233 (1967).
2. American Public Health Association, *Standard Methods for the Examination of Water and Wastewater,* 13th ed., Washington, D.C., 1971, pp. 69–73.
3. American Society for Testing and Materials, *1974 Annual Book of ASTM Standards, Part 12,* Philadelphia, pp. 37–39.
4. *Ibid.,* pp. 99–100.
5. American Society of Testing and Materials, *1973 Annual Book of ASTM Standards, Part 23,* ASTM Designation D 3082-72T, Philadelphia, pp. 255–257.
6. Andrew, T. R., and P. N. R. Nichols, *Analyst,* **91**, 664 (1966).
7. Archer, V. S., F. G. Doolittle, and L. N. Young, *Talanta,* **15**, 864 (1968).
8. Austin, C. M., and J. S. McHargue, *J. Assoc. Offic. Agri. Chem.,* **31**, 284, 427 (1948).
9. Baron, H., *Z. Anal. Chem.,* **143**, 339 (1949).
10. Bell, D., and K. McArthur, *Analyst,* **93**, 298 (1968).
11. Berger, K. C., and E. Truog, *Ind. Eng. Chem., Anal. Ed.,* **11**, 540 (1939).
12. Bhargava, O. P., and W. G. Hines, *Talanta,* **17**, 61 (1970).
13. Borchert, O., *Chem. Tech.* (Berlin), **7**, 554 (1955).
14. Bovalini, E., and A. Casini, *Ann. Chim.* (Rome), **41**, 745 (1951).

15. Bovalini, E., and M. Piazzi, *ibid.*, **48**, 305 (1958).

16. Bramen, R. S., and T. N. Johnston, *Talanta,* **10**, 810 (1963).

17. Brewster, D. A., *Anal. Chem.,* **23**, 1809 (1951).

18. Brown, R. S., *Anal. Chim. Acta,* **50**, 157 (1970).

19. Bunton, N. G., and B. H. Tait, *J. Amer. Water Works Assoc.,* **61**, 357 (1969).

20. Bush, G. H., and D. G. Higgs, *Analyst,* **76**, 683 (1951).

21. Calkins, R. C., and V. A. Stenger, *Anal. Chem.,* **28**, 399 (1956).

22. Callicoat, D. L., and J. D. Wolszon, *ibid.*, **31**, 1434 (1959).

23. Callicoat, D. L., J. D. Wolszon, and J. R. Hayes, *ibid.*, **31**, 1437 (1959).

24. Campbell, R. H., and M. G. Mellon, *ibid.*, **32**, 50 (1960).

25. Capelle, R., *Anal. Chim. Acta,* **25**, 59 (1961).

26. Cerrai, E., and G. Testa, *Energia Nucleare* (Milan), **5**, 824 (1958).

27. Chirnside, R. C., H. J. Cluley, and B. M. C. Proffitt, *Analyst,* **82**, 18 (1957).

28. Codell, M., and G. Norwitz, *Anal. Chem.,* **25**, 1446 (1953).

29. Cogbill, E. C., and J. H. Yoe, *ibid.*, **29**, 1251 (1957).

30. Coursier, J., J. Hure, and R. Platzer, *Anal. Chim. Acta,* **13**, 379 (1955).

31. Coursier, J., J. Hure, and R. Platzer, *Comm. Energie Atomique* (France), **Rappt. No. 404**, (1955).

32. Crawley, R. H. A., *Analyst,* **89**, 749 (1964).

33. Danielsson, L., *Talanta,* **3**, 138 (1959).

34. Danielsson, L., *ibid.*, p. 203.

35. Devoti, A., and A. Sommariva, *J. Iron Steel Inst.,* **189**, 227 (1958).

36. Dible, W. T., E. Truog, and K. C. Berger, *Anal. Chem.,* **26**, 418 (1954).

37. Ducret, L., *Anal. Chim. Acta,* **17**, 213 (1957).

38. Ducret, L., and P. Sequin, *ibid.*, p. 213.

39. Dyrssen, D. W., Y. P. Novikov, and L. R. Uppstrom, *ibid.*, **60**, 139 (1972).

40. Dyrssen, D. W., L. R. Uppstrom, and M. Zangen, *ibid.*, **46**, 55 (1969).

41. Eberle, A. R., and M. W. Lerner, *Anal. Chem.,* **32**, 146 (1960).

42. Ellis, G. H., E. G. Zook, and O. Baudisch, *ibid.*, **21**, 1345 (1949).

43. Elwell, W. T., and D. F. Wood, *Analyst,* **88**, 475 (1963).

44. Everest, D. A., and W. J. Popiel, *J. Chem. Soc.,* **1956**, 3183.

45. Fleet, M. E., *Anal. Chem.,* **39**, 253 (1967).

46. Freegarde, M., and J. Cartwright, *Analyst,* **87**, 214 (1962).

47. Fukushi, N., and Y. Kakita, *Bunseki Kagaku,* **15**, 553 (1966).

48. Gaestel, C., and J. Hure, *Bull. Soc. Chim. France,* **1949**, 830.

49. Gagliardi, E., and E. Wolf, *Mikrochim. Acta,* **1968**, 140.

50. Gautier, J. A., and R. Pignarel, *ibid.*, **1951**, 793.

51. Gooch, F. A., *Amer. Chem. J.*, **9**, 23 (1887).

52. Gooch, F. A., and L. C. Jones, *Amer. J. Sci.*, **7**, 34 (1899).

53. Gorfinkiel, E., and A. G. Pollard, *J. Sci. Food Agr.*, **5**, 136 (1954).

54. Goward, G. W., and V. R. Wiederkehr, *Anal. Chem.*, **35**, 1542 (1963).

55. Green, H., *B.C.I.R.A. J.*, **10**, 56 (1962).

56. Greenhalgh, R., and J. P. Riley, *Analyst*, **87**, 970 (1962).

57. Gregordzyk, S., and J. Mrozinski, *Hutnick*, **33**, 426 (1966).

58. Grindstead, R. R., and S. Snider, *Analyst*, **92**, 532 (1967).

59. Grizo, V. A., and E. N. Poluektova, *Zh. Anal. Khim.*, **13**, 434 (1958).

60. Grob, R. L., J. Cogan, J. J. Mathias, S. M. Mazza, and H. P. Piechowski, *Anal. Chim. Acta*, **39**, 115 (1967).

61. Grob, R. L., and J. H. Yoe, *Anal. Chim. Acta*, **14**, 253 (1956).

62. Gupta, H. K. L., and D. F. Boltz, *Anal. Lett.*, **4**, 161 (1971).

63. Gupta, H. K. L., and D. F. Boltz, *Mikrochim. Acta*, **1971**, 577.

64. Gupta, H. K. L., and D. F. Boltz, *ibid.*, **1974**, 415.

65. Harrison, T. S., and W. D. Cobb, *Analyst*, **91**, 576 (1966).

66. Hatcher, J. T., *Anal. Chem.*, **32**, 726 (1960).

67. Hatcher, J. T., and L. V. Wilcox, *Anal. Chem.*, **22**, 567 (1950).

68. Hayes, M. R., and J. Metcalfe, *Analyst*, **87**, 956 (1962).

69. Hill, D. L., E. I. Gipson, and J. F. Heacock, *Anal. Chem.*, **28**, 133 (1956).

70. Hill, W. H., and M. S. Johnston, *Anal. Chem.*, **27**, 1300 (1955).

71. Hillebrand, W. F., G. E. F. Lundell, H. A. Bright, and J. I. Hoffman, *Applied Inorganic Analysis*, 2nd ed., Wiley, New York, 1953, p. 299.

72. Hulthe, P., L. Uppstrom, and G. Ostling, *Anal. Chim. Acta*, **51**, 31 (1970).

73. Isozaki, A., and S. Utsumi, *Nippon Kagaku Zasshi*, **88**, 741 (1967).

74. Jones, A. H., *Anal. Chem.*, **29**, 1101 (1957).

75. Karpen, W. L., *ibid.*, **33**, 738 (1961).

76. Komarovskii, A., and N. Poluetov, *J. Appl. Chem.*, **7**, 831 (1934).

77. Korenman, I. M., and L. V. Sidorenko, *Tr. Khim. Khim. Tekhnol*, **1968**, 127.

78. Kramer, H., *Anal. Chem.*, **27**, 144 (1955).

79. Kuemmel, D. P., and M. G. Mellon, *ibid.*, **29**, 378 (1957).

80. Lang, K., *Z. Anal. Chem.*, **163**, 241 (1958).

81. Lel'chuk, Y. L., and V. A. Ivashina, *Izv. Tomsk, Politekh. Inst.*, **148**, 157 (1967).

82. Lishka, R. J., *J. Amer. Water Works Assoc.*, **53**, 1517 (1961).

83. Logie, E., *Chem. Ind.*, **1957**, 225

84. Luedemann, K. F., R. Zimmermann, and H. Wemme, *Neu. Huette*, **11**, 755 (1966).

85. Luke, C. L., *Anal. Chem.*, **27**, 1150 (1955).

86. Luke, C. L., *ibid.*, **30**, 1405 (1958).

87. Luke, C. L., and S. S. Flaschen, *ibid.*, **30**, 1406 (1958).

88. Lundell, G. E. F., and H. B. Knowles, *J. Res. Nat. Bur. Stand.*, **3**, 91 (1929).

89. Macdougall, D., and D. A. Biggs, *Anal. Chem.*, **24**, 566 (1952).

90. Maeck, W. J., M. E. Kussy, B. E. Ginther, G. V. Wheeler, and J. E. Rein, *ibid.*, **35**, 62 (1963).

91. Martin, J. R., and J. R. Hayes, *ibid.*, **24**, 182 (1952).

92. Materova, E. A. and T. I. Rozhanskaya, *Zh. Neorg. Khim.*, **6**, 177 (1961).

93. Maurice, J., *Ann. Agran.*, **19**, 699 (1968).

94. Melton, J. R., W. L. Hoover, P. A. Howard, and J.L. Ayers, *J. Assoc. Offic. Anal. Chem.*, **53**, 682 (1970).

95. Morrison, G. H., and C. L. Rupp, *Anal. Chem.*, **29**, 892 (1957).

96. Mrozinski, J., *Chem. Anal.* (Warsaw) **12**, 93 (1967).

97. Muto, S., *Bull. Chem. Soc. Japan*, **30**, 881 (1957).

98. Muzzarelli, R. A., *Anal. Chem.*, **39**, 365 (1967).

99. Naftel, J. A., *Ind. Eng. Chem.*, *Anal. Ed.*, **11**, 407 (1939).

100. Nemodruk, A. A., and Z. K. Karalova, *Analytical Chemistry of Boron*, Academy Sci., U.S.S.R., Moscow 1964 ed., pp. 57–58.

101. Nicholson, R. A., *Anal. Chim. Acta*, **56**, 147 (1971).

102. Nishimura, M., and S. Nakoya, *Bunseki Kagaku*, **18**, 154 (1969).

103. Norwitz, G., and M. Codell, *Anal. Chim. Acta*, **11**, 233 (1954).

104. Offner, H. G., *Anal. Chem.*, **37**, 370 (1965).

105. Onishi, H., N. Ishiwatari, and H. Nagai, *Bull. Chem. Soc. Japan*, **33**, 830 (1960).

106. Onishi, H., and H. Nagai, *Bunseki Kagaku*, **17**, 345 (1968).

107. Onishi, H., and H. Nagai, *ibid.*, **18**, 164 (1969).

108. Oven, E. C., *Analyst*, **71**, 210 (1946).

109. Pakalns, P., *Met. Metal Form.*, **39**, 308 (1972).

110. Pasztor, L. C., and J. D. Bode, *Anal. Chem.*, **32**, 1530 (1960).

111. Pasztor, L. C., and J. D. Bode, *Anal. Chim. Acta*, **24**, 467 (1961).

112. Pasztor, L. C., J. D. Bode, and Fernando, Q. *Anal. Chem.*, **32**, 277 (1960).

113. Poluetkov, N. S., L. I. Kononenko, and R. S. Lauer, *Zh. Anal. Khim.*, **13**, 396 (1958).

114. Porter, G., and R. C. Shubert, *Colorimetric Determination of Nonmetals*, 1st ed., D. F. Boltz, Ed., Interscience, New York, 1958, p. 351.

115. Puphal, K. W., J. A. Merrill, G. L. Booman, and J. E. Rein, *Anal. Chem.*, **30**, 1612 (1958).

116. Radley, J. A., *Analyst*, **69**, 47 (1944).

117. Reynolds, C. A., *Anal. Chem.*, **31**, 1102 (1959).

118. Ripley-Duggan, B. A., *Analyst,* **78**, 183 (1953).

119. Rosenbladt, T., *Z. Anal. Chem.,* **26**, 21 (1887).

120. Rudolph, G. A., and L. C. Flickinger, *Steel,* **112**, 114 (1943).

121. Ryabchikov, D. I., and G. E. Kuril'chikov, *Zh. Anal. Khim.,* **19**, 1495 (1964).

122. Rynasiewicz, J., M. P. Sleeper, and W. R. Ryan, *Anal. Chem.,* **26**, 935 (1954).

123. Samuels, J. K., III, and D. F. Boltz, private communication.

124. Sandell, E. B., *Colorimetric Metal Analysis,* 3rd ed., Interscience, New York, 1959, p. 83.

125. Schafer, H., and A. Sieverts, *Z. Anal. Chem.,* **121**, 161 (1941).

126. Schlesinger, H. I., H. C. Brown, D. L. Mayfield, and J. R. Gilbreath, *J. Amer. Chem. Soc.,* **75**, 213 (1953).

127. Schutz, E., *Mitt. Gebiete Lebensm. u. Hyg.,* **44**, 213 (1953).

128. Shanina, T. M., N. E. Gel'man, and V. S. Mikhailovskaya, *Zh. Anal. Khim.,* **22**, 782 (1967).

129. Silverman, L., and K. Trego, *Anal. Chem.,* **25**, 1264 (1953).

130. Smith, G. S. *Analyst,* **60**, 735 (1935).

131. Smith, W. C., Jr., A. J. Goudie, and J. N. Sivertson, *Anal. Chem.,* **27**, 295 (1955).

132. Spicer, G. S., and J. D. H. Strickland, *Anal. Chim. Acta,* **18**, 231 (1958).

133. Spicer, G. S., and J. D. H. Strickland, *J. Chem. Soc.,* **1952**, 4644.

134. Spicer, G. S., and J. D. H. Strickland, *J. Chem. Soc.,* **1952**, 4650.

135. Srivastava, R. D., P. R. Van Buren, and H. Gesser, *Anal. Chem.,* **34**, 209 (1962).

136. Sudo, E., and S. Ikela, *Bunseki Kagaku,* **17**, 1197 (1968).

137. Sugawara, K. F. *AEC Accession No. 4064, Rept. RTD-TDR-63-4082.*

138. Talk, A., W. A. Tap, and W. A. Ligerak, *Talanta,* **16**, 111 (1969).

139. Thierig, D., and F. Umland, *Z. Anal. Chem.,* **211**, 161 (1965).

140. Towndrow, E. G., and H. W. Webb, *Analyst,* **85**, 850 (1960).

141. Trentelman, J., and A. Welthuijsen, *Chem. Weekbl.,* **5**, 829 (1958).

142. Trinder, N., *Analyst,* **73**, 494 (1948).

143. Uppstrom, L. R., *Anal. Chim. Acta,* **43**, 475 (1968).

144. Vasilevskaya, A. E., and T. K. Lenskaya, *Zh. Anal. Khim.,* **20**, 747 (1965).

145. Vernon, F., and J. M. Williams, *Anal. Chim. Acta,* **51**, 533 (1970).

146. Vinkovestskaya, S. Y., and V. A. Nazarenko, *Zavod. Lab.,* **32**, 1202 (1966).

147. Wakamatzu, S., *Bunseki Kagaku,* **9**, 22 (1960).

148. Wolszon, J. D., J. R. Hayes, and W. H. Hill, *Anal. Chem.,* **29**, 829 (1957).

149. Yoe, J. H., and R. L. Grob, *ibid.,* **26**, 1465 (1954).

BROMINE

E. R. WRIGHT, R. A. SMITH, AND B. G. MESSICK

Texas Division, The Dow Chemical Co., Freeport, Texas

Bromine is a common element quite widely distributed in nature. At a concentration of 67 ppm, it is the seventh most abundant mineral element in seawater. It also occurs in many salt and brine deposits. Although its role in the human body has not been precisely delineated, bromine is found in blood plasma to the extent of about 10 ppm and in the thyroid gland at somewhat higher concentrations (27). Most plants, including foods, contain a few parts per million bromine (25). Considerable investigational work has been done to ascertain the bromine content of foods fumigated with methyl bromide or ethylene dibromide (8,31,34,35).

The choice of the best analytic method for determination of bromine is not a simple one because many diverse methods are available. Of the noncolorimetric methods, analysis by neutron activation (5,6) has the greatest sensitivity (limit of detection about 0.001 μg Br). Analysis by x-ray emission is highly specific and is sensitive to about 1 ppm in brines (9). Titrimetric methods depending upon oxidation to bromate (Van der Meulen method) can be made sensitive to about 1 ppm with acceptable accuracy (18). Null-point potentiometry (45) is sensitive to about 1 ppm but requires a physical separation of bromide from chloride.

Of the colorimetric methods, those which measure bromine directly are generally useful in the range of 1 ppm or above. Indirect colorimetric methods which depend upon the catalytic properties of bromide ion upon some chemical system are somewhat more sensitive but are not specific for bromine. Among the various direct colorimetric methods, only the phenol red method is reasonably specific for bromine. The other less specific methods require either a physical separation of the bromine by volatilization or elimination of chlorine interference by the Van der Meulen method.

I. SEPARATIONS

A process of general usefulness is one in which bromide is oxidized to free bromine or cyanogen bromide and volatilized in a current of air or nitrogen. This process accomplishes the purpose of putting the bromine in a readily measurable form and of separating it from excess oxidant or other interfering substances. Chromic acid, permanganate, nitric acid, or persulfate may be used as oxidants. The separation is completely successful, however, only in cases where the chloride ion can be held to a concentration of a few milligrams per milliliter or less. In other cases, such as brines or seawater where the chloride concentration is higher, some contamination of the volatilized bromine by free chlorine invariably occurs. This type of interference can be avoided by a repetition of the oxidation process (26). However, in such cases, it is much better to avoid the separation entirely by use of the bromate process devised by Van der Meulen (39,40). For a discussion of this process, see Section IC.

A. VOLATILIZATION AS FREE BROMINE (46)

The following procedure is essentially that of Neufeld (27) and results in the quantitative volatilization of up to 900 μg bromine with little contamination by chlorine if 5 mg or less chlorine is present. Iodine is converted to iodate and is

TABLE I. Interference of Chloride in Volatilization of Bromine(32)

Bromide present, μg	Iodide present, μg	Chloride present, μg	Bromine found, μg, by method of	
			Neufeld	Van Pinxteren
0	10	1	0	0.6
0	10	5	1.3	2.2
0	10	10	4.5	4.9
60	10	10	65	63
0	10	100	212	112

not volatilized (see Table I). The volatilized bromine can be determined by absorbing it in methyl orange and measuring the extent of bleaching.

Procedure

Dissolve the sample containing 3 to 900 μg bromine in 7.0 ml water in a 100-ml Erlenmeyer flask fitted with a two-hole stopper. Incline the flask at an angle of 60° for maximum aeration. The stopper carries two tubes, one slightly drawn out and reaching the bottom of the inclined flask, the other bent to 60° and attached to a small bubbler containing the collecting medium. Add 2.5 ml concentrated sulfuric acid down the side of the flask very slowly and with constant cooling in tap water. This operation should take at least 10 min. Add 4 ml chromic acid–sulfuric acid mixture (20 g chromic acid, 40 ml concentrated sulfuric acid, and 120 ml water). Insert the stopper and aerate for 1 hr at a rate of about 150 ml air per min. Change receiver and aerate a further 2 hr. If halogen comes over in the second period, this indicates that too much chloride is present to make a sharp separation by this method.

B. VOLATILIZATION AS CYANOGEN BROMIDE

This procedure is that of Van Pinxteren (42). Limitations of the separation process are shown in Table I.

Special Solutions

Potassium Cyanide. Dissolve 6.5 g KCN in 100 ml water.

Benzidine Hydrochloride, water solution, saturated. To prepare a colorless solution, dissolve 2 g benzidine hydrochloride in 100 ml hot water, add 50 mg activated carbon, and filter hot. Filter again after cooling.

Ethanol, 96 and 50%.

Chromic acid. Dissolve 150 g CrO_3 in 80 ml water.

Procedure

Use the apparatus shown in Fig. 1. Place in tube A 2–5 ml of the sample made $1N$ in sulfuric acid and containing 1 to 20 μg bromine. Add 3 ml potassium cyanide solution and 2–5 ml chromic acid. CAUTION: Since copious fumes of hydrogen cyanide are evolved in this and subsequent operations, it is essential that this entire procedure be carried out in a well-ventilated hood with

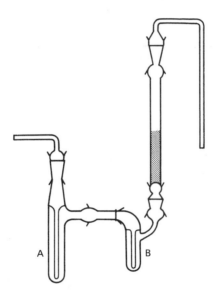

Fig. 1. Apparatus for volatilization of bromine as cyanogen bromide. (Reproduced by permission of the Society for Analytical Chemistry, *Analyst,* 77, 371, 1952.)

proper safety precautions. Add to tube B 4 ml benzidine hydrochloride solution previously mixed with 1 ml pyridine and 1 ml 96% alcohol. Wet the glass beads with this solution also. A pool of liquid should remain with the beads. Place a water bath at 100°C around tube A and an ice bath around tube B. Connect the tubes and pass a current of nitrogen through the apparatus for 8 min at about 75 ml per min. Complete the determination as described in the last paragraph of Section IIIC.

C. OXIDATION TO BROMATE

Although this process is not a physical separation of bromine, it is discussed in this section because the oxidation to bromate and subsequent reactions of the bromate serve the same purpose as a separation by volatilization or other physical process. The method (originally devised by Van der Meulen) (39,40) may be carried out in the presence of large concentrations of chloride ion. Thus, the limitations of the volatilization processes are avoided.

Bromide is oxidized to bromate with excess hypochlorite (or other source of available chlorine) at about pH 6 in a solution containing as much as 10% chloride:

$$Br^- + 3\,OCl^- \longrightarrow BrO_3^- + 3Cl^-$$

The excess hypochlorite is destroyed with a mild reducing agent such as sodium formate under conditions which do not reduce bromate. The bromate can then be reacted with iodide to liberate free iodine which can be determined colorimetrically or by titration, or the bromate can be reacted with bromide to generate free bromine:

$$BrO_3^- + 5\,Br^- + 6H^+ \longrightarrow 3Br_2 + 3H_2O$$

The free bromine can be measured colorimetrically from its bleaching action (methyl orange, Section IIB) or its reaction to give a change in hue of a chromogen (rosaniline, Section IID). One distinct advantage of this bromate procedure lies in the fact that each microgram of bromide originally present is converted by the oxidation process to 6 μg free bromine. The sensitivity is thus increased sixfold compared to oxidation directly to free bromine, for example, by chromic acid.

Van der Meulen used hypochlorite as the oxidant in a weakly acidic solution buffered with boric acid and containing about 10% sodium chloride. Hydrogen peroxide was used to destroy the excess hypochlorite. Later investigators have shown that sodium chloride need not be present if the bromine content is in the order of 10 mg or less (15,22). In colorimetric analysis, therefore, the addition of sodium chloride may be omitted. This omission reduces the blank.

Sodium formate is now generally used instead of hydrogen peroxide for reducing the excess hypochlorite since formate does not affect the later reactions of the bromate and need not be removed from the solution (2,22). According to Willard and Heyn (44), the oxidation is quantitative at pH 5.5 to 7.0, and they recommend an acetate buffer. Phosphate (2,22) and calcium carbonate (7) have also been used for pH control.

The reaction between bromate and bromide must be carried out in acid solution. If the acidity is too high, chlorate (formed during boiling of the hypochlorite) will also release free bromine from bromide. For this reason the acidity should preferably not exceed 3N. With the methyl orange method, Smith (32) found maximum sensitivity at a pH of 1.1. Hunter (17) used 2.9N acid and a considerably lower bromide concentration in the bromination of rosaniline. Under these conditions, 1 μg or less bromine as bromate can be measured with satisfactory accuracy. Details are given in Sections IIIB and IIID.

Iodine is converted to iodate. During the subsequent reaction of bromate with bromide, the iodate also reacts slowly to liberate bromine. Consequently, samples containing iodide in gross excess to bromide require removal of the iodine prior to oxidation of bromide to bromate. Iodide up to 60 mg may be removed from a 10-ml sample by adding 1 ml $0.5M$ sodium nitrite and 2 ml $1N$ sulfuric acid and boiling to drive off the iodine.

All reagents must be examined carefully for traces of bromine. Commercial hypochlorite is ordinarily not useable for determinations in the lower range of bromine content. A satisfactory hypochlorite solution can be made as described in IIIB, using ACS reagent hydrochloric acid as a source for chlorine of low bromine content. Organic matter, including indicators, must be excluded because of the danger of loss of bromine through formation of organic bromine compounds.

II. METHODS OF DETERMINATION

DIRECT COLORIMETRIC METHODS

The more sensitive reagents respond to free chlorine as well as free bromine. Among these are methyl orange, o-tolidine, and the pyridine–benzidine–cyanide reagent of König. The use of these reagents, therefore, requires a complete separation from free chlorine by one of the methods listed under Section I. Other reagents such as phenol red or fluorescein can be employed in the presence of traces of free chlorine, but they are less sensitive.

A. PHENOL RED METHOD

The phenol red method is probably the most commonly used colorimetric method for bromide. In spite of its limitations, it has won acceptance because of its simplicity and tolerance for large amounts of chloride. As modified by Goldman and Byles (12), it is the standard method for examination of water and wastewater (1). Their version is suitable for measuring 5 to 50 μg bromide in a 50-ml water sample.

The method has few interferences. Reducing agents, including ammonium ion, interfere through reaction with the oxidant used in the method. Free iodine or iodide reacts like bromine and must be removed (by oxidation with nitrite); but in most types of samples, iodide will be present in much lower concentrations than bromide. Chloride ion does not interfere in ratios of chloride to bromide as high as 10,000 to 1; consequently, the method can be used on brines or other materials containing a high chloride content.

In the original method of Stenger and Kolthoff (33), bromide is oxidized to hypobromite by hypochlorite in a solution containing phenol red and

TABLE II. Halogenated Compounds of Phenol Red

Indicator	Synonym	pH Range	Color change
Phenol red	phenolsulfonephthalein	6.8–8.4	yellow–red
Chlorophenol red	dichlorophenolsulfo-nephthalein	4.8–6.4	yellow–red
Bromophenol red	dibromophenosulfo-nephthalein	5.2–6.8	yellow–red
Bromophenol blue	tetrabromophenolsulfo-nephthalein	3.0–4.6	yellow-blue
Bromochlorophenol blue	dibromodichlorophenol-sulfonephthalein	3.0–4.6	yellow–blue
Chlorophenol blue	tetrachlorophenolsulfo-nephthalein	3.0–4.6	yellow–blue

buffered at pH 8.8. Under these conditions, phenol red reacts preferentially with hypobromite provided that the excess hypochlorite is removed after a short time. The major product is tetrabromophenolsulfonephthalein (the indicator bromophenol blue) instead of the chlorine analog, although lesser amounts of other di- and tetrahalogenated phenol red compounds as shown in Table II are formed. After adjusting to pH 5.0–5.4, where unreacted phenol red and dihalogenated products are in the yellow form, the absorbance of the blue color due to bromophenol blue is measured at 590 nm.

The main limitation of the method is the necessity for controlling the ratio of phenol red to bromine within narrow limits to keep the formation of halogenated products other than bromophenol blue at a minimum. Beer's law is not obeyed, the reaction being less sensitive for the extreme ratios of bromine to phenol red.

Larson and Ingber (23) improved the precision of the method by carrying out the oxidation of bromide and bromination of phenol red in two steps rather than simultaneously. Beer's law is closely approximated in the range of 4 to 16 μg bromide per 10-ml sample.

A more widely used modification, however, has been the substitution of chloramine-T for hypochlorite as the oxidant. Balatre (4) found that with chloramine-T, the oxidation of the bromide to hypobromite can be carried out simultaneously with the bromination of phenol red at pH 5. In this case, sodium thiosulfate is used to stop the reaction. Houghton (16) and later Goldman and Byles (12) determined the optimum conditions for the reaction. Beer's law is approximated in the range of 5 to 50 μg bromide per 50 ml solution. Above 50 μg bromide, the ratio of phenol red to bromine is too low and the bromophenol blue product is partially bleached out. Below 5 μg, the ratio of phenol red to bromine is too high, favoring the formation of the dibrominated product (bromophenol red). Details of application of the method as revised by Goldman and Byles are given in Section IIIA.

B. METHYL ORANGE METHOD

This reagent was advocated by Taras (36,37) for the determination of free chlorine in water in the presence of chloramine. Two moles free chlorine react with one mole methyl orange to form colorless products, apparently according to the reaction

$$(CH_3)_2 NC_6 H_4 N:NC_6 H_4 SO_3 Na + 2Cl_2 \longrightarrow$$
$$(CH_3)_2 NC_6 H_4 NCl_2 + Cl_2 NC_6 H_4 SO_3 Na$$

Free bromine reacts with methyl orange in a similar manner. Since the products are colorless, the bromine content is measured by the amount of bleaching of the methyl orange. Visual or photometric measurements may be used. Permanent standards are stable for at least a month.

Bromide in the sample can be converted to free bromine either by Neufeld's method (Section IA) or Van der Meulen's method (Section IC). The Van der Meulen separation gives greater sensitivity.

Smith (32) developed the combination of methyl orange finish after Van der Meulen oxidation for use with scrubber solutions from air pollution sampling. This method is suitable for the range of 1 to 10 μg bromide per 10 ml solution. Almost complete bleaching of the methyl orange indicates that 10 μg Br has been exceeded, because of either too large a sample or contamination of the reagents. After suitable reagents have been prepared using water with low bromine demand and hypochlorite with low bromine content, the analysis is simple. Its principal advantage is its combination of high sensitivity with a tolerance for large amounts of chloride. Application of Smith's method to the determination of bromide in an ashed blood sample is given in Section IIIB.

In using methyl orange, the pH must be kept constant, preferably at 2.0 or below, to avoid variations in hue with pH. Taras stated that the reaction with chlorine was slow in the presence of sulfuric acid, but Smith (32) was unable to confirm this effect for the reaction with bromine. Manganic manganese interferes by reacting like free halogen. Ferric iron up to 25 ppm does not react with methyl orange appreciably. Iodine does not interfere as such. However, if the bromate separation method is used, the iodide is converted to iodate. Iodate will react slowly with bromide to release bromine and bleach the methyl orange. In this method, iodide appears as about 7% of an equal weight of bromide.

C. KÖNIG'S REACTION METHOD (3,24,42,43)

According to W. König (20,21), cyanogen bromide or chloride forms an addition product with pyridine, which in turn reacts with an aromatic amine with rupture of the pyridine ring to form an intensely colored dianilide of glutaconic dialdehyde:

$$C_5H_5N \text{ (pyridine)} + CNBr \longrightarrow C_5H_5 NCNBr$$
$$C_5H_5NCNBr + 2RNH_2 \longrightarrow$$
$$RNH_2 BrCH:CHCH:CHCH:NR + CNNH_2 \text{ (cyanamide)}.$$

Milton (24) suggested this reaction using benzidine as a colorimetric method for free bromine or chlorine. The method has a sensitivity about double that of o-tolidine, enabling about 2 μg bromine in a 5-ml sample to be measured. The reaction can also be used for determination of cyanide.

The colored compound is reddish in hue, with a maximum absorption at 530 nm. Beer's law is followed. Reproducibility is poor unless conditions are carefully standardized. For this reason, it is best to distill the cyanogen bromide as described in Section IB to eliminate the effect of other ions present. The colored compound is stable only for about 1 hr under ordinary laboratory conditions.

Application of the method in determining bromide in flour is given in Section IIIC.

D. ROSANILINE METHOD

Rosaniline reacts in acid solution with free bromine to form a red-colored compound described as either pentabromorosaniline or tetrabromorosaniline. The maximum absorbance is near 570 nm. Beer's law is followed. Since the brominated product is insoluble in aqueous solutions, a solvent must be employed or the colored compound may be extracted with an immiscible solvent such as benzyl alcohol. The latter procedure has the advantage of separating the brominated compound from rosaniline itself, which has a slight color.

The reaction is carried out in $3-7N$ sulfuric acid to minimize color due to rosaniline. If a solvent such as butanol is used, the ratio of rosaniline to bromine must be carefully controlled for best results.

Free chlorine must be avoided. Turner (38) used persulfate to release bromine preferentially to chlorine. Smith (32) found chloride interference by this method to the extent of 1 mg chloride appearing as 5 μg bromide.

Hunter (17) used the bromate separation to avoid interference of chloride. The method gives satisfactory measurement of 1 μg of bromine in 5 ml solution. The color is stable for at least several hours. Further details of this method are given in Section IIID.

E. FLUORESCEIN METHOD (14,41)

The reaction between fluorescein and free bromine to form eosin has been used for the colorimetric detection and determination of bromide in the presence of chloride. Traces of free chlorine do not interfere, but iodine does interfere. Most investigators have agreed that the nature of this reaction is such that only semiquantitative results can be attained. Pinching and Bates (29) give a simple semiquantitative test for bromide in sodium chloride by the fluorescein method. Because of its simplicity, this type of test may be useful in some cases. In general, however, other colorimetric methods are superior where quantitative results are desired.

F. GOLD CHLORIDE METHOD (13,23)

In the presence of bromide ion, gold chloride assumes a brownish hue. The method is widely used in clinical laboratories for analysis of blood of patients undergoing bromide therapy. The method, although rapid and simple, is neither accurate nor sensitive.

G. CRESOL RED METHOD

Cresol red has been used by Fennell and Webb (10) for the submicrodetermination of bromine in organic compounds containing no chlorine or iodine. After decomposition of about 50 μg sample by Kirsten's (19) combustion method, bromide is oxidized to free bromine using bromate as the oxidant in about 15 ml acidic solution. The free bromine reacts with cresol red, present during the oxidation, to reduce the absorbance at 519 nm. The method gives good precision in the 2- to 40μg bromide range. Chloride or iodide interfere at the 15-μg level.

INDIRECT COLORIMETRIC METHODS

H. OXIDATION OF IODINE CATALYZED BY BROMIDE METHOD

Methods based on the catalytic effect of bromide on the oxidation of iodine to iodate enables 0.01 μg bromide in 10 ml water solution to be determined. Chloride interferes so that the methods are best suited for use when the ratio of chloride to bromide is less than 250.

Shiota et al. (30) originally developed the method using permanganate in sulfuric acid as the oxidant at 0°C. The amount of iodine left in solution after a given reaction time is inversely proportional to the amount of bromine present. They determined the iodine by extraction into carbon tetrachloride followed by reaction with mercuric thiocyanate and ferric alum to form the colored ferric thiocyanate complex.

Fishman and Skougstad (11) simplified the method by measuring the absorbance of the iodine–carbon tetrachloride extract at 515 nm without further treatment. Their method is suitable for 0.01–1 μg bromide in 10 ml natural water samples. Chloride interferes, with 25 μg chloride appearing as about 0.1 μg bromide. Oxidizing or reducing substances interfere, but normally they are not present in natural water at levels giving interference.

More recently, Yonehara et al. (47) have recommended ceric ammonium sulfate in combination with dichromate and nitric acid as the oxidant. The method has no advantage over the method with permanganate except that the reaction proceeds favorably nearer room temperature (30°C).

I. CHLORINATION OF AMMONIA INHIBITED BY BROMIDE METHOD

Zitomer and Lambert (48) developed a method based on the "inhibition" effect of bromide on the reaction between hypochlorite and ammonia to form trichloramine. The amount of trichloramine formed is dependent upon the amount of bromide originally present. Trichloramine is measured as the blue starch–triiodide complex after first selectively destroying unreacted hypochlorite with nitrite ion. Absorbance decreases linearly with increased bromide concentration.

The determination is rapid and applicable in the range of 1–60 μg bromide per 50-ml sample. Chloride interferes at ratios of chloride to bromide above about 500. Ammonia and iron also interfere, but they may be removed by a simple ion exchange technique described by Zitomer and Lambert.

III. APPLICATIONS

A. BROMINE IN SEAWATER OR BRACKISH WATER

This is an example of the use of the phenol red method. It is applicable to seawater or dilutions of seawater as great as 600-fold.

Special Solutions

Acetate buffer solution. Dissolve 68 g sodium acetate trihydrate, $NaC_2H_3O_2 \cdot 3H_2O$, in distilled water. Add 30 ml glacial acetic acid and make up to 1 liter. The pH should be 4.6–4.7.

Phenol red solution. Dissolve 0.021 g phenolsulfonephthalein sodium salt in 100 ml distilled water. If the sodium salt is not available, dissolve 0.02 g phenolsulfonephthalein in 2.84 ml 0.02N NaOH and dilute to 100 ml with distilled water.

Chloramine T solution. Dissolve 0.05 g chloramine-T, and dilute to 100 ml with distilled water. Store in a dark bottle and refrigerate or prepare a fresh solution daily.

Sodium thiosulfate solution, approximately 2M. Dissolve 49.6 g $Na_2 S_2 O_3 \cdot 5H_2 O$ and dilute to 100 ml with distilled water.

Bromide standard solution. Dissolve 0.149 g ACS reagent potassium bromide and dilute to exactly 100 ml. Dilute 5 ml of this solution to 1 liter to give a solution containing 5 μg bromide per ml.

Procedure

Pipet a sample expected to contain 5 to 50 μg bromide into a 100-ml flask or beaker. Add distilled water, if necessary to give 50 ml total volume. Add 2 ml buffer solution, 2 ml phenol red solution, and 0.5 ml chloramine-T solution. Mix thoroughly. Exactly 20 min after the chloramine-T addition, dechlorinate by adding, with mixing, 0.5 ml sodium thiosulfate solution. The pH at this point should be 5.0–5.4.

Read the color using a filter with maximum transmittance at 590 nm. Determine the amount of bromide by use of a standard calibration graph relating bromide content to photometric reading. Prepare the curve by using as samples amounts of bromide in 5-μg intervals from 0 to 50 μg.

If, during the oxidation with chloramine-T, a sample turns purple and then fades, too much bromide is present. Repeat the test with a smaller portion of sample.

B. TOTAL BROMINE IN BLOOD

This is an example of the Van der Meulen separation combined with the methyl orange finish. The procedure is applicable to samples containing 2 ppm or more bromide and gross amounts of chloride.

Reagents

Redistilled water. Make distilled water approximately 0.05% in potassium permanganate and sodium hydroxide and distill. Discard the first 10% distilled over. Use the redistilled water for reagent preparation and throughout the Van der Meulen oxidation amd methyl orange bleaching.

Acidified water, two solutions. Adjust the pH to 1.1 and 2.0 with 6N hydrochloric acid.

Sodium formate, 85 g in 250 ml solution.

Methyl orange stock solution, 0.05%.

Methyl orange, 0.002%. Twenty milliliters methyl orange stock diluted to 500 ml. Adjust the pH of this solution to 1.1.

Sodium molybdate, 1%.

Potassium bromide, 20%; pH 1.1.

Sodium hydroxide, A.C.S. reagent pellets.

Potassium bromide standard, 1 ml equals 1 μg bromide. Dissolve 74.5 mg potassium bromide in 500 ml water. Transfer 5 ml of the solution to a second 500-ml flask and dilute to the mark.

Sodium hypochlorite. In a two-necked flask place 40 g manganese dioxide and 130 ml ACS reagent hydrochloric acid. Fit a relief tube in one neck of this flask and fit the other neck with a tube running to the bottom of a second two-necked flask containing about 100 ml water. Connect this second flask with a tube to the bottom of a third two-necked flask containing 500 ml 1.1N soduim hydroxide made from ACS reagent pellets. Immerse the flask of sodium hydroxide in an ice bath. Generate chlorine in the first flask by gentle heating. Scrub the chlorine through water in the second flask and absorb the chlorine in the third flask. From time to time, remove 1 ml of the caustic solution, add 5 ml 3% hydrogen peroxide to destroy the sodium hypochlorite, and titrate with 0.1N acid. Stop the chlorine generation when about 1 ml 0.1N acid is required. Transfer the hypochlorite to a brown bottle. The solution is stable indefinitely when stored in a refrigerator.

Procedure

Dissolve a pellet of sodium hydroxide in a 25-ml nickel crucible, using a few drops of water. Pipet 1 ml blood into the crucible and mix the sample with the alkali solution. (Blood samples can be stored without clotting by making them 0.5% sodium citrate.) Cautiously evaporate the sample to dryness on a hot plate. Add 2 g sodium hydroxide pellets to the crucible and heat on a hot plate or burner to decompose the sample. This operation is preferably carried out under a hood because of the disagreeable odor produced. After frothing of the sample has ceased and a mobile liquid remains, place the crucible in a muffle furnace at 600°C. After bubbling ceases, remove the crucible from the furnace and cautiously add a few milligrams sodium peroxide. Return the crucible to the muffle for a few minutes and then repeat the peroxide addition until no reaction occurs on addition of peroxide. Turn the crucible to the side to rinse down any unreacted sample. When the oxidation is complete, remove the crucible from the furnace and allow the melt to cool on the sides of the crucible. After the melt has cooled, fill the crucible two-thirds full with water and dissolve the fusion residue. Transfer the solution to a 100-ml beaker. Partially neutralize the sodium hydroxide by adding 7 ml 6N hydrochloric acid. Boil the sample for 10 min to decompose peroxides. Filter the solution through No. 42 Whatman paper and make to 50 ml volume in a flask.

Pipet a suitable aliquot (10 ml suggested) of the sample into a 50-ml beaker. Place the sample on a pH meter and add 6N hydrochloric acid until a pH of about 3 is reached. Add 1.0 ml 1N hypochlorite. The pH must not be below 4 after this hypochlorite addition or chlorine will be lost. Add hydrochloric acid to bring the pH below 6. Add about 0.1 g calcium carbonate. Add a boiling chip, cover the beaker, and boil the solution for 8 min. Rinse down the beaker walls and cover glass with redistilled water. Add 2.5 ml formate and boil for 8 min. Cool the sample to room temperature. Adjust the pH to 1.1 with hydrochloric acid. Add 5 ml 0.002% methyl orange, 0.5 ml 1% sodium molybdate, and 2 ml 20% potassium bromide in this order. Mix the solution. Five min after the addition of potassium bromide, adjust the pH of the solution to 2.0. Transfer the solution to a 50-ml volumetric flask and make to volume with water at pH 2.0. Read the absorbance using a green filter. Determine the amount of bromide by use of a standard curve relating bromide content to colorimeter reading. Prepare the curve by carrying amounts of bromide ranging from 0 to 8 μg through the hypochlorite oxidation and methyl orange bleaching. Determine a reagent blank for the blood sample by carrying 2 g sodium hydroxide through the entire procedure.

C. TOTAL BROMINE IN FLOUR

This is an example of the use of the Van Pinxteren separation method (Section IB) with the König colorimetric finish (Section IIC). The procedure is applicable to samples containing 2 ppm or more bromide and less than 0.1% chloride.

Special Solutions
Those reagents described in Section IB plus

Potassium hydroxide, 5% in ethanol.

Sulfuric acid, approximately 1N and 7N solutions.

Procedure

Weigh 1 g sample into a 25-ml nickel crucible. Wet the sample with 2 ml 5% alcoholic potassium hydroxide. If the sample has been freshly fumigated with methyl bromide, allow it to stand 1 hr at room temperature to hydrolyze the organic bromide. Samples not recently fumigated may be dried immediately on a steam bath. Do not evaporate the alcohol in an electric oven because of danger of explosion. After evaporating the alcohol, heat the sample a short time at about 110°C and then add 1 g sodium hydroxide pellets. Heat the sample on a hot plate at such a temperature as to melt the pellets and cause a moderate rate of decomposition of the flour. After bubbling on the hot plate ceases, place the sample in a muffle at 600°C. If the sample boils violently or catches fire, remove it from the muffle until the reaction subsides, then return the sample to the muffle.

After bubbling ceases, remove the crucible from the furnace and cautiously add a few milligrams of sodium peroxide. Return the crucible to the furnace for a few minutes and then repeat the peroxide addition until no reaction occurs on addition of peroxide. Turn the crucible to the side to rinse down any unreacted sample. When the oxidation is complete, remove the crucible from the furnace and allow the melt to cool on the sides of the crucible. After the melt has cooled, fill the crucible two-thirds full with water and dissolve the fusion residue. Transfer the solution to a 100-ml beaker and partially neutralize the sodium hydroxide by adding 3 ml 7N sulfuric acid. Add a boiling chip and boil the sample for 10 min to decompose peroxides. Filter the solution through paper into a 50-ml beaker. Add a boiling chip to the filtrate and evaporate by boiling until a scum appears on the surface. Stop the evaporation at this point to prevent spattering. Cool the sample. Using normal sulfuric acid, acidify the sample to the methyl red color change, then add 1 ml excess.

Transfer the sample or aliquot thereof to tube A of the apparatus shown in Fig. 1 of Section IB. Add 3 ml potassium cyanide solution and 5 ml chromic acid. Connect tube A, tube B, and the bead column. Pour into the bead column 4 ml benzidine hydrochloride solution previously mixed with 1 ml pyridine and 1 ml 96% alcohol. Let air escape from the joint between tubes A and B so that part of the pyridine solution will run down into tube B. The end of the dip pipe in tube B should be covered with liquid and a pool of liquid should remain on the beads. Place a water bath at 100°C around tube A and an ice bath around tube B. Pass a current of nitrogen at 75 ml per min for 8 min.

Rinse the solution from the beads and from tube B into a 50-ml volumetric flask. Make the solution to volume with 50% alcohol. Measure the absorbance within 10 min at 580 nm. Read the amount of bromide from a standard plot. Prepare the standard plot by carrying amounts of bromide from 0 to 20 μg through the volatilization procedure. A reagent blank should be determined for the complete procedure.

D. BROMINE IN BRINE

This is an example of the use of the Van der Meulen separation with the rosaniline colorimetric finish. The procedure is applicable to samples containing

1 μg or more bromide in the presence of up to 450 mg chloride and up to 250 μg iodide.

Reagents

Sodium hypochlorite. Prepared as given in Section IIIB.

Sodium formate. Dissolve 50 g in 100 ml solution.

Buffer solution at pH 6.35. Make 18.2 g sodium dihydrogen phosphate dihydrate and 3.6 g potassium hydroxide to 100 ml with water.

Rosaniline solution. Dissolve 6 mg rosaniline in 100 ml 2N sulfuric acid. The solution keeps for several weeks in a brown bottle.

Bromide-molybdate mixture. Dissolve 0.15 g potassium bromide and 3 g ammonium molybdate in water and dilute to 100 ml.

Sulfuric acid, 14N.

Tertiary butanol, containing 5% absolute ethanol to prevent freezing.

Procedure

Stage 1: Conversion of Bromide to Bromate. Carry through this procedure a series of standards containing up to 25 μg bromide. Also carry through the unknown sample in which up to 750 mg sodium chloride may be tolerated.

To a test tube (16 x 130 mm) add 1 ml buffer solution and then a suitable volume of the bromide solution and make up with water to a total volume of 4.5 ml. Do this for each standard solution and for the sample of unknown bromide content. To each test tube add 0.25 ml hypochlorite solution, mix the contents, and immerse the tube in a bath of boiling water for 10 min. Add 0.25 ml formate solution to destroy the excess of hypochlorite, mix the solution, and replace the tube in the boiling water for a further 5 min. Cool the tube, add water to replace that lost by evaporation, and mix the contents. This solution is used for the colorimetric procedure described in Stage 2.

Stage 2: Colorimetry. *(a) Preparation of the standard curve.* In a test tube or colorimeter tube place 0.1 ml bromide-molybdate reagent mixture, 0.1 ml rosaniline solution per microgram bromine present as bromate in the test solution to be used, 0.4 ml or less water, 0.4 ml 14N sulfuric acid and 1 ml of the solution prepared in Stage I, in that order, and mix them. The total volume at this stage is 2.0 ml, the amounts of bromide–molybdate and sulfuric acid being constant in all tests. Any volume of test solution less than 1.0 ml can be used if the balance is made up with water, which is added before the test solution. Leave the reaction mixture at 20–30°C for 3 min and add 2 ml tertiary butanol and 1 ml 14N sulfuric acid. Mix the solution and measure the absorbance at 570 nm.

(b) Analysis of the brine sample. In two test tubes, place the reagents as in (*a*) on the assumption that 0.1 ml of the brine solution is to be used, and also that one test tube contains 0.05 ml and the other 0.5 ml rosaniline. Now add 0.1 ml of the unknown solution to each test tube and continue as in (*a*). From the two observations, one can determine approximately the amount of bromide present, unless the amount taken for conversion to bromate exceeds about 250 μg. Further test portions are then taken with suitable amounts of rosaniline solution until the absorbance is found to be proportional to the test portions taken. When the liberated bromine is greatly in excess of the rosaniline necessary to absorb it, the readings will be low owing to the bleaching action of the bromine on any bromorosaniline that may be formed.

REFERENCES

1. American Public Health Association, *Standard Methods for the Examination of Water and Wastewater*, 13th ed., Washington, D.C., 1971, p. 75.

2. d'Ans, J., and P. Höfer, *Angew. Chem.*, **47**, 71 (1934).

3. Asmus, E., and H. Garshagen, *Z. Anal. Chem.*, **136**, 269 (1952).

4. Balatre, P., *J. Pharm. Chim.*, **24**, 409 (1936).

5. Bowen, H. J. M., *Biochem. J.*, **73**, 381 (1959).

6. Cosgrove, J. F., R. P. Bastian, and G. H. Morrison, *Anal. Chem.*, **30**, 1872 (1958).

7. Doering, H., *Z. Anal. Chem.*, **108**, 255 (1937).

8. Dudley, H. C., *Ind. Eng. Chem., Anal. Ed.*, **11**, 259 (1939).

9. Dunton, P. J., *Appl. Spectr.*, **22**, 99 (1968).

10. Fennel, T. R. F. W., and J. R. Webb, *Z. Anal. Chem.*, **205**, 90 (1964).

11. Fishman, M. J., and M. W. Skougstad, *Anal. Chem.*, **35**, 146 (1963).

12. Goldman, E., and D. Byles, *J. Amer. Water Works Assoc.*, **51**, 1051 (1959).

13. Gray, M. G., and M. Moore, *J. Lab. Clin. Med.*, **27**, 680 (1942).

14. Hahn, F. L., *Mikrochemie*, **17**, 222 (1935).

15. Haslam, J., and G. Moses, *Analyst*, **75**, 343 (1950).

16. Houghton, G. U., *J. Soc. Chem. Ind.* (London), **65**, 277 (1946).

17. Hunter, G., and A. A. Goldspink, *Analyst*, **79**, 467 (1954).

18. Kaplan, D., and I. Schnerb, *Anal. chem.*, **30**, 1703 (1958).

19. Kirsten, W. J., *Microchem. J.*, **7**, 34 (1963).

20. König, W., *J. Prakt. Chem.*, **69**, 105 (1904); *J. Chem. Soc., Abstr.*, **1**, 449 (1904).

21. König. W., *J. Prakt. Chem.*, **70**, 19 (1904); *J. Chem. Soc., Abstr.*, **1**, 816 (1904).

22. Kolthoff, I. M., and H. Yutzy, *Ind. Eng. Chem., Anal. Ed.*, **9**, 75 (1937).
23. Larson, R. P., and N. M. Ingber, *Anal. Chem.*, **31**, 1084 (1959).
24. Milton, R. F., *Nature*, **164**, 448 (1949).
25. Monier-Williams, G. W., *Trace Elements in Foods*, Wiley, New York, 1949.
26. Murphy, T. J., W. S. Clabaugh, and R. Gilchrist, *J. Res. Nat. Bur. Stand.*, **53**, 13 (1954).
27. Neufeld, A. H., *Can. J. Res.*, **14B**, 160 (1936).
28. Paul, W. D., R. W. Knouse, and J. I. Routh, *J. Amer. Pharm. Assoc., Sci. Ed.*, **41**, 205 (1952).
29. Pinching, G. D., and R. G. Bates, *J. Res. Nat. Bur. Stand.*, **37**, 311 (1946).
30. Shiota, M., S. Utsumi, and I. Iwasaki, *Nippon Kagaku Zasshi*, **80**, 753 (1959).
31. Shrader, S. A., A. W. Beshgetoor, and V. A. Stenger, *Ind. Eng. Chem., Anal. Ed.*, **14**, 1 (1942).
32. Smith, R. A., and Sarah Black, unpublished work from the files of the Dow Chemical Co.
33. Stenger, V. A., and I. M. Kolthoff, *J. Amer. Chem. Soc.*, **57**, 831 (1935).
34. Stenger, V. A., S. A. Shrader, and A. W. Beshgetoor, *Ind. Eng. Chem., Anal. Ed.*, **11**, 121 (1939).
35. Tanada, A. F., H. Matsumoto, and P. J. Scheuer, *J. Agr. Food Chem.*, **1**, 453 (1953).
36. Taras, M., *Anal. Chem.*, **19**, 342 (1947).
37. Taras, M., *J. Amer. Water Works Assoc.*, **38**, 1146 (1946).
38. Turner, W. J., *Ind. Eng. Chem., Anal. Ed.*, **14**, 599 (1942).
39. Van der Meulen, J. H., *Chem. Weekblad*, **28**, 82 (1931).
40. Van der Meulen, J. H., *Chem. Weekblad*, **28**, 238 (1931).
41. Van der Meulen, J. H., *Chem. Weekblad*, **36**, 702 (1939).
42. Van Pinxteren, J. A. C., *Analyst*, **77**, 367 (1952).
43. Van Pinxteren, J. A. C., *Pharm. Weekblad*, **88**, 489 (1953).
44. Willard, H. H., and A. H. A. Heyn, *Ind. Eng. Chem. Anal. Ed.*, **15**, 321 (1943).
45. Winefordner, J. D. and M. Tin, *Anal. Chem.*, **35**, 382 (1963).
46. Yates, E. D., *Biochem. J.* (London), **27**, 1763 (1933).
47. Yonehara, N., S. Utsumi, and I. Iwasaki, *Bull. Chem. Soc. Japan*, **38**, 1887 (1965).
48. Zitomer, F., and J. L. Lambert, *Anal. Chem.*, **35**, 1731 (1963).

CHAPTER

3

CARBON

LARRY G. HARGIS

Department of Chemistry, University of New Orleans, New Orleans, Louisiana

CARBON DIOXIDE

Very few colorimetric methods for the determination of carbon dioxide are available. This may be due in part to inherent properties of carbon dioxide that are not particularly suitable to colorimetric methods and to the suitability of other, noncolorimetric methods. In air, the relative inactivity of carbon dioxide is a barrier to the development of direct colorimetric methods. In water, the equilibria between several different forms of carbon dioxide creates numerous difficulties. It is interesting, however, that the most popular colorimetric method utilizes these species in equilibrium in water. In addition to being a natural constituent of air and water, carbon dioxide is a common product of chemical and biochemical oxidations. The determination of carbon dioxide output is frequently helpful in gaining an understanding of the various processes that take place during such changes.

I. METHODS OF DETERMINATION

A. PHENOLPHTHALEIN METHOD (39)

This method is based on the conversion of the colored, basic form of phenolphthalein (pK_a 9.7) to the colorless, acidic form that accompanies the pH change brought about by dissolving carbon dioxide in the solution. A standard dilute solution of sodium hydroxide is prepared and colored with phenolphthalein indicator. As the solution absorbs carbon dioxide from the gas sample, the pH is decreased and part of the indicator is converted to the colorless acidic form. The decrease in absorbance, measured at 515 nm, is proportional to the amount of carbon dioxide. It is important that a known volume of the standard solution and the gas sample be brought into intimate contact in such a way as to ensure that all the carbon dioxide is absorbed by the solution. This may be accomplished by means of bubblers or by drawing or forcing the gas sample into the solution and then shaking thoroughly.

Since it is desired that a small amount of neutralization cause a substantial change in pH, the sodium hydroxide must be very dilute. Concentrations in the vicinity of $10^{-4}M$ are suitable for samples containing carbon dioxide in the range of 0.0005 to 0.032%. Lower concentrations of sodium hydroxide can be used to gain increased sensitivity provided extreme caution is used in guarding against contamination of the standard solution. The standard sodium hydroxide—phenolphthalein solution may change slightly in absorbance over a period of several days, necessitating a recalibration. Beer's law is followed for carbon dioxide concentrations in the range of 0.0005 to 0.032%. Acid- and base-forming gases such as sulfur dioxide and ammonia interfere and must be removed prior to analysis.

Reagents

Sodium Hydroxide—Phenolphthalein Solution. Dissolve 4 g sodium hydroxide in 1 liter freshly distilled water. Transfer a 10-ml aliquot of this solution to a 1-liter flask and dilute to volume with freshly distilled water. This solution is approximately $10^{-4}M$. Add phenolphthalein indicator until the solution absorbance is approximately 1.0 in a 10-cm cell.

Standard Carbon Dioxide Mixtures. Accurate carbon dioxide mixtures may be obtained from various companies specializing in gases and gas products. Alternatively, various volumes of "normal" air may be used. The carbon dioxide content of air in areas free of heavy industry remains fairly constant at about 0.031% throughout most of the United States.

Special Apparatus

Absorption Vessel. The reaction vessel may be an ordinary gas-sampling tube of known volume (about 300 ml) fitted at the top with a three-way stopcock and at the bottom with a two-way stopcock and graduated in such a way that the volumes of solution and gas in it can be read.

Procedure

Purge the absorption tube with carbon dioxide-free nitrogen. By means of the stopcock, introduce 100 ml of the standard sodium hydroxide solution plus an additional amount equal to the volume of the gas sample to be used (typically 50 to 200 ml). Connect one end of the absorption vessel to the gas sample and drain out sodium hydroxide through the other end until the predetermined volume of gas has been sucked in. Shake vigorously for 3 min, then vent the absorption tube to the atmosphere through a soda-lime absorption tube and drain the solution into a 10-cm absorption cell previously purged with nitrogen Measure the absorbance at 515 nm against a distilled water blank. Unknowns may be determined directly from standard curves of absorbance versus concentration prepared from known standards using this procedure.

B. OTHER METHODS

A method somewhat similar to the phenolphthalein method using methyl red, pK_a 5.1, as the indicator has been reported (42). Under conditions where less than half of the indicator is converted to the acid form, the concentration is inversely proportional to the transmittance. The method appears useful in the range of 0.1 to 8% carbon dioxide.

Absorption of carbon dioxide in dilute sodium hydroxide and titration of the excess hydroxide by a spectrophotometric titration employing phenolphthalein as the indicator has been reported (24). Complete absorption of carbon dioxide was accomplished for flow rates up to 100 ml per min. The sensitivity is about 1 ppm carbon dioxide for 20 liters of gas.

Direct spectrophotometric titration of carbon dioxide has been reported based on its ultraviolet absorptivity at 235 nm (40). Kinetic conditions attending the formation of carbonic acid from carbon dioxide and water were overcome by addition of the enzyme carbonic anhydrase, which catalyzes the reaction.

II. APPLICATIONS

A. CONTINUOUS DETERMINATION OF CARBON DIOXIDE IN GAS STREAM (26)

This procedure may be used for carbon dioxide concentrations in the range of 0.06 to 12%.

Reagents

Sodium Hydroxide, $10^{-4} M$.

Phenol Red, 0.02%.

Standard Carbon Dioxide Mixtures. These may be obtained commercially from several companies specializing in compressed gases and gas products.

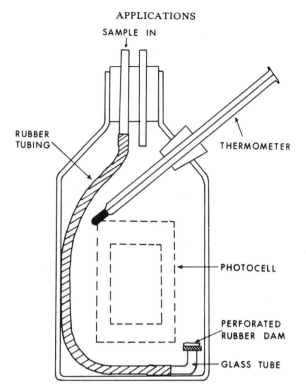

Fig. 1. Schematic diagram for absorption cell.

Apparatus

The cell is diagrammed in Fig. 1, having dimensions of approximately 3.5 x 7 x 13 cm. The components must be carefully arranged so that the optical path is not obstructed by objects or bubbles.

Procedure

Sodium hydroxide solution of known concentration is added to the cell up to a carefully noted level. The sodium hydroxide is used for convenience and is converted to a carbon dioxide-sodium bicarbonate buffer when carbon dioxide is added during the normal course of the determination. The colorimeter or spectrophotometer is adjusted to read 100% transmittance. Add the phenol red indicator until a transmittance of 14.8% is reached. This corresponds to a partial pressure of CO_2 (pCO_2) of zero (see ref. 26 for theory). Pass continuously a gas stream of known carbon dioxide concentration through the absorption cell at a rate of less than 500 ml per min. When equilibrium is established, read the % transmittance. Pass continuously the unknown gas sample through the solution and, when equilibrium is established, measure the % transmittance.

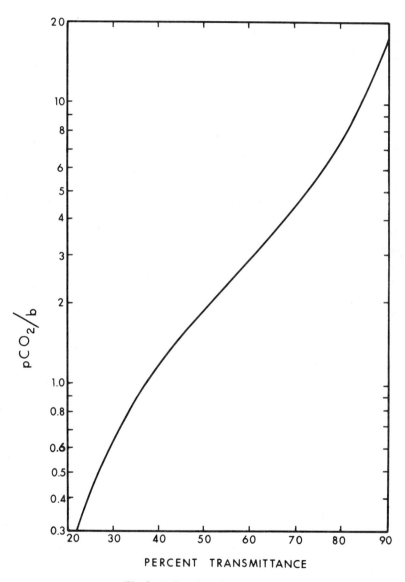

Fig. 2. Calibration plot for apparatus.

Calculations

From the theoretical relationships described in the original paper, the working curve shown in Fig. 2 can be constructed. Using the transmittance of the known carbon dioxide sample, the constant b, which depends on the concentration of the indicator, may be determined. Knowing b, the unknown may be similarly determined from the graph.

CARBON MONOXIDE

Carbon monoxide is the most frequently encountered toxic gas in industrial and domestic daily life. As a product of oxidation of fossil fuels, it is a potential hazard in mines and around fuel-burning equipment and heavy industry, particularly in enclosed or protected areas which are not adequately ventilated. In cities, the principal sources of this gas are the exhaust products from motor vehicles. The toxic effect of carbon monoxide is associated with its ability to complex hemoglobin. The affinity of hemoglobin for carbon monoxide is approximately 200 times greater than its affinity for oxygen. Thus, 0.1% carbon monoxide in air complexes the same amount of hemoglobin as does oxygen, which is present at about 20% in air. Colorimetric methods for carbon monoxide are limited. Why this is the case is not obvious, but certainly it is not entirely for the same reasons as for carbon dioxide. Beatty has summarized the various analysis methods, including colorimetric methods, for carbon monoxide determination (8).

I. METHODS OF DETERMINATION

A. SILVER p-SULFAMINOBENZOATE METHOD (37)

This method is based on the formation of a colloidal suspension of silver from the reaction of carbon monoxide with the silver salt of p-sulfaminobenzoic acid. Depending on the wavelength at which the measurements are made, as little as 2 ppm and as much as 1800 ppm CO can be determined with a precision of about 5%. Acetylenes, olefins, and aldehydes interfere, if present in high concentrations, by reducing the silver p-sulfaminobenzoate. Hydrogen sulfide in high concentration interferes by precipitating silver sulfide.

Reagents

p-Sulfaminobenzoic Acid. Dissolve 2 g in 100 ml of $0.1M$ NaOH

Silver Nitrate, 0.1M.
Sodium Hydroxide, 1.0M.

Procedure

Add 20 ml of the *p*-sulfaminobenzoic acid and 20 ml of the silver nitrate reagents to a 125-ml Erlenmeyer flask. Mix and add 10 ml of the sodium hydroxide solution with shaking. The final solution should be clear and colorless.

Place 10 ml of this solution in a 125-ml ground-glass Erlenmeyer flask fitted with a stopcock and evacuate until the solution begins to bubble. Close the stopcock and cover the exit tube with a rubber septum. Introduce the gas sample via a gas-tight syringe, inserting the needle through the septum. Draw the sample into the flask by opening the stopcock.

Bring the flask to atmospheric pressure by opening the stopcock in a CO-free atmosphere. Close the stopcock immediately and shake on a wrist shaker for 2 hr ± 5 min. Measure the absorbance in a 1-cm cell against a reagent blank. For concentrations below 400 ppm CO, use 425 nm; for concentrations above 400 ppm CO, use 600 nm as the measurement wavelength. Prepare a calibration plot by measuring the absorbance of a series of solutions of known CO concentration.

B. PALLADIUM CHLORIDE—MOLYBDOPHOSPHORIC ACID METHOD (31)

This method is capable of determining 0.002 to 0.06% (by volume) carbon monoxide in 50-ml air or gas samples. Higher concentrations can be determined by suitable gas dilutions. The method is based on the reduction of molybdophosphoric acid to a heteropoly blue in the presence of palladous chloride. Acetone increases and stabilizes the color. The method is not specific for carbon monoxide. Other reducing gases may interfere, causing reduction of the molybdophosphoric acid reagent. Ethylene, acetylene, and hydrogen sulfide show particularly strong interferences. Hydrogen also interferes if its concentration exceeds that of the carbon monoxide.

Reagents

Palladium Chloride. Add 500 mg reagent-grade palladium(II) chloride to 100 ml distilled water containing 2 ml concentrated hydrochloric acid. Cover the beaker with a watch glass and heat until the salt is dissolved. Transfer the solution to a 250-ml volumetric flask, add 3.5 ml concentrated hydrochloric acid, and, after cooling, dilute to volume with distilled water.

Molybdophosphoric Acid. Dissolve 5 g of the reagent in distilled water and dilute to 100 ml.

Acetone, Reagent Grade.

Sulfuric Acid, 3N.

Mixed Carbon Monoxide Reagent. Mix equal volumes of the palladium chloride, molybdophosphoric acid, and sulfuric acid solutions and allow the mixture to stand at room temperature for 48 hr. Some darkening of the reagent may occur on standing, but the color can be stabilized by storing the reagent in a refrigerator.

Procedure

Collect the air sample in, or transfer it to, a 50-ml glass-stoppered test tube or absorption cell designed to fit in the cell compartment of the colorimeter or spectrophotometer. The sample may be introduced by water displacement or any other suitable means. Add 3 ml of the mixed carbon monoxide reagent and then 3 ml acetone. Quickly stopper the tube. When multiple samples are being analyzed, add the mixed carbon monoxide reagent to the tubes first, and then add the acetone. Equilibrate the sample by agitating in a water bath at $60 \pm 1°C$ for 60 ± 1 min. After equilibration, cool the tubes to room temperature and clean and dry them before measuring in the colorimeter. Measure the absorbance against a reagent blank at 820 nm or, if this exceeds the capabilities of the instrument, at 650 nm. Measurements should be made within 24 hr. Prepare a standard curve using the same procedure with a series of standard solutions.

C. PALLADIUM CHLORIDE–PALLADIUM IODIDE METHOD (3,12)

This method is derived from an old test for carbon monoxide based on the following reaction:

$$PdCl_2 + CO + H_2O \longrightarrow Pd + CO_2 + 2HCl$$

The remaining palladium chloride is determined colorimetrically after conversion to the red palladium(II) iodide. It is necessary to know approximately the carbon monoxide concentration in order to determine the required amount of palladous chloride. This can be calculated from the fact that 0.1261 ml carbon monoxide reduces 1.0 mg palladium(II) chloride. The absorbance maximum of greatest intensity is located at 408 nm. However, the absorbance at this wavelength gradually increases with time and exhibits a dependence upon the potassium iodide concentration. Consequently, the absorbance is measured at a somewhat weaker absorbance band whose maximum is located at 490 nm. This latter band does not appear to be affected by the problems associated with the band at 408 nm.

Reagents

Palladium(II) Chloride. Add 100 ml 0.012M hydrochloric acid to a weighed amount of reagent-grade palladium chloride (about 125 mg). Cover the container and allow to stand overnight until the salt dissolves. Transfer to a 250-ml volumetric flask and dilute to volume.

Aluminum Sulfate, 10%.

Potassium Iodide. Dissolve 12 g potassium iodide in 100 ml water.

Standard Carbon Monoxide Mixtures. Accurate carbon monoxide mixtures may be obtained from various companies specializing in gases and gas products. Alternatively, standard mixtures may be prepared from pure carbon monoxide and nitrogen by suitable gas-handling techniques.

Special Glassware

A necessary piece of glassware consists of a three-way stopcock modified in such a way that one arm contains the male portion of a ground-glass joint which fits the standard taper of the volumetric vessels to be used. The apparatus is shown in Fig. 3.

Procedure

Add a known volume of the palladium chloride solution which contains at least 10% more reagent than the expected amount of carbon monoxide and one drop of the aluminum sulfate solution to a 40-ml glass-stoppered graduated cylinder that is connected to a mercury manometer and a water pump. The apparatus is evacuated to a desired extent, and the mercury level in the manometer is noted. The special stopcock is used to close the cylinder and open the manometer to room air. The carbon monoxide sample in a Douglas bag or rubber bladder is connected to the vertical arm of the three-way stopcock, and a rubber tube from an inverted 1-liter graduated cylinder filled with water is connected to the horizontal arm. By pressing the bag and opening its screw clamp, a sufficient volume of gas is expelled to flush the connections. The volume expelled should be at least 10 times the volume of the connections. Turn the stopcock to permit the gas to enter the cylinder containing the palladium chloride. After 15 sec, close the cylinder and note the barometric pressure and room temperature.

Gently rotate the reaction cylinder several times and lay it horizontally without permitting liquid to enter the stopcock capillary. After 2 hr at room temperature, filter the mixture and collect the filtrate in a 50-ml volumetric flask containing 25 ml distilled water. Wash the cylinder with two 3-ml portions of water and add to the filter. Add about 3 ml of a 10-ml aliquot of the

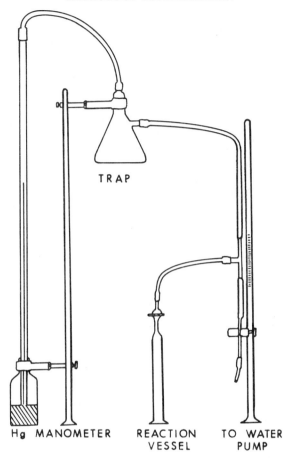

Fig. 3. Apparatus for introduction of gas sample into the reaction vessel.

potassium iodide solution to the filter, and add the remainder directly to the volumetric flask. Wash the filter with several milliliters of water and dilute the flask to volume. Measure the absorbance in a 1-cm cell at 490 nm against a water blank.

Prepare three identical standard carbon monoxide solutions and treat them in a similar manner. The volume of carbon monoxide (normal temperature and pressure, dry) in the unknown sample is computed as follows:

$$0.1261 \left[Q - \left(\frac{A_u}{A_s} \right) \times Q \right]$$

where Q refers to the milligrams of palladium(II) chloride originally present, and

A_u and A_s refer to the absorbances of the unknown sample and the standard, respectively. The volume of the gas sample drawn into the reaction vessel is computed as follows:

$$(C - V) \times \frac{\Delta P}{P} \times \frac{P - p}{760} \times \frac{273}{273 + T}$$

where C is the capacity (ml), V is the volume (ml) of the reagent solutions in the reaction vessel, ΔP is the rise in mercury (mm) in the monometer, P is the temperature-corrected atmospheric pressure, p is the vapor pressure of water at room temperature, and T is the temperature (°C).

D. OTHER METHODS

Lambert (23) has devised a method for carbon monoxide based on the formation of a soluble, colored complex rather than a colloidal dispersion. This method utilizes the reduction product of the reaction between $PdCl_4^{-2}$ and carbon monoxide, which is probably colloidal Pd, to reduce iron(III)–EDTA to iron(II)–EDTA. The iron(II)–EDTA undergoes ligand exchange with 2,2'-bipyridine or 1,10-phenanthroline to form a colored complex. The method is sensitive to about 25 ppm CO.

CYANIDE

Cyanide is a common constituent of sewage and metal finishing and refinery waste waters. Because of its toxicity, it is a very important constituent that has been quite thoroughly studied. The problem of toxicity, analysis, and control is complicated by the fact that the cyanides show varying degrees of chemical activity. It is convenient to consider two types of cyanides: total and free. Total cyanide is taken to mean all inorganic cyanide, present as hydrogen cyanide, ionic cyanide, or inert metal complexes such as ferricyanide. Free cyanide is taken to mean cyanide found as hydrogen cyanide, ionic cyanide, and certain metal cyanides that are in equilibrium with hydrogen cyanide. The alkali cyanides and certain metal cyanides such as those of cadmium, nickel, and zinc are readily converted to hydrogen cyanide during distillation with acid. Other metal cyanides, such as those of iron, cobalt, thallium, mercury, and silver, show much greater resistance to decomposition into simple ions.

I. SEPARATIONS

The principal separations involve the removal of sulfide from cyanide samples and the removal of cyanide from thiocyanate. Both reagents interfere in the common methods employed for cyanide. The chemical pretreatment for determination of total cyanide normally requires separation of the cyanide as hydrogen cyanide.

A. DISTILLATION METHOD

Total cyanide may be separated as hydrogen cyanide by distillation from concentrated phosphoric acid (21,22). Phosphoric acid is especially suitable because it is nonvolatile, even when concentrated; is strong enough to decompose metal complexes; and complexes several metal cations. Ethylene-diaminetetraacetic acid may be added to the distillation mixture to aid in decomposing the more stable metal cyanides of cobalt and copper. The distilled hydrogen cyanide is collected in a dilute sodium acetate solution.

Hydrogen cyanide may also be distilled from dilute sulfuric acid containing cuprous chloride (14) or mercuric chloride plus magnesium chloride (4,36). In these procedures, the distillate is collected in dilute sodium hydroxide. This process is apparently not as efficient as distillation from phosphoric acid, and several repetitive distillations may be necessary.

Free cyanide may be separated from ferricyanide by distillation at reduced pressure in the presence of zinc acetate (33). Free cyanide may be separated from thiocyanate by a simple aeration procedure which takes advantage of the large difference in volatility of hydrocyanic acid and hydrothiocyanic acid in aqueous acidic solution (6,7,36).

B. EXTRACTION METHOD

Free cyanide may be separated from metals, thiocyanate, and complex cyanides by extraction as hydrogen cyanide from acetic acid solutions with isopropyl ether (22). Thiocyanate concentrations below 200 ppm show no interference, and higher concentrations introduce only small errors. The hydrogen cyanide is reextracted from the organic solvent with sodium hydroxide. The distribution coefficient is not large, and as many as four extractions may be necessary for complete separation. Fatty acids commonly present in sewage may cause undesired foaming during extractions. This type of interference may be removed by extraction of the neutralized sample with isooctane, hexane, or chloroform (21,22).

C. PRECIPITATION METHOD

Sulfide may be removed by addition of small amounts of solid lead carbonate to an alkaline solution of the cyanide (4). Black, insoluble lead sulfide is formed and can be filtered. A large excess of lead carbonate should be avoided.

            CARBON

D. DIFFUSION METHOD

Hydrogen cyanide in tartaric acid has been separated from other sample constituents by microdiffusion in a Conway dish (29,30). The diffused hydrogen cyanide is collected in mercuric nitrate. Sulfide interferes because it diffuses as hydrogen sulfide. Tartaric acid prevents the slow air oxidation of sulfide to polysulfides which in turn react with cyanide to produce thiocyanate.

II. METHODS OF DETERMINATION

A. PYRIDINE–PYRAZOLONE METHOD (4,14,15,21,25,36)

This method employs the König reaction whereby the pyridine ring is opened by the action of cyanogen chloride followed by a condensation of the resulting glutaconic aldehyde with 1-phenyl-3-methyl-5-pyrazolone to give a highly colored product. Absorption spectra of the aqueous and extracted product are shown in Fig. 4. Thiocyanate interferes seriously. Numerous cations may precipitate as hydrous oxides in the approximately neutral solution used for

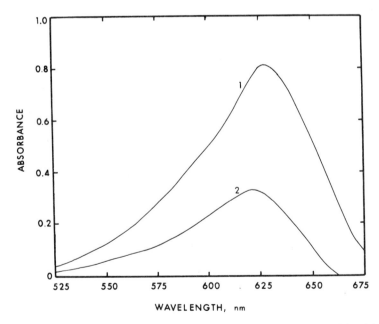

Fig. 4. Absorption curves for pyridine–pyrazolone method: (1) extracted in 1-butanol; (2) aqueous.

determination of free cyanide, but the precipitates may be removed by filtering prior to color development.

Reagents

Pyridine—Pyrazolone. Add 1-phenyl-3-methyl-5-pyrazolone to 125 ml hot water to form a saturated solution. Cool and filter. To the filtrate, add 25 ml redistilled pyridine containing 25 mg bis(1-phenyl-3-methyl-5-pyrazolone). The pyridine and aqueous pyrazolone should be freshly prepared and mixed just before use. The bispyrazolone can be prepared by adding 17.4 g 3-methyl-1-phenyl-5-pyrazolone and 25 g phenylhydrazone to 100 ml 95% ethanol in a 250-ml distilling flask and refluxing for 24 hr. Filter the insoluble bispyrazolone every few hours and wash with hot ethanol. Finally, wash with isopropanol to remove any hydrogen cyanide.

Chloramine-T, 1%. Prepare fresh each day.

Phosphate Buffer. Dissolve 14.3 g disodium hydrogen phosphate dodecahydrate and 13.6 g potassium dihydrogen phosphate in distilled water and dilute to 1 liter. The pH of this solution is 6.8.

Standard Cyanide Solution. Dissolve that weight of cyanide, assayed titrimetrically by the Liebig method, equivalent to 1.884 g pure sodium cyanide in distilled water and dilute to 1 liter. Dilute 10 ml of this stock solution to 1 liter with distilled water to obtain a standard cyanide solution containing 1 μg cyanide per ml.

Procedure

Direct. Transfer 10-, 5-, 3-, 2-, and 1-ml aliquots of the standard cyanide solution to 50-ml volumetric flasks. Add 5 ml of the buffer and 0.3 ml chloramine-T solution, mix, and allow to stand 1 min. Add 15 ml of the pyridine—pyrazolone reagent, dilute to volume, mix, and allow to stand stoppered for 30 min. Measure the absorbance at 620 nm in 1-cm cells against a reagent blank. The sample, containing 1—10 μg cyanide, is neutralized to pH 6—7 with acetic acid or sodium hydroxide and treated in the same manner.

Extraction. Follow the above procedure until the color is fully developed. Transfer with rinsings to a 125-ml separatory funnel containing exactly 10 ml 1-butanol. Stopper and mix. If the emulsion at the interface does not break within 3 min, add another few milliliters of the phosphate buffer and mix again. Transfer a portion of the 1-butanol layer to a 1-cm absorption cell and measure the absorbance at 630 nm against a reagent blank. The extraction procedure is reported to offer slightly better sensitivity (4).

B1. PYRIDINE–BENZIDINE METHOD (1,2,10,11,17,25,28)

This method is very similar to the pyridine–pyrazolone method. It is based on the König reaction where the pyridine ring is opened by cyanogen bromide and the product condensed with benzidine to form the final colored product. The method is slightly less sensitive than the pyridine–pyrazolone method. The final colored solution is not stable, and the measurement times must be carefully controlled. Thiocyanate and sulfide interfere seriously.

Reagents

Benzidine Solution. Dissolve 0.5 g benzidine hydrochloride in 50 ml $0.5M$ hydrochloric acid.

Benzidine–Pyridine Solution. Dissolve 15 ml pyridine in 10 ml of $1M$ hydrochloric acid. Add 10 ml of the benzidine reagent and mix thoroughly. This solution should be prepared fresh daily.

Bromine Water, Saturated.

Standard Cyanide Solutions. Prepare according to the directions listed for the pyridine–pyrazolone method.

Arsenious Acid, 2%.

Procedure

Direct. Transfer the neutralized sample containing $1-10$ μg cyanide to a 25-ml volumetric flask. Add 1 ml bromine water and allow to stand for 5 min with occasional shaking. Remove the excess bromine by dropwise addition of the arsenious acid. When no bromine color can be observed, add 0.2 ml in excess. Add 5 ml of the pyridine–benzidine reagent, mix thoroughly, and allow the solution to stand 15 min. Measure the absorbance at 530 nm against a reagent blank. Prepare a standard curve using the same procedure with standard cyanide solutions.

Extraction. Follow the same procedure used for extraction in the pyridine–pyrazolone method. Measure the absorbance at 480 nm against a reagent blank.

B2. PYRIDINE–PHENYLENEDIAMINE METHOD (7)

This method is essentially the same as the pyridine–benzidine method but avoids the use of an active carcinogen such as benzidine. The method is applicable with slight modifications over the concentration range of 0.005 to 100 ppm cyanide.

Reagents

p-Phenylenediamine. Dissolve 100 mg of the reagent in 50 ml 0.5M hydrochloric acid.

Pyridine Solution. Dissolve 18 ml pyridine in 15 ml 2.5M hydrochloric acid.

Pyridine–Phenylenediamine Reagent. Prepare immediately prior to use by mixing three parts of the pyridine solution with one part of the phenylenediamine solution.

Bromine Water, Saturated.

Arsenious Acid, 2%.

Standard Cyanide Solutions. Prepare according to the directions listed for the pyridine–pyrazolone method.

Procedure

Transfer 40 ml of the sample containing between 0.4 and 20 μg cyanide to a 50-ml volumetric flask. Add 2 ml bromine water and 0.1 ml concentrated hydrochloric acid. Mix and allow the solution to stand 5 min. Add 0.4 ml arsenious acid and dilute to 50 ml. Treat a 5-ml aliquot of this solution with exactly 4 ml freshly mixed pyridine–phenylenediamine solution. Mix and allow to stand for 30 min. Measure the absorbance at 515 nm against a reagent blank. Prepare a standard curve using the same procedure with standard cyanide solutions.

C. PICRIC ACID METHOD (16,19)

Picric acid reacts with cyanide at pH 7.8–10.2 to form a dark-red color. The exact nature of the reaction is unknown. The color is best developed by heating for 10 min at 70–85°C. Thiocyanate does not interfere, but many metals interfere by causing pH changes or forming turbid solutions.

Reagents

Picric Acid, 1%.

Standard Cyanide Solutions. Prepare according to the directions listed for the pyridine–pyrazolone method.

Procedure

Transfer 10 ml of the sample containing 0.1–0.5 mg cyanide in sodium carbonate. Add 5 ml of the picric acid reagent, mix, and heat in a boiling water bath for exactly 5 min. Remove the volumetric flask and immediately dilute to 100 ml with room-temperature water to stop further reaction. Cool the entire solution to room temperature, adjust the volume to the mark, and measure the absorbance at 520 nm against a reagent blank. Prepare a standard curve using the same procedure with standard cyanide solutions.

D. PHENOLPHTHALEIN METHOD (32)

This method was once very widely used in cyanide analysis but has yielded to the generally superior methods based on the König reaction. The method is based on the oxidation of phenolphthalin to phenolphthalein by cyanide in the presence of copper. The oxidation probably results through the reduction of copper(II) by cyanide to form cuprous cyanide. Addition of base converts the phenolphthalin to the red form for measurement. The reaction is slightly temperature dependent. The sensitivity is reported to be 0.25 μg per ml. The method suffers from numerous interferences from substances such as ferricyanide, ferrocyanide, halogens, sulfides, and phenols.

Reagents

Phenolphthalin–Copper Sulfate. Mix 0.5 ml 1% phenolphthalin in absolute ethanol with 99.5 ml 0.01% copper sulfate pentahydrate. This solution is stable for two days at room temperature and several months if stored in a refrigerator.

Potassium Hydroxide, 0.05%.

Standard Cyanide Solutions. Prepare according to the directions listed for the pyridine–pyrazolone method.

Procedure

Transfer a 5-ml sample of the cyanide solution to a 10-ml volumetric flask and add 2 ml of the potassium hydroxide solution and 2 ml of the phenolphthalin–copper sulfate reagent. Dilute to volume, mix thoroughly, and immediately measure the absorbance at 548 nm against a reagent blank. Prepare a standard curve using the same procedure with standard cyanide solution.

E. OTHER METHODS

Cyanide reacts with sodium p-toluene sulfochloroamide to form a cyanohalide which then reacts with pyridine and barbituric acid to form a colored derivative (5). Neutral dicyanobis(1,10-phenanthroline)iron(II) has been extracted into chloroform and measured at 597 nm (35). The method is sensitive to 2 μg cyanide. Cyanide has been determined based on its ability to mask mercury(II) in its reactions with *p*-dimethylaminobenzylidenerhodanine (30). The mercuric complex is highly colored, and the loss in absorbance can be related to the cyanide concentration.

III. APPLICATIONS

A. TOTAL CYANIDE IN SEWAGE (22)

Reagents

Ethylenediaminetetraacetic Acid, 10%. Dissolve 10 g of the free acid in 85 ml water containing 5.6 g sodium hydroxide.

Phosphoric Acid, 85%.

Chloramine-T, 1%. Dissolve 0.2 g of the reagent in 20 ml water.

Bis(1-phenyl-3-methyl-5-pyrazolone). Dissolve 17.4 g 1-phenyl-3-methyl-5-pyrazolone in 95% ethanol. Add 25 g freshly distilled phenylhydrazine and reflux for 4 hr. Filter and wash the precipitate, which is bis(1-phenyl-3-methyl-5-pyrazolone), with hot methanol.

Pyridine–Pyrazolone Reagent. Prepare a saturated solution of pyrazolone in 125 ml water and filter the excess solid. To the solution, add 25 ml redistilled pyridine containing 25 mg bispyrazolone. Prepare this solution fresh just prior to use.

Sodium Acetate, 10%.

Standard Cyanide Solutions. Prepare according to the directions listed for the pyridine–pyrazolone method in Section II.

Procedure

Remove salt and insoluble materials by filtering and wash the precipitate thoroughly. Add about 10 ml of the ethylenediaminetetraacetic acid solution to 250 ml of the sample which should contain 0.03 to 10 mg cyanide. Disregard any precipitate formation.

Add 0.3 g citric acid and 35 ml phosphoric acid to 100 ml water in a 1-liter distilling flask. Fit a 125-ml separatory funnel to the distilling flask and close the stopcock. Add water to the receiving flask until the end of the adapter tube is covered by at least 0.5 cm. Add 10 ml sodium acetate to the receiving solution. Heat the solution in the flask to boiling. Next, allow water to circulate around the condenser, which will cause the column of water in the delivery tube to rise to about 30 cm.

Slowly add all the sample from the separatory funnel to the boiling solution. Maintain a reduced pressure in the distilling flask. When the sample has been added, close the stopcock, rinse the container and separatory funnel with distilled water, and add these washings to the distilling flask. Boil until the solution becomes sirupy (about 40 ml). Turn off the heat and admit air through the separatory funnel.

If any organic materials are present in the distillate, they may be removed by extraction with isooctane, hexane, or carbon tetrachloride. Adjust the pH of the distillate to 5–8 with acetic acid or sodium acetate, transfer to a 500-ml volumetric flask, and dilute to the mark.

Transfer an aliquot containing 0.001 to 0.010 mg cyanide to a 50-ml volumetric flask and add 0.3 ml chloramine-T. Allow the solution to stand 1 min and add 15 ml of the pyridine–pyrazolone reagent. Let the solution stand 30 min and then adjust the volume to 50 ml and measure the absorbance of the blue color at 620 nm against a reagent blank. Prepare a standard curve using the same procedure with a series of standard cyanide solutions.

B. CYANIDE IN REFINERY GASES AND SEPARATOR WATER CONTAINING SULFIDE (6)

Reagents

Benzidine Solution. Dissolve 0.5 g benzidine in 50 ml hot $0.5N$ hydrochloric acid. Cool, filter, and store in a dark bottle.

Benzidine–Pyridine Solution. Dissolve 18 ml pyridine in 12 ml water and add 3 ml concentrated hydrochloric acid. Add 10 ml of the benzidine solution and mix well. Prepare this solution daily.

Sulfide Solution. Dissolve 0.75 g sodium sulfide nonahydrate in water and dilute to 100 ml. One milliliter of this solution contains 1 mg sulfide.

Bromine Water, Saturated.

Arsenious Acid, 2%.

Standard Cyanide Solutions. Prepare according to the directions listed for the pyridine–pyrazolone method in Section II.

Procedure

Calibration. Transfer appropriate volumes of the standard cyanide solution to give 0, 1, 3, 5, 7, and 10 µg cyanide to 25-ml volumetric flasks containing 1 ml 1% sodium hydroxide. Add 2.5 ml of the sulfide solution to each flask and acidify with glacial acetic acid, adding 0.5 ml in excess. Add immediately 2 ml bromine water and mix thoroughly. Allow the solution to stand 10 min with

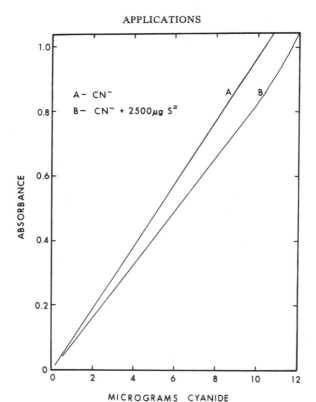

Fig. 5. Calibration plots with and without added sulfide. Cell length, 1 cm; wavelength, 530 nm.

occasional shaking. If a precipitate of free sulfur appears or if all the bromine is consumed, add additional bromine water in 0.2-ml increments until an excess is present as indicated by its red color. When the solutions clear, add the arsenious acid solution dropwise with mixing until the excess bromine is reacted, then add 0.2 ml in excess. Add 4 ml of the pyridine–benzidine reagent and mix. Wait 30 sec, add 5 ml ethanol, and dilute to volume with water. After 15 min, measure the absorbance of each solution at 530 nm against a reagent blank containing approximately the same concentration of sulfide. The color is stable for approximately 30 min. Calibration plots with and without sulfide vary somewhat. Typical plots are shown in Fig. 5.

Unknown. Determine the sulfide concentration of the unknown by an appropriate method. Select a sample containing less than 2500 μg sulfide. If the sample contains less than 50 μg sulfide, add sufficient sulfide solution to the 25-ml volumetric flask to give 100 to 2500 ppm in the final solution. Proceed as described above and determine the cyanide concentration from the calibration curve.

THIOCYANATE

The major applications of spectrophotometric methods for thiocyanate are in the analysis of sewage and industrial effluents and in biological materials. The iron(III) thiocyanate method has been widely used in the past, probably because of its great simplicity coupled with its relatively good sensitivity. The instability of the iron(III) thiocyanate complex and the need for better sensitivity and specificity have resulted in the development of several other good methods. As one might expect, several of the methods developed for cyanide, such as the pyridine–benzidine and pyridine–pyrazolone methods, are also suitable for thiocyanate.

I. SEPARATIONS

Instead of isolating thiocyanate from the other constituents in a sample, it is far more common to separate the various potential interferences. Thus, hydrocyanic acid can be removed by extraction with isopropyl ether (21,22) or volatilization (4,6,7,14,21,22,36). Complex cyanides and cyanates are usually decomposed first to hydrocyanic acid for subsequent removal. Iodide can be oxidized to iodine and extracted with carbon tetrachloride. Sulfide can be precipitated with lead carbonate (4).

II. METHODS OF DETERMINATION

A. COPPER–PYRIDINE METHOD (13,21)

Thiocyanate reacts with copper and pyridine to form the complex $Cu(Py)_2(CNS)_2$. The neutral complex is isolated and concentrated by a chloroform extraction. The method is not quite as sensitive as the pyridine–benzidine and pyridine–pyrazolone methods, but it is more specific and employs only stable reagents that are easily prepared. The color-developing reaction is carried out between pH 2.5 and 4. The copper–pyridine complex is destroyed if the solution is too acidic, and copper hydroxide precipitates if it is too basic. The volume of the sample affects the final absorbance; consequently, the standards and sample volumes should not differ by more than 3%. The color system conforms to Beer's law from 2 to 40 ppm thiocyanate but shows a negative deviation at concentrations of less than 2 ppm. The molar absorptivity is 790 liter $\cdot$ mole^{-1} $\cdot$ cm^{-1} at 407 nm.

Reagents

Copper(II) Solution. Dissolve 250 mg copper metal in dilute nitric acid and dilute to 1 liter.

Standard Thiocyanate Solution. Dissolve 0.194 g potassium thiocyanate in water and dilute to 1 liter. Standardize by titration with standard silver nitrate. One milliliter of the thiocyanate solution contains approximately 0.116 mg thiocyanate.

Procedure (13)

Select a sample containing less than 800 μg thiocyanate in a total volume not exceeding 10 ml. Transfer the sample to a 125-ml separatory funnel, and, if necessary, add water to bring the volume to approximately 10 ml. Add 20 ml of the copper solution and 1 ml of the pyridine solution and mix. Add 20 ml chloroform and shake for 6 min on an automatic shaker. Allow the phases to separate and collect the green organic layer in a polyethylene bottle. Measure the absorbance at 407 nm in a 1-cm cell against chloroform. Prepare a standard curve using solutions of known thiocyanate concentration.

B. PYRIDINE–PYRAZOLONE METHOD (4,15,19)

The method is based on the conversion of thiocyanate to cyanogen chloride by chloramine-T. The cyanogen chloride reacts with pyridine and the product condenses with 3-methyl-1-phenyl-5-pyrazolone in a König-type reaction to produce the final colored product. The system conforms to Beer's law in the concentration range of 0.04 to 0.25 ppm thiocyanate. Cyanide is a serious interference and may be removed by volatilization from acid solution.

The reagents and procedure are identical to those listed for cyanide in the previous section.

C. PYRIDINE–BENZIDINE METHOD (1,2,19,41)

This method is very similar to the pyridine–pyrazolone method except that benzidine is used in place of pyrazolone in the condensation reaction. For additional information on the nature of the reaction and interferences, refer to the earlier section describing the pyridine–benzidine method for cyanide. The same reagents and procedure may be used for thiocyanate.

D. IRON(III) THIOCYANATE METHOD (9,18,19,38)

Despite the greater sensitivity of the pyridine–pyrazolone and pyridine–benzidine methods and the superior stability of the copper–pyridine method, the iron(III) thiocyanate method is still in wide use, mainly because the reagents are very readily available and the procedure is simple and fast. Thiocyanate forms a series of consecutive complexes with iron(III) (34), each with a slightly different absorption spectrum. At low thiocyanate concentrations, the major species is the red $[Fe(CNS)]^{2+}$ complex.

The color intensity and stability of the iron—thiocyanate system depends on many factors such as concentration and type of acid employed, presence of oxidizing agents, and temperature (34). Fluoride, iodide, meta- and pyrophosphate, oxalate, and phenols interfere. When present in large quantities, phosphate and sulfate may also interfere. Mercury(I) and silver interfere by precipitating thiocyanate. Bismuth, cadmium, copper, mercury(II), molybdenum, titanium, and zinc are other common metals that interfere by forming thiocyanate complexes.

Reagents

Iron(III) Nitrate. Dissolve 50 g iron(III) nitrate, $Fe(NO_3)_3$, in 25 ml concentrated nitric acid and dilute to 1 liter with water.

Standard Thiocyanate Solution. Prepare according to the directions listed for the copper—pyridine method.

Procedure (19)

Transfer a 5-ml aliquot of the neutralized sample, containing between 0.05 and 0.5 mg thiocyanate, to a test tube or small Erlenmeyer flask. Add exactly 1 ml of the iron(III) nitrate reagent, mix, and after 5 min measure the absorbance at 550 nm against a reagent blank. Prepare a calibration curve using this same procedure with standard thiocyanate solutions.

E. OTHER METHODS

Asmus and Garschagen employed a König-type method using barbituric acid (5). The thiocyanate is converted to cyanogen chloride at pH 2 to 10 with chloramine-T and ferric ion as a catalyst. Barbituric acid in pyridine is added and the absorbance is measured at 578 nm. Concentrations of 1 to 25 ppm thiocyanate can be determined.

Methylene blue forms a 1:1 complex with thiocyanate that is extracted into dichloroethane from an acid solution (20). Methylene blue is only very slightly extracted. The absorbance of the complex is measured at 657 nm. Concentrations of 0.012 to 0.35 ppm thiocyanate can be determined.

Tetrathionate has been determined by decomposing it with alkaline cyanide (27) as shown by the following reaction:

$$CN^- + S_2O_3^{2-} \longrightarrow CNS^- + SO_3^{2-}$$

The thiocyanate is determined colorimetrically employing a ferric thiocyanate procedure.

REFERENCES

1. Aldridge, W. N., *Analyst,* **69**, 262 (1944).
2. Aldridge, W. N., *Analyst,* **70**, 474 (1945).
3. Allen, T. H., and W. S. Roote, *J. Biol. Chem.,* **216**, 309 (1955).
4. American Public Health Association, *Standard Methods for the Examination of Water and Wastewater,* 13th ed., Washington, D.C., 1971, pp. 397–406.
5. Asmus, E., and H. Garschagen, *Z. Anal. Chem.,* **138**, 414 (1953).
6. Baker, M. O., R. A. Foster, B. G. Post, and T. A. Hielt, *Anal. Chem.,* **27**, 448 (1955).
7. Bark, L. S., and H. G. Higson, *Talanta,* **11**, 621 (1964).
8. Beatty, R. L., *U.S. Bur. Mines Bull.,* No. **557** (1944).
9. Bowler, R. G., *Biochem. J.,* **38**, 385 (1944).
10. Brooke, M., *Anal. Chem.,* **24**, 583 (1952).
11. Bruce, R. B., J. W. Howard, and R. F. Hanzal, *Anal. Chem.,* **27**, 1346 (1955).
12. Christman, A. A., W. D. Block, and J. Schultz, *Ind. Eng. Chem., Anal. Ed.,* **9**, 152 (1937).
13. Danchik, R. S., and D. F. Boltz, *Anal. Chem.,* **40**, 2215 (1968).
14. Elly, C. T., *J. Water Poll. Contr. Fed.,* **40**, Part I, 848 (1968).
15. Epstein, J., *Anal. Chem.,* **19**, 272 (1947).
16. Finkel'shtein, D. N., *Zh. Anal. Khim.,* **3**, 188 (1940).
17. Fisher, F. B., and J. S. Brown, *Anal. Chem.,* **24**, 1440 (1952).
18. Iwaski, I., S. Utsumi, and T. Ozawa, *J. Chem. Soc. Japan, Pure Chem. Sect.,* **78**, 474 (1957).
19. Karchmer, J. H., *The Analytical Chemistry of Sulfur and Its Compounds,* Part I, Wiley, New York, 1970.
20. Koh, T., and I. Iwasaki, *Bull. Chem. Soc. Japan,* **40**, 569 (1967).
21. Kruse, J. M., and M. G. Mellon, *Anal. Chem.,* **25**, 456 (1953).
22. Kruse, J. M., and M. G. Mellon, *Sewage Ind. Wastes,* **23**, 1402 (1951).
23. Lambert, J. L., *Anal. Lett.,* **4**, 745 (1971).
24. Loveland, J. W., R. W. Adams, H. H. King, Jr., F. A. Nowak, and L. J. Cali, *Anal. Chem.,* **31**, 1008 (1959).
25. Ludzack, F. J., W. A. Moore, and C. C. Ruckhoft, *Anal. Chem.,* **26**, 1784 (1954).
26. Maxon, W. D., and M. J. Johnson, *Anal. Chem.,* **24**, 1541 (1952).
27. Nietzel, O. A., and M. A. DeSesa, *Anal. Chem.,* **27**, 1839 (1955).

28. Nusbaum, I., and P. Skupeko, *Sewage Ind. Wastes,* **23**, 875 (1951).

29. Ohlweiler, O. A., and J. O. Meditsch, *Anal. Chim. Acta,* **11**, 111 (1954).

30. Ohlweiler, O. A., and J. O. Meditsch, *Anal. Chem.,* **30**, 450 (1958).

31. Polis, B. D., L. B. Berger, and H. H. Schrenk, *U.S. Bureau Mines, Report of Investigations, No. 3785,* Washington, D.C., 1944.

32. Robbie, W. A., *Arch. Biochem.,* **5**, 49 (1944).

33. Roberts, R. F., and B. Jackson, *Analyst,* **96**, 209 (1971).

34. Sandell, E. B., *Colorimetric Determination of Trace Metals,* 3rd ed., Interscience, NY, 1959.

35. Schilt, A. A., *Anal. Chem.,* **30**, 1409 (1958).

36. Serfass, E. J., R. B. Freeman, B. F. Dodge, and W. Zabban, *Plating,* **39**, 267 (1952).

37. Smith, R. G., R. J. Bryan, M. Felstein, B. Levadie, F. A. Miller, E. R. Stephens, and N. G. White, *Health Lab. Sci.,* **Suppl.**, **7**, 75 (1970).

38. Snell, F. D., C. T. Snell, and G. A. Snell, *Colorimetric Methods of Analysis,* Vol. II, Van Nostrand, Princeton, N.J., 1949.

39. Spector, N. A., and B. F. Dodge, *Anal. Chem.,* **19**, 55 (1947).

40. Underwood, A. L., and L. H. Howe, III, *Anal. Chem.,* **34**, 693 (1962).

41. Wagner, F., *Z. Anal. Chem.,* **162**, 106 (1958).

42. Winzler, P. J., and J. P. Baumberger, *Ind. Eng. Chem., Anal. Ed.,* **11**, 371 (1939).

CHAPTER

4

CHLORINE

D. F. BOLTZ
Department of Chemistry, Wayne State University, Detroit Michigan

W. J. HOLLAND
Department of Chemistry, Windsor University, Ontario, Canada

J. A. HOWELL
Department of Chemistry, Western Michigan University, Kalamazoo, Michigan

The colorimetric determination of small amounts of chlorine is very important in the analysis of waters that have been chlorinated to destroy pathogenic organisms. The chlorine that is not consumed by reacting with impurities remains as free residual chlorine or combined residual chlorine. The residual chlorine content at specified times after treatment of the water with chlorine is used as a measure of the effectiveness of the disinfecting process. The addition of an ammonium salt, or anhydrous ammonia, with the addition of chlorine results in the formation of monochloramine and dichloramine. When this disinfection method is followed, the permissible residual chlorine content of

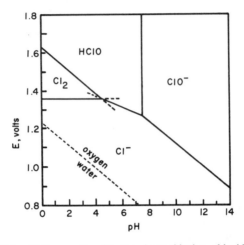

Fig. 1. Potential–pH diagram for chlorine–hypochlorite–chloride system (56).

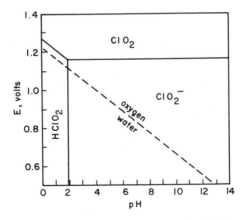

Fig. 2. Potential–pH diagram for chlorine dioxide–chlorite system (56).

water is higher. In water analysis, the total residual chlorine present, regardless of type, is called "total chlorine residual." The term "free available chlorine residual" refers to hypochlorous acid and/or hypochlorite ions, and the term "combined available chlorine residual" indicates chlorine combined with ammonia nitrogen or other nitrogenous compounds (2).

Chlorine dioxide is a water disinfectant sometimes used in contending with taste and odor problems. Because of the higher cost of chlorine dioxide, an admixture of chlorine and chlorine dioxide is employed. The analytical problem of determining chlorine dioxide residuals is especially difficult because of the chemistry of chlorine dioxide (30). For example, chlorine dioxide produces chlorite upon reduction, as illustrated by the following equation:

$$ClO_2 + e^- \rightleftharpoons ClO_2^-$$

It is also possible for this chlorite to be reduced to chloride, especially in solutions of low pH:

$$ClO_2^- + 4H^+ + 4e^- \rightleftharpoons Cl^- + 2H_2O$$

Some chlorous acid may also form and disproportionate to give chlorine dioxide, chloride, and chlorate (16,47):

$$ClO_2^- + H^+ \rightleftharpoons HClO_2$$

$$4HClO_2 \rightleftharpoons 2ClO_2 + H^+ + Cl^- + HClO_3 + H_2O$$

Pourbaix (56) potential–pH diagrams for the chloride–chlorine–hypochlorous acid system and the chlorine dioxide–chlorite–chlorous acid system illustrate the dependence of redox potentials on pH and the acid–base equilibria of these systems (Fig. 1 and 2). Hence, in order to determine the chlorine dioxide concentration, it is necessary that the interference due to free and combined chlorine and chlorite either must be eliminated or compensated for.

I. SEPARATIONS

Because of the nonselectivity of the colorimetric methods used in the determination of chlorine and chloride, any other halogen or halide present in the sample must be removed. The separation of chloride from the matrix of the sample is often of extreme importance.

A. PRECIPITATION METHOD

Precipitation of silver chloride in acidic solution permits the separation of chloride from borate, phosphate, arsenate, arsenite, and oxalate which precipitate as insoluble silver salts only in neutral solution. Bromide, iodide, cyanide, and sulfide would also precipitate from acidic solution upon addition of silver ions.

B. DISTILLATION METHOD

When bromide and iodide are both present with chloride in the sample, a preferential oxidation of the bromide and iodide to the free halogen is possible, the free halogen then being volatilized. A number of oxidizing agents have been used for this purpose, with control of acidity and the removal of excess oxidizing agent being of principal concern. Potassium iodate in $0.2-0.4M$ nitric acid solutions has been recommended by Andrews (1) as being a suitable oxidizing agent to oxidize iodide and bromide in the presence of relatively small amounts of chloride. Phosphorous acid was used to reduce the excess iodate. Permanganate has also been recommended in oxidizing appreciable amounts of bromide in the presence of small amounts of chloride, the addition of manganese(II) sulfate preventing the oxidation of the chloride (42). The critical factors affecting the separation of traces of halides by the preferential oxidation volatilization method have not been delineated.

One way of removing chloride from a sample consists of using the Watters and Orlemann nitrogen-swept cell method. According to this procedure, the sample is dissolved in sulfuric acid and the liberated hydrogen chloride is removed with a stream of nitrogen. The hydrogen chloride passes through a sulfuric acid–lead dioxide oxidizer, which results in the liberation of free chlorine. The chlorine can then be absorbed in the *o*-tolidine or other colorimetric reagent used (58). Another method of removal involves the distillation of hydrogen chloride from acidic solution at a temperature of $150°C$. The absorption of the hydrogen chloride in a silver nitrate solution has been used in the nephelometric determination of traces of chloride (59). Chlorine can also be liberated by the addition of permanganate, dichromate, or other oxidizing agents to the acidic chloride solution in the outer cell of a Conway microdiffusion cell. The inner cell contains $0.1N$ potassium hydroxide to absorb the chlorine (60). The liberated chlorine diffuses into the inner cell and is absorbed by the caustic solution. This diffusion method seems to be preferable to the hydrogen halide distillation method. A special microdistillation cell has been described (15).

C. EXTRACTION METHOD

The halogens are extractable with immiscible inert solvents, such as carbon tetrachloride, carbon disulfide, and chloroform. Thus, iodine when formed by the action of a mild oxidizing agent on an iodide can be extracted with carbon tetrachloride or carbon disulfide, which has a much higher distribution ratio.

Either 3% hydrogen peroxide or nitrite is a suitable oxidant. Bromine can be liberated by a controlled oxidation of bromide with nitric acid or permanganate and then extracted with carbon tetrachloride. If the permanganate oxidation is used, it is advisable to add a small amount of nitrite to remove any manganese dioxide which may have formed before proceeding with the determination of chloride. Because of the lower distribution ratio for bromine, as compared to iodine, several extractions may be required to completely remove the bromine.

D. ION EXCHANGE METHOD

The use of anion exchange resins in the separation of halides has been investigated by Zalevskaya and Starobinets (76) who determined the distribution isotherms for the halides between the resin phase and $0.1-1M$ hydroxide solutions. A procedure was developed using Dowex 1-X4 (100−200 mesh) in OH form, in which the solution containing the halide ions is added to the column and elution is started 30 min later with a series of basic solutions. A constant elution rate of 1 ml per min was maintained with a $0.045N$ potassium hydroxide solution used to elute the fluoride. The chloride was not eluted with a $0.32N$ potassium hydroxide solution. By using in sequence $0.70N$, $1.63N$, and $1.73N$ potassium hydroxide solutions, reasonable separations of bromide, iodide, and thiocyanate of the 1-mg equivalent level for each ion were obtained.

Another chromatographic method for the fractionization of halides using a thin-layer film of silica gel−gypsum and a solvent mixture consisting of 65 ml acetone, 20 ml n-butanol, 10 ml ammonium hydroxide, and 5 ml distilled water has been described by Seiler and Kaffenberger (63). The order of ion migration is $I^- > Br^- > Cl^- > F^-$.

Weiss and Stanbury (73) used a weakly basic exchanger to isolate perchlorate in biological fluids prior to elution and colorimetric determination of the perchlorate.

E. ELECTRODEPOSITION METHOD

Controlled potential coulometry has been used by Lingane to analyze iodide−chloride and bromide−chloride mixtures. Thus, halide ions react with the silver anode to form a silver halide deposit. The silver halide deposit may be removed by reducing electrolytically in sulfuric acid solution or by dissolution in ammonium hydroxide. The removal of iodide from chloride by using an ammonia−ammonium acetate supporting electrolyte and a controlled potential of −0.06 volts against S.C.E. is especially attractive (39).

II. METHODS OF DETERMINATION

A. *o*-TOLIDINE METHOD FOR CHLORINE

Chlorine in aqueous acid solution oxidizes *o*-tolidine (12,22−24,43,67) forming an intense yellow color. This reaction furnishes one of the most sensitive methods for the determination of trace amounts of chlorine. Organic matter causes serious interferences. Iron(III), manganese(VII) or (IV), and nitrite interfere if present in larger amounts than 0.3, 0.01, and 0.1 ppm, respectively (2). The method is not specific; other oxidizing agents react with the reagent in the same manner. The color produced depends on the pH. The color is an intense yellow if the pH is 0.8−3. In the pH range of 3−6, a bluish-green color develops which fades rapidly to give a colorless solution. Brown-colored solutions are formed when the pH is above 6. *o*-Tolidine has been used in neutral solution, but the color reaction is not as sensitive as in acidic solution (32). At pH 1.6, Beer's law is obeyed at 438 nm in the concentration range of 0.01−1.0 ppm chlorine (Fig. 3). The color intensity reaches a maximum almost immediately and decreases thereafter at a moderate rate. The concentration of the reagent should be at least three times that of the chlorine. This method gives the total free and combined available chlorine.

Reagents

o-**Tolidine.** Dissolve 0.10 g *o*-tolidine dihydrochloride in 15 ml hydrochloric acid and dilute to 100 ml with water. Store in dark bottles.

Standard Chlorine Solution, 0.01 mg Chlorine per ml. Pass a slow stream of chlorine from a cylinder into 500 ml distilled water for 20 min. Pipet 50 ml of the solution into a 250-ml stoppered Erlenmeyer flask. Add 10 ml of a 10% potassium iodide solution. Determine the liberated iodine by titration with $0.1N$ sodium thiosulfate using starch as indicator. Calculate the chlorine content per ml. Measure with a buret that volume of the solution calculated to contain 10 mg chlorine and transfer to a 1-liter flask. Dilute to the graduation mark with distilled water. Each milliliter of the final solution contains 0.01 mg chlorine. This solution is stable for 4 hr.

Procedure

Transfer 5 ml of the *o*-tolidine reagent solution to each of five 100-ml volumetric flasks. Add 50 or 75 ml of the sample solution to one 100-ml volumetric flask. Dilute to the mark with distilled water. Prepare standard solutions in the same manner using 2, 5, and 8 ml of the standard chlorine solution. Prepare the blank solution by diluting 5 ml *o*-tolidine solution to

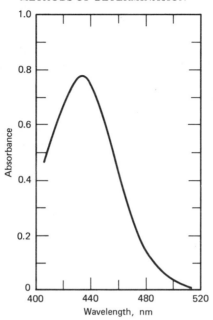

Fig. 3. Absorption spectrum: *o*-tolidine–chlorine oxidation product (1 ppm Cl_2, 1-cm cell).

100 ml. Measure the absorbance or transmittance at 438 nm 9 min after the addition of chlorine solution.

B. METHYL ORANGE METHOD FOR CHLORINE (21,61,64,66)

The decrease in the color to give a colorless product due to the chlorination of methyl orange is the basis of another colorimetric method. The following equation shows the reaction involved:

$$NaO_3S\text{—}\langle\ \rangle\text{—}N{=}N\text{—}\langle\ \rangle\text{—}N(CH_3)_2 + Cl_2 + H_2O \longrightarrow$$

$$(CH_3)_2N\text{—}\langle\ \rangle\text{—}OH + Cl\text{—}\langle\ \rangle\text{—}SO_3H + N_2 + NaCl$$

The color reaction is dependent on the pH. In acidic solution, there is a relatively large decrease in absorptivity at 505 nm proportional to the chlorine concentration. The decrease in color due to the reaction of methyl orange and chlorine is critically dependent on the manner in which the reagents are brought together. Thus, the dropwise addition of the chlorine solution to the methyl orange solution with vigorous stirring causes a much larger decrease in the color intensity than a rapid addition of the chlorine solution without stirring. This effect is probably due to a secondary chlorination of the products of the reaction. The method is not specific; other oxidizing agents such as bromine cause a similar decrease in intensity. Advantages of this method are that interferences due to iron(III) and nitrite in concentrations up to 10 ppm are negligible and that chlorine–ammonia compounds show very slight interference in the absence of bromide and in solutions containing less than 500 ppm chloride ion. The only serious interference is from the oxidized forms of manganese, and this can be overcome by the use of an arsenite modification (64).

The sensitivity of the methyl orange method is approximately 70% that of the *o*-tolidine method. Beer's law is obeyed at 505 nm up to a concentration of 0.6 ppm chlorine.

Solutions

Methyl Orange. Dissolve 0.06 g of methyl orange in water and dilute to 1 liter.

Buffer, pH 2. Add 37 ml 0.2*N* hydrochloric acid and 250 ml 0.2*N* potassium chloride to a 1-liter volumetric flask and dilute to the mark with water.

Standard Chlorine Solution. Prepared as described in Section A (*o*-tolidine method).

Procedure

Add exactly 10.00 ml of the methyl orange solution and approximately 45 ml of the buffer solution to each of four 100-ml beakers. Transfer with dropwise addition 3, 6, and 9 ml of the standard chlorine solution to three of the beakers with vigorous stirring of the solutions. Add dropwise a measured volume of the sample solution, for example, 25 ml, to the fourth beaker with constant stirring of the solution. Transfer the resulting solutions to 100-ml volumetric flasks. Rinse the beakers with the buffer solution and transfer washings to the proper flask. Dilute each flask to the graduation mark with the buffer solution. Measure the absorbance of each solution at 505 nm using distilled water in the reference cell.

C. KÖNIG REACTION METHOD FOR CHLORINE (3,35,45,46,50)

The König reaction refers to a procedure, first described by König, of opening a pyridine ring by the action of cyanogen chloride or cyanogen bromide and the condensation of the resulting glutaconic aldehyde with an aromatic amine. The compounds formed, are intensely colored. The method is rather specific; only

$$KCN + Cl_2 \longrightarrow KCl + CNCl$$

or

$$C_5H_5N + CNCl + 2H_2O \longrightarrow O=\overset{\overset{\displaystyle H}{|}}{C}-\overset{\overset{\displaystyle H}{|}}{C}=\overset{\overset{\displaystyle H}{|}}{C}-CH_2-\overset{\overset{\displaystyle H}{|}}{C}=O + CN-NH_2 + HCl$$

$$+$$

$$\downarrow \quad 2\,ArNH_2$$

$$2H_2O + Ar-N=\overset{\overset{\displaystyle H}{|}}{C}-\overset{\overset{\displaystyle H}{|}}{C}=\overset{\overset{\displaystyle H}{|}}{C}-\overset{\overset{\displaystyle H}{|}}{C}=\overset{\overset{\displaystyle H}{|}}{C}-\overset{\overset{\displaystyle H}{|}}{N}-Ar$$

chlorine, bromine, and substances containing available chlorine or bromine give the reaction. The sensitivity of this method depends upon the aromatic amine used. When barbituric acid is used as the amine, there are no interferences from iron(III) (0.5 ppm), permanganate (0.5 ppm as chlorine), nitrite (0.15 ppm as nitrogen) chlorite (0.5 ppm as chlorine), and iodine (0.5 ppm as chlorine). Bromine, manganese(III), and chloroamines interfere. Comparison of the barbituric acid procedure with nine other colorimetric procedures indicates that it is the most suitable laboratory method for determining free chlorine in the absence of combined chlorine, being reproducible, sensitive, and relatively specific. This procedure is, however, less convenient to use, and the reagent is unstable (49).

The method using sulfanilic acid as the aromatic amine has approximately 50% of the sensitivity of the o-tolidine method. The color formed by the reaction of potassium cyanide, pyridine, chlorine, and sulfanilic acid at 60°C reaches its maximum intensity in 40 min (27). At room temperature, over 2 hr are required for maximum development of the color. Beer's law is obeyed at 395 nm (27) and at pH 8 for chlorine concentrations up to 2 ppm.

Solutions

Sodium Sulfanilate, 0.5%. Adjust to pH 8.3 by the addition of 0.1N sodium hydroxide.

Pyridine, 0.25%.

Potassium Cyanide, 0.25%.

Sodium Hydroxide–Boric Acid Buffer, pH 8.   Add 50 ml 0.1M boric acid to 3.97 ml 0.1N sodium hydroxide and dilute to 100 ml.

Standard Chlorine Solution. Prepare as described in Section A (o-tolidine method).

Procedure

Adjust the sample to pH 5–6 by the addition of a dilute hydrochloric acid or sodium hydroxide solution. Add 1 ml each of sulfanilic acid, pyridine, and potassium cyanide solutions and 70 ml distilled water to each of five 100-ml volumetric flasks. Dilute the first flask to the graduation mark with distilled water. This solution serves as a reagent blank in the reference cell. Transfer 3, 6, and 9 ml respectively, of the standard chlorine solution to three of the remaining flasks and dilute each to the graduation mark with distilled water. Add a measured amount of sample solution to the remaining flask and dilute to the graduation mark with distilled water. Place each flask in a constant-temperature bath maintained at a temperature of 60°C for 30 min and then cool in an ice–water mixture for 3 min. Measure the absorbance of each solution at 395 nm 40 min after the addition of the chlorine solution.

D. OTHER METHODS FOR CHLORINE

Several colorimetric methods for chlorine have been developed using either N,N-dimethyl-*p*-phenylenediamine (11,51) or N,N-diethyl-*p*-phenylenediamine (2,49,52,54) as the aromatic amine, which reacts with chlorine to form a colored oxidation product. The diethyl analog is preferable inasmuch as it gives a better differentiation between free and combined chlorine and reacts rapidly to give a red color. Palin (52) developed a colorimetric procedure for sequential determination of (i) free chlorine and (ii) combined chlorine by (a) immediate measurement of the absorbance of red color developed by adding the sample to the reagent solution containing diethyl-*p*-phenylenediamine oxalate (Eastman # 7102) and a phosphate buffer solution, and (b) measurement of the increase in color when potassium iodide crystals were added to the intial colored system. The increase in color is attributed to the combined chlorine, monochloramine, and dichloramine. Bjorklund and Rand (6) have studied the determination of free residual chlorine in water by this method and rcommend a pH of 6.6 and measurement of absorbance at 540 nm. No interference was found for dissolved oxygen, 10 ppm copper, 0.3 ppm nitrite nitrogen, 10 ppm iron(II), 0.5 ppm manganese(II), and up to 2.4 ppm combined residual chlorine. Manganese(VII) interferes, but this difficulty can be eliminated by the modification using arsenic(III) as reducing agent (64).

4,4′,4″-Methylidynetris (N,N-dimethylaniline), commonly called "leuco crystal violet," under controlled conditions can be used to determine either free chlorine or total chlorine, provided iodide is added in the latter determination (2,7).

3,3′-Dimethylnaphthidine (3,3′-dimethyl-4,4′-diamino-1,1′-dinaphthyl) gives a violet-colored oxidation product upon reaction with chlorine. The reaction is not specific for chlorine (4). According to Nicholson (49), the sensitivity is almost 1.5 times better than that obtained with the acidic *o*-tolidine method.

E. NEPHELOMETRIC METHOD FOR CHLORIDE (41,42,57,74)

The nephelometric measurement of the turbidity resulting from the action of silver nitrate on dilute nitric acid solution of chloride is suitable for the determination of as little as 0.01 mg chloride per 100 ml solution. Interfering substances are those anions, such as bromide and iodide, whose silver salts are insoluble in dilute nitric acid.

Solutions

Generator Solution. Dissolve 1.6 mg sodium chloride in 80 ml water. Add 10 ml of a 1:160 nitric acid solution and 10 ml of a 0.1% solution of silver nitrate.

Standard Chloride Solution. Weigh 0.1649 g reagent-grade sodium chloride and transfer to a 1000-ml volumetric flask. Dilute to the mark with double-distilled water. Pipet 2 ml of the solution into a 200-ml volumetric flask and dilute to the mark with double-distilled water. The final solution contains 0.001 mg chloride ion per ml.

Nitric Acid. Pipet 1 ml concentrated nitric acid into 160 ml double-distilled water.

Silver Nitrate. Dissolve 0.1 g silver nitrate in 100 ml water.

Apparatus

The use of a photoelectric nephelometer is recommended. The data presented in Fig. 4 were obtained with a Fisher Nefluoro-Photometer under the following conditions: 425 center filter, 430 + right filter, 430 + left filter, mercury vapor lamp, 1-in. left cell containing distilled water, ¾-in. black-fritted center cell containing sample, and ¾-in. right cell containing generator solution.

Procedure

Add 10 ml of the nitric acid solution to each of five 100-ml volumetric flasks. Add 50, 40, 25, and 10-ml aliquots of the standard chloride solution to four of these flasks. To the fifth flask add a measured volume of the sample containing 0.02–0.05 mg chloride. Add 10 ml of the silver nitrate solution and sufficient double-distilled water to make the volume exactly 100 ml. The flasks are warmed at 40°C in a constant-temperature bath for 30 min and then cooled rapidly to room temperature in an ice-water mixture. The transmittance scale of

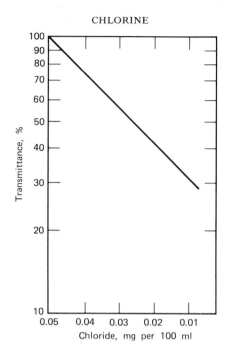

Fig. 4. Effect of chloride concentration (nephelometric method).

the instrument is set at 100% with the most concentrated of the standard solutions and the transmittance of the remaining solutions determined.

F. INDIRECT ULTRAVIOLET SPECTROPHOTOMETRIC DETERMINATION OF CHLORIDE: MERCURY(II) CHLORANILATE METHOD

This method, developed by Bertolacini and Barney (5), is applicable to the determination of 0.1 to 10 ppm chloride. The chemical basis of the method is the displacement of chloranilate upon the addition of solid mercury(II) chloranilate to an acidic chloride solution. The absorbance of the liberated chloranilic acid is measured. Methyl Cellosolve is added to decrease the solubility and dissociation of the mercury(II) chloride:

$$2Cl^- + HgC_6Cl_2O_4 + H^+ \longrightarrow HgCl_2 + HC_6Cl_2O_4^-$$

Interfering cations are removed by a cation exchange resin in the H form prior to development of the absorptive system. Sulfate, acetate, and citrate ions do not interfere. However, phosphate, fluoride, bromide, iodide, iodate and thiocyanate ions must be absent.

Special Solutions

Standard Chloride Solution. Dissolve 0.1887 g ammonium chloride in distilled water in a 1-liter volumetric flask and dilute to the mark. Transfer 100.0 ml of this stock solution to a 1-liter volumetric flask and dilute to volume with distilled water. This standard solution contains 1.25 μg chloride per ml.

Nitric Acid Solution (1M). Transfer 63.3 ml concentrated nitric acid to a 1-liter flask containing about 800 ml distilled water. Dilute to the mark with water.

Methyl Cellosolve (Union Carbide Chemicals Co.).

Mercury(II) Chloranilate. The solid mercury(II) chloranilate is prepared by the following procedure: Add dropwise with constant stirring a 5% solution of mercury(II) chloride which is also 2% in nitric acid to a 0.1% solution of chloranilic acid (2,5-dichloro-3,6-dihydroxy-p-benzoquinone) at 50°C until no additional precipitate of mercury(II) chloranilate forms. Decant the supernate and then wash the precipitate by decantation three times with ethanol and once with diethyl ether. Dry at 60°C in a vacuum oven. Mercury(II) chloranilate is also commercially available.

Procedure

Transfer a 25-ml aliquot of the sample solution containing not more than 0.5 mg chloride to the top of a cation exchanger. (A column 15 cm in length, 1.5 cm in diameter, and containing Dowex 50-X8, 20–50 mesh, in the H form or equivalent resin is satisfactory.) Discard the first 10 ml of effluent and collect the remainder of the effluent. Wash the column with 20–25 ml distilled water. Adjust the pH of the effluent to pH 7. The volume of the effluent should not exceed about 40 ml.

Transfer the effluent to a 100-ml volumetric flask and add 5 ml of the 1*M* nitric acid. Add 50 ml methyl Cellosolve and dilute to mark with distilled water. Add 0.20 g mercury(II) chloranilate and shake the flask for 15 min, preferably with mechanical shaking apparatus. Filter through a fine-porosity filter to remove excess mercury(II) chloranilate, or centrifuge.

Measure the absorbance of filtrate at 305 nm in 1.000-cm cells using a reagent blank, prepared by this same procedure in the reference cell.

Preparation of Calibration Graph. Use a buret to transfer 0, 5, 10, 15, 20, 25, 30, 35, and 40 ml of the standard chloride solution to nine 100-ml volumetric flasks. Add 5 ml 1*N* nitric acid and 50 ml methyl Cellosolve to each flask. Dilute to volume with distilled water. Add 0.2 g mercury(II) chloranilate to each flask and shake the flasks for 15 min on a mechanical shaker. Centrifuge to remove excess reagent. Measure the absorbance of each supernatant solution at 305 nm in 1.000-cm silica cells using the reagent blank solution in the reference cell.

G. OTHER METHODS FOR CHLORIDE

A number of indirect methods have been recommended for the determination of chloride and are summarized briefly.

1. Silver Chromate Method (31,33,38)

The action of chloride in neutral solution on solid silver chromate results in the precipitation of silver chloride and the passage into solution of an equivalent amount of chromate ion. The chromate thus provided is measured spectrophotometrically in ammoniacal solution at 373 nm. The chromate method is less sensitive than the nephelometric method. Beer's law is obeyed in the chloride concentration range of 0.3–1.5 mg per 200 ml. A calibration curve is necessary in the range of 0–0.3 mg per 200 ml. The sensitivity of the method is limited by the relatively high solubility of silver chromate, resulting in large blank readings. Interfering substances are those anions whose silver salts are less soluble than silver chromate, such as bromide, iodide, and phosphate.

2. Silver Phosphate–Heteropoly Blue Method (27)

The action of chloride in neutral solution on solid silver orthophosphate results in the precipitation of silver chloride and the passage into solution of the phosphate ion. The phosphate thus produced is determined by the heteropoly blue method (see Chapter 9). Beer's law is obeyed in the chloride range of 0–5 ppm. The method is less sensitive than the nephelometric method. Its advantage over the silver chromate method lies in the lower solubility of silver phosphate and the high sensitivity of the heteropoly blue method. Interfering ions are those anions whose silver salts are less soluble than silver phosphate in neutral solution or ions which interfere in heteropoly blue method.

3. Mercury(II)–Diphenylcarbazide Method (13,17,34,40,62)

Mercury(II) ions are complexed by chloride ions to give a slightly dissociated mercury(II) complex. The decrease in the mercury(II) concentration is measured colorimetrically with diphenylcarbazide in slightly acidic solution. Beer's law is obeyed for chloride concentrations up to 0.3 mg per 100 ml. Interfering substances are ammonium, cobalt, copper(II), chromate, iron(III), lead(II), zinc, and sulfate ions.

4. Mercury(II) Thiocyanate–Iron(III) Method

Chloride will displace thiocyanate from mercury(II) thiocyanate, and the equivalent thiocyanate is then determined spectrophotometrically as the iron(III) thiocyanate complex (77). This method is discussed in more detail in Section IIIB of this chapter.

5. Silver Iodate–Starch Method

Upon equilibration of a chloride solution with solid silver iodate, an equivalent of iodate ion is liberated. Treatment with the cadmium iodide–linear starch reagent of Lambert (37) results in a blue color.

H. BENZIDINE METHOD FOR CHLORATE (28)

Chlorate in concentrated hydrochloric acid solution oxidizes benzidine, forming an intense yellow color. This reaction furnishes the most sensitive method for the determination of chlorate. The method is not specific; other oxidizing agents interfere. The intensity of the color and the time required to reach maximum intensity depends upon the concentration of the hydrochloric acid. In solutions containing up to 12.6% hydrogen chloride, no color forms within 10 min. In solutions containing 16.8% hydrogen chloride, an immediate color forms reaching maximum intensity within 24 min. In solutions containing 21% hydrogen chloride, maximum intensity is reached within 8 min. In the latter case, Beer's law is obeyed at 434 nm for up to concentrations of 0.05 mg chlorate per 100 ml.

In a procedure developed by Burns (10) to determine trace amounts of chlorate in ammonium perchlorate, solutions containing at least 19% hydrogen chloride were used. Absorbance readings were taken at 438 nm within 12 to 15 min after mixing. For hydrochloric acids containing trace amounts of metal sulfites, a straight line was obtained when absorbance was plotted against chlorate concentration, which intersected the concentration axis. This displacement was attributed to the following reactions:

$$3SO_3^{2-} + ClO_3^- \longrightarrow Cl^- + 3SO_4^{2-}$$

$$SO_3^{2-} + Cl_2 + H_2O \longrightarrow 2Cl^- + HSO_4^- + H^+$$

If the hydrochloric acid contains trace amounts of free chlorine as an impurity, a straight line passing through the origin would be expected if a proper reference reagent blank was used.

Solutions

Benzidine Dihydrochloride. Dissolve 0.5 g benzidine hydrochloride in 100 ml 0.06N hydrochloric acid.

Hydrochloric Acid, 5.8N. Dilute 50 ml concentrated hydrochloric acid to 100 ml with water.

Standard Chlorate Solution. Weigh 0.1468 g potassium chlorate and transfer to a 1000-ml volumetric flask. Dilute with water and shake until dissolved. Pipet 10 ml of the solution into a 100-ml flask and dilute to mark with water. The final solution contains 0.01 mg chlorate ion per ml.

Procedure

Transfer 3 ml of the benzidine solution and 90 ml 5.8N hydrochloric acid to each of four 100-ml volumetric flasks. Add 2, 3, 4, and 5 ml of the standard chlorate solution at 2-min intervals to three of the flasks. Add a measured volume of the solution whose chlorate content is desired to the fourth flask. Dilute the flask to the mark with the hydrochloric acid solution. The Reagent blank to be used in the reference cell consists of 3 ml of the benzidine solution and 97 ml of the hydrochloric acid solution. Measure the absorbance of each solution 11 min after addition of the chlorate at 434 nm using a 1.000-cm cell. Use a reagent blank in the reference cell.

I. NEPHELOMETRIC METHOD FOR CHLORATE

By the reducing of chlorate with sulfurous acid and determining the resulting chloride nephelometrically, it is possible to determine as little as 0.02 mg chlorate ion per 100 ml (8). Interfering substances are those anions whose silver salts are insoluble in dilute nitric acid, such as chloride, bromide, and iodide.

Solutions

Generator Solution. Dissolve 1.6 mg sodium chloride in 80 ml water, add 10 ml of a 1:160 nitric acid solution and 10 ml of a 0.1% silver nitrate solution.

Sulfurous Acid, 6% (reagent grade).

Standard Chlorate Solution. Weigh 0.1467 g potassium chlorate and transfer to a 1000-ml volumetric flask. Dilute to the mark with double-distilled water. Shake until the salt dissolves. Pipet 25 ml of the solution into a 250-ml volumetric flask and dilute to the mark with double-distilled water. Each milliliter of the final solution contains 0.01 mg chlorate ions.

Nitric Acid Solution. Dilute 1 ml concentrated nitric acid to 500 ml with double-distilled water.

Silver Nitrate Solution, 0.1%.

Procedure (27)

Transfer 10, 20, 30, and 40 ml of the standard chlorate solutions to four conical flasks. To a fifth flask, add a measured volume of the solution whose chlorate content is desired. Add sufficient double-distilled water to each flask to make the volume approximately 40 ml. Add 2 ml of the sulfurous acid to each flask and place on a hot plate. Evaporate the solutions in each case to approximately 10 ml and cool to room temperature. After the addition of 25 ml of the nitric acid solution, add 25 ml of the silver nitrate solution. Transfer the solutions to 250-ml volumetric flasks. Dilute the flasks to the mark with double-distilled water and place them in an oven at 35°C for 30 min. Cool the solutions rapidly to room temperature by placing the flasks in an ice–water mixture for 1 or 2 min. Set the transmittance scale of the instrument at 100% using the most concentrated of the chlorate solutions. Determine the transmittance of the remaining solutions (see Section E).

J. INDIRECT SPECTROPHOTOMETRIC METHOD FOR PERCHLORATE

Perchlorate ion in acidic solution forms an ion pair complex with neocuproine, perchloratobis(2,9-dimethyl-1,10-phenanthroline)copper(I), which is extractable with ethyl acetate (14). A method based on the extraction of ferroin perchlorate with n-butyronitrile has also been reported (20). The following procedure is applicable to the determination of 5 to 100 μg perchlorate.

Special Solutions

Standard Perchlorate Solution. Dissolve 0.1232 g reagent-grade "anhydrous" sodium perchlorate in distilled water and dilute to 1 liter. Dilute exactly 25 ml of this stock solution to 100 ml to obtain a standard solution containing 25.0 μg perchlorate per ml.

Neocuproine Reagent Solution. Dissolve 0.4165 g neocuproine (2,9-dimethyl-1,10-phenanthroline) in reagent-grade ethyl acetate and dilute to 1 liter with this solvent.

Copper(II) Sulfate Solution. Dissolve 0.3140 g reagent-grade anhydrous copper(II) sulfate in distilled water and dilute to 1 liter.

Phosphate Buffer Solution. Dissolve 34.0 g potassium dihydrogen phosphate, KH_2PO_4, in distilled water and dilute to 1 liter.

Hydroxylamine Sulfate Solution. Dissolve 50 g hydroxylamine sulfate, $(NH_2OH)_2H_2SO_4$, in 950 ml water.

Procedure

Weigh or measure by volume an amount of sample containing no more than 0.6 mg perchlorate ion. The resulting solution should be adusted to pH 3–5 and diluted to 25 ml.

Transfer 6 ml of the copper(II) sulfate solution to a 25-ml volumetric flask and add 5 ml sample solution. Add 2 ml of the hydroxylamine sulfate solution and 5 ml of the phosphate buffer solution. Dilute to volume with distilled water and mix thoroughly. Add 10 ml of the neocuproine reagent to a 60-ml separatory funnel. Transfer solution from volumetric flask to separatory funnel. Use 1 ml buffer solution to rinse volumetric flask. Add with rinsing to the separatory funnel. Place the separatory funnel on a shaking apparatus and extract for 3 min. Remove the aqueous layer, drain the extract through a cotton pledget placed in the stem of the funnel and collect in a small bottle. Measure the absorbance at 456 nm in 1.000-cm cells using a reagent blank solution in the reference beam. Refer the absorbance reading to a calibration graph obtained using standard perchlorate solutions.

K. OTHER METHODS FOR PERCHLORATE

Perchlorate in the 5- to 500-ppm range can be determined by extraction of the tris(1,10-phenanthroline)–iron(II) perchlorate complex into n-butyronitrile and measurement of the absorbance at 510 nm (20). The formation of iron(II)-2,2'-bipyridine perchlorate and the extraction of the complex with nitrobenzene is the basis of another method for the determination of perchlorate (75).

A novel indirect spectrophotometric method has been proposed by Trautwein and Guyon (68). This method is based on the interference of the perchlorate ion in the development of the color due to the rhenium furildioxime formed when perrhenate is reduced with tin(II) chloride in the presence of α-furildioxime.

Perchlorate in aqueous solution reacts with methylene blue to form a slightly soluble complex (25,36). The methylene blue–perchlorate complex can be extracted with chloroform (27) or 1,2-dichloroethane (65). This method is applicable to the determination of 5 to 50 mg perchlorate when the absorbance is measured at 655 nm in 1.000-cm cells. In another methylene blue method, a known amount of methylene blue is added to the aqueous perchlorate solution, and the exces methylene blue in the filtrate is determined spectrophoto-metrically, following removal of the methylene blue perchlorate percipitate by filtration (48). Other dyes which have been used include brilliant green, malachite green (71), and neutral red (69,70).

The decrease in absorbance following the precipitation of tetrapyridine-copper(II) perchlorate is the basis of another spectrophotometric method (9). A redox reaction in which perchlorate was reduced to chloride by vanadium(III) to give an equivalent amount of vanadium(IV) has been utilized (78). Absorbance measurements must be made at two wavelengths, 400 and 700 nm, so that a correction can be made for the absorbance due to the excess vanadium (III). Perchlorate can be extracted into o-dichlorobenzene as tetrabutylphosphonium perchlorate. When this extract is contacted with an aqueous solution containing iron(III) thiocyanate, an iron(III) thiocyanate—tetrabutylphosphonium complex forms in the organic layer. The absorbance is measured at 510 nm (19).

L. ACID CHROME VIOLET METHOD FOR CHLORINE DIOXIDE

This method, developed by Masschelein (44), is based on measuring the decrease in absorbance at 550 nm of 1,5-bis(4-methylphenylamino-2-sodium sulfonate)-9,10-anthraquinone (acid chrome violet K) in an ammonia—ammonium chloride buffer solution of pH 8.1—8.4. This method is presumably free of interference due to free chlorine, hypochlorites, chlorites, chlorates, fluoride, and aluminum. Two moles chlorine dioxide react with 1 mole of the dye.

By utilizing this colorimetric method for the determination of chlorine dioxide and two iodometric-potentiometric titrations, Myhrstad and Samdal (47) determined chlorine dioxide, chlorite, and chlorine in water. Palin also determined chlorine dioxide and chlorite using N,N-diethyl-p-phenylenediamine and several concomitant determinations for differentiation (53,55). The color developed by the reaction of chlorine dioxide and tyrosine has also been used to determine chlorine dioxide in water (26).

Special Solution

Acid Chrome Violet K. Suspend 175 mg of the reagent in an aliquot of distilled water, add 20 mg sodium hexametaphosphate, 48.5 g ammonium chloride, and 0.4 ml concentrated ammonium hydroxide (sp. gr. 0.90, 28% NH_3), dilute to 1 liter with distilled water, and allow to stand 24 hr. This solution when stored in an amber bottle is suitable for use for at least 30 days.

Procedure

Transfer 10.0 ml of the acid chrome violet K reagent to a 100-ml flask. Add an aliquot of the standard ClO_2 solution or sample solution containing less than 1.25 mg/liter ClO_2, with the special precaution that this solution drop directly into the reagent without running down the inside of the flask in order to avoid loss of ClO_2 before it reacts with the reagent. Dilute the solution to 100 ml with distilled water. Use 5-cm absorption cells and measure the absorbance at 550 nm against distilled water in the reference cell for both the decolorized sample solution and the reagent blank solution. This difference in absorbance, ΔA, is a measure of chlorine dioxide.

M. INDIRECT SPECTROPHOTOMETRIC METHOD FOR CHLORINE DIOXIDE

This method is applicable only to solutions combining chlorine dioxide and chlorine but no chlorite. The method is based on measuring the diminution of absorbance of the tris-1,10-phenanthroline–iron(II) complex relative to a reagent blank by a differential spectrophotometric technique (29).

Solutions

Iron(II) Perchlorate Solution. Dissolve 326 mg Fe $(ClO_4)_2 \cdot 6H_2O$ in 300–400 ml distilled water, add approximately 1.5 ml concentrated sulfuric acid, and dilute to 1 liter.

1,10-Phenanthroline Solution. Dissolve 1.0 g 1,10-phenanthroline monohydrate in 300 ml water, heating, if necessary, to effect solution, and dilute to 1 liter.

Sulfuric Acid Solution. Dilute 33.3 ml concentrated sulfuric acid (sp. gr. 1.84; 95.5% H_2SO_4 by weight) to 1 liter with distilled water.

Saturated Sodium Acetate Solution. Add 1200 g anhydrous sodium acetate to 500 ml distilled water, dilute to 1 liter, and mix thoroughly to ensure saturation.

Malonic Acid Solution. Dissolve 2.0 g malonic acid in 100 ml distilled water.

Procedure

To a 100-ml volumetric flask containing 30 ml 0.6M sulfuric acid solution, pipet 10.00 ml 0.0009M iron(II) perchlorate solution. Add 1 ml of the malonic acid solution to 100 ml of the sample if free chlorine is also present. Transfer an aliquot of the chlorine dioxide solution containing from 0.7 to 70 μg chlorine dioxide to the flask containing the sulfuric acid and iron(II) perchlorate solutions. Mix thoroughly and allow 2 min for the reaction to reach completion; then add 10 ml saturated sodium acetate solution and 10 ml 0.1% 1,10-phenanthroline solution. Mix thoroughly and dilute to volume. Prepare a reagent blank in a similar manner omitting the aliquot of chlorine dioxide solution.

Measure the differential absorbance at 510 nm by placing the reagent blank in the sample beam and the sample containing the aliquot of chlorine dioxide in the reference beam. The concentration of chlorine dioxide may be read from a calibration graph, previously constructed from data obtained from the standard chlorine dioxide solution, or by computation from the Beer-Lambert relationship. It should be noted, however, that the quantity measured and plotted is the differential absorbance, ΔA, and is directly proportional to the chlorine dioxide concentration.

III. APPLICATIONS

A. CHLORINE IN WATER

The following method for the determination of free chlorine in water is based on the use of a stabilized neutral o-tolidine reagent and is sometimes referred to as the SNORT method. Johnson and Overby (32) have made a thorough study of this method and developed a procedure which eliminates interference due to iron(III) and nitrite and minimizes the interference due to chloramines. Instead of the development of a yellow color as with the acid o-tolidine method (IIA), a blue color is produced and stabilized by an anionic surface active agent, perhaps due to ion pair formation. Control of pH to the 7 ± 0.5 range, the use of large excess of o-tolidine (at least a 8-to-1 ratio of o-tolidine dihydrochloride to chlorine), and the addition of the water sample to the reagents are important aspects of this method. Monochloramine and dichloramine can be determined by a differential technique in which the absorbances are measured after the addition of iodide to a neutral solution of the sample and to an acidic solution of the sample and development of the blue color by the SNORT method (2,32). When trichloramine is in the sample, it produces an absorbance equivalent to 48% of the trichloramine present.

Special Solutions

o-**Tolidine Reagent.** Add 5.0 ml concentrated hydrochloric acid to 100 ml chlorine demand-free water. Transfer 5.0 ml of this acidic solution to a 1-liter volumetric flask and add chlorine demand-free water almost to volume. Dissolve 1.5 g o-tolidine dihydrochloride in this acidic solution, and dilute to volume with chlorine demand-free water. Store solution in a glass-stoppered amber glass bottle.

Buffer–Stabilizer Reagent. Dissolve 25.9 g disodium hydrogen phosphate, Na_2HPO_4, 12.6 g potassium dihydrogen phosphate, KH_2PO_4, and 8.0 g sodium bis(2-ethylhexyl) sulfosuccinate (Aerosol OT, American Cyanamide Co.) in a solution of 600 ml chlorine demand-free water (2,32) and 200 ml diethylene glycol monobutyl ether. After dissolution is complete, dilute to volume with chlorine demand-free water in a 1-liter volumetric flask.

Sodium Arsenite Solution. Dissolve 100 g sodium arsenite, $NaAsO_2$, in distilled water and dilute to 1 liter.

Procedure

Transfer 5 ml buffer–stabilizer reagent and 5 ml o-tolidine reagent to a 250-ml beaker. Use a magnetic stirrer and, while stirring the solution, add 100 ml of the water sample.

Prepare a blank reference solution by adding 1 ml of the sodium arsenate to 100 ml water sample and then adding this solution to 5 ml of the buffer–stabilizer and o-tolidine reagents with gentle stirring.

Measure the absorbance against a blank reference solution. Prepare a calibration graph by using 100-ml aliquots of standard solutions containing 0.6 to 6 μg chlorine per ml. Commercial bleach, which may contain 30 to 50 g chlorine per liter, can be standardized iodometrically and used to prepare these standard solutions.

B. CHLORIDE IN WATER

This indirect method is based on the reaction of mercury(II) thiocyanate with chloride to form mercury(II) chloride, with the displaced thiocyanate being determined as the Fe(SCN)$^{2+}$ complex (72,77). Bromides, iodides, cyanides, and thiosulfates interfere, and small amounts of fluorides, nitrates, nitrites, sulfates, and phosphate do not interfere. Appreciable amounts of sulfate, phosphate, and nitrite decolorize the solution.

Special Solutions

Standard Chloride Solution. Transfer 0.4205 g dry, reagent-grade potassium chloride to a 1-liter volumetric flask and dilute to volume with distilled water. Transfer 50.00 ml of this stock solution to a 1-liter volumetric flask and dilute to volume. This standard solution contains 10.0 μg Cl$^-$ per ml.

Mercury(II) Thiocyanate Reagent. Dissolve 0.700 g mercury(II) thiocyanate in 1 liter distilled water. Prepare a fresh solution after one week. Store in a glass-stoppered amber bottle.

Iron(III) Perchlorate. Dissolve 11.6 g hydrated ferric perchlorate, Fe(ClO$_4$)$_3$ · 6H$_2$O, in 100 ml 4N perchloric acid.

Procedure

Transfer 10 ml water sample to a 50-ml volumetric flask. Add 5 ml 60% perchloric acid, 1 ml mercury(II) thiocyanate, and 2 ml of the iron(III) perchlorate. Dilute to volume with distilled water and mix thoroughly. After 10 min, measure the absorbance at 460 nm using 1.000-cm cells against a reagent blank solution in the reference cell. The use of 5.000-cm cells is recommended for low-absorbance solutions.

Prepare a calibration graph by using 1.00 to 25.0 ml of the standard chloride solution instead of the 10-ml water sample and use this same procedure for color development.

REFERENCES

1. Andrews, L. W., *J. Amer. Chem. Soc.,* **29,** 275 (1907).
2. American Public Health Association, *Standard Methods for the Examination of Water and Wastewater,* 13th ed., Washington, D.C., 1971, pp. 107–144.
3. Asmus, E., and H. Garschagen, *Z. Anal. Chem.,* **136,** 269 (1952).
4. Belcher, R., A. J. Nutten, and W. I. Stephen, *Anal. Chem.,* **26,** 772 (1954).
5. Bertolacini, R. J., and J. E. Barney, II, *ibid.,* **30,** 202, 498 (1958).
6. Bjorklund, J. G., and M. C. Rand, *J. Amer. Water Works Assoc.,* **60,** 608 (1968).
7. Black, A. P., and G. P. Whittle, *ibid.,* **59,** 607 (1967).
8. Blattner, N., and J. Brasseur, *Chem. Ztg. Rept.,* **24,** 793 (1900).
9. Bodenheimer, W., and H. Weiler, *Anal. Chem.,* **27**; 1293 (1955).
10. Burns, E. A., *ibid.,* **32,** 1800 (1960).
11. Byers, D. H. and M. G. Mellon, *Ind. Eng. Chem., Anal. Ed.,* **11,** 202 (1939).
12. Chamberlain, N. S., and J. R. Glass, *J. Amer. Water Works Assoc.,* **35,** 1065 (1943).
13. Clarke, F., *Anal. Chem.,* **22,** 553 (1950).
14. Collinson, W. J., and D. F. Boltz, *ibid.,* **40,** 1896 (1968).
15. Elsheimer, H. N., A. L. Johnston, and R. L. Kochen, *ibid.,* **38,** 1684 (1966).
16. Feigl, F., and F. Neuber, *Z. Anal. Chem.,* **62,** 370 (1923).
17. Feuss, J. V., *J. Amer. Water Works Assoc.,* **56,** 607 (1964).
18. Fogg, A. G., C. Burgess, and D. T. Burns, *Analyst,* **96,** 854 (1971).
19. Fogg, A. G., D. T. Burns, and E. H. Yoowart, *Mikrochim. Acta,* **1970,** 974.
20. Fritz, J. S., J. E. Abbink, and C. A. Campbell, *Anal. Chem.,* **6,** 2123 (1964).
21. Gad, G., and E. Priegnitz, *Gesundh.-Ingr.,* **68,** 174 (1947).
22. Gilcreas, F. W., and F. J. Hallinan, *J. Amer. Water Works Assoc.,* **36,** 1343 (1944).
23. Grant, J., *Chemist Analyst,* **26,** 59 (1937).
24. Haase, L. W., *Von Wasser,* **11,** 116 (1936).
25. Hahn, F. L., *Z. Angew. Chem.,* **39,** 451 (1926).
26. Hodgen, H. W., and R. S. Ingols, *Anal. Chem.,* **26,** 1224 (1954).
27. Holland, W. R., H. S. Brand, and D. F. Boltz, unpublished research.
28. Horvorka, V., and Z. Holzbecher, *Collect. Czech Chem. Commun.,* **14,** 490 (1949).
29. Howell, J. A., G. E., Linington, and D. F. Boltz, *Microchem. J.,* **15,** 598 (1970).
30. Ingols, R. S., and G. M. Ridenour, *J. Amer. Water Works Assoc.,* **40,** 1207 (1948).

31. Isaacs, M. L., *J. Biol. Chem.*, **53**, 17 (1922).

32. Johnson, J. D. and R. Overby, *Anal. Chem.*, **41**, 1744 (1969).

33. Kishima, I., *J. Japan. Biochem. Soc.*, **20**, 147 (1948).

34. Kolthoff, I. M., *Chem. Weekblad.*, **21**, 20 (1924).

35. König, W., *J. Prakt. Chem.*, **69**, 105 (1904); *ibid.*, **70**, 17 (1904); *ibid.*, **83**, 325 (1911).

36. Kurger, D., and E. Tschirch, *Z. Anal. Chem.*, **85**, 171 (1931).

37. Lambert, J. L., and S. K. Yasuda, *Anal. Chem.*, **27**, 444 (1957).

38. Letonoff, T. V., *J. Lab. Clin. Med.*, **20**, 12 (1935).

39. Lingane, J. J., and L. A. Small, *Anal. Chem.*, **21**, 1119 (1949).

40. Litvinenko, P., and M. Rozivskii, *Gigienai Sanit,* **1954** (No. 1), 52.

41. Lure, Y. Y., and Z. V. Nikolaeva, *Zavod. Lab.*, **12**, 161 (1946).

42. McAlpine, R. K., *J. Amer. Chem. Soc.*, **51**, 1065 (1929).

43. Marks, H. C., and R. R. Joiner *Anal. Chem.*, **20**, 1197 (1948).

44. Masschelein, W., *ibid.*, **38**, 1839 (1966).

45. Milton, R. F., *Nature,* **164**, 448 (1949).

46. Morris, P. K., and H. A. Grant, *Analyst,* **76**, 492 (1951).

47. Myhrstad, J. A., and J. E. Samdal, *J. Amer. Water Works Assoc.*, **61**, 205 (1969).

48. Nabar, G. M., and C. R. Ramachandran, *Anal. Chem.*, **31**, 263 (1959).

49. Nicholson, N. J., *Analyst,* **90**, 187 (1965).

50. Nusbaum, I., and P. Skupeko, *Anal. Chem.*, **23**, 1881 (1951).

51. Palin, A. T., *Analyst,* **70**, 203 (1945).

52. Palin, A. T., *J. Amer. Water Works Assoc.*, **49**, 873 (1957).

53. Palin, A. T., *ibid.*, **62**, 483 (1970).

54. Palin, A. T., *J. Inst. Water Eng.*, **21**, 537 (1967).

55. Palin, A. T., *Water & Sewage Works,* **107**, 457 (1960).

56. Pourbaix, M., *Atlas of Electrochemical Equilibria in Aqueous Solutions,* Pergamon Press, New York, 1966, pp. 590–603.

57. Richards, T. W., and R. C. Wells, *J. Amer. Chem. Soc.,* **31**, 235 (1904).

58. Rodden, C. J., *Analytical Chemistry of the Manhattan Project,* McGraw-Hill, New York, 1950, p. 299.

59. *Ibid.,* p. 295.

60. *Ibid.,* p. 297.

61. Schmidt, M. P., *J. Prakt. Chem.*, **85**, 235 (1912).

62. Scott, A., *J. Amer. Chem. Soc.*, **51**, 3351 (1929).

63. Seiler, H. and T. Kaffenberger, *Helv. Chim. Acta,* **44**, 1282 (1961).

64. Sollo, F. W., Jr., and T. E. Larson, *J. Amer. Water Works Assoc.,* **57**, 1575 (1965).

65. Swasaki, I., S. Utsumi, and C. Kang, *Bull. Chem. Soc. Japan,* **36**, 325 (1963).

66. Taras, M., *Anal. Chem.,* **19**, 342 (1947).

67. Todd, A. R., *J. Amer. Water Works Assoc.,* **30**, 115 (1938).

68. Trautwein, N. L., and J. C. Guyon, *Anal. Chem.,* **40**, 639 (1968).

69. Tsubouchi, M., *Anal. Chim. Acta,* **4**, 143 (1971).

70. Tsubouchi, M., and Y. Yamamoto, *Bunseki Kagaku,* **19**, 966 (1970).

71. Uchikawa, S., *Bull. Chem. Soc. Japan,* **40**, 798 (1967).

72. Utsumi, S., *J. Chem. Soc. Japan, Pure Chem. Sect.,* **73**, 835, 838 (1952).

73. Weiss, J. A., and J. B. Stanbury, *Anal. Chem.,* **44**, 619 (1972).

74. Wells, R. C., *J. Amer. Chem. Soc.,* **35**, 99 (1906).

75. Yamamoto, Y., and K. Kotsuji, *Bull. Chem. Soc. Japan,* **37**, 594 (1964).

76. Zalevskaya, T. L., and G. L. Starobinets, *Zh. Anal. Khim.,* **24**, 721 (1969).

77. Zall, D. M., D. Fisher, and M. Q. Garner, *Anal. Chem.,* **28**, 1665 (1956).

78. Zatko, D. A., and B. Kratochvil, *ibid.,* **37**, 1560 (1965)

FLUORINE

STEPHEN MEGREGIAN

Sanitation Director (retired), U.S. Public Health Service

The methodology of fluoride colorimetric analysis has undergone little change in recent years, although many papers have been published on this subject. This is due primarily to the lack of color-producing reactions wherein the fluoride ion is one of the primary color-producing reactants. Consequently, most colorimetric methods are based on the fluoride ion acting upon a color-producing reaction, usually between a metal and an organic dye. Nearly all such reactions result in an inverse relationship — a decrease in color intensity with increasing fluoride concentration. A few systems exist wherein the fluoride reaction with the metal either releases a dye which imparts a color in direct proportion to fluoride concentration or forms a colored ternary complex with the dye.

To the analyst, this methodology results in problems affecting the precision of the fluoride determination. In addition to the care normally required during analytic manipulations, the addition of the color-forming reagents must be controlled with extra precision, and the composition of the color-forming reagents accurately reproduced. Therefore, the analyst should be fully aware of these extra sources of error inherent in this methodology.

Notwithstanding these limitations, several new methods have been developed that offer some improvement and have been substituted for other methods presented in the first edition.

The newer zirconium–SPADNS method (9) retains nearly all of the features of the earlier zirconium–alizarin method plus a significant reduction in time and effects from interferences. A new distillation technique developed by Bellack (8) has been added. This technique results in a total transfer of fluoride without dilution of the original sample concentration, thereby improving the precision of the distillation.

A significant development in fluoride analysis has been the development of a fluoride electrode by Frant and Ross (22). This selective ion technique is under investigation and is being used by many laboratories. It shows promise as a versatile adjunct to the colorimetric measurement of fluoride, with a special advantage possibly in situations requiring continuous monitoring or with large concentration differences. Because the electrode responds to the activity of the fluoride ion, the electrode response depends upon the ionic strength of the test solution (21).

I. SEPARATIONS

A. PRELIMINARY TREATMENT OF SAMPLE

Before fluorine can be measured quantitatively by colorimetric procedures, it must be present in solution as the fluoride ion essentially free from organic matter and ionic interferences. This is necessary because there is no color-forming reagent available that reacts specifically with fluoride. Consequently, the most time-consuming and tedious operations in fluoride analysis are the ashing and the distillation of the sample in preparation for its determination.

Samples containing organic matter, such as urine, body fluids, and plant materials, are first ashed under alkaline conditions in order to remove the organic matter. The usual ashing agent is calcium oxide, specially prepared to be free from fluoride. Samples containing organic matter in covalent combination with fluorine cannot be ashed directly but must be decomposed by special procedures such as bomb oxidation or combustion at high temperature in special apparatus.

The fluoride in minerals has been liberated as hydrogen fluoride by a pyrohydrolytic method: a flux of bismuth trioxide, sodium tungstate, and vandium pentoxide is used, and the hydrogen fluoride is absorbed by a basic solution prior to spectrophotometric determination (14).

Another preliminary treatment that has been suggested as being applicable to the microdetermination of fluoride in natural waters, salt solutions, urine, and plant extracts consists of the adsorption of fluoride on pure calcium phosphate and collection of the resulting product by centrifugation (63). The fluoride adsorbed on the calcium phosphate is then removed by the microdiffusion method (23,56).

B. DISTILLATION METHOD

After the sample has been ashed, the fluoride is removed from interferences by distillation. The basic procedure for this removal was described by Willard and Winter (65). This procedure is based on the volatilization of fluoride as H_2SiF_6 from a perchloric or sulfuric acid solution at a temperature of $135°C$. Distilled water was used as a sweeping agent, and the fluoride was separated from nonvolatile interferences such as sulfate, phosphate and metals. The procedure has been studied by many investigators. Complex apparatus (24,32,49) has been designed to ensure optimum temperature control. Compressed air (61) and steam (24) have been substituted for distilled water as sweeping agents. Studies on the effect of various interferers such as silica and aluminum (15) on fluoride recovery have been made. The effect of acid volume, glassware size, distillation rate, distillation temperature, and kind of acid used has also been studied (15,43). From these many investigations, the following factors appear to be essential to quantitative fluoride recovery.

Distillation apparatus

Figure 1 is an illustration of a fluoride distillation apparatus and steam generator that meet the requirements for quantitative fluoride recovery and that can be assembled from common laboratory glassware.

More elaborate stills are available in which refinements such as ground-glass connections, jacketed distillation flasks to maintain constant boiling temperature, or electrically heated, thermostat-regulated flasks can be incorporated. However, since the primary consideration in the design of this apparatus is the quantitative transfer of fluoride to the distillate, free from interferences and in a reasonable time, any apparatus that is capable of satisfying these requirements can be used. The following details should be considered in its assembly:

The distillation flask capacity should be as small as practicable for the volume of sample and acid used in the distillation. A 250- to 300-ml flask is usually

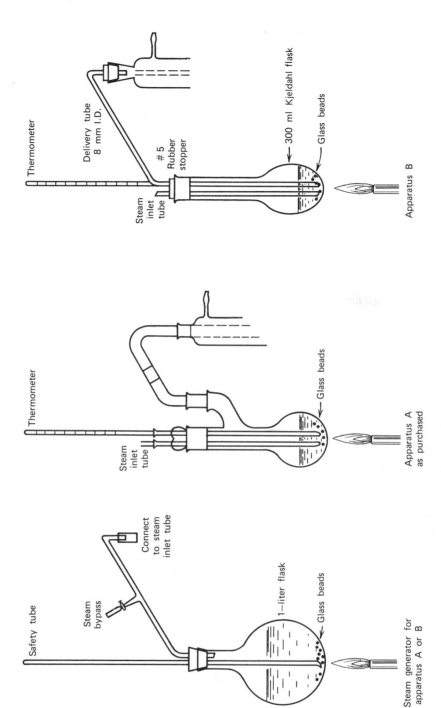

Thermometer

Delivery tube 8 mm I.D.

Steam inlet tube

#5 Rubber stopper

300 ml Kjeldahl flask

Glass beads

Apparatus B

Thermometer

Steam inlet tube

Glass beads

Apparatus A as purchased

Safety tube

Connect to steam inlet tube

Steam bypass

1–liter flask

Glass beads

Steam generator for apparatus A or B

Fig. 1. Fluoride distillation apparatus.

112

ample for most distillations. Where very small quantities of fluoride are being distilled, use of a 125-ml flask will be advantageous.

The bore of the delivery tube between the flask and the condenser should be as large as possible in order to prevent condensate from forming a vapor seal across the tube and being carried over with the distillate. The end of the delivery tube that protrudes into the distillation flask should be cut at an angle to permit free fall of condensate returning to the flask. Tubing of less than 8 mm bore is not suitable for this purpose.

The steam inlet tube and thermometer bulb must remain immersed in the boiling liquid throughout the distillation. All connections must be leakproof. Considerable amounts of fluoride can be lost through leaky connections.

The heat under the distillation flask must not touch any portion of the flask not in contact with the liquid inside. Local superheating will decompose the distilling acid and result in the contamination of the distillate. This can be prevented by seating the bottom of the flask into a hole cut in a piece of Transite or asbestos sheet, thus shielding the exposed glass surfaces from direct contact with the flame.

A distillation apparatus using either steam or distilled water introduced through a capillary has been used successfully as the sweeping agent for fluoride separation. If distilled water is used, extra precautions (such as spray traps) are needed to prevent acid carry-over. In general, distillations should not be made at a rate faster than 5 ml per min with this type of apparatus, as this seems to be the upper limit for the prevention of acid carry-over. However, steam distillation can be carried out at much faster rates, the rate depending only on the efficiency of the condenser used to cool the distillate.

Distillation procedure

This general procedure is for the distillation of fluoride samples using the assembly shown in Fig. 1. The sample is transferred quantitatively to the distillation flask; 15 to 50 ml water should accompany the transfer. Eight to 10 glass beads are added, sufficient silver sulfate (or perchlorate) is added to precipitate any chloride which may be present in the sample, and 15 ml concentrated sulfuric (or 60% perchloric) acid is added. The flask is connected to the condenser and to the steam generator, a receiver is placed under the condenser, and the contents of the flask are then mixed by gentle swirling. Heat is applied to the flask and to the steam generator with the bypass valve open. When the temperature of the liquid in the flask reaches 130–135°C, steam is introduced by closing the bypass valve, and the distillation is continued until at least 100 ml have distilled over after the temperature has reached the optimum. During the distillation, flask temperature must be maintained between 130° and 150°C (between 130° and 140°C if perchloric acid is used) by controlling the heat under the flask. Distillation rate should be at least 5 ml per min and can be much faster without danger of acid carry-over. When the appropriate amount of distillate has been collected, the heat is removed from beneath the distillation flask and the steam by-pass is opened. The distillate is now ready for colorimetric measurement. A fresh sample may be introduced into the flask and distilled without changing the acid, unless other considerations such as accumulation of aluminum, silica, or silver chloride require the acid to be discarded.

Factors Affecting Quantitative Recovery. Samples containing aluminum or silica, which will precipitate in gelatinous form on acidification, require either a larger distillate volume or higher distillation temperature for complete fluoride recovery. Aluminum forms a complex with fluoride and retards its distillation. Therefore, with samples containing over 20 mg aluminum, the usual procedure is to distill first from a sulfuric acid bath at a temperature of $160-170°C$ and then to redistill the distillate at $135°C$ in order to eliminate the sulfate carry-over from the first distillation (1). Grimaldi, (26) recommends a single distillation from a perchloric–phosphoric acid mixture at $140°C$ with the aluminum complexed by phosphoric acid. Some phosphate (about $40-60$ μg) does distill or is carried over as spray. However, this quantity would not interfere with most of the colorimetric procedures. With samples high in silica, the ashed sample is subjected to a sodium hydroxide fusion prior to distillation. This procedure, as described by Remmert (48) and also by Rowley (51), is necessary to release the fluoride entrapped in the silica.

With samples containing chloride, silver salts (perchlorate, sulfate) are added to the distillation flask in order to hold back the chloride. The presence of chloride or other volatile acids in the distillate is undesirable because they interfere with most of the colorimetric procedures for fluoride determination. Also, the hydrogen ions accompanying the volatile anion produce even greater interference if not neutralized.

Perchloric, sulfuric, or phosphoric acid has been used as the distillation acid. Preference has been given to perchloric acid because many laboratories determine the fluoride in the distillate by the thorium-titrimetric method, which cannot tolerate traces of sulfate or phosphate in the distillate. In colorimetric fluoride determinations, traces of sulfate which may be carried over do not produce a significant interference; therefore, sulfuric acid is preferred when fluoride samples are to be determined colorimetrically.

Optimum distillation temperature has been determined to be $130-150°C$ when sulfuric acid is used. Temperature control is not critical, provided the upper limit is not exceeded. High-temperature distillates will contain excessive amounts of the distilling acid and must be redistilled at the optimum temperature. Fluoride will distill over at temperatures below $130°C$, but a larger volume of distillate must be collected to ensure complete transfer.

Glass beads are placed in the distillation flask to reduce bumping and to provide a source of silica for the fluoride to transfer as H_2SiF_6. Their presence also reduces the attack of fluoride on the walls of the flask.

The rate of distillation is not critical. The distillation may be carried out as fast as the system will permit, so long as spray from the distilling flask is not mechanically carried over into the condensate. Distillation rates as high as 30 ml per min have been observed without carry-over of acid spray. Slow rates of distillation, under 3 ml per min, are to be avoided, as they provide no benefits and merely consume time.

The volume of distillate to be collected is dependent on factors previously cited such as temperature and the presence of aluminum or silica in the sample. The following generalization should apply when these interferers are at a minimum and the distillation is carried out at optimum temperature. All of the distillate must be collected. After the distillation flask has reached the optimum temperature and steam has been introduced, at least an additional 100 ml distillate must be collected at the optimum temperature. Fluoride will begin to distill over as soon as boiling commences, and a considerable quantity will be transferred even though the temperature of the distillation is below optimum. The major portion of the fluoride will transfer in the first 50 ml collected after the steam has been introduced. The additional 50 ml distillate will sweep out the system, transferrring the remaining traces and preparing the setup for subsequent distillations. Distillate volumes totaling up to 500 ml are recommended when alumium and silica are present in order to assure complete recovery.

Most colorimetric methods for the determination of fluoride require a concentration range of 0.05 to about 1.50 mg fluoride per liter. Consequently, the optimum quantity of fluoride in a distillate volume of 200 ml should be between 10 and 300 μg. If less than this amount of fluoride is collected, the distillate must be concentrated, under alkaline conditions, to a volume that will yield the minimum concentration. If a large quantity of fluoride is present, the distillate must be diluted and the result multiplied by the appropriate dilution factor. In actual practice, it has been found that when low fluoride values are being determined, poor recoveries are frequently obtained because of adsorption or entrapment of fluoride in the distillation apparatus. It is recommended, therefore, that a sample size be selected for distillation such that the quantity of fluoride distilled falls within the above range and, whenever possible, each distillation contains the same quantity of fluoride. With samples known to contain only small quantities of fluoride, if the sample size must be limited by other considerations, it is best, then, that the sample not be distilled in an apparatus through which a high fluoride sample has been previously run. In situations of this type, it is good practice to reserve one apparatus for use with samples low in fluoride and another for high fluoride samples. It is also good practice, especially when the apparatus is first assembled, to run through a blank

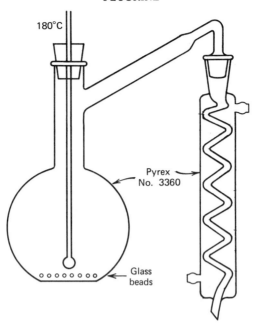

180°C

Pyrex
No. 3360

Glass
beads

Fig. 2. Bellack fluoride still.

distillation, sweeping out the apparatus, and removing any fluoride or other interference which may be present.

Where many samples of similar composition are analyzed, the distillation apparatus developed by Bellack (8) can be used effectively. Automatic apparatus using this principle has also been developed whereby the analyst adds a sample to the acid mixture and then retrieves it as distillate after about 1 hr completely free from interference.

The Bellack distillation apparatus, Fig. 2, consists of a flask containing a minimum of 400 ml 1:1 sulfuric acid, connected to a condenser. The flask also requires glass beads or chips to control bumping and a 180°C thermometer. A preliminary boilout with distilled water is performed first. Heat is applied to the acid mixture until a flask temperature of exactly 180°C is reached. The mixture is then cooled to below 120°C. A 300-ml sample is then poured into the flask. After thorough mixing, the apparatus is connected and heat is applied again until the flask temperature reaches exactly 180°C. The resulting distillate will measure 300 ml and, after thorough mixing, will represent the concentration of the original sample with all interferences removed. An excess of silver sulfate should be present in the acid mixture to prevent chloride transfer. Many repeated distillations can be performed without changing the acid. No steam, water, or other sweeping agent is necessary. About 30–50 min is needed to complete a distillation, but very little attention by the analyst is necessary.

C. OTHER METHODS OF SEPARATION

Prior to 1933, when Willard and Winter described the technique of separating fluoride as H_2SiF_6, most separations were carried out by volatilizing fluoride as SiF_4. This method was not completely satisfactory and has been abandoned in favor of the much simpler and more reliable Willard and Winter distillation. Attempts to separate the fluoride by volatilization as HBF_4 (65) and as BF_3 (18,19) have also been described.

Direct distillation of fluoride with sulfuric acid before ashing has also been recommended. Smith and Gardner (57) used this technique with blood samples, claiming that iron fluoride will volatilize if blood is ashed by the usual ashing procedure. Venkateswarlu and Rao (62) also utilized this technique and claimed superior recoveries with samples of tea. Also, rather than submit the first distillate to ashing followed by a second distillation, they adsorb the fluoride on magnesium oxide, separate by filtration or centrifugation, and distill again by the usual technique.

According to the Association of Official Analytical Chemists (3), preliminary sulfuric acid distillation is also advantageous when fatty materials of low fluoride content are being analyzed. Use of a larger still is recommended as well as a small pellet of paraffin to reduce foaming. A larger volume of distillate is collected, and after treatment with 30% H_2O_2, it is evaporated to dryness with excess $Ca(OH)_2$ in a platinum dish, ashed, and redistilled in the usual manner.

Singer and Armstrong (55) described a procedure based on the diffusion of hydrofluoric acid in the confined atmosphere of a small polyethylene bottle. In this technique, the HF is volatilized by incubation at 60°C in a perchloric acid solution and is collected on a polyethlene strip containing sodium hydroxide. The procedure is particularly applicable to very small quantities of fluoride, in the range of 1–20 μg. The incubation requires 20 hr. However, a large number of samples can be run concurrently. Organic matter prevents complete fluoride recovery. The effect of excessive quantities of volatile acids can be overcome by maintaining a sufficient quantity of sodium hydroxide on the recovery strip. Previous attempts to separate fluoride as HF by simple distillation had not proved satisfactory except when a special inert apparatus was used.

Wharton (64), in separating microamounts of fluoride from materials containing calcium and orthophosphate, recommends a time of 16 to 24 hr and a temperature of 60°C for the isolation of small amounts of fluoride (4 μg) by the microdiffusion method using a Conway cell. It has been reported that the microdiffusion separation can be accelerated by adding hexamethyldisiloxane to the sample compartment. When the volatilized trimethylfluorosilane is collected by a basic solution, the fluoride is liberated (60).

Separation of fluoride by ion exchange (34) followed by elution has received some attention but very little acceptance. The principal drawback seems to be incomplete elution of the fluoride adsorbed on the resin. Interfering cations have been removed by a cation exchange resin prior to using spectrophotometric determination (11).

II. METHODS OF DETERMINATION

Most colorimetric methods which have been developed are indirect in that they are dependent upon the effect of fluoride on color systems such as ferric–thiocyanate (20) ferric-acetylacetone (4), zirconium–alizarin (16,38,53), zirconium–quinalizarin (58), aluminum–hematoxylin (46), and thorium–alizarin (59). The early investigators described procedures whereby low fluoride concentrations could be estimated visually by comparison with a series of standards. These methods all have many disadvantages. They are all subject to various interfering ions such as chloride, sulfate, phosphate, aluminum, iron, hydrogen or hydroxyl, and oxidizing agents. Because of the empirical nature of the reactions involved, these methods require careful control of reaction conditions such as time, temperature, pH, and quantities of reagents added, and they result in weak colors necessitating long light paths in order to obtain sufficient color intensity to distinguish fluoride concentration intervals of 5.0 to 10.0 μg per 100 ml. Consequently, they are unsuitable for use in spectro-photometric analysis.

Studies by several investigators have resulted in the adaptation of some of these methods to photometric measurement. Boonstra (10), Bumstead and Wells (12), and Megregian and Maier (42) studied the zirconium–alizarin methods and recommended reagents that can be used in photometric analysis. Richter (50) adapted the aluminum–eriochrome cyanine R color system to photometric measurement. Monnier (44) described a method based on the titanium–peroxide color system; Icken and Blank (33), on thorium and alizarin; Megregian (41), on the zirconium–eriochrome cyanine R system; and Bellack and Schouboe (9), on zirconium with SPADNS. None of these methods is specific for the fluoride ion. Therefore, all require a preliminary separation, especially with complex samples, before the determination can be completed. Also, many of these methods have comparable precision and accuracy. There-fore, if the analyst were to select one of these procedures as best for photo-metry, he would be guided in his choice by other considerations, such as sensitivity, stability, simplicity, and rapidity.

With these parameters as guide, the zirconoum–eriochrome cyanine R method of Megregian (41) has been selected as the one most generally suited for photometric analysis. This choice is based on the following advantages of this method. It is very sensitive to small fluoride concentration differences; a concentration difference of 0.02 μg fluoride per ml can be accurately measured in the range of 5.0–120.0 μg per 100 ml (0.05–1.2 mg fluoride per liter). By scaling down the sample quantities and reagent volumes proportionately, it should be possible to determine as little as 0.25 μg in a 5.0-ml sample. The colored system follows Beer's law for the concentration range cited. When no interfering ions such as aluminum or phosphate are present, the color reaction is immediate and reproducible, requiring no specific time interval before measurement. The reagents are easy to obtain and to prepare and are indefinitely stable, requiring no special care other than normal protection from direct sunlight and evaporation. The procedure is very simple, no manipulations other than the measurement of the reagent and sample volumes are necessary, and its fluoride sensitivity is about 0.5 absorbance units per cm light path and μg fluoride per ml, thus allowing photometric measurements to be made accurately with the short light paths commonly available in most spectrophotometers.

The one disadvantage of this method arises when water samples are analyzed without prior distillation. Interference due to sulfate is quite high. Therefore, when undistilled samples are analyzed directly, the sulfate concentration must be known and a correction applied. A nomogram for sulfate correction has been provided and also a simple method for determining sulfate concentration to a precision of ±5 mg/liter, thus providing a means for circumventing this disadvantage.

A. ZIRCONIUM–ERIOCHROME CYANINE R METHOD (21)

Zirconyl ions form both monomolecular and bimolecular colored complexes with eriochrome cyanine R in hydrochloric acid solution, depending on the acidity and the quantity of dye present. The structures of the complexes formed are believed to be as represented below:

and ... Bimolecular

The absorption spectra of the two complexes are very similar, and when an excess of the dye is present in solution, both complexes may be present. When a limited quantity of fluoride is present in the reaction mixture, the zirconyl ion reacts preferentially with it, forming a complex of the composition ($ZrOF_2$). This reaction withdraws zirconyl from the colored complex and results in a reduction of the total color of the system. This system, like the other color systems discussed previously, involves complex equilibria in which acidity, zirconyl ions, eriochrome cyanine R, fluoride, temperature, and interfering ions all contribute to the amount of color formed. Therefore, no simple stoichio-metric relationship exists between the fluoride, zirconium, and dye.

Consequently, with any reaction based on these considerations, careful control of all parameters is necessary in order to achieve a quantitative relation-ship with any one of them. In the procedure to be described, the reaction conditions have been so selected that an inverse relationship exists between fluoride ion concentration and the color intensity of the solution. With samples freed from interference, the precision and accuracy of the measurements are subject only to the precision inherent in the absorbance measurement, a temperature control of $\pm2°C$ with reference to the standard graph, and the precision of the volumetric measurements involved in carrying out the determination.

Two procedures are described. The first is a simplified method for use with samples that have been previously distilled. The second procedure has been developed for fluoride solutions containing known interferences or for water samples which must be run routinely. The apparatus and reagents required for both procedures are identical.

Special Solutions

Reagent A. Dissolve 1.800 g eriochrome cyanine R dissolved in distilled water and dilute to 1 liter.

Reagent B. Dissolve 0.265 g zirconyl chloride octahydrate or 0.220 g zirconyl nitrate dihydrate in 50 ml water. Add 700 ml concentrated hydrochloric acid, reagent grade (sp. gr. 1.19), to the zirconyl solution and dilute the mixture to 1 liter with distilled water. This reagent should be cooled to room temperature before use.

The above zirconyl salts have a tendency to lose water and acid on storage. A preferred technique in their use is to prepare a stock solution of the zirconyl salt, assay it for zirconium by precipitation with ammonia, and then use an aliquot of the stock solution when preparing repeated batches of reagent B. Reagents A and B cannot be combined as a single reagent because prior mixing results in formation of a precipitate on prolonged storage.

Reference Solution. Add 10.0 ml reagent A to 100 ml distilled water. Add 10 ml of a hydrochloric acid solution prepared by diluting 7.0 ml of the concentrated acid to 10 ml with distilled water. This solution will keep indefinitely when stoppered.

Sodium Arsenite (0.1N). Dissolve 6.5 g $NaAsO_2$ in 1 liter water.

Standard Sodium Fluoride Solution. Dissolve 0.221 g NaF in 1 liter water (1 ml = 0.1 mg F). This solution may be diluted to provide working solutions of lesser F content.

Procedure

Fluoride Distillates. The following procedure is recommended for fluoride distillates or solutions containing no interference. The volumes indicated below may be scaled down proportionately with the volume of sample taken if only small sample volumes are available.

Transfer a 50.0-ml portion of distillate or aliquiot diluted to 50.0 ml and containing no more than 60 μg fluoride (1.20 mg per liter) to a 150-ml beaker and adjust to a standard temperature ±2°C. Add 5.0 ml reagent A followed by 5.0 ml reagent B to the sample and mix well. Adjust the spectrophotometer to 0 absorbance (at 525–530 nm) with the reference solution and measure the absorbance of the sample. Determine the fluoride value of the sample aliquot from a calibration plot prepared by using standard solutions of fluoride and the above procedure (see Fig. 3). Standards should be selected in the range of 0.0 to 60.0 μg fluoride in 50.0 ml (0.0–1.20 mg per liter). The resultant plot should be a straight line of negative slope. Because the color reaction reaches equilibrium rapidly and the color is stable at a standard temperature, absorbance readings can be taken immediately after mixing with the reagents or at any convenient time thereafter. A new plot must be prepared each time a fresh batch of reagents A or B is prepared, and frequent checks of the standard graph should be made during the life of the reagents since concentration changes due to evaporation will occur as the reagents are being used.

The reference solution is necessary because of the high absorbance due to

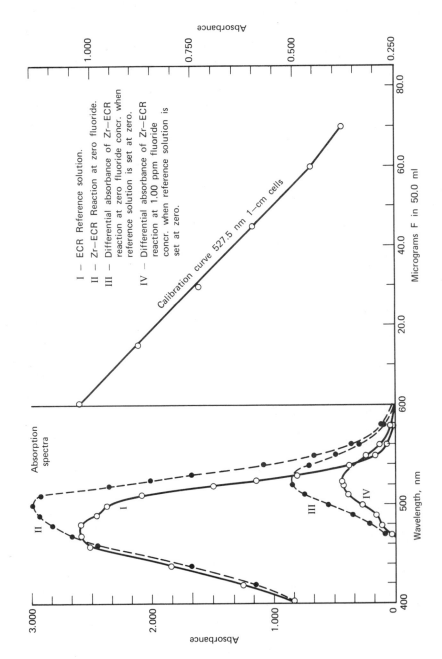

Absorbance

Micrograms F in 50.0 ml

Calibration curve 527.5 nm 1—cm cells

I — ECR Reference solution.
II — Zr–ECR Reaction at zero fluoride.
III — Differential absorbance of Zr–ECR
    reaction at zero fluoride concr. when
    reference solution is set at zero.
IV — Differential absorbance of Zr–ECR
    reaction at 1.00 ppm fluoride
    concr. when reference solution is
    set at zero.

Absorption
spectra

Wavelength, nm

Absorbance

Fig. 3. Absorption spectra and standard curve for zirconium—eriochrome cyanine R reaction.

122

unreacted dye in the sample solution. Any stable colored solution or standard colored glass absorbing light at 525–530 nm may be used as reference, and the standard graph should be prepared from it. The need for a reference solution is evident from the absorption spectra of the system shown in Fig. 3. From this figure, it is evident that maximum sensitivity can be obtained only with photometric systems employing relatively pure monochromatic radiant energy. Thus, ordinary filter photometers will not provide as good a sensitivity as spectrophotometers that are capable of providing a much narrower spectral bandwidth.

Water Samples Containing Interferences. Take 50.0 ml water sample or an aliquot diluted to 50.0 ml containing no more than 0.06 mg fluoride and adjust

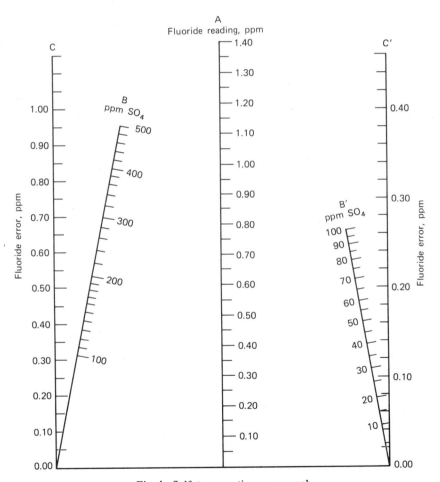

Fig. 4. Sulfate correction nomograph.

the temperature to within ±2.0°C of that used in preparing the standard graph. If the sample contains free chlorine, add two drops 0.1N arsenite solution for each milligram chlorine present per liter of reagent A, followed by 5.0 ml reagent B, and mix well.

Set the spectrophotometer at 0 absorbance (at 525–530 nm) with the reference solution, and measure the absorbance of the sample within 5 min after the reagents have been added to the sample solution. If the absorbance reading of the sample falls beyond the range of the standard graph, the combined sulfate and fluoride concentrations in the sample are too high. In this case, the procedure must be repeated with a smaller sample aliquot.

If the presence of aluminum ion is suspected in the water sample, allow the reaction to continue an additional 15 min and measure the absorbance. An appreciable decrease in absorbance indicates that aluminum is present as an interference. At this point, holding the reacted sample for 2 hr before making the final absorbance reading will eliminate the interference effect of up to 5.0 mg aluminum per liter. The distillation procedure can also be used to separate the fluoride from the interference, and the distillate can then be analyzed by the procedure outlined previously for distillates.

The correction for sulfate concentration in the sample aliquot is made as follows: The fluoride value of the absorbance reading is obtained from the standard graph; this value is applied to the nomogram.in Fig. 4 and the sulfate error determined. The sulfate error is then subtracted from the first value to obtain the corrected fluoride value of the aliquot, and this value is multiplied by the aliquot factor to obtain the fluoride content of the original sample.

Interferences

As previously indicated, distillates can be analyzed directly by this method. The complete analysis of a distillate need not take longer than 5 min. Most distillates do not need to be neutralized or concentrated, and buffer solutions must not be added. The only possible interference would be the presence of large quantities of volatile anions such as nitrates or unprecipitated chloride, which would come over during the distillation and produce high results in the subsequent analysis. If this is suspected, a portion of the distillate can be neutralized to the phenolphthalein endpoint with 1.0N sodium hydroxide and the neutralized aliquot taken for the fluoride determination. The effect of large quantities of volatile anions on the reagent is to increase the acidity at which the reaction takes place, resulting in a reduction of the total color produced.

For water samples that are subjected to direct analysis, the tolerance to various interferences is shown in Table I. No interference was noted with calcium ion concentrations of 200 mg per liter or magnesium ion concentrations of 100 mg per liter. The reagent can also tolerate large quantities of alkali metal

TABLE I. Interferences Normally Present in Potable Waters

Interfering substance	Concentration, mg/liter, necessary to produce fluoride error equivalent to	
	0.02 mg/liter F	0.10 mg/liter F
Alkalinity (as $CaCo_3$)	3300.0	undetermined
Cl^-	3500.0	undetermined
Fe^{3+}	10.0	40.0
PO_4^{3-}	2.5	6.0
$(NaPO_3)_6$	1.0	2.0
Al^{3+}	5.0^a	—
Free chlorine	reduce with arsenite	
Turbidity, color	compensate	

^{a}When reaction is permitted to continue for 2 hr.

ions and barium. The effect of very large concentrations of bicarbonate, carbonate, or hydroxide ions (alkalinity) is to reduce the acidity of the reaction, thereby increasing the total color produced by the reagent and resulting in low fluoride values. Large concentrations of chloride influence the reaction in the same direction. Iron(III) in high concentration produces a color reaction with eriochrome cyanine R, thus increasing the apparent color of the system and resulting in low fluoride values. Phosphate and hexametaphosphate (also other polyphosphates) act on the reagent similarly to fluoride in that they form a complex with the zirconyl ion and inhibit the color-forming reaction. Water samples containing phosphates in excess of that listed in Table I should be distilled prior to the determination of fluoride. Aluminum does not affect the primary color reaction under the acid conditions specified. Aluminum reacts with fluoride to form a complex similar to the complex formed between zirconyl and fluoride ions. Therefore, if no fluoride is present, aluminum does not interfere. When both fluoride and aluminum are present in a sample that has been treated with the reagents, the fluoride is distributed between the zirconyl and aluminum ions. Under the reaction conditions imposed, the zirconium slowly removes the fluoride from the aluminum complex, and after a sufficient lapse of time (about 2 h), the aluminum effect is no longer evident.

If many samples containing up to 5.0 mg aluminum per liter are to be determined, the 2-hr waiting period can be reduced to 15 min by a slight modification of the above procedure. A few drops of $1.0N$ sodium hydroxide, sufficient to make the sample alkaline and to convert the aluminum into the aluminate anion, are added. The reagents are then added as previously outlined. The effect of the added alkalinity is negligible. A reasonable explanation for this phenomenon would be that fluoride does not form a complex with aluminum existing as the aluminate ion. When the reagent is added, the fluoride present immediately complexes with the zirconyl ions. By the time aluminum ions are released from aluminate, there is very little fluoride remaining to form the aluminum complex; consequently, much less time is required for the reaction to reach equilibrium.

Free chlorine and other strong oxidizing agents attack eriochrome cyanine R, thereby reducing the total color of the system. These oxidants can be easily reduced with sodium arsenite. A slight excess of arsenite does not affect the fluoride reaction. If desired, arsenite can be incorporated in reagent A.

Sulfate Error and Correction

Zirconyl ions in acid solution complex with sulfate ions to produce the complex $[ZrO(SO_4)_2]^{2-}$ anion. Although the reaction conditions have been adjusted to reduce this effect, the zirconium sulfate complex is stable enough to prevent some of the zirconium from reacting with eriochrome cyanine R to produce the full color of the reaction. Therefore, the total color reduction produced in a water sample containing both sulfate and fluoride ions is the result of the sum of the two complex-forming reactions. Figure 5 shows how the presence of both fluoride and sulfate affects the fluoride reading. The data obtained in this figure were used to construct the nomograph in Fig. 4, which can be used to determine the sulfate error in a given sample.

Figure 4 is used as follows: Obtain the fluoride reading of the sample aliquot, A. Determine the sulfate concentration of the aliquot, B. Determine the fluoride

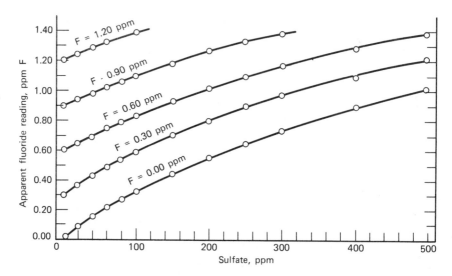

Fig. 5. Effect of sulfate interference.

error of the sulfate concentration by running a line through A and B to C. Subtract the value obtained on line C from A to give the fluoride value of the aliquot. Multiply this value by the aliquot factor to obtain the fluoride content of the original sample. Lines B' and C' are expansions of A and B and should be used when the sulfate concentration is in the range of 0–100 ppm.

Any procedure requiring additional manipulations to remove interferences would nullify the advantages gained by using a rapid method for determining fluoride concentration. Thus, sulfate might be removed from the sample by barium precipitation or by the usual distillation procedure, but the advantages gained in using the rapid method would be counterbalanced by the time required to effect a separation by these conventional procedures. Another possibility would be to diminish the sulfate effect on the reagent by adding a known amount of sulfate to the reagent. When this is done, the reagent becomes less sensitive to fluoride and the measurable fluoride range is reduced. Large quantities of sulfate in the reagent render it practically insensitive to fluoride, and this cannot be offset by increasing the zirconium concentration of the reagent.

Fortunately, a method exists whereby the sulfate concentration of a water sample can be quickly and accurately determined; and by use of Fig. 4, a correction can be applied which eliminates the necessity for time-consuming separations without materially sacrificing the accuracy of the rapid method. A slightly modified procedure based on a turbidimetric method for sulfate is described below.

Reagents

Barium Chloride Dihydrate Crystals, Reagent Grade, 20–30 mesh.

Acid–Salt Solution. Dissolve 240 g sodium chloride in 900 ml water. Add 20 ml concentrated hydrochloric acid and dilute the solution to 1 liter.

Procedure

Transfer a 50-ml water sample or aliquot which, when diluted to 50 ml, will contain no more than 5.0 mg sulfate (100 mg/liters) to a 250-ml beaker or conical flask. Add 10 ml of the acid–salt solution and mix. Add approximately 0.4 ml barium chloride crystals measured dry with a small spoon. Swirl the container immediately and continue for 30 sec. Allow the sample to stand for 5 min and then transfer the resulting suspension to an absorption cell and measure the absorbance at 525 nm.

The sulfate concentration of the aliquot is obtained from a standard plot, which is prepared by subjecting solutions of known sulfate concentration in the range of 0.0 to 5.0 mg sulfate (0 to 100 mg/liter) to this same procedure.

B. ZIRCONIUM–SPADNS METHOD (5)

This method provides a single, stable reagent, an instantaneous reaction, a high tolerance to sulfate as well as to other interferences, and a satisfactory fluoride sensitivity for most purposes. It has been included as a standard method in the 13th edition (1971) of *Standard Methods* (2). It can be used directly on many water samples without need for prior distillation. Its chemistry is very similar to that of the eriochrome cyanine R method. SPADNS is the acronym for sodium 2(*p*-sulfophenylazo)-1,8-dihydroxy-3,6-naphthalene disulfonate.

Special Solutions

SPADNS Solution. Dissolve 0.958 g sodium 2(*p*-sulfophenylazo)-1,8-dihydroxy-3,6-naphthalene disulfonate (SPADNS) in distilled water and dilute to 500 ml.

Zirconium(IV) Solution. Dissolve 0.133 g zirconyl chloride octahydrate, $ZrOCl_2 \cdot 8H_2O$, in about 25 ml distilled water. Add 350 ml concentrated hydrochloric acid and dilute to 500 ml with distilled water.

Mixed Reagent Solution. Mix equal volumes of the SPADNS and zirconium(IV) solutions.

Reference Solution. Add 10.0 ml of the SPADNS solution to 100 ml distilled water. Dilute 10.0 ml concentrated hydrochloric acid with 10.0 ml distilled water and add to the diluted SPADNS solution.

Standard Fluoride Solution. See Section IIA.

Sodium Arsenite Solution. See Section IIA.

Procedure

Take a 50.0-ml aliquot of the fluoride distillate or a water sample, dechlorinated with a few drops of sodium arsenite solution, and adjust to the temperature used in preparing the standard graph. Add exactly 10.0 ml of the mixed (zirconium-SPADNS) reagent solution to the sample. Mix well after addition of this reagent.

Set spectrophotometer at 570 nm, transfer the reference solution to both absorption cells, and adjust the spectrophotometer to zero absorbance. Transfer the sample solution to the matched sample absorption cell and read the absorbance.

Determine the fluoride value of the sample aliquot by reference to a standard graph prepared using standard fluoride solutions. A straight line of inverse slope is obtained in the fluoride concentration range of 0.00 to 1.4 mg/liter. A standard graph must be prepared each time a fresh batch of zirconium—SPADNS reagent is prepared. The zirconium—SPADNS reagent has been known to remain stable for at least two years.

Interferences

Table II summarizes the tolerance to interferences commonly present in water samples. The tolerance is based on a fluoride error of 0.1 mg/liter at a fluoride level of 1.0 mg/liter. Interference effects vary with the concentration of fluoride present.

TABLE II. Effect of Interferences on the Zirconium—SPADNS Method for Fluoride[a]

Interferer	Concentration mg/liter	Effect on fluoride value mg/liter
Alkalinity (as $CaCO_3$)	5000	decreases 0.1
Al^{3+}	0.1[b]	decreases 0.1
Cl^-	7000	increases 0.1
Fe^{3+}	10	decreases 0.1
$(NaPO_3)_6$	1	increases 0.1
PO_4^{3-}	16	increases 0.1
SO_4^{2-}	200	increases 0.1

[a]Residual chlorine must be completely removed. Use arsenite reagent. Color and turbidity must be removed or compensated.
[b]Effect on immediate reading. Tolerance increases to 3.0 after 2 hr of standing and 30 after 4 hr of standing.

C. ALIZARIN FLUORINE BLUE METHOD

A ternary complex containing equimolar amounts of a dye (alizarin fluorine blue) (39), lanthanum(III), and fluoride exhibits maximum absorbance at 625 nm. This method was developed initially by Belcher, Leonard, and West (5,6,7) who used cerium instead of lanthanum. A comprehensive study of this method has shown that improved sensitivity and rate of color formation can be achieved by using a succinate buffer and a water—acetone solution (25,27,36,45,66).

Most cations interfere seriously, although the tolerance to sodium and potassium ions exceeds 1000 ppm, and as much as 20 ppm ammonium, lithium, barium, calcium, and magnesium ions are permissible. Chloride, bromide, iodide, perchlorate, and nitrate do not interfere at the 1000 ppm level. The tolerances to sulfate (200 ppm), acetate, citrate, silicate, and tartrate (20 ppm) are often acceptable. Carbonate and sulfide interfere very seriously. Errors due to sulfate can be corrected (35). Unless a preliminary separation is used, the standard addition technique has been suggested. The optimum concentration range is about 0 to 0.8 ppm. Therefore, this method is suitable for the determination of 4 to 40 μg fluoride in a 20-ml sample solution assuming a final dilution volume of 50 ml.

Special Solutions

Alizarin Fluorine Blue Solution. Transfer 0.44 g alizarin fluorine blue (1,2-dihydroxyanthraquinon-3-ylmethylamine-N,N-diacetic acid) to a 400-ml beaker, add 5 ml 0.5M sodium hydroxide, and triturate. Dilute to 200 ml and add 50 ml of the succinate buffer solution. Measure the pH with a pH meter and adjust the pH to the range of 4.5–4.8, if necessary. Dilute this solution to 500 ml and filter before storing in a polyethylene bottle in a refrigerator.

Lanthanum Nitrate Solution. Dissolve 0.43 g lanthanum(III) nitrate hexahydrate in distilled water, dilute to 500 ml, and store in a polyethylene bottle.

Lanthanum–Alizarin Fluorine Blue Reagent Solution. This reagent solution should be prepared each day. Mix equal volumes of the alizarin fluorine blue and lanthanum nitrate solutions.

Succinate Buffer Solution. Dissolve 5.9 g succinic acid in about 300 ml distilled water and add 0.5M sodium hydroxide until the pH is 4.6 as monitored with a pH meter. Dilute to 500 ml. This solution should be prepared each day.

Stock Standard Fluoride Solution. Dissolve 2.2103 g sodium chloride, previously dried for 1 hr at 140°C, in distilled water and dilute to 1 liter in a 1-liter volumetric flask. Transfer to polyethylene bottle for storage (1 ml = 1.00 mg fluoride).

Standard Fluoride Solution. Transfer 10.00 ml of the stock solution of fluoride to a 1-liter volumetric flask and dilute to volume (1 ml = 10.0 μg fluoride).

Procedure (45)

Transfer a 20-ml sample solution which has been previously adjusted to pH 4 and which contains 2–40 μg fluoride to a 50-ml volumetric flask. Add 5.0 ml of the succinate buffer solution and mix thoroughly. Add 10.0 ml of the lanthanum–alizarin fluorine blue reagent solution and mix thoroughly before adding 10.0 ml acetone and again mixing. Dilute to the mark with distilled water, mix very thoroughly, and allow to stand for 30 min. Measure the absorbance in 1-cm absorption cells at 625 nm against a reagent blank solution prepared by this same procedure using distilled water instead of the sample solution.

Prepare a calibration graph by transferring 0.5, 1, 2, 3, and 4 ml of the dilute standard fluoride solution by means of a microburet to 50-ml volumetric flasks, carefully follow the recommended procedure, and plot absorbance versus μg fluoride.

D. OTHER METHODS

In an aqueous methyl Cellosolve solution at pH 4.5, thorium chloranilate reacts with fluoride ion to form thorium difluorochloranilate, $ThF_2C_6Cl_2O_4$, and liberates chloranilic acid whose absorbance can be measured at 330 or 540 nm (31).

Solochrome cyanine R and zirconium have been used in a spectrophotometric method for the determination of 0.25 to 2.5 μg fluoride (17).

A procedure utilizing the thorium—alizarin color system has been developed by Icken and Blank (33) to replace the thorium titration technique commonly used on distillates from food samples. The authors indicate that 0.05 to 1.0 μg fluoride/ml can be measured accurately by this procedure utilizing standard 1-cm absorption cells. The procedure requires a period of 2 hr for equilibrium to be reached. The lanthanum—alizarin spectrophotometric method has been automated (29).

Many satisfactory methods for the determination of fluoride based upon visual comparison of color intensity have been described (42,52,54,59,61). These methods have now been generally discarded in favor of methods adapted to photometric measurement.

A kinetichromic method for fluoride has been based on the catalytic action of fluoride on the relatively slow reaction between zirconium(IV) and methylthymol blue and was used to determine 0.5—4.75 μg fluoride (30). The catalytic effect of fluoride on the zirconium(IV) and xylenol orange has been used to determine 0.005—0.05 ppm fluoride (13,37).

Powell and Saylor (47) describe a procedure based on the quenching by fluoride of the fluorescence produced in the aluminum—eriochrome red B (or superchrome garnet Y) complex when irradiated with ultraviolet radiant energy. Reagent concentrations can be varied to cover fluoride ranges of 0—5.0 μg in 50 ml or 0—70.0 μg in 50 ml. Samples and standards are prepared and run concurrently, and a waiting period of 1—3 hr is necessary depending on the dye used to prepare the reagent. This procedure appears to be more tolerant to phosphate interference; therefore, it might be used to advantage with distillates containing traces of phosphoric acid. Traces of fluoride can also be determined spectrofluorimetrically by measuring the increase in fluorescence intensity of the zirconium—calcein blue complex resulting from the formation of a 1:1:1 ternary complex (28). This method is applicable to the 0.01—0.06 ppm fluoride range.

III. APPLICATIONS

A. BIOLOGICAL SAMPLES

1. Urine.

The following procedure for the fixing and ashing of urine samples is essentially that which has been used at the National Institute of Dental Research (40).

Procedure

Select a volume of sample which will contain not less than 10 μg and not more than 250 μg fluoride. Add 1.0 g of especially purified $Ca(OH)_2$ to maintain the sample alkaline during drying and evaporate the mixture to dryness in a loosely covered platinum dish on the steam bath. Dry at 110°C in a hot-air oven. Then place the dish in a cold muffle furnace and increase the temperature to 350°C. Forty min is required for this preliminary ashing. Cool the sample, wet down with distilled water, triturate, and dry as before. Ash a second time to a temperature of 450°C for 20 min. Repeat this procedure twice with the exception that the ashing temperature is 550°C for 45 min. A greyish ash should result from this treatment. Transfer the ash to the distillation flask with a minimum quantity of distilled water, rinse the platinum dish with 1–2 ml of the distilling acid, followed by a final distilled-water rinse. Total rinse should not exceed 50 ml. Distill the sample according to the procedure given in Section IB and collect a volume of 150–200 ml. Dilute to volume, take a 50-ml aliquot, and determine the fluoride content by the zirconium-eriochrome cyanine R procedure (Section IIA).

Prepare a blank using distilled water and the same quantities of all of the reagents.

2. Teeth and Bones.

Samples can be weighed out and ashed without the addition of fixative.

Procedure

Weigh a dry, fat-free sample previously pulverized in a Wiley mill, sufficient to give about 100 μg fluoride. Ash at 550°C for 3 hr and transfer directly to the distillation flask with a minimum of distilled water. Rinse the platinum dish with 1–2 ml of the distilling acid and follow with a distilled water rinse. The total rinsing volume should not exceed 100 ml. Distill the sample according to the procedure given in Section IB, and then use the zirconium—eriochrome cyanine R procedure (Section IIA).

3. Blood and Soft Tissues.

In the determination of fluoride in blood and soft tissues, the results obtained have not been entirely satisfactory. Because the usual fluoride content of these materials is very low, extra large samples are required in order to obtain sufficient fluoride for analysis. The high organic content of these materials interferes in the fixing of the sample and in the subsequent ashing. Various laboratories have reported poor recoveries in the analysis of these materials, and no two agree as to the best procedure to be followed. The following procedure is essentially that used by the National Institute of Dental Research (40).

Procedure

Take a sample, up to 100 g, after mincing thoroughly in a Waring blender or food chopper. Mix thoroughly with $1-2$ g $Ca(OH)_2$ and dry on a steam bath or with an infrared lamp, partially at $250°C$ and terminating at $550°C$. Increase the muffle temperature in $50°$ steps, each increase occurring after the sample has stopped emitting smoke from the previous step. If necessary, repeat the ashing after cooling, wetting, and redrying the sample. Transfer the ash to the distillation flask with distilled water, using a total rinse volume of not more than 50 ml. After most of the ash has been transferred, rinse the dish with $1-2$ ml of the distilling acid and follow with a final distilled-water rinse. Distill the sample according to the procedure given in Section IB and determine the fluoride in the distillate by the zirconium—eriochrome cyanine R procedure (Section IIA). Prepare a blank using distilled water, the same quantities of all of the reagents, and this same procedure. Special care must be exercised to use reagents of very low fluoride content since the fluoride content of soft tissues is extremely low and blanks may run higher than the fluoride content of the sample. The distillate may be concentrated to a lesser volume by evaporation under alkaline conditions, and the concentrate can be analyzed (Section IIA) using proportionately reduced volumes of sample and reagents, if necessary.

B. AGRICULTURAL SAMPLES

The procedure for ashing and fusion of the sample is essentially that described by Remmert (48). The sample should be in a finely divided state, if possible.

Procedure

Transfer $5-10$ g dry matter to a platinum or nickel dish and cover with a suspension of $Ca(OH)_2$ sufficient to maintain the sample alkaline during the drying and ashing procedure. Dry the sample and char on a hot plate or infrared heater. Place the dish in a muffle furnace and ash at $550-600°C$. If the ash is high in silica or if the silica content is unknown, add 5 g sodium hydroxide and fuse the mixture over a low flame. Continue the fusion for 5–10 min at 650–700°C. Cool the melt, dissolve in a minimum volume of water, and transfer to the distillation flask using no more than 50 ml total volume to effect the transfer. Add sufficient concentrated sulfuric acid to provide an excess of 15 ml over that needed to neutralize the alkalinity of the fusion mixture. Distill the sample according to the procedure outlined in Section IB. For samples with high silica, it may be necessary to collect up to 500 ml distillate in order to obtain complete recovery. The distillate can be concentrated, if necessary, by evaporation under alkaline conditions in a platinum or porcelain dish.

C. WATER SAMPLES

Samples of drinking water can usually be determined directly without preliminary treatment as outlined in Sections IIA, IIB, or IIC. Those samples which require a distillation can be processed by the procedure given in Section IB.

Industrial wastes, sewage, and other highly contaminated water should be subjected to a preliminary ashing prior to the distillation. A procedure identical with that for urine, Section IIIA1, should be followed.

A blank should be run on all reagents used in the procedures described in this section. The blank should be carried through identical steps with the sample being analyzed. In order to ensure a minimum blank, the following recommendations regarding the reagents and apparatus used should be followed.

Reagents

Calcium Hydroxide. Only specially purchased, low-fluoride lime should be used. This must be further purified by the following procedure: Slake about 60 g low-fluoride calcium oxide with 250 ml water. Add 250 ml 60% $HClO_4$ slowly with stirring. Add a few glass beads and evaporate the mixture to fumes of $HClO_4$. Cool the mixture and repeat the evaporation twice after adding 200 ml water each time. Dilute the mixture to about 2 liters and filter through sintered glass to remove silica. Pour the clear solution into a 1-liter solution of 10% NaOH, allow the precipitate to settle, siphon off the supernatant, and wash the precipitate with distilled water at least five times with thorough mixing of the precipitate each time to remove dissolved sodium salts. Finally, shake the precipitate with 2 liters water and store in a polyethylene container. A 100-ml sample of this suspension should have a blank of 2 μg or less fluoride.

Distilled Water. All distilled water used in the above procedures should be redistilled from an alkaline solution pink to permanganate. Discard the first 100 ml.

Sulfuric Acid. Equal volumes of water and reagent-grade sulfuric acid are carefully mixed and boiled down to fumes. This process is repeated, and the cooled acid is stored as is or, if the analyst desires, a 1:1 dilution with water can be made and stored. If a 1:1 acid is used for distillations, 30 ml should be added to the distillation flask in place of the recommended 15 ml.

Perchloric Acid. Reagent-grade 60% $HClO_4$ is added to 3–4 volumes distilled water and boiled down to the original volume. This process is repeated, and, after cooling, the acid is stored in borosilicate bottles.

Distillation Apparatus.

The Association of Official Analytical Chemists (3) recommends that the apparatus used for fluoride distillation undergo the following treatment. After each run, the flask and attachments are rinsed out with distilled water, then treated with hot 10% NaOH, followed by thorough rinsing with distilled water. Also, if the equipment has stood idle for any length of time, 20 ml 1 : 1 H_2SO_4 is added and boiled down to fumes. This is followed by thorough rinsing after cooling, and then the NaOH treatment is applied. Experience by other investigators indicates that when low-fluoride samples are run through stills so treated, the system has a tendency to hold back some fluoride, probably by adsorption, and only after repeated runs through the same setup will an equilibrium be reached and complete recovery achieved. Therefore, it would be advisable to carry out the A.O.A.C. recommendation only for high-fluoride samples.

REFERENCES

1. Aluminum Company of America, *Determination of Fluorine*, April 1947, p. 914.
2. American Public Health Association, *Standard Methods for the Examination of Water, Sewage and Wastewater*, 13th ed., Washington, D.C., 1971, pp,168–178.
3. Association of Official Analytical Chemists, *Methods of Analysis*, 10th ed., Washington, D.C., p. 363.
4. Armstrong, W. D., *Ind. Eng. Chem., Anal. Ed.*, 5, 300 (1933).
5. Belcher, R., M. A. Leonard, and T. S. West, *J. Chem. Soc.*, **1959**, 3577.
6. Belcher, R., M. A. Leonard, and T. S. West, *Talanta*, 2, 92 (1959).
7. Belcher, R., and T. S. West, *Talanta*, 8, 853, 863 (1961).
8. Bellack, E., *J. Amer. Water Works Assoc.*, 50, 530 (1958).
9. Bellack, E., and P. J. Schouboe, *Anal. Chem.*, 30, 2032 (1958).
10. Boonstra, J. P., *Rec. Trav. Chim.*, 70, 325 (1951).

11. Brownley, F. I., Jr., and C. W. Howle, Jr., *Anal. Chem.*, **32**, 1330 (1960).

12. Bumstead, H. E., and J. C. Wells, *Anal. Chem.*, **24**, 1595 (1952).

13. Cabello-Tomas, M. L., and T. S. West, *Talanta*, **16**, 781 (1969).

14. Clements, R. L., G. A. Sergeant, and P. J. Webb, *Analyst*, **96**, 51 (1971).

15. Dahle, D., and H. J. Wickman, *J. Offic. Agr. Chem.*, **19**, 313 (1936).

16. De Boer, H. L., *Rec. Trav. Chim.*, **44**, 1071 (1925).

17. Dixon, E. J., *Analyst*, **95** 272 (1970).

18. Fasano, H. L., *Rev. Fac. Cienc. Quim. (Univ. Nacl., La Plata)*, **21**, 69 (1946).

19. Fasano, H. L., *An. Soc. Cient. Argentina*, **144**, 473 (1947).

20. Foster, M. D., *Ind. Eng. Chem., Anal. Ed.*, **5**, 284 (1933).

21. Frant, M. S., and J. W. Ross, Jr., *Anal. Chem.*, **40**, 1169 (1968).

22. Frant, M. S., and J. W. Ross, Jr., *Science*, **154**, 1553 (1966).

23. Frere, F. J., *Anal. Chem.*, **33**, 644 (1961).

24. Gilkey, W. K., H. L. Rohs, and H. V. Hansen, *Ind. Eng. Chem., Anal. Ed.*, **8** 150 (1936).

25. Greenhalgh, R., and J. P. Riley, *Anal. Chim. Acta*, **25**, 179 (1961).

26. Grimaldi, F. S., B. Ingram, and F. Cuttita, *Anal. Chem.*, **27** 918 (1955).

27. Hall, R. J., *Analyst*, **88**, 76 (1963).

28. Har, T. L., and T. S. West, *Anal. Chem.*, **43**, 136 (1969).

29. Hargreaves, J. A., G. S. Ingram, and D. L. Cox, *Analyst*, **95**, 177 (1970).

30. Hems, R. V., G. F. Kirkbright, and T. S. West, *Talanta*, **17**, 433 (1970).

31. Hensley, A. L., and J. E. Barney, II, *Anal. Chem.*, **32**, 828 (1960).

32. Huckaby, W. B., E. T. Welch, and A. V. Metler, *Anal. Chem.*, **19**, 154 (1947).

33. Icken, J. M., and M. B. Blank, *Anal. Chem.*, **25**, 1741 (1953).

34. Kelso, F. S., J. M. Matthews, and H. P. Kramer, *Anal. Chem.*, **36**, 577 (1964).

35. Kempf, T., *Z. Anal. Chem.*, **244**, 113 (1969).

36. Kletsch, R. A., and F. A. Richards, *Anal. Chem.*, **31**, 1435 (1970).

37. Knapp, G., *Mikrochim. Acta*, **1970**, 467.

38. Lamar, W. L., *Ind. Eng. Chem., Anal. Ed.*, **17**, 148 (1945).

39. Leonard, M. A., and T. S. West, *J. Chem. Soc.*, **1960**, 4477.

40. McCann, H. G., and I. Zipkin, Natl. Inst. of Dental Research, Bethesda, Md., private communication, 1955.

41. Megregian, S., *Anal. Chem.*, **26**, 1161 (1954).

42. Megregian, S., and F. J. Maier, *J. Amer. Water Works Assoc.*, **44**, 239 (1952).

43. Megregian, S., and I. Solet, *J. Amer. Water Works Assoc.*, **45**, 1110 (1953).

44. Monnier, D. R., R. Vaucher, and P. Wenger, *Helv. Chim. Acta*, **33**, 1 (1950).

45. Newman, E. J., R. W. Fennel, E. J. Dixon, J. K. Foreman, G. S. Goff, R. J. Hall, W. C. Hanson, R. F. Milton, and J. W. Ogleby, *Analyst,* **96** 384 (1971).

46. Okuna, H., *J. Chem. soc. Japan*, **63**, 23 (1942).

47. Powell, W. A., and J. H. Saylor, *Anal. Chem.*, 25, 960 (1953).

48. Remmert, L. F., T. D. Parks, A. M. Lawrence, and E. H. McBurney, *Anal. Chem.*, 25, 450 (1953).

49. Richter, F., *Z. Anal. Chem.*, **124**, 161 (1942).

50. Richter, F., *Chem. Tech. (Berlin)*, 1, 84 (1949).

51. Rowley, R. J., J. G. Grier, and R. L. Parsons, *Anal. Chem.*, 25, 1061 (1953).

52. Rubin, L., *J. New England Water Works Assoc.*, **66**, 97 (1952).

53. Sanchis, J. M., *Ind. Eng. Chem., Anal. Ed.*, 6, 134 (1934).

54. Scott, R. D., *J. Amer. Water Works Assoc.*, 33, 2018 (1941).

55. Singer, L., and W. D. Armstrong, *Anal. Chem.*, **26**, 904 (1954).

56. Singer, L., and W. D. Armstrong, *Anal. Chem.*, **31**, 105 (1959).

57. Smith, F. A., and D. E. Gardner, *Arch. Biochem.*, 29, 311 (1950).

58. Smith, O. M., and H. A. Dutcher, *Ind. Eng. Chem., Anal. Ed.*, 6, 61 (1934).

59. Talvitie, N. A., *Ind. Eng. Chem., Anal. Ed.*, 15, 620 (1943).

60. Taves, D. R., *Talanta*, 15, 969 (1968).

61. Thrun, W. E., *Anal. Chem.*, 22, 918 (1950).

62. Venkateswarlu, P., and D. N. Rao, *Anal. Chem.*, 26, 766 (1954).

63. Venkateswarlu, P., and P. Sita, *Anal. Chem.*, 43, 758 (1971).

64. Wharton, H. W., *Anal. Chem.*, 34, 1296 (1962).

65. Willard, H. H., and O. B. Winter, *Ind. Eng. Chem., Anal. Ed.*, 5, 7 (1933).

66. Yamamura, S. S., M. A. Wade, and J. H. Sikes, *Anal. Chem.*, 34, 1308 (1962).

IODINE

BENNIE ZAK

Department of Pathology, Wayne State University School of Medicine and Detroit General Hospital, Detroit, Michigan

Changes in methodology for the determination of trace amounts of iodine are occurring at an accelerated rate. Much of the newer technology has been developed in clinical chemistry where the need, especially in the ultramicro range, has been maximal. Automation by either on-stream treatment and spectrophotometry or robotized treatment (discrete sample analysis) of the sample has been developed and improved to a point where a parts-per-billion determination is now easily possible. Much of this chapter is primarily oriented toward the trace determinations of iodine in analytical biochemistry. However, this treatment of the analytical area is deliberate, because these same methods or minor modifications thereof could easily be applied to many other materials especially when ultramicro trace analysis is the required objective.

I. SEPARATIONS

At present, there are several procedures which are used for the isolation of iodine from any media, thereby freeing the element for reaction in a particular spectrophotometric method. However, it is not always necessary to separate iodine from its milieu if the latter can be removed by destructive alkaline fusion (12,27,32,68,94,95,135,138,140,180,182,183,228), oxidative acid ashings, or other means (60,88,95,115,155,193,194,203,239—241), thus leaving iodine in a form suitable for quantitative absorptiometric analysis. These ashings are carried out on iodine-containing organic materials, most often those found in, or pertaining to, biologic systems. Separatory processes for inorganic iodine may include distillation, extraction, or diffusion, and those for organic iodine may include chromatography prior to rupture of the carbon—iodine bond or complete destruction of the organic moiety in order to make the iodine available for measurement by one of several spectrophotometric techniques.

A. DISTILLATION METHOD

The distillation of iodine may have to be carried out after the destructive fusion of the organic material present with the concomitant oxidation of the iodine to a higher nonvolatile valence state. The large amount of salts remaining as residue requires distillation in order to determine the trace amount of the desired constituent. The steps are effected by addition of comparatively large amounts of acid oxidizing mixtures such as chromic—sulfuric or permanganate—sulfuric acids (102). After rigorous heating, reduction to hydriodic acid or iodine (10,15,33,34,36,47,83,98,104,108,114,134,144,202,205,214,217,220, 222,233) enables one to quantitatively distill the iodine into an alkaline scrubber, where it can be further treated in a manner dictated by the analytic procedure to be followed thereafter.

For example, if the catalytic ceric arsenite procedure is the method of choice, then the sample need not be treated by an oxidation process since the iodide—iodine couple is the reactive form. In addition, the presence of excess arsenite ensures that the higher oxidation states will not exist. However, if the iodine—starch complex method is preferable (67,82,109,201,213,215,228,236), or if the method involving the ultraviolet or visible measurement of the absorbance of the triiodide ion in water or organic solvent is selected (41,54,191,195,197), the distillate must be treated with an oxidizing agent such as bromine (31), chlorine (159), or permanganate (30,66) before spectrophotometric determination can be carried out on the iodine liberated from the iodate formed. Such treatment expands the sensitivity obtained, since six equivalents of iodine are obtained for each iodate ion formed.

Other distillation devices or train apparatuses exist which were used more extensively in the past than are the comparatively simple, much smaller, and more compact stills employed more recently (142,151,179,209,212,235). These distillation procedures and train devices were developed because of the importance of iodine determinations. Therefore, extensive efforts were made to solve this microanalytic problem (92) which now seems to have been reasonably accomplished by modern methodology. The efficacy of distillation has also been tested by the employment of tracer amounts of radioactive iodine (153).

B. MICRODIFFUSION METHOD

Another mode of separation somewhat more delicate than the distillation previously described and sometimes called "isothermal distillation" (92) involves the use of the Conway diffusion cell (37). The iodine-containing liquid is placed in the outer compartment of a Conway cell (Fig. 1), and a scrubbing medium such as 20% potassium iodide or starch is placed in the central well. Potassium dichromate and sulfuric acid are added to the outer compartment, which

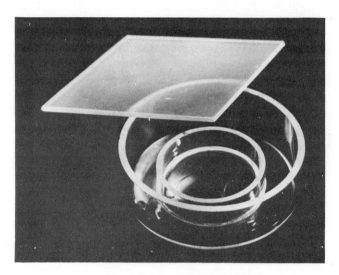

Fig. 1. Conway cell and cover.

contains the iodine, and a cover glass is sealed on with a petroleum jelly and paraffin mixture. Gentle rocking mixes the reactants in the outer well, and liberated iodine diffuses over through the vapor phase into the scrubbing solution of the center well. The constant upsetting of the liquid and vapor phase with respect to the iodine content of the latter by absorption into the scrubber ensures the completeness of diffusion (37,92,175).

A more sophisticated approach to isolation by diffusion was developed by Spitzy (80,121,204). In this elegant technique, a vacuum was created in a sealed apparatus which caused inorganic iodine to diffuse into an alkaline trap contained in a small procelain boat. The latter step was carried out after the iodine has been removed from its organic bonds by oxidative disruption with hot dichromic–sulfuric acids in a silicone heating bath and then liberated from this residue by diffusion after the addition of a strong reducing agent. This procedure appears to be the only diffusion technique successfully developed for the ultramicro determination of organic-bound iodine from physiologic samples (148,149).

C. EXTRACTION METHOD

Sometimes, separation of elemental iodine can be effectively carried out by the use of organic solvents to extract trace amounts of iodine from an aqueous medium. Methods involving isolation of elemental iodine by means of extraction date back more than a century when Chatin (35), using a method based on the procedure of Rabourdin (169), established the presence of iodine in air by passing the latter through an alkaline scrubber, converting it to alkali iodide, converting to iodine with nitrous acid, and finally extracting the iodine with carbon disulfide to give a purple color. This method was shown to be sensitive to several parts per million of iodine.

The solvents used for the extraction of iodine include chloroform (51,123,168,234), carbon disulfide (3,137,140,216,237), toluene (41), benzene (41), ethanol (41,54,192,201), xylol (150), carbon tetrachloride (4,17,56,65,81,205), and cyclohexane (115). However, a 5% potassium iodide solution has a partition coefficient which favors extraction into the aqueous instead of the organic phase. Custer and Natelson (41) have shown how the partition coefficient favors linear extraction with one pass for several of these solvents without great difficulty (Figs. 2 and 3).

Organic-bound iodine, which has biologic (diagnostic) significance, can be differentially extracted by butanol or an isoamyl alcohol—trimethylpentane mixture and thereby separated from inorganic iodine or nondiagnostic organic-

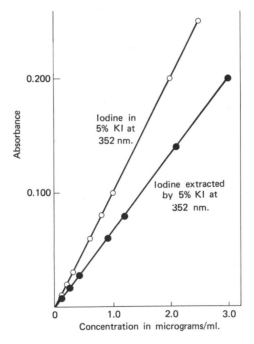

Fig. 2. Calibration plots of potassium iodide solutions of iodine with and without an extraction step.

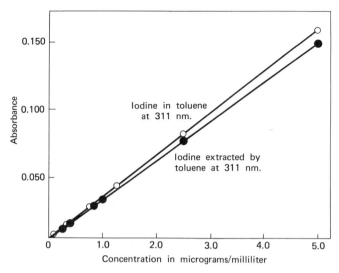

Fig. 3. Calibration plots of toluene solutions of iodine with and without an extraction step.

bound iodine. This has most commonly been carried out by extraction from acidified samples, a process which favors partition into the organic phase, removal of the washed organic phase by careful evaporation, and determination of the iodine present after destruction of the iodo-carbon bond (18,100,126,132,147,162,163).

D. DESTRUCTION OF THE IODINE-CONTAINING MEDIUM METHOD

There are several methods which can be used for the destruction of the medium containing the iodine and/or the iodine-containing compound leaving the iodine in a form suitable for quantitative analysis. Alkaline fusions using sodium carbonate (10,32,94,138,228), sodium carbonate–sodium hydroxide (182), potassium hydroxide (135,228), or sodium hydroxide–potassium nitrate (183) are capable of destroying organic matter when a muffle furnace with a controlled temperature of 600–650°C is employed. The alkali iodide formed will not be lost because it has not yet reached its melting point. The acidified solution of the ash gives a clear medium capable of being used as is if a procedure for the determination of iodide is involved (27,106,138,182). If the iodide is oxidized to iodate by the Groak-permanganate (66) or bromine procedure, an ultraviolet measurement or the starch–triiodide absorptiometric method may be employed after reaction of the iodate with iodide to liberate iodine.

Acid ashing, which obviates the employment of muffle oven techniques, can be carried out with chloric (9,193,194,239) or chloric–chromic acids (114,155,240,241), where the choice of either depends on the amount of iodine to be determined. The destructive agent can be removed by evaporation, leaving a small amount of soluble residue and the iodine in the form of iodate. The iodate can be reduced with arsenite and then treated with cerium(IV) for a catalytic method of spectrophotometric determination, or six equivalents of iodine can be liberated from the iodate at an adjusted pH if the starch–triiodide or the ultraviolet spectrophotometric method is to be used (239).

Another application of the "liquid–fire" reaction (199,200) for the determination of trace amounts of iodine has been developed for a biologic medium such as serum (129). This system employs perchloric acid containing a small amount of vanadium for the rapid destruction of a small amount of organic material prior to determination of the iodine content.

A third liquid–fire system that is also very effective for the complete destruction of organic materials and the concomitant conversion of organic iodine to iodate is a mixture of nitric, perchloric, and sulfuric acids (20,128). Still another and recent application to a mild destruction technique for removing iodine from its carbon bond involves the use of bromine, from bromine water or homogeneously generated on acidifying a mixture of bromate–bromide for the purpose (189). Chlorine and hypochlorite have also been used (132,161).

Thus, the organic medium in which the iodine is present may be destroyed by oxidizing acids or alkaline fusion, and the subsequent determination is made on the fusion residues without prior removal of the iodine as previously described for the distillation, diffusion, or extraction methods of separation. Some procedures for aqueous media exist where the iodine is not removed or the medium destroyed, but the determination is carried out by direct treatment of the solution being tested (93).

The use of radioactive iodine has borne out the fact that the alkaline fusion technique does not lead to loss (42). The chloric acid procedure has also been tested for loss by indirect means by comparison to a modification of this alkaline fusion procedure (242).

E. CHROMATOGRAPHIC METHOD

An iodo-organic compound of primary importance and diagnostic significance is thyroxine. It is normally present in human serum at about 0.05 μg/ml in terms of its iodine content. A simple way to isolate it is to pass buffered serum through an ion exchange resin which retains thyroxine. This compound can then be removed from the resin very simply by use of the proper eluant. From this point on, thyroxine can be treated under mild oxidative conditions to liberate its iodine as iodate which can then be determined by its catalytic effect on the cerium(IV)—arsenic(III) reaction (1,2,19,20,22,81,87,91,96,97,99,101,119,128, 156,164,210,232,238). Sephadex columns and ultrafiltration have also been used for the separation process (61,122,146).

Isolation of the iodine compounds of serum has been carried out using paper chromatography or thin-layer chromatography as the medium for separation (78,174,215). The location reagent may then be a variation of the cerium(IV)—arsenic(III) reagent or the ferric ferricyanide—arsenic(III) reagent, and the concentration of material in the spot may be densitometrically evaluated for its iodine content (62). Other chromatographic separations of significant interest have also been reported (16,116,152,165,231).

II. METHODS OF DETERMINATION

There are a number of absorptiometric methods available which are well suited for the determination of trace amounts of iodine. The most important in terms of extensive practical use and the greatest sensitivity are the procedures in which the catalytic property of iodine is used. For the most part, this has been the kinetic mediation of the reaction between cerium(IV) and arsenic(III) by iodide.

A. IODIDE AS A CATALYST METHOD

The estimation of iodide by the measurement of its catalytic action on the reaction taking place between cerium(IV) and arsenic(III) in acidic solution, described by Sandell and Kolthoff (184,185) and adopted by Chaney (33) and later by other investigators (26,32,53,73,93,106,111,138,144,175,182, 183,210), is utilized most commonly. The color intensity of the cerium(IV) ion is measured in the presence of arsenic(III) after undergoing reaction under rigidly controlled conditions of time, temperature, and acidity. The following reaction requires many hours to go to completion without the aid of heat or a catalyst, in spite of the fact that the difference of potential seems to favor spontaneous reaction:

$$2Ce(IV) + As(III) \rightleftharpoons 2Ce(III) + As(V)$$
$$\text{(orange)} \quad \text{(colorless)} \qquad \text{(colorless)} \quad \text{(colorless)}$$

In the presence of iodide, the rate of decrease of color increases in proportion to the amount of iodide present, and this enables one to effectively determine amounts of iodide in the range of 0.005 to 0.020 μg per sample with a comparatively high degree of accuracy.

One hypothesis which may explain the mechanism of catalytic reaction is the Schaffer equivalence change theory of ionic oxidation—reduction reactions (186) in which it is postulated that there is only a possibility of certain reactions taking place that are in themselves dependent on the equal valence states of the reacting compounds. For if the redox change, that is, electrons gained and lost for both oxidant and reductant, is not equivalent, then a bimolecular reaction could not occur even though bimolecular collision had taken place. Therefore, successful reaction is involved with the factor of the equivalence of valence change, and this must determine that portion of the collisions which achieves favorable results. Since reaction velocity can be inhibited by unequivalent valence changes, the introduction of a third substance able to exist in the two valence states required for mediation of the reaction in consecutive steps should markedly catalyze the reaction. These ideas were challenged when unequivalent reactants were shown to take part in rapid reactions, thereby obviating the equivalence change concept, since rapid reaction between monodelectronator and didelectronator would not fit the theory (172).

This viewpoint found partial support in the Remick version (172) of the Bancroft potential hump theory (8). He believed that if Schaffer's theory was extended by the introduction of the potential hump concept, it would offer a satisfactory explanation of the observed facts. If a substance is capable of acting as a monodelectronator on a didelectronator by successive one-step oxidations, it must have a potential higher than the potential barrier present. If not, "it could only surmount the hump slowly because the Maxwell-Boltzman distribution of energy would only endow relatively few ions at a given moment with sufficiently high potential, whereas a didelectronator would not have to surmount the hump."

Many other mechanistic views concerning this useful kinetic reaction have been described with much elegant evidence to justify the views taken. A listing of these findings rather than an extensive elaboration of each in this chapter should enable readers to further their interests along those lines (11,40,46,76, 81,130,136,154,170,176,177,229,230).

The effect of the several variables, such as reagent concentration, time, temperature, inhibitors, and enhancers, has been investigated. The following ions have been found to interfere in the reaction (33,34,52,77,89,139,144,183,184, 207,210): (1) Cyanide, which forms the iodo-cyanide complex, a compound that does not catalyze the reaction; (2) mercury, which forms insoluble mercuric iodide and inhibits the catalysis entirely; (3) chloride, which enhances the reaction but at a somewhat slower rate than does iodide; the chloride concentration must be controlled and kept constant; (4) bromide, which enhances the reaction to a somewhat greater extent than does chloride; (5) osmium, which catalyzes the reaction in much the same manner as does iodide; (6) silver, which ties up iodide as the insoluble silver iodide. Although never reported as an apparent enhancer of the cerium(IV)–arsenic(III) reaction, it has been shown (48–50,107) that it can form the complex $H[AgCl_2]$ which catalyzes the liberation of chlorine due to an activation of the hydrochloric acid and this process causes decolorization of cerium(IV) ions. Feigl described its use as a spot test for silver (49). (7) Thiocyanate and citrate have been found to interfere with the action of iodine although for the latter, 200 μg must be present to interfere with 0.1 μg iodide.

Other interferences are found in the technical difficulties due to the fact that the reaction is still in progress, necessitating the rapid measurement of the absorbance of the cerium(IV) solution at a moderately elevated temperature (68,178). However, modern instrumentation has the capacity for eliminating such problems, especially since established automated systems are capable of making kinetic measurements in relatively simple fashion (5,6,28,29,72,90, 120,124,221). Also, it will be shown later that the reaction can be stabilized by controlled inhibition to a fixed color representing a defined time of reaction.

There are several controllable variables involved during the reaction which must be considered in order to obtain successful analytic results. These variables include cerium(IV) concentration, arsenic(III) concentration, addition of chloride to enhance the reaction time, temperature, acidity, and oxidation state of the iodine. Several workers (104,113,141) have found that the reaction is

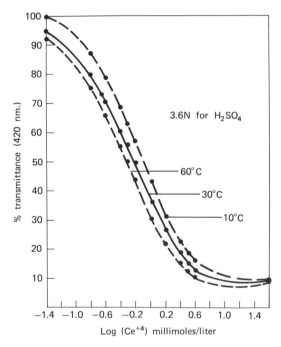

Fig. 4. Relation of % transmittance to log concentration of ceric ion in ceric ammonium sulfate solutions.

complex and follows pseudo first-order kinetics. The absorbance-versus-concentration curves for ceric ammonium sulfate at 420 nm do not always obey Beer's law (64) and a representative curve describing the relationship is shown in Fig. 4, where % transmittance is plotted against the log cerium(IV) concentration expressed in millimoles per liter. Because of the S-shaped nature of the curve, several workers (32,94,113,155) carried out their determinations by mathematical interpretation which only utilized the straight-line portion of the curve. The peak absorbance for ceric sulfate is at 315 nm, although this peak is not usually employed in the determination of iodine.

An increase in the reaction rate for a given quantity of iodide was obtained by Lein and Schwartz (113) who found that a 20:1 initial ratio of arsenic to cerium was most convenient for accelerating the loss of cerium(IV) colors (Fig. 5). The validity of this observation has been confirmed by other groups of workers (138,155).

A somewhat different finding concerning the variation of the arsenic(III): cerium(IV) ratio was described for an experiment wherein the effect of the arsenic ratio was plotted before and after the addition of a trace of mercury (138). When the arsenic concentration of a blank system was increased,

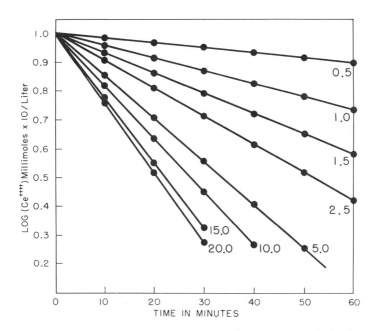

Fig. 5. Rate of change in log concentration of ceric sulfate at various initial concentrations of arsenious acid. Number of curve is ratio of initial molar concentration of arsenious acid to ceric sulfate.

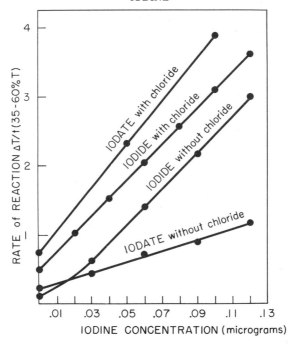

Fig. 6. Effect of iodide concentration on first-order velocity constant.

the rate at which the decolorization occurred also increased. On the addition of a trace of mercury to complex any possible contaminating iodine, it was found that the curve flattened out. This suggested that the arsenic solutions were increasingly contaminated with iodine and that this contamination could perhaps be responsible for the "arsenite effect." In any event, this is a phase of the reaction which could bear reinvestigation.

The empirical calibration curve was shown to be nonlinear with respect to low concentrations of iodine by Sandell and Kolthoff (185), Barker (10), and others (104). Therefore, the addition of varying amounts of sodium chloride to obtain linearity was studied (113), and Fig. 6 shows how a linear relationship can be obtained for iodide concentration versus residual cerium absorbance when 100 mg sodium chloride is used for enhancement. O'Neal and Simms (155) carried out a similar study with iodide and iodate, and their results are depicted in Fig. 7. Since iodine is a natural potential contaminant encountered in sodium chloride, Moran (144) and Leffler (111) showed how the reagents could be systematically tested during their preparation to locate the source of the contamination and to keep the chloride effect more constant in their determinations.

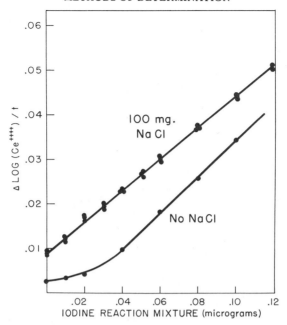

Fig. 7. Iodate and iodide catalysis. Influence of added chloride (50 mg). Iodine concentration represents the amount present in reaction mixture having a total volume of 7 ml.

Another interesting experiment was carried out concerning the iodine content of the chloride added to enhance the cerium(IV)–arsenic(III) reaction characteristics which embodied the same investigative principles described previously for arsenic(III):cerium(IV) ratios, namely, mercury inhibition (137,206). When 0, 50, and 100 mg NaCl was added to each of a set of reagent blanks of the cerium(IV)–arsenic(III) system, a marked but nonlinear increase in reaction was noticed if the residual cerium color was measured against time. When a small amount of mercury (1.0 mg) was preadded to a second set of similar blanks, it was found that the chloride still enhanced the cerium(IV)–arsenic(III) reaction but to a lesser extent, indicating that most of the increase in activity might have been due to iodide contamination of the chloride.

The effect of increasing the temperature has been investigated (155), and several different temperatures from room temperature to 40°C have been reported. Fig. 8 shows how an increase in temperature markedly changes the rate of reaction for a fixed amount of iodide.

Changing the concentration of sulfuric acid has a marked effect on the transmittance of cerium ammonium sulfate solutions. Fig. 9 shows a plateau region which remains constant over a wide range of sulfuric acid concentrations.

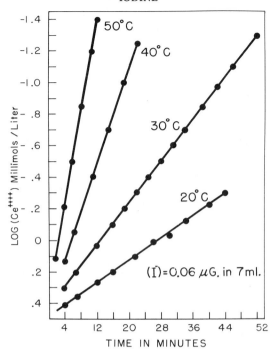

Fig. 8. Reduction of ceric ion by arsenious acid as catalyzed by iodide. Effect of temperature.

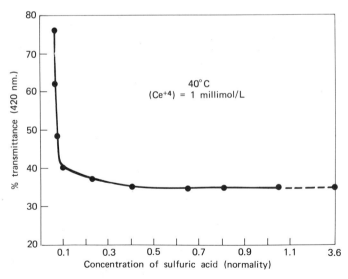

Fig. 9. Relation of acid concentration to % transmittance of ceric ammonium sulfate solution.

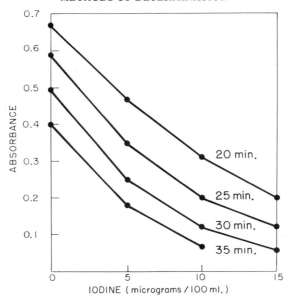

Fig. 10. Shift of iodine calibration plot with variation in time.

The factor of time is clearly shown in Fig. 10, where several calibration curves are made for different reaction times while using the same cuvettes.

The original ceric-arsenite method was chronometric and involved waiting until a completely colorless system was present and then noting the time (185). This procedure proved to be too unwieldly and time consuming for routine determinations. Therefore, procedures for the determination of iodine utilizing the catalytic ceric-arsenite method have been developed based on the employment of an empirical curve prepared under controlled conditions of time, temperature, acidity, concentration, and addition of reagents (33). The original use of this empirical calibration curve with its many subsequent modifications was followed by procedures which used mathematical interpretations as a substitute for the absorbance—concentration curve (113,155). A third approach involved the addition of a fixed amount of iodide to a second aliquot of the analysis solution. The difference in the rate of reactions for the two solutions was then used as a measure of the amount of iodine present as described in Section IIA3 (4). This latter procedure has been studied in detail, and some modifications have been described (32,93).

Chaney was the first to adapt the use of a photoelectric recording colorimeter to the quantitative absorptiometric procedure involved in the kinetic determination of trace amounts of organic-bound iodine (34). He recorded the output of a vacuum-type phototube on a strip chart recorder. The light passed through a lens system and filter and was brought to a secondary focus at the center of a cuvette

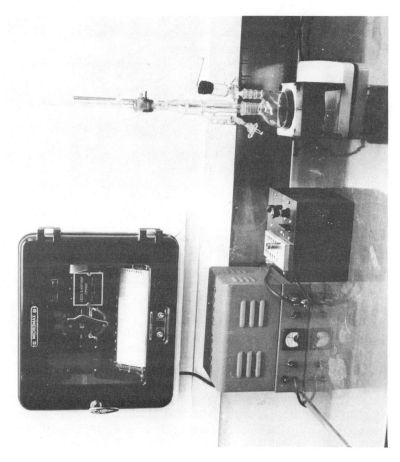

Fig. 11. At left, recording apparatus with incubator for cuvets whose contents are to be measured. At right, Chaney still for the distillation procedure.

154

containing cerium(IV) solution. This transmitted light passed through to the phototube, and its output was then recorded on the strip chart. The test tube housing, which was an aluminum block with a thermostatic control to maintain a constant temperature during the recording of color change, could hold several test tubes in the ready state at one time while they were waiting to be tested. A vacuum-type phototube was used instead of a barrier layer cell because of better spectral response for the measurement of yellow solutions (141). An electrically regulated power supply and dc amplifier whose principle had previously been described (64) provided a stable light source. Fig. 11 shows the entire instrument as well as the distilling apparatus which also has served as the basis for many of the procedures that were later developed. Chaney studied the effect of interfering ions, order of addition of reagents, temperature, pH, reagent purity, mixed reducing agents, and the choice of reducing agents. Man, Siegfried, and Peters used the same instrument with a somewhat different digestion and distillation procedure (126,127). Although the instrument itself is not commercially available, similar and now more sophisticated ones are. The description is included because it seems to be a historical and important prototype of modern instrumentation and kinetic measurements which are presently of great interest.

A later technique akin to Chaney's system which has seen practical application in water analysis is the automatic reaction-rate method of Malmstadt and Hadjiioanou (69,125). In this system, the time required to bring about a small change in absorbance was determined automatically on a modified spectrotitration unit (69) with a temperature-controlled test tube cell and photoconductive detector for greater sensitivity. The slope measurement could then be directly calibrated in terms of iodine concentration.

Although several techniques exist for handling the spectrophotometric data obtained in the ceric-arsenite reaction, the calibration curve technique has proved to be the most durable and popular for both manual and automated procedures. A single representative example of many available techniques follows.

1. Catalytic Calibration Technique (138).

Reagents

Ceric Ammonium Sulfate, 0.02N, containing 40 ml sulfuric acid per liter reagent.

Sodium Carbonate, 4N.

Sulfuric Acid, 7N.

Sodium Arsenite, 0.2N, containing 20 g sodium chloride per liter reagent.

Procedure

Add 1.0 ml 4*N* sodium carbonate and 3.0 ml 7*N* sulfuric acid to 3.0 ml water solution containing the iodide in the range of 0.02 to 0.06 µg. The addition of acid is carefully carried out or loss of iodide will occur because of vigorous effervescence. The alkali is present to approximate the amount that would be present in an alkaline fusion technique or possibly in a distillate. Add 2.0 ml 0.2*N* sodium arsénite to a 3.0-ml aliquot in a cuvette and, after mixing, allow the tubes to achieve temperature equilibrium in a 35°C constant-temperature bath. After 10 min of equilibration, carefully pipet 1.0 ml 0.02*N* ceric ammonium sulfate solution into each tube at exactly 1-min intervals and then immediately replace the tubes in the bath. At the end of exactly 20 min for each tube, measure the absorbance at 420 nm with a spectrophotometer using a water blank. Determine the concentration from a calibration curve constructed with varying amounts of standard iodide and the same amount of the reagents that were present in the sample.

It is also possible to add 1 ml 17% mercuric acetate in 1% acetic acid to arrest the reaction after the fixed time interval of incubation is over. Unhurried measurement of the absorbance can then be made and compared to similarly treated standard within several hours. This kind of measuring system is described later in Section IIA4.

2. Catalytic Computation Technique (111)

Reagents

Ceric sulfate, 0.01*M* in 3.6*N* sulfuric acid.

Arsenious acid, 0.10*M*.

Sulfuric Acid, 7*N*, containing 50 mg sodium chloride.

Sodium carbonate, 2*N*.

Procedure

Add 2.0 ml 2.0*N* sodium carbonate, 2.0 ml 7*N* sulfuric acid, and 2.0 ml 0.1*M* arsenious acid to the unknown iodide solution contained in a cuvette. The cuvettes and contents are brought to the temperature of a 40°C constant-temperature bath. Then, 1.0 ml of the ceric sulfate solution is added. After about 5 min, at time $t = 0$, the transmittance reading is made. These transmittance readings are repeated at several intervals and the transmittance recorded. The transmittance values that fall between 30% and 65% are used to calculate $\Delta T/t$. The equation for the regression line in Fig. 6 obtained with the use of 100 mg sodium chloride was used to derive the following equation:

$$I = (0.0439 \, \Delta T/t) - 0.0244$$

Substituting the value of $\Delta T/t$ obtained in this equation permits one to calculate the iodine content in the range of 0 to 0.12 μg.

3. Kinetic Curve Calculation Technique (12)

Reagents

Ceric Ammonium Sulfate, 0.02N in 1 liter water that contains 230 ml 7N sulfuric acid.

Sodium Arsenite, 0.1N.

Sulfuric Acid, 7N.

Hydrochloric Acid, 2N.

Sodium Carbonate, 4N.

Potassium Iodide, 0.04 μg per ml iodide.

Procedure

Add 1.0 ml sodium carbonate to the iodide-containing solution and evaporate the solution to dryness. Add 2.0 ml 2N hydrochloric acid, 2.0 ml 7N sulfuric acid, and 3.0 ml distilled water, respectively, to the residue. Two 3.0-ml aliquots of the well-mixed solution are pipetted into cuvettes, and 1.0 ml water is added to one while 1.0 ml of the potassium iodide solution containing 0.04 μg per ml is pipetted into the other. Then, add 0.5 ml sodium arsenite to each tube and place the tubes in a 39°C constant-temperature bath for 10 min. Add 1 ml of the ceric ammonium sulfate solution to each tube using an interval of 30 sec. A zero time reading is made for a given batch of reagents by quickly reading a blank immediately after the addition of the cerium(IV) solution. Readings are also obtained at 6 and 12 min, and the lines describing the two reaction rates are drawn as shown in Fig. 12. The concentration of iodine in the sample is determined by the use of the following equation the rationale of which is described in the same figure:

$$\frac{t_2 \times 0.04}{t_1 - t_2} = \text{concentration of iodine}$$

This equation can be used for the calculation of iodine, but the answers obtained are described as being a little lower than those obtained by the empirical standard curve technique with no explained rationale for the difference (12).

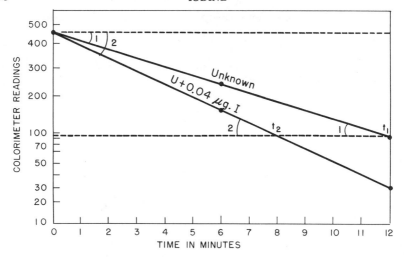

Fig. 12   Method of plotting results for calculation of PBI:

$$\tan \angle 1 = \frac{\text{initial color reading} - \text{final color reading}}{t_1} = \frac{\text{difference in color readings}}{t_1}$$

$$= U \ \mu g \ I \ \text{in sample}$$

difference in color readings $= Ut_1$ (where U = unknown)

In a similar fashion,

$$\tan \angle 2 = \frac{\text{difference in color readings}}{t_2} = U + 0.04 \ \mu g \ I$$

difference in color readings $= (U + 0.04 \ \mu g \ I)t_2$

Since color readings are identical

$$Ut_1 = (U + 0.04 \ \mu g \ I)t_2$$

$$Ut_1 - Ut_2 = 0.04 \ \mu g \ I \times t_2$$

$$U = \frac{t_2 \times 0.04 \ \mu g \ I}{t_1 - t_2}$$

4. Arrested Catalytic Technique

Some variations in the procedure for the catalytic method exist that circumvent making the spectrophotometric measurement at 420 nm of the yellow cerium(IV) system while the reaction is still taking place. Rogina and Dubravcic (178) arrested the catalytic reaction after a fixed time by the addition of excess iron(II), which immediately reduced all residual cerium(IV) ions to the colorless cerium(III) state. Potassium thiocyanate was then added and the extent of the iodine reaction was determined from the intensity of the red color of the ferric thiocyanate which formed. A linear relationship exists between the intensity of the red color of the complex and the decrease in iodide concentration. Their rationale for the employment of the arrested catalytic reduction of ceric ions involved the fact "that the catalytic reaction takes place in a thermostatically controlled enclosure, whereas the concentration of the remaining cerium(IV) ions is measured elsewhere with a photoelectric absorptiometer. Because of instability of the voltage, frequent adjustments of the photometer are necessary and the measurements are therefore protracted. It is difficult to carry out the measurement quickly enough to avoid inaccuracies arising from both the reaction being still in progress and from changes in temperature. This results in errors in the determination of iodides."

Reagents

Sulfuric Acid, 60% w/w.

Arsenious Acid, $0.1N$ in $0.01N$ sulfuric acid.

Ceric Ammonium Sulfate, $0.02N$ in 1.6N sulfuric acid.

Ferrous Ammonium Sulfate, 1.5% in 0.6% sulfuric acid.

Potassium Thiocyanate, 4% w/v.

Iodide Standards, 1.0 μg per ml and 0.1 μg per ml.

Procedure

The high-range reaction (0.1–1.0 μg per 8.0 ml sample solution) is carried out at 20°C for 8 min, while the lower-range reaction 0.01–0.1 μg per 8 ml sample solution) is carried out at 30°C for 20 min. Pipet 3.0 ml iodide-containing solution in the each test tube, followed by 0.5 ml arsenious acid. After 20 min, add 1 ml ceric ammonium sulfate solution to each tube at exactly 1-min intervals. After 8 or 20 min, depending on the amount of iodine being determined, arrest the reaction by the addition of 1 ml ferrous ammonium sulfate. Add 1 ml potassium thiocyanate solution, which forms a red color. Measure the absorbance after 45 min in 0.5-cm cells using a green filter. An instrument capable of measuring solutions with very high absorbance is often necessary.

5. Miscellaneous Catalytic Techniques

Other techniques that arrest the color reaction involve the addition of mercuric ion to tie up the iodide after the fixed time interval of incubation is over (43,87,138,157,207,227). Unhurried measurement of the absorbance can then be made and compared to similarly treated standards within a lengthened time period. The addition of brucine to the ceric-arsenite mixture, like ferrous iron, stops the reaction by reducing all of the cerium to colorless cerous ions (68,87,157). At the same time, the brucine is oxidized to an orange color whose absorbance values are related indirectly to iodine content. The advantage of this type of inhibited catalysis is found in an expansion of sensitivity because equivalent amounts of oxidized brucine have much greater absorbance in the visible range than do ceric ions. The same kind of sensitivity expansion is true for ferric thiocyanate.

Arsenious acid and permanganate in acidic solution undergo reaction at a very slow rate because of the intermediate production of manganic salt interfering with the rate and stoichiometry of the following equation (47):

$$5As_2O_3 + 4MnO_4^- + 12H^+ \longrightarrow 5As_2O_5 + 4Mn^{2+} + 6H_2O$$

However, in the presence of trace amounts of iodide, the reaction stated goes rapidly and stoichiometrically to completion because of the following reaction:

$$2Mn^{3+} + 2I^- \longrightarrow I_2 + 2Mn^{2+}$$

The iodine formed reacts with arsenious acid to regenerate iodide:

$$As^{3+} + I_2 \longrightarrow As^{5+} + 2I^-$$

and this maintains the cyclic nature of the reaction. The principles of the Schaffer equivalence change theory (186) for the manganic—arsenite reaction along with the potential hump concept seem to be similar here to the ceric-arsenite system, with the manganic ion acting as the monodelectronator in place of cerium while the arsenious acid remains the didelectronator. Boguth and Schaeg determined the iodine content of rat thyroids after wet ashing with dichromate and 30% hydrogen peroxide using the manganese—arsenite catalytic system (23).

Gmelin and Virtanen (62) determined that the I^-–I_2 couple could be used as a catalyst for an iron(III)—arsenic(III) system as well as for a ferricyanide—arsenic(III) system. In addition, when iron(III) and ferricyanide were used together, insoluble Prussian blue was formed in the process, and it was extremely sensitive in showing chromatographic separation of spots representing iodide, iodinated tyrosines, and thyronines. The blue spots formed were quantitative and capable of being analyzed by densitometry in the range of 0.01 to 0.07 μg per spot. The overall reaction was as follows:

$$2Fe(CN)_6^{3-} + 2I^- \longrightarrow 2Fe(CN)_6^{4-} + I_2$$

$$As(III) + I_2 \longrightarrow As(V) + 2I^-$$

$$2Fe(CN)_6^{3-} + As(III) \xrightarrow{(2I^- - I_2)} 2Fe(CN)_6^{4-} + As(V)$$

Jungreis and Gedalia (86) used the catalytic action of the iodide—iodine couple on the reaction between sodium p-toluenesulfonechloramide (chloramine-T) and N,N'-tetramethyldiaminodiphenylmethane (tetra base acetate). The sequence of reactions were described as follows:

$$H_3C-\!\!\!\left<\!\!\bigcirc\!\!\right>\!\!-SO_2\,NCl^- + H_2\,O \rightleftharpoons CH_3-\!\!\!\left<\!\!\bigcirc\!\!\right>\!\!-SO_2\,NH_2 + ClO^-$$

$$ClO^- + 2I^- + 2H^+ \longrightarrow Cl^- + H_2O + 2I^0$$

$$(CH_3)_2\,N-\!\!\!\left<\!\!\bigcirc\!\!\right>\!\!-CH_2-\!\!\!\left<\!\!\bigcirc\!\!\right>\!\!-\overset{+}{\underset{H}{N}}(CH_3)_2 + 2I^0 \longrightarrow$$

$$(CH_3)_2\,N-\!\!\!\left<\!\!\bigcirc\!\!\right>\!\!-CH=\!\!\!\left<\!\!\bigcirc\!\!\right>\!\!=\overset{+}{N}(CH_3)_2 + 2I^- + 2H^+$$

A disadvantage of the reaction is that the blue quinoid compound is unstable and quickly changes to a green to green-yellow color.

Lundgren (118) determined the iodide content of sulfite solutions by measuring the catalytic effect on the reaction between sulfite and methylene blue. Herries and Richards (75) found that on addition of ICN to iodide, iodine was liberated rapidly according to the following equation:

$$H^+ + I^- + ICN \longrightarrow HCN + I_2$$

The iodine could be measured spectrophotometrically or extracted with carbon tetrachloride for measurement.

Bobtelsky and Kaplan (21) measured the rate of decolorization of permanganate as catalyzed by iodine. Iwasaki, Utsumi, and Ozowa (84) used the fading of the red color of ferric thiocyanate as a measure of the catalytic activity of iodine.

Fuchs et al (57) used p-aminophenol and peroxide as reagents for the catalytic determination of small amounts of iodide or iodate according to the following reaction scheme:

$$\text{iodide} + \text{peroxide} \xrightarrow{p\text{-aminophenol}} \text{iodate}$$

$$\text{iodate} + p\text{-aminophenol} \longrightarrow \text{iodide} + \text{indamine dye}$$

$$\text{iodide} + \text{peroxide} \xrightarrow{p\text{-aminophenol}} \text{iodate}$$

This process was reported to be useful for concentrations as low as 2.5 ppm iodide (57).

Funahashi et al. (58) developed a technique for the determination of trace amounts of iodine by measuring its effect as a catalyst on the substitution reaction of the mercury 4-(2-pyridylazo)resorcinol complex with 1,2-cyclohexanediamine-N,N,N′,N′-tetraacetic acid (CyDTA). As an analytical application, they used the procedure for the determination of iodide in rain water. This procedure is very sensitive and said to be capable of determining $10^{-8} M$ iodide concentrations, whereas chloride and bromide do not interfere at much higher concentrations of $5 \times 10^{-4} M$ and $10^{-5} M$, respectively. Interfering cations were eliminated by passing the sample through Amberlite IR 120-B.

B. STARCH–IODINE METHODS

The determination of an inorganic nonmetal such as iodine is also possible in analytic situations where it is capable of undergoing formation of a spectro-photometrically measurable adsorption compound. Here, an organic material functions as the adsorbent in the formation of the inorganic–organic adsorption compound (48). Several excellent studies have been carried out on the iodo-starch complex (180,195,211) as well as the amylose–iodine reaction (59,187,196,218,219) and all can, therefore, not be discussed here.

The analysis of iodine by starch was used as early as 1851 by Grange (63). Now, soluble starches of several kinds such as Lintner (117) or Zulkowski starches (243) which readily form colloidal solutions with water are used. Purified linear starch containing cadmium iodide (7,105,106), which does not readily undergo oxidation by atmospheric oxygen, has begun to receive more attention as a colorimetric reagent for the determination of various oxidizing agents such as iodic acid. It fulfilled a need for the preparation of an iodo-starch complex capable of being used for the absorptiometric determination of minute quantities of iodine without contamination from iodine liberated by action of oxygen (189). Lambert began the use of a specially purified "linear starch" which contained cadmium iodide dissolved in it and which, by virtue of the low dissociation constant of the latter, was capable of being used under much greater acidity conditions than had previously existed. The pH of the cadmium iodide–linear starch reagent is approximately 6, regardless of its age. The recommended pH for precipitating and recrystallizing the linear A fraction starch to prevent hydrolysis is also 6. An additional feature is that the toxic property of cadmium with respect to microorganisms effectively prevents their growth and increases the stability of the starch–iodide mixture to a much greater extent than is usually shown by soluble starch–iodide mixtures. The specially prepared linear starch may be isolated according to the procedures of Krishnaswamy and Sreenvasan (102) Schoch (188), or Lambert (105,106).

The spectral characteristics of the starch–iodine chromogen have been thoroughly investigated and found to calibrate linearly (67,225) when excess iodide is present. But there is some question as to whether the chromogen obeys Beer's law because of the phenomenon known as "the threshold." Recent studies agree that the calibration is linear (44). However, when the calibration line is extrapolated, it has been found to have an intercept on the abscissa a little away from zero. Lambert and Zitomer (107) agree with the theory that the complex is formed initially by iodine penetration, followed by adsorption on the outer surface of the amylose (or starch) helix (79). So, a little more than one triiodide ion per starch helix would be required to give a color. Drey (44) believed and showed that the iodine—starch threshold was not overcome by increasing the iodine or starch concentration or by altering acidity. He questioned the penetration of the helix theory (107) on the basis that the threshold would vary with starch content, and it did not. He theorized that iodine hydrolyzed according to the following scheme:

$$I_2 + H_2O \rightleftharpoons HOI + H^+ + I^-$$

and that a certain small amount of hypoiodite was necessary before iodine can react with starch.

Some fortification for this concept later appeared in the findings of Moknach and Rusakova (145) who found that ultraviolet spectrophotometry of the starch—iodine complex showed a 360 nm band indicative of hypoiodite. Their postulated sequence of reaction schematic was as follows:

$$I_2 + H_2O \rightleftharpoons OI^- + 2H^+ + I^-$$

$$I_2 + I^- \rightleftharpoons I_3^-$$

$$I_3^- + starch \longrightarrow starch{-}I_3^- \text{ complex}$$

The starch—iodine complex is a sensitive color system and enables one to effectively determine iodine present in trace quantities as low as 0.02 ppm in the final dilution of iodine. Several interesting applications for iodide and/or iodate determinations using starch have been described (133,143,208,223,224).

1. Method of Gross et al. (67)

Reagents

Sulfuric Acid, 1.4N and 0.1N

Sodium Hydroxide, 1N.

Potassium Permanganate, Saturated.

Sodium Nitrite, 1.5M.

Urea Solution, 5M.

Potassium Iodide, 1g per 6 ml, freshly prepared.

Arrowroot Starch Solution. Two grams starch is ground to a thin paste with water and poured into 800 ml boiling water. This is boiled for 15 min and 1 g salicylic acid is added as a preservative. Stability time is about two weeks.

Procedure

Add aliquots of a solution of potassium iodide containing iodine in the range of 1 to 14 µg to about 15 ml distilled water which contains 1 ml 1N NaOH and boil. Perform a Groak oxidation (66) by adding two drops of the permanganate solution to the boiling solution and boil for 5 min. Add 1 ml 1.4N sulfuric acid to neutralize the sodium hydroxide and to provide the proper acidity for the final liberation of iodine. If the pink permanganate color fades out at this point, add another drop. After 2 min, decolorize the solution by adding sodium nitrite dropwise and then boil the solution for 5 min. After cooling to room temperature, transfer the mixture to a 50-ml volumtric flask and add 3.0 ml potassium iodide solution and 4.0 ml starch solution. Dilute this solution to the mark, mix, and measure the absorbance after 30 min at 575 nm using a blank made up from 4 ml 0.1N sulfuric acid, 3.0 ml potassium iodide, and 4.0 ml starch, respectively, diluted to 50 ml in a volumetric flask.

The simplicity of the previous procedure was used in a modification which made it feasible to determine small amounts of iodine. The volume of solution was decreased to one tenth, the amount of iodide employed was lessened to obviate air oxidation, phosphoric acid was substituted in place of sulfuric to maintain proper pH conditions, the temperature of the final reaction was that of an ice bath which effected a stability of the iodo-starch color for an indefinite period, and potato starch gave a sol of greater transparency than arrowroot starch.

2. Method of Houston (82)

Reagents

Sodium Hydroxide, 0.2N.

Potassium Permanganate, 0.2N.

Phosphoric Acid, 28%.

Potassium Iodide, 5%.

Potato Starch, 0.25%.

Procedure

Transfer the iodide-containing solution (0.1–5.0 µg) to a 50-ml beaker and add 1 ml sodium hydroxide. Add one or two drops permanganate and reduce the volume to 2–3 ml by boiling. Transfer the contents of the beaker quantitatively to a calibrated tube graduated at 5 ml. Ice for 2 min and then treat with 0.05 ml of the potassium iodide solution and 0.05 ml of starch solution. Mix, dilute to the mark, and measure the absorbance after 2 min in a 1-cm cuvette at 575 nm.

3. Cadmium Iodide-Linear Starch Method (105–107)

Reagents

Starch—Cadmium Iodide (105–107). Dissolve 11 g cadmium iodide in 300–400 ml distilled water and boil gently for 15 min to expel iodine, adding water to maintain the volume. Add distilled water to a volume of 800 ml and slowly add 15 g superlose to the gently boiling, stirred solution. Boil and stir 10 min longer, then filter with suction through fine paper while maintaining the unfiltered solution at 65°C. If five to six changes of filter paper are made to speed filtration, this step should be completed in 1 hr or less, or one can use high-speed centrifugation to avoid filtration altogether. Dilute the filtrate to 1 liter.

Sulfuric acid, 3M.

Procedure

Add enough water to an aqueous solution of iodate to bring the volume up to 18.0 ml and then add 1 ml of the 3M sulfuric acid and 1.0 ml of the starch solution. After 15 min has elapsed to allow full color formation, measure the absorbance at 615 nm using 1-cm cells. The concentration range is about 0 to 2 μg iodate per ml.

C. TRIIODIDE ION METHODS

One of the more sensitive methods for the determination of iodine involves the measurement of the triiodide ion in the ultraviolet region of the spectrum. Studies have been carried out which indicate that iodine in various solvents shows a varied state which depends entirely on the nature of the solvent (25,26). Iodine dissolved in solvents where it does not undergo solvation, such as carbon tetrachloride, carbon disulfide, and chloroform, is purple and has a spectrum similar to that of vapor-state iodine (25). Brown colors, however, are obtained in solvents that form complexes with iodine, such as water, potassium iodide, and ethanol, yielding spectra of solvated or molecular addition compounds of iodine and solvent and exhibiting visible and strong ultraviolet peaks. Several of these absorption spectra are shown for both solvated and unsolvated solutions in Figs. 13 and 14. The addition of a small amount of ethanol to a purple solution of iodine in chloroform changes the color to brown, which indicates that the resonating molecule is apparently dissolved in the ethanol and that the ethanol—iodine complex formed dissolves in the chloroform (25). Studies have been made on iodine in the atomic state and in many polar and nonpolar solvents (25,26,65).

Ultraviolet absorption due to the triiodide ion was investigated by

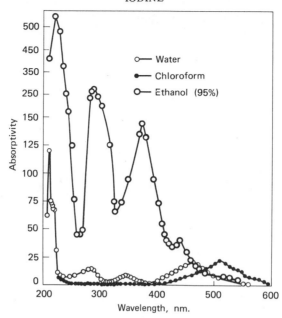

Fig. 13. Absorption spectra of iodine dissolved in water, chloroform, and 95% ethanol.

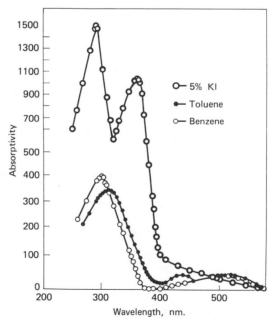

Fig. 14. Absorption spectra of iodine dissolved in toluene, benzene, and 5% aqueous potassium iodide.

Brode (25,26) whose early work on alkali halides showed a band due to free iodine present. He confirmed this by observing an increase in absorbance of potassium iodide solutions by letting then stand for several days in the light and then removing the intensified band by the addition of thiosulfate.

1. Method of Custer and Natelson (41)

Reagents

Potassium Permanganate, 1%.

Sodium Hydroxide, 1%.

Hydrogen Peroxide, 6%.

Sulfuric Acid, 5%.

Toluene.

Potassium Iodide, 1% and 5%.

Procedure

Pipet 0.2 ml of the test potassium iodide solution into graduated centrifuge tubes (5 ml) and treat with 0.1 ml sodium hydroxide and 0.1 ml potassium permanganate. If organic material is present, dissolve in 1% sodium hydroxide, dry at 95°C, and ash in a muffle oven at 625°C. Treat a 0.2-ml aliquot of the residue in the same manner as the potassium iodide solution. Place the tube containing the permanganate in a boiling water bath for 0.5 hr and then cool it in a refrigerator until it reaches its temperature. Place the tube in an ice bath and add cold hydrogen peroxide dropwise to decolorize the permanganate. Incubate the tube at 37°C for 1 hr and allow it to stand at room temperature overnight to decompose the excess peroxide. Dilute the solution to 2 ml and centrifuge it for 15 min. Remove a 1.5-ml aliquot and treat according to one of the following procedures.

Procedure A. Transfer the aliquot to a 12-ml glass-stoppered centrifuge tube and add 0.1 ml 1% potassium iodide and 0.2 ml 5% sulfuric acid. Add 1.8 ml toluene and shake for 10 min. Centrifuge and transfer the toluene layer to 1-cm quartz cuvettes and read in a spectrophotometer at 311 nm.

Procedure B. Transfer the aliquot to a glass-stoppered centrifuge tube and treat it with potassium iodide and sulfuric acid solutions as described in Procedure A. Add 1.8 ml chloroform and shake the mixture for 10 min. Centrifuge 3 min at 1500 rpm, aspirate off the upper layer, transfer 1.5 ml of the chloroform solution to the quartz cuvette, and read the absorbance at 352 nm.

2. Method of Shahrokh and Chesbro (195)

Reagents

Alkaline Sulfite Solution, 0.25 g sodium sulfite per 250 ml 0.5N sodium hydroxide.

Chlorine–Carbon Tetrachloride, bubble in chlorine until the solution is bright green.

Hydrochloric Acid, 2N.

Potassium Iodide, 0.1%.

Procedure

To an Erlenmeyer flask containing iodine trapped in alkaline sulfite following distillation from an acid digestion mixture, add 1 ml 2.0N hydrochloric acid followed by 0.5 ml chlorine solution. Boil the sample until less than 3.0 ml remains. Transfer the solution in the flask to a test tube calibrated at 4.0 ml and dilute to the mark with distilled water. Pipet 1 ml 0.1% potassium iiodide solution into the tube. The absorbance of the triiodide ion formed is measured at 353 nm against a water blank.

3. Chloric Acid Method (239)

Reagents

Chloric Acid, approximately 25% (111,155,239,240).

Potassium Iodide, 1%.

Procedure

Transfer a weighed sample of an organic iodine compound or pharmaceutical material such as desiccated thyroid to a 150-ml breaker. Add 10–25 ml of the chloric acid digestion reagent. Evaporate at the low temperature of the hot plate to the formation of dense fumes of perchloric acid. It is important that all the chlorine is driven off and that no excess chlorate remains. Dilute the digested sample to a fixed volume which is dependent upon the iodine content of the sample as well as the amount of sample used. Add 1 ml 1% potassium iodide solution to a 2-ml aliquot. The absorbance of the liberated iodine is measured at 290 nm or 353 nm, with the lower absorption peak at 353 nm used for the more concentrated solutions. The iodine-containing digest can be neutralized and treated as in Section B1 if the iodo-starch method of determination is preferable (67).

D. OTHER METHODS

There are a multitude of procedures aside from those previously described that are available to be used in the absorptiometric determination of iodine. Alpha-naphtholflavone reacts with iodine to form a spectrophotometrically measurable blue compound (171). Iodine forms nitrous acid when reacted with hydroxylamine, which then diazotizes sulfanilic acid, and the latter couples with alpha-naphthylamine to produce a red dye (45). o-Tolidine reacts with iodine to yield a blue-green color (110). Iodide has been determined after reaction with dioxane (181). In acidic solution, iodate oxidizes pyrogallol to purpurogallin, forming a sensitive reddish-brown color (201). The decrease in fluorescence of fluorescein has been used because the diiodo derivative is nonfluorescent and its formation is controlled by the pH of the solution (71). The measurement of the turbidity of silver iodide has been successfully used for small amounts of iodide (190). The adsorption of mercurous iodide on mercuric chloride (93,150) and the formation of palladous iodide (131) have been used in the determination of iodine. Poziomek and Reger (167) used 4-cyano-1-methylpyridinium perchlorate in a charge-transfer complexation technique for the spectrophotometric determination of iodide in water.

III. APPLICATIONS

A. IODINE IN PHYSIOLOGIC MEDIA (SERUM)

1. Wet Ash (Manual) Method

Three primary wet ashing systems for complete destruction of organic material with subsequent iodine determination in the residue have been described. Although many variations of these "liquid fire" reactions may exist (112,160,198,226), three procedures have been selected to detail these techniques for manual, semiautomated, and completely automated determinations of the organic-bound iodine of serum because this determination has diagnostic significance.

Wet ashing is in general the best approach for the determination of iodine in most media in which it is difficult to determine iodine without destruction or at least partial destruction. The three wet ashing liquid mixtures described here are similar in their actions and in the residues in which final analytic determinations are carried out. For serum analyses, they all really stem from the chloric–chromic acid technique of almost two decades ago (241).

Reagents

Digestion Reagent. Prepare a solution of vanadic acid (0.25%) in 72% perchloric acid.

Ceric Ammonium Sulfate Solution. Dissolve 6 g per liter of 27% sulfuric acid.

Arsenious Acid Solution. Dissolve 9.0 g anhydrous arsenious oxide in 8.2% sulfuric acid and dilute the mixture to 1 liter with the same solvent.

Iodate Stock Standard. Prepare a solution containing 100 μg iodine per 100 ml water. Dilute this to 0–20 μg per 100 ml for working standards.

Anion Exchange Resin. Use analytical-grade anion exchange resin such as IRA-400 of large particle size for rapid settling.

Procedure

Deiodinate the serum of inorganic iodine by adding resin to serum in a 1:10 ratio and shake for a fixed period (10 min) before the clear serum is decanted. Some believe it is best to pass each serum through a small column of ion exchange resin to avoid hydrolytic action on the iodo-carbon bonds of the molecule which may liberate iodine and cause low results (39). Pipet 0.1 ml of each serum into 19 x 150 mm cuvettes which will serve as the container for the entire process. Add 2 ml digestion reagent from an automatic pipet. Place the tubes in an aluminum electrical heating block (Fig. 15) preheated to 230 ± 5°C.

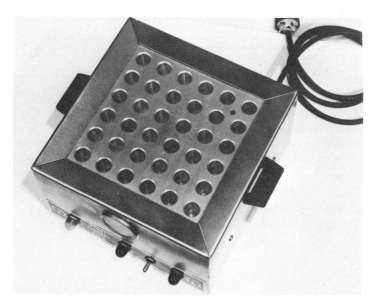

Fig. 15. Thirty-six-place aluminum heating block drilled to hold 19 by 36 mm tubes at 230° ± 5°C.

The block can hold 36 tubes, and they can be placed in the block at 10-sec intervals. This takes exactly 6 min. In 6 min (at 10-sec intervals), remove the tubes for cooling. Removal of 36 tubes takes exactly 6 min. Add 2.0 ml ceric ammonium sulfate to each tube, place them in a 37°C water bath to come to temperature, and then add 2.0 ml arsenious acid solution to these tubes at timed intervals. At the end of the catalytic incubation period of 20 min, remove each tube at timed intervals and read its absorbance at 420 nm against a water blank. In order to obtain a reading of approximately 0.70 for the blank in this regressing calibration system, the wavelength can be varied between 420 and 450 nm. This assures adequate sensitivity in terms of slope of the calibration curve used.

An automated version of this procedure has been developed by Hycel, Inc. Houston, Texas and is commercially available on their multiphasic screening system.

2. Wet Ash (Semiautomatic) Method

It is possible to automate the catalytic reaction step used in the determination of protein-bound iodine or the inorganic iodine of other kinds of samples, including nonbiochemical ones. The procedure to be described involves a manual destruction step followed by an automatic analytic step using an instrument devised for on-stream spectrophotometry with automatic addition of the reagents used (18). If the analyst prefers or lacks the necessary equipment, the final colorimetric step can still be carried out manually because the procedure is derived from a manual method and makes no change in the reagents. A second liquid-fire reaction is used involving a chloric—chromic—perchloric acid mixture as the destruction agent (241).

Reagents

Chloric—Chromic Acid Solution. Dissolve 500 g potassium perchlorate and 200 mg sodium chromate in approximately 1 liter iodine-free water by heating on a hot plate. Carefully and slowly add 375 ml 72% perchloric acid to the hot solution and stir continuously. After it cools overnight in a refrigerator, filter off the precipitated potassium perchlorate from the cold solution through Whatman 41-H filter paper (rapid filtering) and store the clear solution in the refrigerator.

Ceric Ammonium Sulfate Solution. Dissolve 25.3 g ceric ammonium sulfate in several hundred milliliters distilled water, add 104 ml sulfuric acid, and dilute the solution to 1 liter.

Arsenious Acid Solution. Dissolve 14 g sodium hydroxide in approximately 800 ml distilled water, add 19.6 g arsenious oxide, 56 ml sulfuric acid in excess of neutrality, and 50 g sodium chloride, and dilute to 2 liters with water.

Stock Standard Solution. Weigh out 168.5 mg potassium iodate, dissolve, and dilute to 1 liter with distilled water.

Working Standard Solution. Prepare 0–20 μg iodine per 100 ml in distilled water using 5 μg per 100 ml increments.

Procedure

Pipet 0.5 ml of each serum sample into 10 ml 5% trichloracetic acid, allow to stand for 30 min, centrifuge at 3000 for 10 min, and decant the supernatant fluid. Add 5 ml chloric–chromic acid solution and three granules of anti-bumping stones to each tube and heat the tubes in an aluminum thermo-regulated heating block at 160°C to an orange residue of about 0.5 ml. Add 2.0 ml water to each residue to dissolve the residues and pour the well-mixed contents into sample cups for the automatic step. The schematic for analysis of 40 samples per hour is shown in Fig. 16. The sample is picked up, partitioned with air, and mixed with arsenious acid solution. This solution is then mixed, ceric ammonium sulfate solution is added, and the final solution is mixed again. This reaction mixture is then delayed in a single glass coil (40 ft) at 37°C, passed into the flow cuvette of the colorimeter, read with a 420-nm interference filter, recorded, and discarded. If a 15-mm "N" cuvette is used instead of the gravity

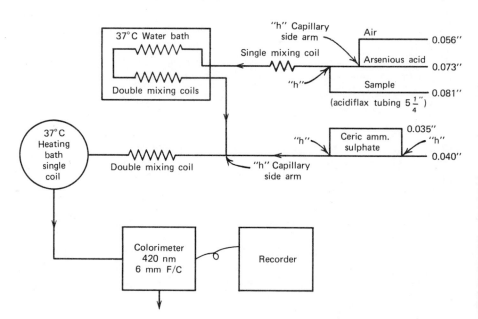

Fig. 16. Manifold schematic for carrying out automatic cerium(IV)–arsenic(III) reaction under the catalytic influence of iodine.

flow cuvette shown in Fig. 16, the manifold can be easily modified. A 0.081-in.-inner-diameter tube would then be added to the manifold as a pull-off line for the cuvette. The concentration of the cerium reagent must be cut to 6.3 g per liter because the "N" cuvette used is approximately 15 mm long, or about 2.5 times as long as the cuvette it replaces.

3. Wet Ash (Automatic) Method

This is the, complete automation procedure for iodine wherein all steps are automatic except the preliminary deiodination with anion exchange resin for removal of inorganic iodine (39,173). It is really a two-channel system hooked together so that the destruction step is completed on one section of the instrument before the colorimetric section of the instrument automatically samples the residue for treatment in the catalytic step. The cuvette differs from the semiautomatic system previously described in that it is much smaller and requires debubbling prior to feeding liquid into it. In addition, the sampler has the capability of washing out the sample line after it picks up the sample, and this results in less interaction between samples. It is mechanically simple to convert the colorimeter of the semiautomatic procedure described and replace the sampler if one wished to do so.

Reagents

Digestion Mixture. Mix 75 ml nitric acid with 300 ml 70% perchloric acid. Prepare weekly and store in a brown bottle.

Arsenious Acid Solution. Dissolve 19.6 g arsenious oxide and 15 g sodium hydroxide in about 800 ml distilled water. Neutralize to phenolphthalein endpoint with concentrated sulfuric acid and dilute the solution to 2 liters. Add 50 g sodium chloride, dissolve, and mix well.

Ceric Ammonium Sulfate Solution. Dissolve 12.6 g ceric ammonium sulfate in about 1400 ml water containing 104 ml concentrated sulfuric acid. Dilute to 2 liters with water and allow the solution to stand overnight. If a precipitate forms, filter through glass wool.

Sulfuric Acid Diluent, 5.2% v/v

Iodate Stock Standard. Prepare the same solution as described for the semiautomated procedure. Some workers prefer to add albumin (protein) to the standards because they feel that inorganic iodate standards tend to adsorb to the tubing, yielding lower results for the standards and higher results for the samples.

Screening Standard. Dilute the stock standard to 50 and 500 μg iodide per 100 ml.

Ion Exchange Resin. Use the same anion exchange resin as described for the manual procedure. (Technicon Corporation calls this anion exchanger Iobeads.)

Procedure (Screening)

Put a 50-μg standard in the first cup of the sampler, follow it with a water cup, and aspirate samples and standards at the screening rate of 120 per hr with the special screening cam in the sampler and a water cup between all cups sampled. Use the ruler provided with this procedure to determine the values of samples that are contaminated. Reject all samples whose values produce a peak at this speed, for only severely contaminated samples can decolorize cerium under these circumstances.

Procedure (Samples)

Add about 300 μg resin to 1.5–2.0 ml serum in a throwaway plastic tube, mix well for 3 min, allow the resin to settle, and decant the deiodinated serum into a sample cup. Load the standards, quality controls, and samples onto the sample plate eliminating the water cup between each of the cups. Put the 20 sample-per-hour cam into the sampler.

The procedure then takes place as follows: Figure 17 shows all of the modules in place. These are (right to left) sampler (40 places) at right front, proportioning pump at the right rear, digestion apparatus with its heated (350°C) glass digestion vessel, second proportioning pump (color reaction takes place here), 55°C delay coil (8 min), colorimeter (410 nm), and recorder. This

Fig. 17. AutoAnalyzer setup showing modular arrangement for iodine determination.

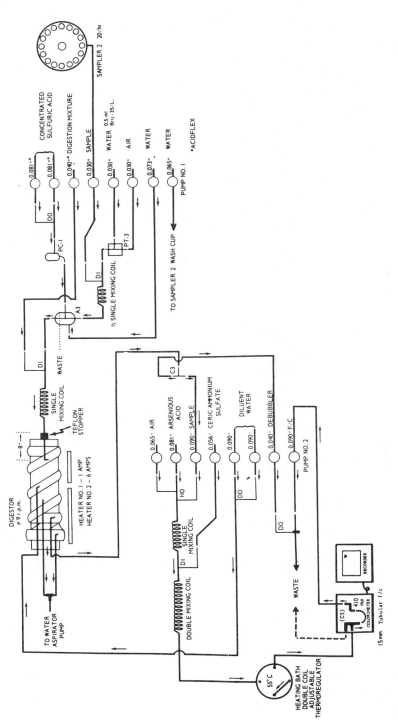

Fig. 18. Manifold schematic for AutoAnalyzer setup of Fig. 17.

175

really might be considered two channels – a destruction channel and an analytic channel. The sample is picked up from the sample cup and mixed with the digestion reagents by the aspiration caused by the pump rollers. The volumes are determined by the diameters of the tubing on the pump manifold. This mixture is passed through the heated helical coil for several minutes, and this destroys the protein and liberates iodine as iodate. Diluent is then added to the digest, and an aliquot is picked up for mixing with arsenious acid and ceric ammonium sulfate solutions. This mixture is then delayed to allow catalysis to take place, and it is then passed through the colorimeter for reading and recording. The schematic of the entire operation is shown in Fig. 18.

4. Method of Bromine Oxidation After Chromatographic Separation (74)

It is possible to isolate thyroxine by column chromatography (161) or organic extraction (132) prior to carrying out a mild oxidation with bromine to liberate iodide in order to subject this to determination by its catalytic reaction in the cerate–arsenite system.

Reagents

Ceric Ammonium Sulfate Solution. Dissolve 20 g ceric sulfate to 1 liter with a solution containing 48.6 ml sulfuric acid.

Arsenious Acid Solution. Dissolve 1.44 g arsenious acid in 15 ml $0.5N$ sodium hydroxide. Warm to effect solution. Dilute this solution to approximately 850 ml with distilled water, add 28.0 ml hydrochloric acid, and dilute to 1 liter.

Bromide–Bromate Solution. Prepare a $0.1M$ $KBrO_3$–$0.6M$ KBr solution in distilled water.

Saturated Bromine Solution.

Thyroxine Stock Standard Solution. Dissolve 5.77 mg hydrated sodium L-thyroxine $(11H_2O)$ in 50 ml methanol–ammonium hydroxide solution (99:1).

Thyroxine Working Standard Solution. Dilute the stock standard 1 ml to 50 ml with 50% (v/v) acetic acid (2 μg thyroxine per ml). Finally prepare working standards by diluting the 2 μg/ml standard with 50% aqueous acetic acid.

Acetic Acid, $0.2M$.

Sodium Acetate, $0.2M$.

Acetate Buffer, pH 4.0. Prepare a solution in the ratio of 4.16:1 of $0.2\,M$ acetic acid and $0.2M$ sodium acetate.

Acetic Acid Solution, pH 2.2. Prepare a solution of 11.5:1 of water and glacial acetic acid. If necessary, adjust the pH of the solution to 2.2.

Acetic Acid Solution, pH 1.4. Prepare a 1:1 solution of glacial acetic acid and water. If necessary, adjust to pH to 1.4.

Resin Column (Dowex AG1-X2, 200–400 mesh, Cl⁻ form). Wash the resin three to four times with pH 1.4 acetic acid solution. Repeat the washing with pH 4 buffer until the washing has the pH of the buffer.

Procedure

Use a stopcock column with a 10-mm internal diameter and a glass wool plug and pour in the resin to 4–5 cm. Keep the level of the liquid above the edge of the resin, and when the column is to be used, draw off the liquid to the other side of the resin edge. This is the same pH as the transfer liquid for applying serum to the column. (These columns are presently purchasable as disposables from several companies such as Curtis Nuclear and Bio-Rad of California, and others.) Add 1.5 ml of pH 2.2 acetic acid to 3 ml serum, then transfer the sample to the column with the use of the pH 4 buffer solution. Also prepare a blank column to be used without serum. Remove liquid from the column until the level is just below the resin surface and then add 25 ml of pH 4.0 buffer to the column. Draw off solution again until the level is just below the resin surface. Pass two successive 25-ml volumes of pH 2.2 acetic acid solution through the column until gravity will not remove more liquid. Pass three successive 10-ml aliquots of pH 1.4 acetic acid solution through the column separately, closing the stopcock on adding the liquid and opening the stopcock until the liquid stops flowing. Place 2-ml aliquots of each of the three fractions into cuvettes. Add one drop bromine water to impart a yellow color to the solutions and to oxidize the thyroxine releasing its iodine. After 5 min, add 4 ml arsenious acid solution to each cuvette to reduce iodate and incubate the cuvettes to temperature in a 37°C water bath. At timed intervals, add 1 ml ceric ammonium sulfate solution to each cuvette, and after an appropriate wait, read the absorbance of the solution in each cuvette at 420 nm.

5. Microdiffusion Method

The procedure to be described is the only kind of microdiffusion technique applicable to the determination of serum protein-bound iodine at the present time (204).

Reagents

Iodine-Free Water. Double distill water in a glass apparatus from potassium carbonate.

Chromic Acid, 60%. Dissolve 600 g chromic acid (Merck) to 1 liter with distilled water.

Boiling Stones (1 mm). Clean 1-mm sieved fire clay beads by heating with sodium hydroxide solution, washing with water, boiling with chromic–sulfuric acid, washing with water, and drying.

Sulfuric Acid, 83%. Prepare a solution containing 730 ml sulfuric acid and 270 ml water. Heat the solution for several hours, and dilute it back to its original volume. The heating step purifies the solution of contaminating iodide.

Phosphorous Acid, 60%. Dissolve 600 g phosphorous acid to 1 liter, including the heating step described for the sulfuric acid solution.

Sodium Hydroxide Solution, 10%.

Ceric Sulfate Solution, 0.045N. Dissolve 9.098 g ceric sulfate to a volume of 500 ml with 3.5N sulfuric acid.

Arsenite Solution, 0.15N. Dissolve 3.71 g resublimed arsenious oxide in 50 ml 1N potassium hydroxide, dilute the solution to 250 ml with water, neutralize with 83% sulfuric acid, add 54 ml more acid and 3.125 g sodium chloride, and dilute this solution to 500 ml with water.

Iodide Stock Standard, 100 μg per ml.

Trichloracetic Acid, 68%.

Procedure

Mix 2.5 ml water and 0.25 ml trichloracetic acid into 0.5 ml serum in a combustion tube and let the mixture sit for 10 min before centrifuging and discarding the fluid. Repeat this step with the same additions, again discarding the supernatant fluid. Add 2.0 ml water, 6 ml 83% sulfuric acid, and 0.6 ml chromic acid to each precipitate and mix to solubilize. Connect the tubes to clamps in a 130°C silicone bath (Fig. 19) and raise the temperature within 40 min to 210°C. After 5–10 min at temperature, remove the rack holding the tubes, place it in ice water, and add 1 ml 60% phosphorous acid to each tube. Keep this in the ice bath until all parts are prepared for connection. Pipet 0.15 ml 10% sodium hydroxide onto quartz glass wool in the absorption spoon. Carefully connect the three parts of the diffusion apparatus (Fig. 20), open the stopcock, create a vacuum with a sink aspirator, and then close the stopcock. Carefully rotate the ashing tube to pour its contents below the spoon, place the chambers in a 50°C oven, and allow 12 hr for diffusion to be completed. The time involved to achieve a plateau for complete diffusion is shown in Fig. 21. At the end of the diffusion period, release the vacuum, dismantle the diffusion chambers, and wash the alkali-trapped iodine with a jet of water to a final volume of 5 ml. Add 0.4 ml 0.15N arsenious acid and incubate the mixtures at 40°C for 15 min to attain temperature equilibrium. At regular intervals

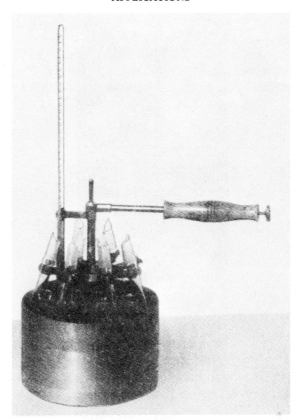

Fig. 19. Silicone heating bath containing rack which can hold 12 digestion tubes for the partial destruction step using dichromic–sulfuric acids to liberate iodine as iodate.

(30–60 sec), add 0.5 ml cerium solution and incubate again for 30 min. Using the same intervals, remove the tubes and read them at exact times at 430 nm in 2-cm cuvettes. Prepare standards containing 0.0–0.1 μg iodide in 0.2 ml 10% sodium hydroxide, dilute them to 5.0 ml with water, and process them through the same color reaction as described for the samples.

6. Manual Analysis by Alkaline Fusion Method (55)

Reagents

Ceric Ammonium Sulfate. Add 12.65 g ceric ammonium sulfate to 730 ml solution containing 230 ml 7N sulfuric acid. Boil the solution, cool, and dilute to 1 liter with water. The solution is stable in a brown bottle for one month.

Fig. 20. Isothermal distillation apparatus consisting of destruction tube at right, scrubber and evacuation system at upper left, and central portion of the diffusion apparatus at lower left.

Sodium Arsenite solution. Dissolve 4.95 g arsenious oxide in 25 ml $1N$ sodium hydroxide and then dilute the solution to 1 liter with water.

Stock Iodide, 100 μg per ml. Dilute 118.1 mg dried NaI to 1000 ml with distilled water.

Standard Iodide Solutions. Dilute stock iodide to 0.1 μg per ml. Prepare standards by diluting 5–30 ml of the 0.1 μg/ml solution to 50 ml with distilled water.

Sulfuric–Hydrochloric Acid Mixture. Prepare a solution of $0.65N$ in hydrochloric acid and $7N$ in sulfuric acid.

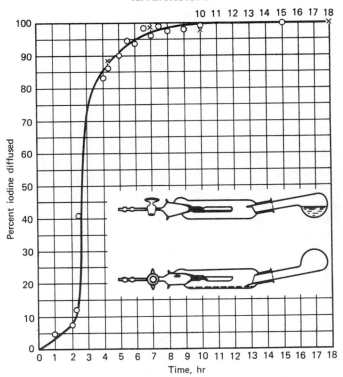

Fig. 21. Plot relating % iodine diffused vs. time in hr. Also shown is apparatus before and after destruction tube is turned to place reaction mixture below the scrubber in the diffusion chamber.

Potassium Hydroxide, 2N.

Sodium Hydroxide, 0.5N.

Zinc Sulfate, 10%. Titrate this solution against the 0.5N NaOH and adjust so that 10.8–11.2 ml sodium hydroxide neutralize 10 ml zinc sulfate.

Procedure

Dilute duplicate 1-ml aliquots of serum to 8 ml with distilled water. Add 1 ml 10% zinc sulfate, mix well, and then add 1 ml 0.5N NaOH with mixing to effect the protein precipitation. After the mixture has stood for 10 min, centrifuge out the proteins at 2000 rpm. Decant off the supernatant fluid, resuspend the proteins in 10 ml iodine-free distilled water, and recentrifuge. Repeat the washing process two more times, decanting each supernate. Dissolve the protein in 1 ml 2N potassium hydroxide and put the tube in a 115°C oven to dry overnight. Carry three blanks through the entire process at the same time. Transfer the dried tubes in a rack to a muffle oven and adjust the temperature (within ½ hr) to 600°C. Continue heating the tubes for a full hour at this temperature. During this 1-hr period, open the door of the oven for 15 sec at 5, 20, and 40 min. Remove the tubes, allow them to cool, and then dissolve the residues in 10 ml distilled water. Centrifuge out any undissolved material such as silica and remove 4 ml of the clear supernatant fluid to a cuvette. Add 0.5 ml sodium arsenite solution and 1 ml of the sulfuric–hydrochloric acid reagent. Place the tubes in a 37°C water bath for 10 min and then add 0.5 ml of the cerium reagent at timed 30-sec intervals. After 20 min, read each tube at 420 nm against a water blank.

Prepare standard solutions by pipetting 1 ml of each standard into cuvettes followed by 0.4 ml 2N potassium hydroxide, 0.5 ml sodium arsenite, 2.6 ml water, and 1 ml of the sulfuric–hydrochloric acid solution. Incubate these tubes at 37°C for 10 min. Add cerium solution and process the tubes spectrophotometrically as described for the samples. These standard results represent 0–15 μg iodine per 100 ml.

7. Distillation Method (92,214)

Although this is the least common of all the techniques described for the determination of serum protein-bound iodine, it is still used somewhat and will therefore be described. Its importance for the purpose is questionable, certainly fleeting, and its duration of existence for the future in analytical biochemistry may definitely be limited.

The original distillation apparatus described by Chaney (33) and successfully modified by others serves as a comparatively simple means for the digestion of organic material by permanganic (103) or chromic acid (10,33,34).

Procedure

Add 25 ml 70% sulfuric acid and 2 ml 60% chromic acid to a 300-ml double-necked, round-bottom flask, containing enough sample to provide 0 to 0.25 μg iodine. Add silicon carbide chips as antibump material and allow the digestion to proceed until fumes have filled the flask. Cool the flask, add 15 ml water, and heat the flask (as previously described). Add 25 ml distilled water. Attach the flask to the condenser and again heat until condensation appears in the trap, which contains 12.0 ml 1% sodium hydroxide added at the top of the condenser. Add slowly 3 ml 50% phosphorous acid from the dropping funnel and distill for a fixed period of 7 to 10 min. The time varies somewhat in the many modifications of this procedure that exist. The sodium hydroxide in the trap has been found by some (144,217) to work more effectively if there is a little arsenious acid present, whereas the phosphorous acid is found to be an improved reducing agent if it contains arsenious acid (34). The distillate can be treated by one of the catalytic procedures described in Section IIA1, 2, 3, or 4.

B. IODINE IN ROCKS, SOIL, PLANTS, AND ANIMAL TISSUE

1. Rock Samples, Alkaline Fusion Method (38)

Reagents

Sodium Chloride Solution, 0.1N. Treat with a few drops of 0.1N silver nitrate to remove iodide present.

Bromine Buffer Solution, pH 2.8. Prepare an acetate buffer by adding 0.5N sodium acetate and 0.5N acetic acid in the ratio of 1:50. Add 5 ml bromine per 100 ml solution and shake it vigorously until it is saturated.

Cadmium Iodide Solution, 5%. Dissolve the cadmium iodide crystals in water, boil them for 5–10 min, dilute to volume, and store in a brown bottle. The boiling eliminates contaminating iodine adsorbed on the crystals.

Starch Solution, 0.5%. Triturate 1 g of crude potato starch and about 1 mg mercuric iodide preservative to a paste with a few milliliters water. Add this to 200 ml near-boiling water and continue the warming for 10 min. Cool the solution, centrifuge it at 3000 rpm for 10 min, and filter it through Whatman #42 paper into a brown glass bottle.

Procedure

Heat sodium hydroxide (three to five times the sample weight) in a 250-ml nickel crucible to fusion. Add the powdered rock sample (5–10 g) to the cooled flux and then heat, moderately at first and then to a temperature of about 500°C, until the contents appear to be a viscous fluid. Cool, then extract the melt with distilled water, and transfer to a 400-ml beaker. Neutralize the extract with 6N hydrochloric acid and then add an excess of 0.5–1.0 ml acid. Filter this through previously washed fine paper and wash the residue several times with distilled water. Add 0.5 ml 1% sodium sulfite to the filtrate and dilute to 200 ml. Add 5 ml 0.1N silver nitrate to this solution. Add 3N sulfuric acid in the ratio of 1:1000 to hasten coagulation and allow the mixture to stand overnight.

Siphon off the supernatant and wash the precipitate twice with distilled water, twice with 1:100 nitric acid, and twice with water. Transfer the precipitate to a 100-ml beaker, add 2.0 ml 0.1N sodium chloride, and bring the volume to 15–20 ml. Cool the solution to 15–20°C and add 10 ml of the bromine–acetate buffer solution. Warm the beaker to expel bromine and remove the iodate from the silver chloride precipitate by filtration (Whatman #42). Wash the residue twice with water. Warm the solution to expel residual bromine, adjust the volume to 15 ml, then add 0.2 ml 1% sodium sulfite to destroy hypobromous acid and to reduce iodate to iodide. Repeat the bromine oxidation step using 10 ml bromine—acetate buffer solution.

Warm the solution to eliminate bromine, then add water to a volume of 10 ml. Transfer the solution to a 25-ml volumetric flask and cool to 15°C in a water bath. Add 1 ml of the 0.5% starch solution and 1 ml of the cadmium iodide solution. Shake the solution for a few seconds, then allow it to stand in the dark for 30 min. Adjust the pH to 4.2 by adding 0.3 ml 25% sodium acetate solution. Dilute the solution to the mark and measure the absorbance at 580 nm. Determine the concentration against a standard calibration curve.

2. Plant Material, Wet Ash Method (24)

Reagents

Sulfuric—Nitric Acids, 1:1. This is called Neumann's acid.

Perchloric Acid.

Arsenious Acid Solution. In sequence, dissolve 9.8 g arsenious oxide in 14 ml $10N$ sodium hydroxide, add 600 ml water, neutralize the solution with $10N$ sulfuric acid, add 42 ml concentrated sulfuric acid, and dilute to 1 liter.

Dilution Solution, 3.0N Sulfuric Acid.

Ceric Ammonium Sulfate Solution. Dissolve 5 g ceric ammonium sulfate in 70 ml $5N$ sulfuric acid by heating gently, filter and dilute to 100 ml with distilled water. Dilute 25 ml 100 ml with 3.5N sulfuric acid.

Iodate Standard Solution, 1 mg as iodide per liter of water.

Procedure

Dry 200 mg plant material at 105°C and transfer it to a 50-ml flask. Add 7.5 ml Neumann acid and 1 ml perchloric acid. Heat on a hot plate for 1 hr at 160—190°C. Raise the temperature at 275°C until the solution turns yellow (about 30 min more). Cool, dilute the digest to 40 ml, and add 1 ml arsenite solution. Allow the solution to stand at room temperature for one to two days and remove 5-ml aliquots into colorimeter tubes containing 0.4 ml arsenite solution. Incubate the equilibrium in a 30°C water bath, add 1 ml ceric ammonium sulfate to each tube, and read each tube at 30 sec and 20 min at 420 nm, using timed intervals of 1 min between the cerium addition to each tube. Calibrate by carrying standards through the same process in amounts of 0—0.5 ml of the standard iodate solution.

3. Distillation Method

Procedure

Add saturated chromium trioxide solution (10–30 ml) to 1–5 g sample in the flask of an iodine distillation apparatus (93), followed by 5 ml concentrated sulfuric acid for each milliliter chromium trioxide used. Add the chromium trioxide dropwise until the vigorous reaction subsides, and then add in large amounts. Heat the solution to 220°C for 5 min, cool to 100°C, add 50 ml distilled water, and mix well. Connect the flask to the still with a 50-ml beaker containing 1 ml 1N sodium hydroxide as the receiver. When the beaker contents are boiling, add dropwise 10–15 ml 30% phosphorous acid to the distilling apparatus and distill until 40 ml distilled water has been collected in the beaker. After evaporating the alkaline solution to 15 ml, treat the contents as in Section IIB1 using the iodine–starch method of determination.

C. IODINE IN WATER

1. Alkaline Ash Method (205)

Use a 50-liter sample of water made alkaline with sodium carbonate and evaporate to 1 liter. Remove the solids by filtration. Evaporate the filtrate to 50 ml, filter into a nickel or platinum crucible, evaporate to dryness, and heat to 110°C. Transfer the residue to a nickel boat and place in the center of a Pyrex combustion tube of a McClendon apparatus where it should be heated until the ash is white. Transfer the ash to a 100-ml breaker with alkali rinsings, evaporate to 20 ml, filter, and neutralize with phosphoric acid. Add a drop of arsenious acid and transfer the contents to a 30-ml separatory funnel. Add 1 ml carbon tetrachloride and one drop nitrosylsulfuric acid, shake the mixture, and remove the carbon tetrachloride. The nitrosylsulfuric acid is made by placing starch paste, made from 50 g starch and an equal amount of water, in a sidearm distilling flask heated by a water bath. Nitric acid is run in dropwise on the hot starch rapid enough to keep a steady stream of nitrogen oxides coming over into 30 ml sulfuric acid. The nearly saturated solution of nitrosylsulfuric acid is stable indefinitely in a closed bottle. Repeat the extraction and compare the color of the combined extracts with the color of standards in the range of 0.5 to 15 μg per 20 ml that have been prepared in a similar manner.

The same procedure can be carried out without the use of the McClendon apparatus (216). In this case, treat 50 liters water, 1 to 10 liters mineral water, or 100 ml brine with 6 g sodium carbonate and evaporate to a viscous consistency. Cool this solution and add 100 ml ethanol. Heat the mixture, stir, and filter. Reextract the solids by the same process and combine the filtrates.

Add a few drops concentrated sodium hydroxide and evaporate to a viscous consistency. Treat the solids again with alcohol, heat, stir, and filter as previously described. Add a few drops sodium hydroxide solution and evaporate the solution to dryness. Transfer the solution quantitatively to a 250-ml separatory funnel using 100 ml water. Add 5 ml concentrated phosphoric acid, 10 ml carbon disulfide, and 20 ml hydrogen peroxide. After shaking the carbon disulfide layer, compare the color with standards prepared by adding iodide solutions to the separatory funnel and extracting in the same manner as the sample.

2. Direct Determination Method

A more recent determination of iodide or iodate in water made use of a very sensitive color reaction involving the reaction of iodate with p-amino-phenol (57). In the case of iodide, the reaction became catalytic by the introduction of an oxidizing agent, hydrogen peroxide, which caused cyclization of reactions by the continued oxidative regeneration of iodate to oxidize p-aminophenol.

Reagents

Aqueous p-Aminophenol Hydrochloride, 0.05%.

Hydrogen Peroxide.

Iodate Standards.

Procedure (Iodate)

Mix 2 ml test solution with 2 ml of the color reagent, aqueous p-aminophenol. If the quantity of iodate is less than 20 ppm, measure the absorbance at 540 nm after 90 min. If the quantity of iodide is greater than 20 ppm, measure the absorbance after 60 min.

Procedure (Iodide)

Mix 2 ml neutral or slightly acidic sample with 2 ml of the aqueous p-aminophenol color reagent. Add two drops hydrogen peroxide solution, then measure the absorbance at 540 nm after 1 hr. This is a much more sensitive procedure for iodide than it is for iodate because of the cyclic reaction as follows:

$$I^- + H_2O_2 \xrightarrow{p\text{-aminophenol}} IO_3^-$$

$$IO_3^- + p\text{-aminophenol} \longrightarrow I^- + \text{indamine dye}$$

In the presence of p-aminophenol, hydrogen peroxide easily oxidizes iodide to iodate, which then oxidizes *p*-aminophenol to the indamine dye through a quinoneimine which condenses with unchanged *p*-aminophenol. The minimum amount of iodide that can be determined is 2.5 ppm in the final solution for the procedure as described.

3. Method for Seawater (13,14)

Reagents

Arsenious Acid. Dissolve 19.8 g arsenious oxide in 200 ml distilled water containing 8 g sodium hydroxide, dilute the solution to 800 ml with water, acidify with 94 ml 18N sulfuric acid, and then dilute to 1 liter.

Ceric Sulfate Solution. Prepare a 0.016N solution by dissolving 10.1 g ceric ammonium sulfate dihydrate in concentrated sulfuric acid. Dilute to 1 liter and then filter through a millipore filter after the solution has aged three days.

Brucine Reagent. Dissolve 2.5 g brucine in 500 ml distilled water containing 14 ml 36N sulfuric acid (stable two to three weeks in brown bottle).

Procedure

Filter samples of seawater through a millipore filter, determine their chlorinities, and dilute 25 ml of each with water to a chlorinity of 15‰. Pipet 25 ml of the diluted sample into a large test tube followed by 20 ml arsenious acid reagent. Place the tubes in a 30°C water bath for 30 min. Then add 5 ml of the cerium reagent to start the reaction. Remove four 5-ml aliquots at predetermined intervals and pipet them into 1 ml brucine solution to quench the reaction. Read the absorbances of the quenched aliquots at 440 nm. Pipet a fifth aliquot after the cerium color has disappeared, add it to brucine, and use it as a reagent blank. Prepare calibration by plotting reaction rate constant versus total iodine, which gives a linear correlation. Prepare standards by making up artificial iodine-free seawater and adjusting this to 15‰Cl. Then add various known amounts of iodine.

REFERENCES

1. Ahuja, J. N., and A. Kaplan, *Clin. Chem.*, **14**, 809 (1968).
2. Ahuja, J. N., A. Kaplan, and P. Van Dreal, *Clin. Chem.*, **14**, 664 (1968).
3. Allott, E. M., J. A. Dauphinee, and W. H. Hurtles, *Biochem. J.*, **26**, 1665 (1932).
4. Almquist, H. J., and J. W. Givens, *Ind. Eng. Chem., Anal. Ed.*, **5**, 254 (1933).
5. Anderson, N. G., *Amer. J. Clin. Pathol.*, **53**, 778 (1970).

6. Anderson, N. G., *Anal. Biochem.*, **28**, 545 (1969).

7. Arthur, P., T. E. Moore, and J. L. Lambert, *J. Amer. Chem. Soc.*, **71**, 3260 (1949).

8. Bancroft, W. O., and J. E. Magoffin, *J. Amer. Chem. Soc.*, **57**, 2561 (1935).

9. Banick, W. W., and G. F. Smith, *Anal. Chim. Acta*, **16**, 464 (1957).

10. Barker, S. B., *J. Biol. Chem.*, **173**, 715 (1948).

11. Barker, S. B., *Biochem. J.*, **90**, 214 (1964).

12. Barker, S. B., M. J. Humphrey, and M. H. Soley, *J. Clin. Invest.*, **30**, 55 (1951).

13. Barkley, R. A., and T. G. Thompson, *Anal. Chem.*, **32**, 154 (1960).

14. Barkley, R. A., and T. G. Thompson, *Deep. Sea. Res.*, **7**, 24 (1960).

15. Bauman, E. J., and N. J. Metzger, *J. Biol. Chem.*, **121**, 231 (1937).

16. Bellabarba, D., R. E. Peterson, and K. Sterling, *J. Clin. Endocrinol*, **28**, 305 (1968).

17. Bellucci, I., *Gazz. Chim. Ital.*, **72**, 501 (1942).

18. Benotti, J., S. Pino, and H. Gardyna, *Clin. Chem.*, **12**, 491 (1966).

19. Bjorksten, F., R. Grasbeck, and B. A. Lamberg, *Acta Chem. Scand.*, **15**, 1165 (1961).

20. Blanquet, P., M. Croizet, A. M. Moura, F. Dumora, and A. Baghdiantz, *Science*, **158**, 381 (1967).

21. Bobtelsky, M. M., and O. Kaplan, *Z. Anorg. Chem.*, **172**, 196 (1928).

22. Bognar, J., and L. Nagy, *Mikrochim. Acta* (Wien), **1969**, 108.

23. Boguth, W., and W. Schaeg, *Mikrochim. Acta* (Wien), **1967**, 658.

24. Borst Pauwels, G. W. F. H., and J. Ch. Van Wesemael, *Anal. Chim. Acta*, **26**, 532 (1962).

25. Brode, W. R., *Chemical Spectroscopy*, 2nd ed., Wiley, New York, 1947.

26. Brode, W. R., *J. Amer. Chem. Soc.*, **48**, 1877 (1926).

27. Brown, H., A. M. Reingold, and M. Samson, *J. Clin. Endocrinol. Metab.*, **13**, 444 (1953).

28. Burtis, C. A., J. C. Mailen, W. F. Johnson, C. D. Scott, T. O. Tiffany, and N. G. Anderson, *Clin. Chem.*, **18**, 752 (1973).

29. Bush, I. E., *Amer. J. Clin. Pathol.*, **53**, 755 (1970).

30. Butler, A. Q., and R. A. Burdett, *Ind. Eng. Chem.*, *Anal. Ed.*, **11**, 237 (1939).

31. Campbell, R. B., and E. G. Young, *Can. J. Res.*, *Sect. F*, **27** (8), 301 (1949).

32. Caraway, W. T., *J. Clin. Endocrinol.*, **12**, 1215 (1952).

33. Chaney, A. L., *Ind. Eng. Chem.*, *Anal. Ed.*, **12**, 179 (1940).

34. Chaney, A. L., *Anal. Chem.*, **22**, 939 (1950).

35. Chatin, A., *Compt. Rend.*, **32**, 669 (1851).

36. Connor, A. C., R. E. Swenson, C. W. Park, E. C. Gangloff, R. Lieberman, and G. M. Curtis, *Surgery*, **25**, 510 (1949).

37. Conway, E. J., *Microdiffusion Analysis and Volumetric Error*, Crosby, Lockwood, London, 1947.

38. Crouch, W. H. Jr., *Anal. Chem.*, **34**, 1698 (1962).

39. Crowley, L. J., and D. R. Jensen, *Clin. Chim. Acta*, **12**, 473 (1965).

40. Csanyi, L. J., *Disc. Faraday Soc.*, **29**, 146 (1960).

41. Custer, J. J., and S. Natelson, *Anal. Chem.*, **21**, 1003 (1941).

42. Decker, J. W., and H. S. Hayden, *Anal. Chem.*, **23**, 798 (1951).

43. Deman, J., *Mikrochim. Acta* (Wien), **1964**, 67.

44. Drey, R. E. A., *Anal. Chem.*, **36**, 2200 (1964).

45. Endres, G., and L. Kaufman, *Z. Physiol. Chem.*, **243**, 144 (1936).

46. Everett, K. G., and D. A. Skoog, *Anal. Chem.*, **43**, 1541 (1973).

47. Fashena, G. J., and V. Trevorrow, *J. Biol. Chem.*, **114**, 351 (1936).

48. Feigl, F., *Chemistry of Specific Selective and Sensitive Reactions*, Academic Press, New York, 1949, pp. 127–8.

49. Feigl, F., *Spot Tests, Vol. 1, Inorganic Applications*, Elsevier, New York, 1954, p. 59.

50. Feigl, F., and E. Frankel, *Ber.*, **65**, 544 (1932).

51. Fellenburg, Von, T., *Biochem. Z.*, **139**, 371 (1923); *ibid.*, **152**, 116 (1924).

52. Fisher, A. B., R. B. Levy, and W. Price, *New England J. Med.*, **273**, 812 (1965).

53. Fisher, D. A., M. D. Morris, H. Lehman, and C. Lackey, *Anal. Biochem.*, **7**, 37 (1964).

54. Flox, J., I. Piteskey, and A. S. Alving, *J. Biol. Chem.*, **142**, 147 (1942).

55. Foss, Q. P., L. V. Hankes, and D. D. Van Slyke, *Clin. Chim. Acta*, **5**, 301 (1960).

56. Fraps, G. S., and J. F. Fudge, *J. Assoc. Offic. Agr. Chem.*, **23**, 164 (1940).

57. Fuchs, J., E. Jungreis, and L. Ben-Dor, *Anal. Chim. Acta*, **31**, 187 (1968).

58. Funahashi, S., M. Tabata, and M. Tanaba, *Anal. Chim. Acta*, **57**, 311 (1971).

59. Gaillard, B. D. E., H. S. Thompson, and J. Morak, *Carbohydr. Res.*, **11**, 505 (1969).

60. Gasner, F. X., *Ind. Eng. Chem., Anal. Ed.*, **12**, 120 (1944).

61. Gimlette, T. M. D., *J. Clin. Pathol.*, **20**, 170 (1967).

62. Gmelin, R., and A. I. Virtanen, *Acta Chem. Scand.*, **13**, 1469 (1959).

63. Grange, M., *Compt. Rend.*, **33**, 627 (1851).

64. Greenwood, I. A., J. V. Hodam, and D. Macrae, Jr., *Electronic Instruments*, McGraw-Hill, New York, 1948, pp. 548–67.

65. Grimaldi, F. S., and M. M. Schnepfe, *Anal. Chim. Acta*, **53**, 181 (1971).

66. Groak, B., *Biochem. Z.*, **270**, 291 (1934).

67. Gross, W. G., L. K. Wood, and J. S. McHargue, *Anal. Chem.*, **20**, 900 (1948).

68. Grossman, A., and G. F. Grossman, *J. Clin. Endocrinol. Metab.*, **15**, 354 (1955).

69. Hadjiioannou, T. P., *Anal. Chim. Acta*, **30**, 488 (1964).

70. Hamilton, R. H., *Clin. Chem.*, **8**, 194 (1962).

71. Harlay, V., *Ann. Pharm. Franç.*, **5**, 81 (1947).

72. Hatcher, D. W., and N. G. Anderson, *Amer. J. Clin. Pathol.*, **52**, 645 (1969).

73. Heerspink, W., and G. J. Op De Weegh, *Clin. Chim. Acta*, **39**, 327 (1972).

74. Henry, R. J., *Clinical Chemistry, Principles and Techniques*, Hoeber-Harper, New York, 1964.

75. Herries, D. G., and F. M. Richards, *Anal. Chem.*, **36**, 1155 (1964).

76. Hoch, H., and C. G. Lewallen, *Clin. Chem.*, **15**, 204 (1969).

77. Hoch, H., S. L. Sinnett, and T. H. McGavack, *Clin. Chem.*, **10**, 799 (1964).

78. Hollingsworth, D. R., M. Dillard, and P. K. Bondy, *J. Lab. Clin. Med.*, **62**, 347 (1963).

79. Hollo, J. J., *Szejtli (Polytech. Univ. of Budapest) Periodica Polytech*, **1**, 223 (1957).

80. Honetz, N., and R. Kotzaurek, *Klin. Wochenschr.*, **38**, 494 (1960).

81. Horn, K., T. Ruhl, and P. C. Scriba, *Z. Klin. Chem. u. Klin. Biochem.*, **10**, 99 (1972).

82. Houston, F. G., *Anal. Chem.*, **22**, 493 (1950).

83. Hovorka, V., *Collect. Czech. Chem. Commun.*, **2**, 559 (1930).

84. Iwasaki, I., S. Utsumi, and T. Ozowa, *Bull. Chem. Soc. Japan*, **26**, 108 (1953).

85. Jacobs, M. B., *Chemical Analyses of Foods and Food Products*, 2nd ed., Van Nostrand, New York, 1951, p. 757.

86. Jungreis, E., and I. Gedalia, *Mikrochim. Acta* (Wien), **1960**, 145.

87. Jüngst, D., and L. Strauch, *Z. Klin. Chem. u. Klin. Biochem.*, **7**, 636 (1969).

88. Karns, G. M., *Ind. Eng. Chem., Anal. Ed.*, **4**, 299 (1932).

89. Keller, H. E., B. Gloebel, B. Hoffsummer, E. Oberhausen, and W. Leppla, *Clin. Chem.*, **14**, 844 (1968).

90. Kelley, M. T., and J. M. Jansen, *Clin. Chem.*, **17**, 701 (1971).

91. Kessler, G., and V. J. Pileggi, *Clin. Chem.*, **14**, 811 (1968).

92. Kirk, P. L., *Quantitative Ultramicroanalysis*, Wiley, New York, p. 114.

93. Klein, E., *Biochem. Z.*, **36**, 9 (1954).

94. Klugerman, M. R., *Amer. J. Clin. Pathol.*, **24**, 490 (1954).

95. Knapheide, M. D., and A. R. Lamb, *J. Amer. Chem. Soc.*, **50**, 2121 (1928).

96. Knapp, G., H. Spitzy, and H. Leopold, *Anal. Chem.*, **46**, 724 (1974).

97. Kologlu, S., H. L. Schwartz, and A. C. Carter, *Endocrinology*, **78**, 231 (1966).

98. Kolnitz, H., Von, and R. E. Remington, *Ind. Eng. Chem.*, *Anal. Ed.*, **5**, 38 (1933).

99. Kono, T., L. Van Middlesworth, and E. B. Astwood, *Endocrinology*, **66**, 845 (1960).

100. Kotzaurek, R., *Klin. Wochenschr.* (Berlin), **38** 995 (1960).

101. Kreutzer, E. K. J., *Clin. Chim. Acta*, **37**, 519 (1972).

102. Krishnaswaym, K. G., and A. Sreenivasan, *J. Biol. Chem.*, **176**, 1253 (1948).

103. Kydd, D. M., E. B. Man, and J. P. Peters, *J. Clin. Invest.*, **29**, 1033 (1950).

104. Lachwer, F., and J. Leloup, *Bull. Soc. Chim. Biol.*, **31**, 1128 (1949).

105. Lambert, J. L., *Anal. Chem.*, **23**, 1247 (1951), **23**, 1251 (1951).

106. Lambert, J. L., P. Arthur, and T. E. Moore, *Anal. Chem.*, **23**, 1101 (1951).

107. Lambert, J. L., and Zitomer, *Anal. Chem.*, **35**, 405 (1963).

108. Lang, R., *Z. Anorg. u. Allgem. Chem.*, **152**, 197 (1926); *Ber.*, **60**, 1389 (1927).

109. Lange, B., *Kolorimetrische Analyse*, Verlag Chemie, G.m.b.H., Weinheim, 1952, p. 211.

110. Lange, N. A., and L. A. Ward, *J. Amer. Chem. Soc.*, **47**, 1000 (1925).

111. Leffler, H. H., *Amer. J. Clin. Pathol.*, **24**, 483 (1954).

112. Leffler, H. H., *Amer. J. Clin. Pathol.*, **41**, 95 (1964).

113. Lein, A., and N. Schwartz, *Anal. Chem.*, **23**, 1507 (1951).

114. Leipert, T., *Biochem. Z.*, **261**, 436 (1933).

115. Levenson, G. I. P., *J. Soc. Chem. Ind.* (London), **66**, 198 (1947).

116. Lewallen, C. G., Chromatographic Fractionation of Iodinated Compounds in Biologic Fluids, in *Evaluation of Thyroid and Parathyroid Functions*, Sunderman, F. W., and F. W. Sunderman, Jr., Eds., Lippincott, Philadelphia, 1963, p. 37.

117. Lintner, C. J., *J. Prakt. Chem.*, **34**, 378 (1886).

118. Lundgren, H. P., *J. Amer. Chem. Soc.*, **59**, 413 (1937).

119. MacLaglen, N. F., C. H. Bowden, and J. H. Wilkinson, *Biochem. J.*, **67**, 5 (1957).

120. Maclin, A., *Clin. Chem.*, **17**, 707 (1971).

121. Magee, R. J., and H. Spitzy, *Mikrochim. Acta* (Wien), **1959**, 101.

122. Makowetz, E., K. Muller, and H. Spitzy, *Microchem. J.*, **10**, 194 (1966).

123. Malijarov, K. L., and W. B. Matskievich, *Mikrochemie*, **13**, 85 (1933).

124. Malmstadt, H. V., C. J. Delaney, and E. A. Cordos, *CRC Crit. Rev. Anal. Chem.*, **2**, 560 (1972).

125. Malmstadt, H. V., and T. P. Hadjiioannou, *Anal. Chem.*, **35**, 2157 (1963).

126. Man, E. B., D. M. Kydd, and J. P. Peters, *J. Clin. Invest.*, **30**, 531 (1951).

127. Man, E. B., and D. A. Siefried, *J. Biol. Chem.*, **168**, 119 (1947).

128. Mandl, R. H., and R. J. Block, *Arch. Biochem. Biophys.*, **81**, 25 (1959).

129. Manual No. 5004, Hycel Inc., *Cuvette PBI Determination*, Houston, Texas 1967.

130. Mark, H. B., and G. A. Rechnitz, *Kinetics in Analytical Chemistry*, Interscience, New York, 1968, p. 292.

131. Martens, F. F., *Verhandl. Deut. Phys. Ges.*, **4**, 138 (1930).

132. Mason, J. M., *Amer. J. Clin. Pathol.*, **48**, 561 (1967).

133. Matthews, A. D., and J. P. Riley, *Anal. Chim. Acta*, **51**, 295 (1970).

134. Matthews, N. L., G. M. Curtis, and W. R. Brode, *Ind. Eng. Chem., Anal. Ed.*, **10**, 612 (1938).

135. McCullagh, D. R., *J. Biol. Chem.*, **107**, 35 (1934).

136. McCurdy, W. H., and G. G. Guilbault, *J. Phys. Chem.*, **64**, 1825 (1960).

137. McHargue, J. S., D. W. Young, and R. K. Calfee, *Ind. Eng. Chem., Anal. Ed.*, **6**, 318 (1934).

138. Meyer, K. R., R. C. Dickenman, E. G. White, and B. Zak, *Amer. J. Clin. Pathol.*, **25**, 1160 (1955).

139. Meyers, J. H., and E. B. Man, *J. Lab. Clin. Med.*, **37**, 867 (1951).

140. Michel, O., and G. Deltoux, *Bull. Soc. Chim. Biol.*, **31**, 1125 (1949).

141. Miller, R. M., *Ind. Eng. Chem., Anal. Ed.*, **11**, 12 (1939).

142. Milton, R. F., and W. A. Waters, *Methods of Quantitative Micro-Analysis*, Edward Arnold, London, 1949, pp. 84–86.

143. Miyake, Y., and S. Tsunogai, *La Mer, Bull. Soc. Fr.-Jap. Oceanogr.*, **4**, 65 (1966).

144. Moran, J. J., *Anal. Chem.*, **24**, 378 (1952).

145. Moknach, V. D., N. M. Rusokova, *Dokl. Akad. Nauk USSR*, **145**, 1290 (1963); Consultants Bureau Translation p. 753.

146. Mougey, E. H., and J. W. Mason, *Anal. Biochem.*, **6**, 223 (1963).

147. Mougey, E. H., and J. W. Mason, *J. Lab. Clin. Med.*, **59**, 672 (1962).

148. Muller, K., *Clin. Chim. Acta*, **17**, 21 (1967).

149. Muller, K., *Mikrochim. Acta* (Wien), **1967**, 585.

150. Muller, H., *Jod-, Chlor- und Calciumbestimmungen an normalen und an kropfig veränderten Shilddrüsen,* Inaug. Diss. Univ. of Zurich, 1923.

151. Munster, W., *Microchemie*, **14**, 23 (1933).

152. Nauman, J. A., A. Nauman, and S. C. Wermer, *J. Clin. Invest.*, **46**, 1346 (1967).

153. Nesh, F., and W. C. Peacock, *Anal. Chem.*, **22**, 1573 (1950).

154. Offner, H. G., and D. A. Skoog, *Anal. Chem.*, **38**, 1520 (1966).

155. O'Neal, L. W., and E. S. Simms, *Amer. J. Clin. Pathol.*, **23**, 493 (1953).

156. Orth, G. W., and O. Fenner, *Ärztl. Lab.*, **17**, 334 (1971).

157. Paletta, B., *Mikrochim. Acta* (Wien), **1969**, 956.

158. Pierson, G. G., *Ind. Eng. Chem., Anal. Ed.*, **6**, 437 (1934); *ibid.*, **11**, 86 (1939).

159. Pihar, O., and S. Vohmout, *Casopis Lekaru Ceskych*, **87**, 1182 (1948).

160. Pjck, J., J. Gillis, and J. Hoste, *Int. J. Appl. Rad. Isotopes*, **10**, 149 (1961).

161. Pileggi, V. J., and G. Kessler, *Clin. Chem.*, **14**, 339 (1968).

162. Posner, I., *Anal. Biochem.*, **23**, 492 (1968).

163. Posner, I., *J. Lab. Clin. Med.*, **57**, 314 (1961).

164. Postmes, Th., *Acta Endocrinol.*, **12**, 153 (1963).

165. Postmes, T. J., *Clin. Chim. Acta*, **10**, 581 (1964).

166. Postmes, T. J., and J. M. Coenegracht, *Clin. Chim. Acta*, **38**, 313 (1972).

167. Poziomek, E. J., and D. W. Reger, *Anal. Chim. Acta*, **58**, 459 (1972).

168. Rabin, M. S., *Anal. Chem.*, **22**, 480 (1950).

169. Rabourdin, S. M., *Ann. Chem. Liebig's*, **76**, 375 (1850).

170. Rechnitz, G. A., *Anal. Chem.*, **36**, 453R (1964).

171. Reith, J. F., *Pharm. Weekblad*, **66**, 1097 (1929).

172. Remick, A. E., *J. Amer. Chem. Soc.*, **69**, 94 (1947).

173. Riley, M., and Gochman, N. A., *Technicon Co. Bulletin 62*, Technicon Instruments Corporation, Chauncey, New York, 1964.

174. Roche, J. Lissitzky, S., and Michel, R., in *Methods of Biochemical Analysis*, Vol. 1, Glick, D., Ed., Interscience, New York, 1954, p. 243.

175. Rodden, C. J., *Analytical Chemistry of the Manhattan Project*, 1st ed., McGraw-Hill, New York, 1950, pp. 287–302.

176. Rodriguez, P. A., and H. L. Pardue, *Anal. Chem.*, **41**, 1369 (1969).

177. Rodriguez, P. A., and H. L. Pardue, *Anal. Chem.*, **41**, 1376 (1969).

178. Rogina, B., and M. Dubravcic, *Analyst*, **78**, 594 (1953).

179. Roth, H., *Quantitative Organic Microanalysis of Fritz Pregl*, 3rd ed., Churchill, London, 1937, pp. 113–16.

180. Rundle, R. E., and D. French, *J. Amer. Chem. Soc.*, **65**, 558 (1943).

181. Saifer, A., and J. Hughes, *J. Biol. Chem.*, **118**, 241 (1937).

182. Salter, W. T., and M. W. Johnson, *J. Clin. Endocrinol. Metab.*, **8**, 911 (1948).

183. Salter, W. T., and E. A. McKay, *Endocrinology*, **35**, 380 (1944).

184. Sandell, E. B., and I. M. Kolthoff, *J. Amer. Chem. Soc.*, **56**, 1426 (1934).

185. Sandell, E. B., and I. M. Kolthoff, *Microchim. Acta*, **1**, 9 (1951).

186. Schaffer, P. A., *J. Amer. Chem. Soc.*, **55**, 2169 (1933); *idem, J. Phys. Chem.*, **40**, 1021 (1936).

187. Schnepfe, M. M., *Anal. Chim. Acta*, **58**, 83 (1972).

188. Schoch, T. J., *Advances in Carbohydrate Chemistry*, Vol. 1, Academic Press, New York, 1945, pp. 247–77.

189. Scott, A. F., *J. Amer. Water Works Assoc.*, **26**, 634 (1934).

190. Scott, A. F., and F. H. Hurley, *J. Amer. Chem. Soc.*, **56**, 333 (1934).

191. Sendroy, J., Jr., *J. Biol. Chem.*, **130**, 605 (1939).

192. Sendroy, J., Jr., and A. S. Alving, *J. Biol. Chem.*, **142**, 159 (1942).

193. Shahrokh, B. K., *J. Biol. Chem.*, **147**, 109 (1943).

194. Shahrokh, B. K., *J. Biol. Chem.*, **154**, 517 (1944).

195. Shahrokh, B. K., and R. M. Chesbro, *Anal. Chem.*, **21**, 1003 (1949).

196. Simerl, L. E., and B. L. Browning, *Ind. Eng. Chem., Anal. Ed*, **11**, 125 (1939).

197. Smaller, B., and J. F. Hall, *Ind. Eng. Chem., Anal. Ed.*, **16**, 64 (1944).

198. Smith, G. F., *The Wet Chemical Oxidation of Organic Compositions Employing Perchloric Acid*, G. Frederick Smith Chemical Co., Columbus, Ohio, 1965.

199. Smith, G. F., *Anal. Chim. Acta*, **8**, 397 (1953).

200. Smith, G. F., and Diehl, H., *Talanta*, **3**, 41 (1959).

201. Snell, F. D., and C. T. Snell, *Colorimetric Methods of Analysis*, 3rd ed., *Vol. II*, Van Nostrand, New York, 1949, p. 740.

202. Sobel, H., and S. Sapsin, *Anal. Chem.*, **24**, 1829 (1952).

203. Spielholtz, G. I., and H. Diehl, *Talanta*, **13**, 991 (1966).

204. Spitzy, H., M. Reese, and H. Skrube, *Mikrochim. Acta* (Wien), **1958**, 488.

205. *Standard Methods for the Examination of Water and Sewage*, 8th ed., American Public Health Association, New York 1936.

206. Stevens, C. S., *J. Lab. Clin. Med.*, **22**, 1074 (1937).

207. Strickland, R. D., and C. M. Maloney, *Anal. Chem.*, **29**, 1870 (1957).

208. Sugawara, K., T. Koyama, and K. Teroda, *Bull. Chem. Soc. Japan*, **28**, 494 (1955).

209. Sundberg, O. E., and G. L. Roger, *Ind. Eng. Chem., Anal. Ed.*, **18**, 719 (1948).

210. Sunderman, F. W., Jr., *CRC Crit. Rev. Clin. Lab. Sci.*, **1**, 551 (1970).

211. Swanson, M. A., *J. Biol. Chem.*, **172**, 825 (1948).

212. Swift, E. H., *A System of Chemical Analysis for the Common Elements*, Prentice-Hall, New York, 1939, pp. 448–521.

213. Talbot, N. B., A. M. Butler, A. H. Saltzman, and P. M. Rodriguez, *J. Biol. Chem.*, **153**, 479 (1944).

214. Taurog, A., and I. L. Chaikoff, *J. Biol. Chem.*, **163**, 313 (1946).

215. Taurog, A., and I. L. Chaikoff, in *Methods in Enzymology, Vol. 4*, Colowick, S. P., and N. O. Kaplan, Eds., Academic Press, New York, 1957, p. 856.

216. Theroux, E. R. T., E. F. Eldrige, and W. L. Mallmawn, *Laboratory Manual for Chemistry and Bacteriology. Analysis of Water and Sewage*, 3rd ed., McGraw-Hill, London, 1943, p. 16.

217. Thomas, J. W., L. A. Shinn, H. G. Wiseman, and L. A. Moore, *Anal. Chem.*, **22**, 726 (1950).

218. Thompson, J. C., and E. Hamori, *Biopolymers*, **8**, 689 (1969).

219. Thompson, J. C., and E. Hamori, *J. Phys. Chem.*, **75**, 272 (1971).

220. Thompson, J. J., and U. O. Oakdale, *J. Amer. Chem. Soc.*, **52**, 1195 (1930).

221. Tiffany, T. O., D. O. Chilcote, and C. A. Burtis, *Clin. Chem.*, **19**, 908 (1973).

222. Trevorrow, V., and G. J. Fashena, *J. Biol. Chem.*, **110**, 29 (1935).

223. Tsunogai, S., *Anal. Chim. Acta*, **55**, 444 (1971).

224. Tsunogai, S., and T. Henmi, *Chikyu Kagaku*, **3**, 14 (1969).

225. Turner, R. G., *J. Amer. Chem. Soc.*, **52**, 2768 (1930).

226. Ueitwiller, Von A., *Z. Klin. Chem. u. Klin. Biochem.*, **4**, 45 (1968).

227. Vass, L., *Ärtzl. Lab.*, **17**, 334 (1971).

228. Waleszek-Piotrowski, L. J., and F. C. Koch, *J. Biol. Chem.*, **194**, 427 (1952).

229. Weisz, H., and H. Ludwig, *Anal. Chim. Acta*, **60**, 385 (1972).

230. Weisz, H., and K. Rothmaier, *Anal. Chim. Acta*, **68**, 93 (1974).

231. West, C. D., V. J. Chaure, and M. Wolfe., *J. Clin. Endocrinol. Metab.*, **25**, 1189 (1965).

232. Wiener, J. D., and E. T. Backer, *Clin. Chim. Acta*, **20**, 155 (1968).

233. Willard, H. H., and J. J. Thompson, *J. Amer. Chem. Soc.*, **52**, 1895 (1930).

234. Wintersteiner, E., *Z. Physiol. Chem.*, **104**, 54 (1918).

235. Winton, A. L., and K. B. Winton, *The Analysis of Foods*, Wiley, London, 1945, p. 271.

236. Woodward, H. Q., *Ind. Eng. Chem., Anal. Ed.*, **6**, 331 (1934).

237. Wright, C. H., *Soil Analysis*, Thomas Murby, London, 1934, p. 166.

238. Zak, B., and E. S. Baginski, in *Gradwohl's Clinical Laboratory Methods and Diagnosis*, Frankel, S., S. Reitman, and A. C. Sonnewirth, Eds., Mosby, St. Louis, p. 220 (1970).

239. Zak, B., and A. J. Boyle, *J. Amer. Pharm. Assoc., Sci. Ed.*, **41**, 260 (1952).

240. Zak, B., A. M. Koen, and A. J. Boyle, *Amer. J. Clin. Pathol.*, **23**, 603 (1953).

241. Zak, B., H. H. Willard, G. B. Myers, and A. J. Boyle, *Anal. Chem.*, **24**, 1345 (1952).

242. Zieve, L. M., Dahle, and A. L. Schultz, *J. Lab. Chem. Med.*, **44**, 374 (1954).

243. Zulkowski, K., *Ber.*, **13**, 1395 (1880).

NITROGEN

D. F. BOLTZ

Department of Chemistry, Wayne State University, Detroit, Michigan

and

MICHAEL J. TARAS

*Research Project Director, American Water Works Association,
Research Foundation, Denver, Colorado*

Nitrogen constitutes 78.09% by volume of dry, unpolluted air but is a minor element in respect to its abundance in the earth's crust. However, the fixation of atmospheric nitrogen to give ammonia and other useful compounds is important industrially and in the natural nitrogen cycle. The decay of nitrogenous organic material to produce ammonia, nitrites, nitrates, and nitrogen by the action of various bacteria is an important facet of the nitrogen cycle. Thus, the nitrogen in most samples to be analyzed exists in the form of ammonium, nitrite and nitrate salts, and organically bound nitrogen. However, hydrazine, hydroxylamines, and many derivatives of ammonia are frequently determined spectrophotometrically. The solubilization of samples, the conversion to the desired species, and the isolation of the desired compound prior to measurement of the desired constituent by a spectrophotometric method are important analytic considerations.

One of the classical methods for the determination of total nitrogen is based on the Kjeldahl method (122,124) of digestion and distillation in which the nitrogen compound is converted to an ammonium salt and the ammonia is distilled after addition of a strong base. In the digestion step, the organic compound is decomposed by treatment with hot, concentrated sulfuric acid and the nitrogen is converted to ammonium sulfate. The Kjeldahl digestion is of sufficient analytic importance to justify further discussion of this technique.

The nature of the sample material largely defines the digestion technique and conditions to be applied in the particular circumstance. Kirk (122) has summarized the factors related to the success of a Kjeldahl digestion both on the macro and micro scale. Many of the macro practices are applicable with suitable modification to the micro field.

Sulfuric acid is used in almost every Kjeldahl digestion procedure. Phosphoric acid can also be used to advantage for the digestion of proteins, provided the proper proportion of phosphoric and sulfuric acids is maintained (71).

Potassium sulfate is often added to sulfuric acid to raise the boiling point and increase the rate of digestion (89). A temperature of about $350°C$ is necessary for the decomposition of refractory nitrogen compounds. This temperature can be attained when a mixture of 2 ml concentrated sulfuric acid and 1.3 g potassium sulfate is kept at its boiling point (239). An excess of the salt is inadvisable, especially if the composition of the digestion mixture should approach that of potassium acid sulfate. In such instances, significant losses of ammonia might result. Therefore, it is good practice to add extra sulfuric acid in the middle of a protracted digestion (122).

A catalyst is usually added for the purpose of decreasing the digestion time. The prevailing opinion is that mercury as the metal, oxide, or sulfate is the best single catalyst (30,122). On the basis of a critical study, Hiller, Plazin, and Van Slyke concluded that mercury is the only Kjeldahl catalyst capable of yielding protein nitrogen values as high as the Dumas dry combustion method (102). A system for recovering the mercury catalysts used in Kjeldahl digestions has been developed (63). It should be remembered that thiosulfate or sulfide is added prior to distillation in order to precipitate the mercury(II) ions and prevent aminocomplex formation. Selenium and copper(II) sulfate have been frequently used as catalysts for specific applications (29,114,136).

The indiscriminate use of oxidizing agents for hastening digestion may promote the oxidation of ammonia. Therefore, the careful addition of small quantities of oxidizing agents can be justified only when restricted to digests still rich in carbonaceous reducing agents. Minimal amounts of hydrogen peroxide have been used successfully in micro-Kjeldahl work (58,125,149). Since hydrogen peroxide often contains a nitrogen-bearing preservative, the nitrogen blank must be determined for purposes of correcting the final values.

The duration of digestion is dependent upon the size and chemical character of the sample and the nature of the catalyst and oxidizing agents that may be used. Digestion time may vary from a few minutes to more than several hours. The duration of Kjeldahl digestions can be reduced by sealing the sample in heavy-walled glass tubes. Heterocyclic nitrogen compounds are digested in 7 to 15 min at $470°C$ with concentrated sulfuric acid and mercury(II) oxide catalyst (237). The addition of small amounts of water to the sealed digestion mixture markedly increases the stability of ammonia nitrogen in sulfuric acid and allows somewhat higher temperatures to be used without the loss of ammonia (88).

Although it is common practice to refer to the result obtained by the usual Kjeldahl determination as total nitrogen, the Kjeldahl value actually represents the nitrogen in the trinegative state. Total nitrogen would properly embrace the oxidized as well as trinegative forms of nitrogen in a sample. Unless these oxidized and exceptional nitrogen groups are known to be absent, the term "total nitrogen" is a misnomer.

The Kjeldahl method can be extended to include the determination of oxidized forms of nitrogen. Reducing agents are added prior to digestion in order to convert the oxidized forms to the proper susceptible state. Normally, the determination of organic nitro compounds involves pretreatment with salicylic acid in sulfuric acid solution. The resulting nitrosalicylic acid is then reduced with sodium thiosulfate to form the amino compound, which in turn is converted through digestion to the ammonium salt (204). The use of thiosalicylic acid eliminates the need for thiosulfate (131,142).

The Kjeldahl method has also been adapted to the determination of the nitrogen in such compounds as azines, hydrazones, oximes, and semicarbazones. The material is dissolved in acetic acid and methanol, reduced by the action of zinc and hydrochloric acid, and digested in the typical manner with a catalyst mixture of potassium sulfate, mercury(II) oxide, and selenium metal. The afterboil varies from 30 to 45 min (69).

The nitrogen of nitrile compounds has been determined by the strong reducing action of hydriodic acid as a preliminary to Kjeldahl digestion (76). The hydriodic acid formed by the reaction of sulfuric acid and potassium iodide has also been used to reduce nitrile nitrogen (187). However, a report has been published of the successful determination of nitrogen in nitriles by the use of the ordinary Kjeldahl digestion catalyzed by selenium, without the addition of any reducing agents (221).

Azide nitrogen has been determined by adding sodium thiosulfate to the sample before digestion with sulfuric acid and selenium. The method is based on the observation that hydrazoic acid, in the presence of a reducing agent, liberates one third of the azide nitrogen as ammonia and the remaining two thirds as free nitrogen (175).

Kjeldahl digestion flasks for micro work are available in 10-, 30-, and 100-ml capacities. Routine blood and urine digestions are usually performed in 25 x 200 mm Pyrex test tubes marked at 35 and 50 ml.

The source of heat for digestion may be either electricity or gas. The heater should maintain a mixture of 2 ml concentrated sulfuric acid plus 1.3 g potassium sulfate at a boiling point of $350°C$. At the same time, the heaters should be sufficiently adjustable to enable low-temperature digestion as well as the refluxing of acid into the neck of the flask, without the loss of contents when the flasks are inclined at the normal digestion angle (208).

The digestion of plant material with a sulfuric—perchloric acid mixture has been found to be satisfactory, provided care is taken to avoid conditions of high temperature and a high local concentration of perchloric acid. The addition of a dilute solution of perchloric acid in concentrated sulfuric acid was a convenient approach to circumvent this difficulty (36). A hydrofluoric acid—phosphoric acid—dichromate mixture has been used to dissolve refractory metals and alloys. Most of the hydrofluoric acid was removed by volatilization prior to distillation of ammonia from the basic solution and determination by the indophenol method (116).

I. SEPARATIONS

A. DISTILLATION METHOD

When the acidic Kjeldahl digestion mixture containing ammonium sulfate is treated with an excess of sodium hydroxide, the liberated ammonia may be distilled and collected in an acidic solution. Figure 1 shows a typical steam distillation unit (22). The observance of several precautions will contribute to the success of a distillation. High ammonia values will be averted if the entire distillation assembly is prepared for the analysis by boiling sufficient water in the flask until the distillate is completely freed of ammonia. Each distilling apparatus should be checked with known ammonium standard solutions to observe the volume of distillate which must be collected for complete nitrogen recovery. This volume of distillate should then be collected with unknown samples.

Ammonia losses can be reduced by carefully adding the base down the side of the ordinary Kjeldahl flask so that the alkali forms the lower of two distinct

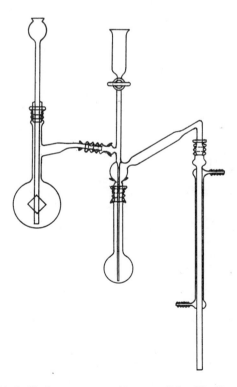

Fig. 1. Micro-Kjeldahl distillation apparatus. (Courtesy Scientific Glass Apparatus Co., Inc.)

layers. After the distilling flask has been properly connected, the layers should be mixed by swirling before distillation is commenced. Glass beads, boiling chips, zinc, and paraffin often regulate the boil and prevent disastrous bumping in the distilling flask. Electrical heating frequently retards bumping and assures smooth operation of distillation units. An efficient trap can minimize entrainment and carry-over of alkali. Carry-over of base is more of a problem in titrimetry than in Nesslerization.

Most of the ammonia distills over during the first few minutes of boiling. The prolongation of distillation ensures the removal of soluble ammonia from the digestion mixture and any ammonia that may have become entrained in the system.

Ammonia levels below 0.2 mg nitrogen can be collected without significant error as the free distillate or by absorption in a small volume of ammonia-free distilled water. If the nitrogen concentration is above 0.2 mg, it is advisable to distill the ammonia into excess acid, with the delivery tube extending below the surface of the acid in order to intercept the volatile ammonia. The amount of acid to be used for capturing the ammonia will depend on the quantity of ammonia distilled. Calculation shows that 1 ml $0.01N$ acid stoichiometrically reacts with 0.14 mg nitrogen. A slight excess of acid is advisable for safety reasons. As little as 10 ml $0.01N$ acid will suffice for nitrogen values of 1 mg. After the ammonia has been absorbed in the acid, the receiver should be lowered and the distillation continued for an additional minute or so. The delivery tube is then disconnected and the outside rinsed into the receiver.

In the determination of nitrate by the 3,4-xylenol method, the nitroxyxylenol is distilled by a micro-Kjeldahl steam distillation technique (100).

B. AERATION METHOD

Another method of isolating ammonia consists of treating the prepared sample or digest with a saturated potassium carbonate or sodium hydroxide solution and sweeping the liberated ammonia by means of an air current into an acidic absorbent. The aeration technique is sometimes used in clinical laboratories. The Van Slyke-Cullen aeration train (222) consists of a wooden block holding three large test tubes connected by glass and rubber tubing. The air first enters the wash tube containing 10 ml $1N$ sulfuric acid which removes the atmospheric ammonia. The air then passes into the tube with the alkali-treated sample, transferring the volatile ammonia to the third tube of the train where the absorption takes place in the acidic receiving solution. A modified aeration train has been described (206).

The aeration method can be used for (a) the removal of ammonia from urine (222); (b) the removal of ammonia formed after the urease incubation of whole blood, Folin–Wu blood filtrate, or urine (71); (c) the removal of ammonia from samples that have been digested in the normal Kjeldahl fashion (71); and (d) the removal of ammonia after the reaction of ninhydrin with alpha-amino nitrogen (201). In the first two instances, the aeration is carried out after the addition of a saturated potassium carbonate solution. In the third and fourth cases, aeration follows the application of a saturated solution of sodium hydroxide.

The usual custom is to perform an aeration at room temperature. Some reports indicate that elevated temperatures of 55–60°C (201) and even 70°C (60) may be necessary for satisfactory recoveries for the 10- to 200-µg nitrogen range.

C. MICRODIFFUSION METHOD

The microdiffusion method of ammonia separation in a closed system is especially applicable to the spectrophotometric determination of small amounts of ammonia. The Conway (56,57) type of microdiffusion cells, Fig. 2, consisting of an outer and inner compartment, has been extensively used (45,55,123,140,181,218). The prepared sample is introduced into the outer well and treated with saturated potassium carbonate solution in order to liberate the ammonia. The cell is then covered, allowing the ammonia to diffuse at room temperature for 1.5 hr into the inner chamber which holds the absorbing acid. The diffusion time can be reduced to 1 hr at 38°C. The absorption solution is transferred by syringe or rubber-bulb pipet to a volumetric flask and Nesslerized, and the color is measured.

Such typical determinations as ammonia, total Kjeldahl nitrogen, and urea nitrogen can be completed at the microgram level by this method. In the urea determination, the urease incubation is allowed to proceed in the outer compartment before the addition of the alkali.

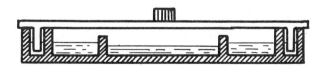

Fig. 2. Schematic diagram of a modified Conway cell (Obrink modification) which can be used to separate traces of ammonia by the microdiffusion method. The outermost chamber is half-filled with a sealing solution having the same composition as the solution in the outer diffusion chamber, with the exception that it contains no sample. The inner diffusion chamber is filled with an absorbing acidic solution.

D. ION EXCHANGE METHOD

Strong cation exchange resins have been used to remove "plasma ammonia" from blood plasma (62,68,150) with subsequent elution of the ammonium ions and spectrophotometric determination of the ammonia nitrogen. Cyanate and cyanide ions have been removed by a strong anion exchange resin before the spectrophotometric determination of ammonia (129). Ammonium ions have also been separated using a column of zirconium phosphate and elution with a cesium chloride solution. The ammonia is determined by the indophenol method (225). O'Neal and Clark (171), in the analysis of fertilizers, employed a cation exchange resin to retain ammonium ions and an anion exchange column to retain nitrate and phosphate, with the urea nitrogen being determined in the effluent. The ammonia nitrogen and nitrate nitrogen were then determined in the eluates from the two resin columns.

In the differential spectrophotometric determination of nitrate and nitrite, Lambert and Zitomer (132) removed nitrite, before reduction of the nitrate, by forming a diazonium cation and using a cation exchange resin and the batch technique.

II. METHODS OF DETERMINATION

Ammonium, nitrite, and nitrate ions are the most common nitrogen-containing species that are determined spectrophotometrically, although gaseous oxides of nitrogen are also determined spectrophotometrically.

In the determination of ammonia, the Nessler, indophenol, and bispyrazolone methods appear to be the dominant methods. The Griess diazotization—coupling reactions are extensively used for the determination of nitrite. A recent trend has been to extract the azo dye into an immiscible organic solvent and to measure the absorbance of the colored extract (73,146) or an aqueous extract obtained by retrograde extraction (141,165). In determining less than 0.3 ppm nitrite, Wada and Hattori (224) concentrated the dye on an anion exchanger and eluted with a small volume of 60% acetic acid in order to improve sensitivity.

Most spectrophotometric methods for the determination of nitrate are based on (a) the nitration of an organic compound, (b) the oxidation of an organic compound, (c) the reduction of nitrate to either nitrite or ammonia, (d) the formation and extraction of an ion-association complex, or (e) the characteristic ultraviolet absorptivity of the nitrate ion. The elimination of the nitrite interference in the determination of nitrate is often accomplished by the addition of sulfamic acid.

Representative general methods for the determination of ammonia, nitrite, and nitrate will be discussed in this section.

A. AMMONIA BY NESSLER METHOD

Nessler first proposed an alkaline solution of mercury(II) iodide and potassium iodide as a reagent for the direct determination of ammonia (160). Nessler reagent ($K_2 HgI_4$) reacts with ammonia to form a reddish-brown colloidal compound with the empirical formula $NH_2 Hg_2 I_3$. The reaction takes place in the following steps (162):

$$2K_2 HgI_4 + 2NH_3 \longrightarrow 2(NH_3 HgI_2) + 4KI$$

$$2(NH_3 HgI_2) \longrightarrow NH_2 Hg_2 I_3 + NH_4 I$$

$$2K_2 HgI_4 + 2NH_3 \longrightarrow NH_2 Hg_2 I_3 + 4KI + NH_4 I$$

Figure 3 reveals that the Nessler–ammonia color system absorbs strongly over a broad wavelength range. Ammonia nitrogen concentrations of 0.2 mg or less absorb sufficiently in the region from 400 to 425 nm to allow photometric measurements at these wavelengths. Higher ammonia nitrogen levels, up to 1 mg,

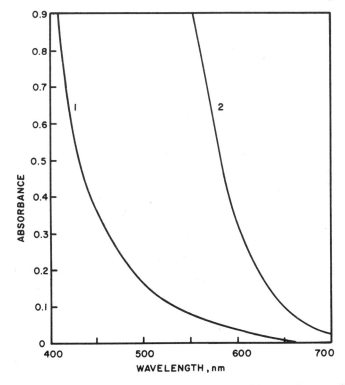

Fig. 3. Absorption spectra of Nessler–ammonia system: (1) ppm nitrogen; (2) 20 ppm nitrogen (1-cm absorption cells).

absorb strongly in the region below 550 nm. Conformity to Beer's law extends to 1.25 mg nitrogen when measurements are made with a spectrophotometer operating at a narrow bandwidth (1 nm), a 1-cm light path, and a wavelength of 580 nm. In this instance, only 1 ml Nessler reagent (preparation A, see below) is added to a 50-ml sample. However, at nitrogen concentrations above 1 mg, an opalescence or turbidity forms when the Nessler reaction is allowed to proceed for more than 15 to 30 min.

Interference in the Nessler reaction generally takes two forms: the appearance of a yellow or green color, and the onset of turbidity. A number of aliphatic and aromatic amines yield yellow colors (137,214). Aldehydes, acetone, alcohols, organic chloramines, and other undefined organic compounds produce turbidity (4,111,214,220). Hydrazine in boiler water has been eliminated by oxidation with potassium iodate prior to use of the Nessler method (59). Ions insoluble in alkaline solution, or producing precipitates with iodide and mercury(II), cause turbidity. Magnesium, manganese(II), iron(III), and sulfide are such ions. Turbidity caused by traces of magnesium can be prevented by adding one or two drops of Rochelle salt solution to a 50-ml sample before Nesslerization. The Rochelle salt solution is prepared by dissolving 50 g potassium sodium tartrate tetrahydrate in 100 ml ammonia-free distilled water, evaporating 20 ml or until all ammonia has been removed, cooling, and restoring the volume with ammonia-free distilled water (8).

Turbidity due to sulfide can be remedied by adding lead carbonate to the flask prior to distillation (5). Off-colors and cloudiness have also been encountered in distillates collected from polluted water samples that have been chlorinated and then dechlorinated (214). No specific procedure can be suggested for eliminating such interferences, particularly when they happen in distillates. However, recourse should be taken to distillation when interference is experienced in direct Nesslerization.

The following procedure has been recommended for reclaiming samples in which turbidity appears after direct Nesslerization (220). The turbid Nesslerized sample is washed into a 125-ml Kjeldahl flask with 25 ml ammonia-free, distilled water. A glass bead and sufficient potassium iodide crystals are added to dissolve the turbidity and convert the orange-colored colloid to a clear-yellow color. An excess of potassium iodide ensures the liberation of all the nitrogen as ammonia. The flask is connected to a Kjeldahl-type distillation apparatus. A 50-ml volumetric flask is arranged so that the outlet of the condenser extends below the surface of the absorbing acid. About 30 ml distillate is collected, and the contents are diluted to the mark. The sample is Nesslerized after proper adjustment of the temperature.

The following variables modify the response of Nessler reagent in a particular application (146,220):

(a) The amounts of the individual chemicals used in the preparation (mercury(II) iodide, mercury(II) chloride, mercury, potassium iodide, iodine, and a base) as well as the impurities introduced by the principal chemicals may affect the sensitivity and reactivity of each Nessler solution.

(b) The condition and stability of the reagent at the time of use are matters of critical importance. The ratio of mercury(II) iodide to potassium iodide in solution bears significantly on Nessler sensitivity and subsequent color development. An appreciable increase in the deposit at the bottom of the reagent bottle should be viewed as visible evidence of a possible alteration in the mercury(II)/iodide ratio.

(c) The chemical agents employed for sample digestion can cause turbidity or off-colors after Nesslerization. Alcohols, added for the suppression of foaming, may contribute a white precipitate upon direct Nesslerization. Sodium sulfate can promote turbidity, particularly as the nitrogen and sodium sulfate levels are increased in the Nesslerized samples.

(d) The alkalinity of the sample plays a vital role in Nesslerization. Color development is inhibited somewhat when the sodium hydroxide concentration falls below $0.15N$ in the final Nesslerized solution. As the concentration of alkali exceeds $0.6N$, a tendency toward turbidity, with increasing nitrogen concentrations, may become evident after Nesslerization.

(e) Raising the sample temperature markedly accelerates Nessler color development. The sample solution temperature is normally adjusted to room temperature, although the $20-40°C$ range is considered satisfactory (151). Higher temperatures are accompanied by a tendency toward formation of turbidity in the presence of increased nitrogen levels. Ice baths may be employed to repress color development, thereby inhibiting the formation of turbidity (179).

(f) The manner of adding Nessler reagent can inhibit or induce turbidity in a Nesslerized sample. Localization of base, resulting from inadequate mixing, favors the production of turbidity at higher nitrogen levels, whereas thorough agitation has an opposite and a more desirable effect. The technique of Nessler addition is often of considerable importance in the direct Nesslerization of digested organic samples (149).

(g) Timing is intimately associated with many of the preceding variables and must be carefully controlled. The longer the Nessler reaction is allowed to continue, the deeper is the color development and the greater is the tendency for turbidity formation. The most rapid color development occurs within the first 10 min. In the case of a relatively insensitive reagent, a reaction period of 30 min may be necessary to demonstrate the presence of nitrogen concentrations of 5 μg or less in 50- or 100-ml volume.

(h) Sometimes, a protective colloid such as gum ghatti (182) is added to biochemical specimens when high nitrogen levels are to be Nesslerized, and a considerable prospect exists for turbidity formation. Other stabilizers, such as gum arabic, have also been used (75). Protective colloids generally decrease color intensity and modify the relationship between color intensity and nitrogen concentration. For this reason they are omitted, if possible.

(i) The calibration graph should be prepared under exactly the same conditions of temperature and reaction times, whether 5, 10, or 30 min or any intermediate interval adopted for the samples. Complete calibration graphs should be constructed after the preparation of each Nessler reagent. Because of the variables that affect a nitrogen determination, the greatest accuracy can be assured only when blanks, samples, and standards are carried simultaneously, and in duplicate, through every step of the digestion, distillation, aeration, microdiffusion, and Nesslerization procedure. The calibration graph should also be rechecked daily for the nitrogen range under investigation.

Reagents

Analytic reagent-grade chemicals, low in nitrogen, and ammonia-free distilled water are required for the preparation of all solutions, with storage preferably in Pyrex glassware. For best results, all dilutions in the ammonia determination should be made with ammonia-free distilled water. All references to concentrated acids in this chapter are understood to mean the reagents regularly available in commerce, namely, sulfuric acid, sp. gr. 1.84; phosphoric acid, sp. gr. 1.69; hydrochloric acid, sp. gr. 1.19; and glacial acetic acid, sp. gr. 1.05.

Ammonia-Free Water. Ammonia-free water can be prepared by distillation or ion exchange methods. Small amounts of ammonia nitrogen may be present in ordinary distilled water as a result of the chloramine treatment of some public water supplies. Several procedures are available for the elimination of this nitrogen when distillation is practiced. Either 1 ml concentrated sulfuric acid or 10 ml alkaline potassium permanganate solution may be added to 1 liter distilled water and the middle portion of the distillate conserved. Ammonia-free distilled water may also be made by treating distilled water with a slight excess of bromine or chlorine water to produce a free halogen residual of about 2–5 mg per liter (6). After the addition of 10–20 ml of pH 7.4 phosphate buffer solution (14.3 g KH_2PO_4 and 68.8 g K_2HPO_4 per liter distilled water), the water is allowed to stand at least 1 hr, preferably overnight. The first 100 ml of the distillate is discarded.

For most work, the ammonia may be removed from ordinary distilled water by shaking 4 liters distilled water with 10 g of a strong cation exchanger (commercially available as Decalso Folin, Folin Permutit, or Ionac C-101 from Ionac Chemical Co., Birmingham, New Jersey) (6). The water is drawn off by siphon or by careful decantation after the resin has settled. Ammonia-free water can also be prepared by passing distilled water through a column of glass tubing 5 cm in diameter and charged with a total of approximately 350 g of the following resins: two parts by volume Amberlite IRA-400 and one part

Amberlite IR-120 (39). Amberlite MB-1, a mixture of two resins, is a satisfactory substitute. Dowex 50, Nalcite HCR, and other similar cation exchange resins are equally applicable (129,130). Water prepared by these exchange processes often retains about 0.01 mg per liter ammonia nitrogen but is satisfactory for most purposes. Daily blanks should be run on the ammonia-free water to guard against unsuspected exhaustion of the exchangers and the attendant release of ammonia to the so-called ammonia-free water.

Standard Stock Ammonium Nitrogen Solution. Dissolve 4.719 g anhydrous ammonium sulfate or 3.819 g anhydrous ammonium chloride in ammonia-free distilled water and dilute to 1.0 liter standard ammonium nitrogen solution (1.0 ml = 1.0 mg nitrogen). Dilute 10.00 ml of the stock standard solution to 1 liter with ammonia-free distilled water (1.0 ml = 10.0 μg nitrogen). Prepare standard solutions of other desired concentrations by appropriate dilution of the stock solution. These dilute standard ammonium nitrogen solutions should be prepared fresh each day (217).

Nessler Reagent. Nessler reagent is an alkaline solution of potassium tetra-iodomercurate, K_2HgI_4. It can be prepared in three ways: from mercury(II) iodide, potassium iodide, and a base; from mercury, iodine, potassium iodide, and a base; or from mercury(II) chloride, potassium iodide, and a base. The special problems involved in the analysis of natural products largely prompted the development of these formulations.

A theoretical ratio of 1.37 g mercury(II) iodide to 1.00 g potassium iodide is required to form K_2HgI_4. The sensitivity of Nessler reagent improves as this theoretical ratio is approached. The merit of preparations A and B is that the ratio of 1.43 to 1.00 offsets the decrease in sensitivity caused by potassium iodide. One milliliter of these reagents is ideal for the determination of small amounts of ammonia in distillates, where no large amount of acidity must be neutralized.

Preparation A (9). Dissolve 100 g mercury(II) iodide and 70 g potassium iodide in 75–100 ml ammonia-free distilled water. Add this mixture slowly, and with stirring, to a cooled solution of 160 g sodium hydroxide pellets that have been dissolved in 700 ml ammonia-free distilled water. Dilute to 1 liter with ammonia-free distilled water. Allow the precipitate to settle, preferably for several days, before using the pale-yellow reagent. A small amount of sediment continues to deposit with the passage of time. The formation of an appreciable quantity of precipitate can affect the reagent's sensitivity and colorimetric reproducibility and, therefore, should be regarded with suspicion. Do not shake the reagent or stir up the sediment. Use only the clear supernatant for color development. The reagent should give a characteristic yellow color within 10 min upon addition to small amounts of ammonia (5 μg nitrogen per 50 ml solution) and should not produce a precipitate within 2 hr. Under optimum conditions, a sensitivity of 1 μg ammonia nitrogen is possible when 1 ml of the reagent is added to a 50-ml sample. Colorimetric response and reproducibility in the range below 5 μg nitrogen may be variable and erratic.

Preparation B. A Nessler reagent identical in all respects to preparation A, except that of alkaline concentration, finds wide acceptance under the name of Bock-Benedict modification (99). Instead of 160 g sodium hydroxide, 100 g sodium hydroxide is used in 1 liter of the reagent. Additional preparations (125,200) are described in the literature. A method of recovering iodine and mercury(II) iodide from Nesslerized solution has been developed (200).

EDTA Solution. Dissolve 50 g disodium dihydrogen ethylenediaminetetraacetate dihydrate in 75 ml distilled water containing 10 g sodium hydroxide. After dissolution is complete, dilute to 100 ml.

Procedure

Transfer 50.0 ml of a neutral sample containing 20 to 250 μg ammonium nitrogen to a 50-ml volumetric flask. Add 0.10 ml of the EDTA solution and mix thoroughly. Add 2.00 ml of the Nessler reagent and mix by inverting the flask eight times. After 10 min, measure the absorbance at 440 nm using 1-cm cells and a reagent blank solution in the reference cell. For less than 50 μg nitrogen, 5-cm cells can be used.

Preparation of Calibration Graph. Transfer 0-, 2-, 5-, 10-, 15-, 20-, and 25-ml aliquots of the dilute standard ammonium solution (1 ml = 10.0 μg nitrogen) to 50-ml volumetric flasks. Dilute to volume with ammonia-free water. Follow the general procedure. Plot absorbance versus μg ammonium nitrogen.

B. AMMONIA BY INDOPHENOL METHOD

The color produced by the reaction of ammonia with phenol and hypochlorite, first described by Berthelet in 1859, is the basis of the indophenol method (40) (Fig. 4). This method has been studied rather extensively, and various modifications in procedure have been suggested. Earlier investigators (74,166,190,199,219,223) noted the high sensitivity of the method but often experienced difficulties in obtaining reproducibility. Russell (190) used manganese(II) ions to catalyze the reaction, while others (52,97,104,139,227,228) have used sodium nitroprusside, sodium pentacyanonitrosylferrate(III), as a catalyst. According to Horn and Squire (106), the actual catalyst is nitritopentacyanoferrate(II) species, $Fe(CN)_5 ONO^{4-}$, formed by the action of the hydroxyl ions and the pentacyanonitrosylferrate(III) ion. Tetlow and Wilson (217) added acetone to increase the

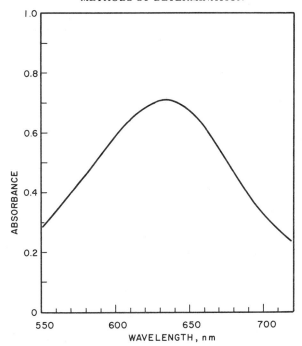

Fig. 4. Absorption spectrum of indophenol blue corresponding to 1 ppm ammonia nitrogen (1-cm absorption cells).

sensitivity of the Berthelet reaction. No catalysts have been employed in several indophenol methods (43,65). Chloramine-T has been substituted for hypochlorite in some procedures (161,189). Thymol (77,83,98,147), o-phenylphenol (245), and m-cresol (246) have been used instead of phenol. The use of a particular catalyst, the relative concentrations of phenol and sodium hydroxide, the temperature, the final pH of the solution, the order of addition of reagents, and timing are factors affecting the ultimate sensitivity and reproducibility obtainable with the indophenol method.

Copper(II) ions cause low results, but this inteference can be eliminated by the addition of EDTA (217). Hydroxylamine, certain amino acids, and urea interfere with the color development, but the latter interferences can be eliminated by a distillation to isolate the ammonia (185). Extraction of the indophenol blue dye into hexanol (161) and isobutyl alcohol (77,83,119,183) have been reported.

A kinetic method in which the maximum value of the reaction rate is directly proportional to the concentration of ammonia in samples has been developed (228). This kinetic method was applied to the determination of blood urea nitrogen.

Weatherburn (227), after studying the factors affecting the reproducibility of the indophenol color, recommended the use of two reagents, a phenol–nitroprusside solution and a basic hypochlorite solution as used earlier by Chaney and Marbach (52). Color measurements were made after allowing color development for 15 min at 37°C. Hoffman (104) used a similar method in determining ammonia in blood. A procedure based on Weatherburn's recommendations will be given in this section. The following equations illustrate the fundamental chemistry of the indophenol method:

$$NH_3 + OCl^- \longrightarrow NH_2Cl + OH^-$$

indophenol

Reagents

Reagent Solution I. Dissolve 10 g phenol and 50 mg sodium nitroprusside, sodium pentacyanonitrosylferrate(III), in ammonia-free water and dilute to 500 ml. Prepare fresh solution on a weekly basis and store in an amber bottle in the refrigerator.

Reagent Solution II. Dissolve 5.0 g sodium hydroxide in ammonia-free water, add 10.0 ml of a sodium hypochlorite solution (5% available chlorine), and dilute to 500 ml. Prepare fresh daily and keep in an amber bottle.

Standard Ammonium Solution. Dissolve 0.3819 g anhydrous ammonium chloride (dried at 100°C) in ammonia-free water and dilute to 1 liter (1 ml = 100 μg N = 122 μg NH$_3$).

Standard Ammonium Solution. Dilute 10.00 ml stock ammonium solution to 1 liter with ammonia-free water (1 ml = 1.0 μg N). Prepare fresh daily.

Procedure

Transfer a 10-ml aliquot of the sample solution containing 0.5 to 10 μg ammonia nitrogen to large test tubes. Add 5.00 ml reagent solution I, cover test tube with polyethylene stoppers, and shake vigorously for 30 sec. Add 5.00 ml reagent solution II. Stopper test tubes and shake vigorously for 30 sec. Immerse test tube in water at 37°C for 15 min. Measure the absorbance at 635 nm against a reagent blank solution, using 1-cm cells. Prepare a calibration graph by preparing standard solutions containing 0.05 to 1.0 μg nitrogen per ml and using 10-ml aliquots instead of sample solution.

C. AMMONIA BY BISPYRAZOLONE METHOD

A pyridine–pyrazolone–bispyrazolone reagent was introduced by Kruse and Mellon (129,130) for the spectrophotometric determination of ammonia in the 0.025- to 2.0-ppm range. Cyanide, thiocyanate, and cyanate also form colors with this reagent and can be removed by an anion exchange resin, for example, Amberlite IRA-400. Chloride, sulfate, nitrate, and sulfide ions do not interfere. Silver, copper(II), zinc, and iron(II) interfere. The following ions have a negligible effect at the designated concentrations: 10,000 ppm potassium; 1000 ppm cadmium, calcium, chromium(III), lead, magnesium, and sodium; 500 ppm mercury(II); 100 ppm aluminum and nickel; and 10 ppm cobalt.

A further study of the ammonia–pyridine–pyrazolone reaction by Lear and Mellon (134) indicated that pyridine was not an essential constituent in the reaction as it evidently was not incorporated in the final product. They applied the use of the method to the determination of both soluble and insoluble nitrogen in steels.

In a critical evaluation of the possible mechanism of the reaction, Prochazkova (180) showed that (a) the main reactants were dichloramine, produced by the reaction of chloramine-T on ammonia, and bispyrazolone; (b) the main product of this reaction was rubazoic acid which decomposes in basic solution to give yellow products that are ultimately extracted; (c) pyridine is not an essential component of the reagent and can be replaced by sodium carbonate; (d) pyrazolone functions to decolorize pyrazolone blue, a side reaction product; and (e) the pyrazolone should be added after the main reaction has taken place. It was also observed that the ammonia–chloramine-T reaction takes place in a buffered solution, pH less than 6.5, and the bispyrazolone–dichloramine reaction takes place at pH 10 ± 0.1. Using this procedure, only iron(II) and reducing agents such as sulfite interfere; cyanide, thiocyanate, zinc, copper(II), and silver do not interfere as was observed with the use of the mixed pyridine–pyrazolone–bispyrazolone reagent of Kruse and Mellon.

Procedures using the Kruse-Mellon reagent, Procedure C1, and the Prochazkova reagents, Procedure C2, are given below.

Reagents for Procedure C1

Aqueous Pyrazolone Solution. Dissolve 0.63 g recrystallized 3-methyl-1-phenyl-5-pyrazolone in 250 ml hot water (75°C) and allow the solution to cool to room temperature. Add five parts of this solution to one part pyridine containing 0.1% bis(3-methyl-1-phenyl-5-pyrazolone), also called 3,3'-dimethyl-1,1'-diphenyl(4,4'-bi-2-pyrazoline)-5,5'-dione, to form the pyridine—pyrazolone reagent. Prepare the pyridine—bispyrazolone solution just prior to the time it is mixed with the aqueous pyrazolone solution. Prepare the final reagent shortly before use.

pH 3.7 Buffer Solution. To 45 ml 10% sodium acetate solution, add sufficient glacial acetic acid to give a pH of 3.7.

Procedure C1

Adjust the pH of a clear 50-ml sample containing 2.5 to 100 μg ammonium nitrogen to about 3.7 and add 10 ml of the pH 3.7 buffer solution. Transfer the sample to a 125-ml separatory funnel, add 0.9 ml 3% chloramine-T, and shake the funnel. After 90 sec, add 30 ml of the pyridine—pyrazolone reagent and mix repeatedly. After about 60 sec for color development, extract the faintly purple solution with 25 ml (if the ammonia concentration is less than 0.1 mg per liter) or 50 ml carbon tetrachloride. The water droplets are removed by filtering the yellow extract through a cotton plug into a volumetric flask. Measure the absorbance of this extract at 450 nm.

Since conformity to Beer's law is not exact, a calibration graph must be prepared at the same time. The carbon tetrachloride extracts often turn cloudy on standing 15 to 30 min, even after filtration through cotton or filter paper. In a prolonged series of determinations, the clouding tendency can be overcome by employing carbon tetrachloride for nulling the instrument. A deduction can then be entered for the color present in the reagent blank.

Reagents for Procedure C2

Chloramine-T Solution. Dissolve 1.0 g chloramine-T in 100 ml ammonia-free water. This reagent is stable for two weeks.

Buffer Solution. Mix 0.1M aqueous citric acid and 0.2M disodium hydrogen phosphate in a volume ratio of 7.91:12.09. Stabilize this solution by adding several drops of a 1:1 toluene:carbon tetrachloride solution.

Bispyrazolone Solution. Dissolve 0.2 g bispyrazolone in 40.0 ml 0.5N sodium carbonate solution at 90°C. Cool and dilute to 100 ml with 0.5N sodium carbonate solution. Prepare fresh daily.

Pyrazolone Solution. Dissolve 0.25 g pyrazolone in 100 ml hot ammonia-free, distilled water.

Hydrochloric Acid, 0.5M.

Procedure C2

Transfer 50 ml of a sample solution, adjusted to pH 6 to 7 and containing 0.5 to 25 μg ammonium nitrogen, to a 125-ml separatory funnel. Add 5 ml of the buffer solution and 2 ml of the 1% chloramine-T solution. Allow to stand for precisely 5 min at 15°C. Add rapidly 6 ml of the bispyrazolone solution and mix thoroughly. After 5 min, add 10 ml of the pyrazolone solution. Add 2 ml 0.5M hydrochloric acid to the decolorized solution. Extract for 3 min with 10 ml carbon tetrachloride. Measure the absorbance at 450 nm in 1-cm absorption cells against reagent blank solutions.

Prepare a calibration graph by using standard solutions containing 1, 5, 10, 15, 20, and 25 μg ammonium nitrogen per 50 ml sample.

D. AMMONIA BY INDIRECT ULTRAVIOLET SPECTROPHOTOMETRIC METHOD

Ammonia is oxidized by hypobromite in basic solution, pH 11.2 to 12.5, to nitrogen, and the decrease in the differential absorbance at 330 nm of the sample relative to that of a reference standard is proportional to the ammonia concentration (107):

$$2NH_3 + 3BrO^- \longrightarrow N_2 + 3Br^- + 3H_2O$$

The optimum concentration is 3 to 23 ppm ammonia. Up to 20 ppm nitrite and 50 ppm nitrate can be tolerated. Iron(III) and many divalent metal ions that form insoluble hydroxides or phosphate salts interfere, as does iodide. Most interferences can be eliminated by a Kjeldahl distillation.

Reagents

Solution A. Prepare a 0.025 to 0.030M sodium hypochlorite solution by taking 23 to 28 ml of a 5.25% commercial bleach, adjust to pH 11.2 to 11.4 with sodium hydroxide, and dilute to 1 liter with distilled water.

Solution B. Prepare a solution that is 2% potassium bromide and 10% dipotassium hydrogen phosphate by dissolving 20 g potassium bromide and 100 g dipotassium hydrogen phosphate in distilled water and diluting to 1 liter.

Solution C. Prepare this reagent shortly before it is needed by thoroughly mixing solutions A and B in a ratio of approximately 2:1 by volume and waiting 3 min for the completion of the reaction.

Solution D. Prepare a 10% tripotassium phosphate solution by dissolving 100 g $K_3PO_4 \cdot XH_2O$ in distilled water and dilute to 1 liter.

Procedure

Transfer an aliquot of ammonia solution containing from 0.05 to 3.0 mg ammonia to a 100-ml volumetric flask. Pipet 15.00 ml of solution C into the flask containing the ammonia solution and 15.00 ml into a volumetric flask containing a volume of ammonia-free distilled water equivalent to the volume of the ammonia solution taken. Shake thoroughly intermittently, and after 3 min, add approximately 10 ml solution D to each flask, mix, and dilute to volume.

Measure the absorbance at 330 nm using the reference standard containing the ammonia-free, distilled water in the sample cell and the sample in the reference cell. Obtain the concentration from a linear calibration graph prepared by using standard ammonia solutions.

E. NITRITE BY MODIFIED GRIESS METHOD

In an acidic solution, nitrite reacts with a primary aromatic amine to produce a diazonium salt. This diazonium salt then couples with an aromatic amine or phenol to form a colored azo dye. The reaction is highly selective and is sufficiently sensitive to detect 1 μg nitrite nitrogen in 1 liter solution. In the original Griess reaction (87), sulfanilic acid was diazotized and then coupled with 1-naphthylamine[1] in sulfuric acid solution to give the characteristic azo color. The following equation represents these two fundamental reactions:

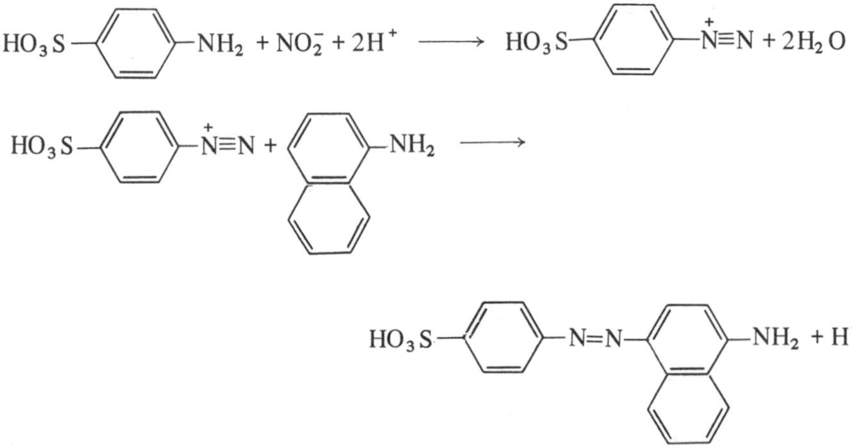

Warington (226) modified the Griess reaction by developing the color in a hydrochloric acid medium. Ilosvay (109) cited the advantages of performing the reactions in acetic acid solution. In a comparative study of the Warington and

[1] 1-Naphthylamine has been classified as a carcinogon (207). Care must be exercised in using this chemical (53).

Ilosvay modifications, Weston (236) reported the Ilosvay method to be more rapid. He also stated that an excess of acetic acid affected reaction sensitivity to a lesser extent than an excess of hydrochloric acid. Weston further suggested that the reagents be made more concentrated for ease in handling, a recommendation which was accepted. The so-called Griess-Ilosvay procedure, as modified by Weston, prevailed for many years.

A reexamination of the pertinent diazotization and coupling reactions, conducted by Rider and Mellon (184) with modern instrumentation, disclosed that four requirements govern the success of the nitrite determination: (a) diazotization should be carried out in strongly acid solution, (b) diazotization should be conducted in as cool a solution as is practicable, (c) coupling should be attempted only after diazoatization is complete, and (d) coupling should be carried out in as low an acidity as is consistent with colorimetric stability. According to Rider and Mellon, the necessary conditions can be met by preparing the sulfanilic acid and 1-napthylamine reagents in hydrochloric acid solution. The proper adjustment of pH for coupling can be achieved with a sodium acetate solution.

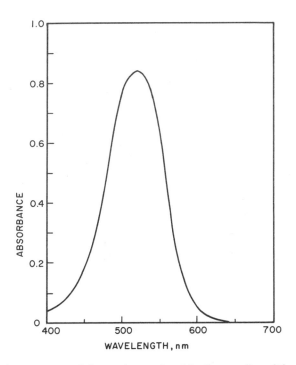

Fig. 5. Absorption spectrum of the azo dye produced by the coupling of the diazonium salt of 4-aminobenzenesulfonic acid and 1-aminonaphthalene (1 ppm NO_2^-, 1-cm absorption cells).

The azo color formed by the Rider and Mellon reagents shows maximum absorption near 520 nm (Fig. 5). Conformity to Beer's law prevails up to 150 μg per liter nitrogen, or 500 μg per liter as nitrite, in a 1-cm light path. Above 300 μg per liter nitrogen, a precipitate forms upon the addition of the 1-napthylamine reagent.

The following ions do not interfere in the Rider-Mellon procedure when present in concentrations 1000 times (400 mg per liter) that of the nitrite (0.4 mg per liter): barium, beryllium, calcium, lead(II), lithium, magnesium, manganese(II), nickel(II), potassium, sodium, strontium, thorium, uranyl, zinc, arsenate, benzoate, borate, bromide, chloride, citrate, fluoride, formate, iodate, lactate, molybdate, nitrate, oxalate, phosphate, pyrophosphate, salicylate, selenate, sulfate, tartrate, tetraborate, and thiocyanate.

Rechecking the Griess-Ilosvay-Weston reagents, Barnes and Folkard (32) found them to be adequate if the diazotization time (3 to 10 min) and coupling conditions of Rider and Mellon are adopted. A coupling time of 30 min is required for full color development. These reagents are prepared by dissolving 0.8 g sulfanilic acid in 100 ml 5N acetic acid and 0.5 g 1-naphthylamine in 100 ml 5N acetic acid. One milliliter of each reagent is used per 50 ml sample volume.

Interference. Interference in the nitrite determination can originate from such sources as amines, oxidizing agents, reducing agents, complex formers, precipitants, agents that disturb the acid—base balance, and colored substances.

Nitrogen compounds such as urea and aliphatic amines can react with nitrite, liberating gaseous nitrogen. Although small amounts of ammonium ion can be tolerated, high concentrations should be avoided. Trichloramine (nitrogen trichloride) produces a false red color with the reagents. When trichloramine is suspected, a check for a free available chlorine and trichloramine residual is advisable by the o-tolidine—arsenite procedure (2). The finding of a free chlorine residual would rule out the presence of nitrite because chemical incompatibility makes the simultaneous existence in a sample of nitrite, free available chlorine, and trichloramine improbable.

Strong oxidants such as permanganate, chlorate, trisulfatocerate, perchlorate, periodate, peroxydisulfate, and tungstate must be absent. On the other hand, 80 mg per liter dichromate can be tolerated. Reductants such as iodide, iron(II), chlorostannite, sulfide, thiosulfate, and sulfite must be absent by virtue of reactivity with nitrite.

The hydrochloric acid content of the reagents can precipitate mercury(I) and silver as chlorides and bismuth(III) and antimony(III) presumably as the oxychlorides. Lead(II) also precipitates as the chloride in concentrated solution but redissolves at room temperature upon dilution, thereby causing no interference. Chloroplatinate, iron(III), gold(II), and metavanadate ions form precipitates with the 1-naphthylamine reagent and must be absent.

Among the agents that can reduce the acidity of the system are the alkali salts of such ions as carbonate, acetate, cyanide, and silicate. The cyanide concentration must be limited to 100 mg per liter, whereas the others can be tolerated in amounts up to 200 mg per liter. The strongly acid solution of chlorostannate must be restricted to 40 mg per liter.

Colored substances that alter the hue of the system must be absent for visual comparison. Permissible concentrations of chromium(III) and cobalt(II) are 40 and 100 mg per liter, respectively, when photometric methods are employed.

Mercury(II) and copper(II) ions should be absent. The first causes high results, whereas the second catalyzes the decomposition of the diazonium salt yielding low results.

Regents

Nitrite-Free Water. Prepare nitrite-free water by either of the following two methods. (a) Add one crystal each of potassium permanganate and barium hydroxide to 1 liter distilled water and redistill in an all-Pyrex glass apparatus (192). Discard the initial 50 ml distillate. Collect the fraction of the distillate which is free of permanganate. Test for permanganate with o-tolidine (192). or any other sensitive reagent. (b) Add 1 ml concentrated sulfuric acid and 0.2 ml 48% manganese(II) sulfate tetrahydrate solution to 1 liter distilled water. Add 1–3 ml 0.04% potassium permanganate solution in order to impart a permanent pink color. Decolorize carefully with 0.09% ammonium oxalate monohydrate solution after 15 min of standing (15).

Sulfanilic Acid Reagent. Dissolve 0.60 g sulfanilic acid (also called 4-aminobenzenesulfonic acid) in 70 ml hot distilled water, cool the solution, add 20 ml concentrated hydrochloric acid, dilute to 100 ml with distilled water, and mix thoroughly.

1-Naphthylamine Hydrochloride Reagent. Dissolve 0.60 g 1-naphthylamine hydrochloride (also called α-naphthylamine hydrochloride or 1-aminonaphthalene hydrochloride), white in color, in distilled water to which 1.0 ml concentrated hydrochloric acid has been added. Dilute to 100 ml with distilled water and mix thoroughly. The reagent discolors and may precipitate on standing after one week in warm surroundings. Discard when the discoloration affects the sensitivity or reproducibility. Storage in a refrigerator extends the life of the reagent to a month or more. If necessary, filter before using.

Sodium Acetate Buffer Solution, 2M. Dissolve 16.4 g anhydrous sodium acetate in distilled water and dilute to 100 ml with distilled water. Filter if necessary.

Stock Sodium Nitrite Solution. Assay the sodium nitrite by pipetting 50.00 ml standard $0.1N$ potassium permanganate, 5 ml 1:5 sulfuric acid, and 25.00 ml approximately $0.1N$ sodium nitrite solution (3.45 g per liter) into a glass-stoppered flask. Shake the stoppered flask frequently and add 2 g potassium iodide after 15 min of reaction time. Titrate with standard $0.1N$ sodium thiosulfate, using starch indicator solution to identify the endpoint (126). Adjust the standardized $0.1N$ sodium nitrite solution so that it contains 0.246 g sodium nitrite in 1.0 liter nitrite-free, distilled water.

Standard Sodium Nitrite Solution. Dilute 10.00 ml of the stock sodium nitrite solution to 1.0 liter with nitrite-free, distilled water (1.0 ml = 0.5 μg nitrogen). Preserve the solution by the addition of 1 ml chloroform and store in a sterilized bottle.

Aluminum Hydroxide Suspension (12). Dissolve 125 g potassium or ammonium alum, $K_2 Al_2 (SO_4)_4 \cdot 24H_2O$ or $(NH_4)_2 Al_2 (SO_4)_4 \cdot 24H_2O$, in 1 liter distilled water. Warm the solution to $60°C$ and slowly add 55 ml concentrated ammonium hydroxide with stirring. Allow the precipitate to settle for 1 hr, then transfer the mixture to a large bottle. Wash the precipitate by the repeated addition, agitation, and decantation of distilled water until the rinse water remains free from ammonia, chloride, nitrite, and nitrate. When freshly prepared, the aluminum hydroxide floc occupies a volume of about 1 liter. Store the liter of suspension in a Pyrex container and shake well before dispensing.

Procedure

When color and suspended solids are present, add 2 ml aluminum hydroxide suspension to 100 ml sample. Stir thoroughly and allow to stand for a few minutes. Filter and discard the first 25 ml of the filtrate. Neutralize a 50.0-ml water sample to pH 6.5–7.5.

Add 1.0 ml sulfanilic acid reagent to 50 ml of the clear, neutralized sample, or an aliquot containing less than 7 μg nitrite nitrogen that has been diluted to 50 ml. Mix thoroughly and let stand 3 to 10 min. Check the pH of the solution electrometrically to see that it is 1.4. Add 1.0 ml 1-naphthylamine hydrochloride reagent and 1.0 ml sodium acetate buffer solution. Mix well. At this stage, the pH of the solution should be 2.0–2.5. Measure the reddish-purple color that develops after 10 to 30 min. Make absorbance measurements at a wavelength of 520 nm and in a 5-cm light path in the nitrogen range of 5–50 μg per liter. With shorter light paths, complete the measurements in proportionate nitrogen ranges up to a maximum of 150 μg per liter beyond which Beer's law fails to apply. Make readings against a reagent blank or distilled water. Run parallel checks with known nitrite standards, preferably in the nitrogen range of the samples. Construct a calibration graph after the preparation of each new batch of reagents.

F. NITRITE BY DIAZOTIZED 4-AMINOBENZENESULFONIC ACID METHOD

Nitrite concentrations in the nitrogen range up to 1.0 mg per liter can be determined without dilution by the ultraviolet spectrophotometric measurement of diazotized sulfanilic acid (also called 4-aminobenzenesulfonic acid) at pH 1.4 (173). The diazotized solutions obey Beer's law from 3 to 50 μg nitrogen per 50 ml sample volume when absorbance measurements are made at 270 nm with a 1-cm light path.

The problem of nitrate interference can be overcome by adding a known amount of nitrate to the reference solution. The inclusion of 5 mg nitrate nitrogen in 50 ml reference solution permits the successful determination of nitrite in samples containing from 0.3 to 10 mg nitrate nitrogen.

Dichromate, molybdate, periodate, permanganate, thiosulfate, tungstate, vanadate, gold(III), iron(III), chloroplatinate, chlorate, and sulfite should be absent. The sample should also be relatively free of organic matter, which may absorb in the ultraviolet region. Iodide, copper(II), uranyl, cyanide, and iron(II) ions can be individually tolerated at concentrations up to 20 mg per liter. The maximum permissible concentration of mercury(II) ion is 50 mg per liter, whereas that of nitrate and oxalate is 60 mg per liter. The maximum for iodate and chromium(III) ions is 100 mg per liter. Selenate up to 200 mg per liter and carbonate, fluoride, persulfate, and pyrophosphate up to 300 mg per liter can be tolerated. Precipitates are formed with mercury(II), bismuth(III), and silver ions.

The following ions exhibit no interference: acetate, arsenite, borate, bromide, chloride, citrate, formate, phosphate, silicate, sulfate, tartrate, tetraborate, thiocyanate, aluminum, ammonium, barium, cadmium, calcium, cobalt(II), lithium, magnesium, manganese(II), nickel(II), potassium, sodium, strontium, thorium, and zinc.

Procedure

A sample containing from 3 to 50 μg nitrite nitrogen is placed in a 50-ml volumetric flask and the pH is adjusted to 1.4. For a previously neutralized unbuffered sample, 1 ml hydrochloric acid solution (20 ml of the concentrated acid diluted to 100 ml) is sufficient. After the addition of 1.0 ml sulfanilic acid reagent (0.60 g per 100 ml), the sample is diluted to volume and the contents are mixed. In the interval between 3 and 15 min, the absorbance of the diazo compound is measured at 270 nm using 1-cm quartz cells. The reading is made against a reagent blank.

G. NITRITE BY ANTIPYRINE METHOD

Nitrous acid reacts with antipyrine (2,3-dimethyl-1-phenyl-3-pyrazolin-5-one) to form 4-nitrosoantipyrine (176,229), which has a molar absorptivity of 7.61×10^3 liter mole^{-1} cm^{-1} at 343 nm. Iron(III), chromium(VI), manganese(VII), sulfite, and thiosulfate interfere. Copper(II), manganese(II), mercury(II), magnesium, nickel, calcium, bromide, chloride, nitrate, oxalate, phosphate, and perchlorate do not interfere. The method is suitable for the determination of 0.1 to 10 ppm nitrite.

Reagents

Standard Nitrite Solution. Dissolve approximately 6.9 g analytic reagent-grade sodium nitrite in distilled water and dilute to 1 liter. Standardize this stock solution by a titrimetric method using a standard $0.1N$ permanganate solution as titrant and a microburet. Use a microburet to measure the appropriate volume of the stock nitrite solution required to prepare a solution containing 25 mg nitrite per liter standard solution.

Antipyrine Reagent Solution. Transfer exactly 5.5 ml concentrated sulfuric acid (sp. gr. 1.84; 98% H_2SO_4) to approximately 700 ml distilled water in a 1-liter volumetric flask. Dissolve 18.823 g dry, recrystallized antipyrine in the acidic solution and dilute to volume. This reagent is $0.1N$ in sulfuric acid and $0.1M$ in antipyrine.

Procedure

Transfer 10 ml of a sample solution containing 0.25 to 30 μg nitrite per ml to a 25-ml volumetric flask. Add 10.0 ml of the sulfuric acid–antipyrine reagent solution and dilute to volume with distilled water. Measure the absorbance at 343 nm in 1.000-cm silica cells using a reagent blank solution in the reference cell. Refer the absorbance readings to a calibration graph prepared using standard solutions.

H. NITRATE BY PHENOLDISULFONIC ACID METHOD

A sulfonated phenol reagent was first proposed for the determination of nitrate by Sprengel in 1863 (230). The method was refined by Grandval and Lajoux in 1885 to a point which is recognizable in current practice (84). Chamot and his co-workers identified the nitration product as the alkaline salt of 6-nitro-1,2,4-phenoldisulfonic acid and gave detailed directions for the preparation of a reagent that consists essentially of 1,2,4-phenoldisulfonic acid (50).

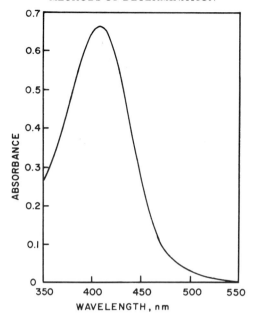

Fig. 6. Absorption spectrum of 1 ppm nitrate nitrogen, phenol–disulfonic acid method (1-cm absorption cells).

Despite its avowed shortcomings in individual situations, the phenoldisulfonic acid reagent can be used to analyze many samples with acceptable precision and accuracy. The high sensitivity of the method is appealing to many analysts. As low as 1 μg nitrate nitrogen yields a faint yellow color upon treatment with phenoldisulfonic acid reagent.

The nitrate–phenoldisulfonic acid color system obeys Beer's law over a broad wavelength range. The greatest sensitivity and accuracy prevail when measurements are made at a wavelength of 410 nm, the point of maximum absorbance (Fig. 6). A 1-cm light path permits measurements of nitrate nitrogen up to approximately 0.2 mg in a 100-ml volume, or 2 mg per liter. A 5-cm light path allows estimations in the nitrogen range of 0.05–0.4 mg per liter. Conformity to Beer's law extends over a concentration interval up to 1.2 mg nitrogen, or 120 mg per liter, when readings are made with a 1-cm light path at 480 nm. At 500 nm Beer's law applies over the nitrogen range up to 70 mg per liter. There is a definite loss in accuracy and sensitivity when measurements are made at these higher wavelengths.

Except for the colored ions, the chief interferences include nitrite, chloride, and organic matter. The separations of these three common interferences from the sample is essential for a successful nitrate determination.

Interference

Organic Materials. Organic matter can interfere in various ways with the nitrate determination, mainly through the off-colors imparted by the materials in the original sample. Evaporation on an electric hot plate or over a gas flame often accentuates charring the sample residue. Additional decomposition of organic substances can be caused by the strongly acidic phenoldisulfonic acid reagent.

Some of the undesirable suspended material can be eliminated during the course of the chloride precipitation with silver sulfate. Sample clarification can also be achieved by the production of a zinc hydroxide floc, a procedure sometimes used in the direct Nesslerization procedure (7). Occasionally, the color contributed by the organic matter can be removed, or reduced in intensity, by treatment with aluminum hydroxide suspension. The preparation of this suspension is described under "Reagents" in Section IIE of this chapter. One or more milliliters of the aluminum hydroxide suspension is used for every 50 ml of sample.

The clarification and decolorization of highly colored samples such as sewage may require the supplementary application of 0.5 g activated carbon in addition to the usual 1 ml aluminum hydroxide suspension for each 50 ml of the sample. Following a thorough mix, the sample is filtered, and the first portion of the filtrate is discarded when a filter paper is employed. The activated carbon used for decolorization purposes should be checked experimentally on known nitrate concentrations to make certain that the adsorption losses of nitrogen are negligible. This step is necessitated by the fact that such carbons as Carbex, Darco, and Norite have been reported to effect appreciable nitrate removals (27,186).

A decolorization method circumventing the use of activated carbon involves the hydrogen peroxide oxidation of the colored organic materials (48,113).

Nitrite also produces a yellow color with phenoldisulfonic acid. Nitrite nitrogen levels in excess of 0.2 mg per liter erratically increase the apparent nitrate concentration. Removal procedures are unnecessary when the nitrite nitrogen concentration falls below 0.2 mg per liter (214).

Two approaches are available for overcoming the nitrite effects. The first attack relies on the reduction of the nitrite to nitrogen by means of such reagents as ammonium compounds (158), urea (231), thiourea (232), and sulfamic acid (105). The second, and more generally accepted, approach involves the oxidation of the nitrite to nitrate by agents such as hydrogen peroxide (38) and potassium permanganate.

Since hydrogen peroxide dissociates rapidly in alkaline solution, oxidation of nitrite to nitrate may be incomplete under alkaline conditions. Therefore, the conversion of nitrite to nitrate is often carried out in an acid environment. One milliliter $1N$ sulfuric acid is added to 100 ml of sample, followed by the dropwise addition of 3% hydrogen peroxide with stirring. The conversion of nitrite is not always instantaneous, occasionally requiring 15 min for some samples.

Potassium permanganate ($0.1N$) can be substituted for hydrogen peroxide under identical experimental circumstances. Completeness of conversion to nitrate is indicated when the sample retains a faint pink hue for at least 15 min in an acid solution. The proper deduction must be made at the end of the nitrate analysis for the amount of nitrite oxidized by hydrogen peroxide or potassium permanganate.

Chloride. Nitrate and chloride are often encountered simultaneously in many laboratory samples. Under the proper acid conditions, the two ions may react with the evolution of volatile oxides of nitrogen. Nitrate losses may range from 12% for a chloride concentration of 5 mg per liter to as much as 30% in the presence of 25 mg per liter chloride (213). Because of this, it is important that the chloride content of a sample be ascertained and the chloride removed before the nitrate determination is undertaken.

Samples vary in the ease with which the chloride can be precipitated. Some samples can be handled by merely adding the equivalent amount of silver sulfate solution (4.40 g per liter, 1 ml = 1 mg chloride), followed by immediate centrifugation. Should centrifugation fail to remove the silver chloride completely, as evidenced by a continuing opalescence in the supernatant, the sample can be given a final clarification by one or more passages through a sintered-glass filter of fine porosity.

Where speed is not critical, and the sample is free of contamination from microorganisms which may seriously disturb the nitrogen balance, substantially complete chloride precipitation can be achieved by adding the equivalent amount of silver sulfate and allowing the precipitate to settle for several hours, preferably overnight, in the dark. The supernatant is then centrifuged or repeatedly poured through a sintered-glass filter of fine porosity to remove any suspended silver chloride. The removal of silver chloride in this way leaves a negligible amount of silver ion in solution to interfere later in the comparison of colors. Evaporation of the treated aliquot on a steam bath will minimize any deleterious browning effects which unprecipitated silver ion may impart to the final yellow color. Such browning is more likely to occur when a sample is evaporated at the uneven and higher heat of an electric hot plate or a gas burner.

In the event that a filter paper is substituted for a sintered-glass filter, a paper free of nitrate or interfering ions should be selected. A known nitrate blank carried through all steps of the procedure will disclose the presence of significant supplemental nitrate concentrations or the presence of interfering ions. A good policy with regard to filter paper is to discard the initial one fourth of the filtrate and conserve the remainder for use in the determination.

The chloride precipitation should be performed with care because some investigators have attributed nitrate losses to the interaction of excess silver ion and organic matter in the sample. The removal of all but 0.1 mg chloride in a 100-ml sample is advised (49,51).

Sodium or potassium hydroxides are interdicted as the color-developing agents in those instances where chloride precipitation must be practiced. No matter how effective the chloride removal, a bare trace of silver ion remains in the sample throughout the procedure. The addition of alkalies other than ammonium hydroxide causes a final brownish off-color due, presumably, to the formation of peptized silver hydroxide. The annoying off-color can only be avoided with ammonium hydroxide. An incidental advantage of ammonium hydroxide is that filtration often becomes unnecessary in the case of samples low in magnesium and iron.

Reagents

Phenoldisulfonic Acid Reagent (49,50). For best results, the phenoldisulfonic acid reagent should be prepared under carefully standardized conditions in order to minimize the concentration of monosulfonic acids. An improperly prepared reagent, especially one containing monosulfonic acids, can change with age, affecting color reproducibility with the passage of time. A dependable reagent will contain little monosulfonic acid or trisulfonic acids. Although sulfonation is usually complete in about 1 hr, a 2-hr heating interval assures the conversion of monosulfonic acids to the disulfonic form. A reliable reagent can only be prepared from phenol crystals free of brown discoloration.

Dissolve 25 g reagent-grade, white phenol crystals in 150 ml concentrated sulfuric acid. Add 75 ml fuming sulfuric acid containing 13 to 15% sulfur trioxide. Mix thoroughly and heat for 2 hr over a boiling water bath or steam bath. The reagent can be prepared directly in the glass-stoppered bottle in which it will be stored. An unchanging and very faint brownish tint usually characterizes an acceptable reagent.

Standard Nitrate Solution. Dissolve 0.722 g anhydrous potassium nitrate and dilute with distilled water to 100 ml (1.0 mg nitrogen or 4.43 mg nitrate per ml). Prepare a standard solution containing 0.01 mg nitrogen per ml by diluting 10.00 ml of the stock solution to 1.0 liter.

Concentrated Ammonium Hydroxide.

Procedure

Remove the nitrite, chloride, and organic matter as indicated in the pertinent paragraphs under "Interference." Transfer the clear and neutralized (pH 7–8.5) sample containing 0.02–0.15 mg nitrate nitrogen to a porcelain casserole or beaker and evaporate to dryness over a boiling water bath or steam bath. Evaporation of the sample over a steam or boiling water bath leaves a residue which is nonvitreous, uncharred, and easier to dissolve in the next step. Add 2.0 ml phenoldisulfonic acid reagent and rub the residue with a glass rod to ensure solution of all solids. If necessary, as a last resort, heat gently on the hot water bath a short time to dissolve a resistant residue, but avoid charring the residue. Dilute with 20 ml distilled water and add sufficient concentrated ammonium hydroxide (approximately 7 ml) to develop maximum color. Remove any suspended matter by passing the colored solution through filter paper into a 100-ml Nessler tube or volumetric flask. Wash the casserole or beaker and filter paper with successive portions of distilled water, dilute to volume, and mix thoroughly. Measure the transmittance in a photometer at 410 nm (blue color filter) using a 5-cm light path for a nitrate nitrogen range of 0.005–0.04 mg and a 1-cm light path for nitrogen levels up to 0.2 mg. Prepare the photometric calibration curve from known amounts of nitrate handled in the same manner as the unknown. Correct the results with a blank carried through all the steps of the procedure.

I. NITRATE BY BRUCINE METHOD

Nitrate reacts with brucine in a sulfuric acid medium, forming a red color which changes rapidly to yellow (66,70,86,112,191,240,241). The yellow color, thought to be an oxidation product, absorbs strongly in the wavelength region of 400–410 nm. Acceptable color development takes place when two volumes of concentrated sulfuric acid are present for each volume of aqueous nitrate solution. The method is sensitive to about 0.1 mg per liter nitrate nitrogen, with the best results prevailing in the range from 1 to 4 mg per liter, where the plot of absorbance versus concentration approaches a straight line.

Two serious shortcomings characterize the brucine color system. One is the failure to conform to Beer's law in the wavelength region of maximum absorbance. The second is the variability that marks color development from one occasion to another. On this account, a series of nitrate standards must always be carried through the entire procedure for exact work.

Interference

A photometric study conducted by Noll (167) on interferences in the brucine–nitrate reaction revealed that nitrite must be absent. The following agents and concentrations can be tolerated in the brucine method: 20 mg per liter iron(II) or iron(III), 50 mg per liter each of ammonium nitrogen and sodium metaphosphate, 100 mg per liter quebracho tannin, 200 mg per liter each of orthophosphate and silica added as sodium silicate), 250 mg per liter each of calcium and magnesium (as calcium carbonate), 1000 mg per liter chloride, 2000 mg per liter total alkalinity (as sodium hydroxide), and 800 units of color.

Since nitrite yields a yellow color similar to that of nitrate, the following procedure has been suggested by Wolf (241) for estimating the interference due to nitrite. A series of nitrite nitrogen standards can be prepared in the range 0.1 to 1.0 mg per liter and treated with brucine reagent in the same manner as the nitrate standards. The nitrate nitrogen equivalent of the nitrite nitrogen is deducted from the total "apparent" nitrate nitrogen in order to obtain the net nitrate nitrogen.

A chlorine residual induces a positive interference and must be absent. The most certain way of eliminating the interference is by evaporating the sample to dryness and taking up the residue with 10.0 ml distilled water before the addition of the brucine reagent. The removal of high chlorine residuals (greater than 5 mg per liter) by natural or artificial ultraviolet irradiation contributes a positive interference. Such common dechlorinants as sodium thiosulfate, sodium sulfite, and phenol are likely to produce a colloid or a decrease in color intensity if applied in excess. Conversely, an excess of sodium arsenite or ammonium sulfate may cause a positive interference.

Reagents

Brucine Reagent. Dissolve 5 g brucine in 100 ml A.R. grade chloroform. Discard when the reagent discolors to the point of affecting the sensitivity. Handle the alkaloid reagent cautiously in view of the extreme toxicity.

Standard Nitrate Solution. Dissolve 0.722 g potassium nitrate, oven dried for 2 hr or more at $100-105°C$, and dilute to 1.0 liter with distilled water (1.0 ml = 0.1 mg nitrogen). Prepare two standard solutions by diluting 5.00 and 25.00 ml of this stock solution to 250 ml with distilled water. The first standard contains 2 μg nitrogen per ml, whereas the second contains 10 μg per ml.

Procedure

Nitrate Nitrogen Range of 0.2–7 mg per liter (90,168). Place two 10.00-ml portions of sample in separate 50-ml Pyrex volumetric flasks. From a 10-ml buret, add 0.20 ml of the brucine reagent to only one of the volumetric flasks. Add 20 ml concentrated sulfuric acid to both flasks, mixing the hot solutions carefully to avoid spattering. Allow the flasks to stand from 5 to 10 min until the color of the brucine-treated sample changes from a reddish cast to yellow. Carefully add distilled water to the 50-ml mark and mix cautiously. Cool quickly to room temperature under a cold water tap or in an ice bath and adjust the volume in both flasks to exactly 50 ml with additional distilled water. Transfer the solutions to cells of a 2- or 1-cm light path. Using the solution from which the brucine was omitted to set the null of the instrument, read the transmittance of the brucine-treated sample in a photometer. Make the reading at a wavelength of 410 nm. Carry 10.0 ml of each of the following nitrate nitrogen standards through the procedure for the preparation of the calibration curve: 0, 2, 6, 10, 20, 40, and 60 μg. In the absence of interference, precision and accuracy are of the order of 0.1 mg per liter nitrogen under the specified conditions.

Nitrate Nitrogen Range of 0.01–0.5 mg per Liter. When interfering substances are absent, accuracy can occasionally be improved by concentrating a low-level nitrate sample. Place two 100-ml portions of the sample in separate evaporating dishes, adjust the pH with sodium or potassium hydroxide to a pH range of 7–8.5, and evaporate to dryness on a hot water bath. Add 10 ml distilled water and rub the residue with a glass rod to dissolve as much of the solid material as possible. To one sample, add 0.20 ml brucine reagent from a buret and mix. To the other sample, add no brucine reagent but continue treatment as indicated hereafter. Add 20 ml concentrated sulfuric acid, mixing the solutions carefully and rubbing any resistant residue with the glass rod. Transfer the solutions to 50-ml Pyrex volumetric flasks. Allow the reaction to proceed for 5 to 10 min. Add distilled water to the mark, mix carefully, and cool quickly to room temperature. On cooling, some water samples yield a heavy turbidity which can be removed by centrifugation or filtration. Transfer to the cells of 2- or 1-cm light path and complete the analysis as described for the nitrogen range of 0.2–7 mg per liter.

Nitrate Nitrogen Range of 0–11 mg per Liter (168). When 5.00 ml sample, 0.20 ml brucine reagent, and 10 ml concentrated sulfuric acid are manipulated in accordance with the directions in the first of the two preceding procedures, nitrogen values up to 11 mg per liter can be measured photometrically at 470 nm with a 1-cm absorption cell.

J. NITRATE BY CHROMOTROPIC ACID METHOD

Chromotropic acid (1,8-dihydroxy-3,6-naphthalenedisulfonic acid) is a sensitive and highly selective reagent in concentrated sulfuric acid medium for nitrogen. This reagent was introduced by West and co-workers (233–235). Two moles nitrate react with one mole chromotropic acid. Iodide, iodate, selenite, selenate, lead, barium, and strontium ions cannot be present. Chromium(III), if present above 20 ppm, interferes. The interference by nitrite and other oxidizing agents can be eliminated by addition of a sulfite–urea solution. The interference caused by chloride is circumvented by the addition of an antimony(III) sulfate solution to complex the chloride. The procedure presented in this section is that recommended by West and Ramachandran (234), although other procedures have been developed (34,37).

Reagents

Standard Nitrate Solution. Dissolve 1.371 g sodium nitrate in distilled water and dilute to 1 liter. Each milliliter is equivalent to 1.00 mg nitrate. Use this solution to prepare standard solutions containing 0.2 to 20 μg nitrate per ml to be used in preparing the calibration graph.

Chromotropic Acid. Dissolve 100 mg purified chromotropic acid (17,234) in analytic reagent-grade concentrated sulfuric acid. Prepare fresh each week.

Sulfite—Urea Solution. Dissolve 5 g urea and 4 g anhydrous sodium sulfite in distilled water and dilute to 100 ml.

Antimony(III) Sulfate Solution. Dissolve 0.5 g antimony metal in 80 ml concentrated sulfuric acid. After heating to effect dissolution, cool the solution and carefully dilute by adding it to 20 ml ice-cold distilled water.

Procedure

Transfer 2.5 ml of a solution containing 0.5 to 50 μg nitrate to a 10-ml volumetric flask. Add one drop of the sulfite—urea solution, mix, and place the flask in a cold-water bath. Add 2 ml of the antimony(III) sulfate solutions and again mix thoroughly. After allowing the solution to stand in the water bath for 4 min, add 1 ml of the chromotropic acid reagent, mix thoroughly, and allow the flask to stand in the water bath for 3 min. Add concentrated sulfuric acid to the graduation mark and mix contents of the flask by inverting four times. Allow to stand at room temperature for 45 min. Adjust to the mark with concentrated acid and mix carefully. After 15 min, measure the absorbance at 410 nm using either distilled water or a reagent blank solution in the reference cell.

Preparation of Calibration Graph. Prepare a series of standard dilute nitrate solutions containing 0.2 to 20 μg nitrate per ml and follow the general procedure using 2.5 ml of each standard nitrate solution.

K. NITRATE BY REDUCTION—DIAZOTIZATION METHOD

Nitrate can be reduced either to nitrite or ammonia with subsequent employment of methods for the determination of nitrite (133,164) or ammonia (44,121). Aluminum (33), Devarda's alloy (24,183), vanadium(II) (163), chromium(II) (94,157), and titanium(III) (242) have been used for the reduction of ammonia. Zinc (54,64,127,159,244), cadmium (72,118,133,154,198), and hydrazine sulfate (155) have been used for the reduction of nitrate to nitrite. The following general procedure involves the reduction of nitrate to nitrite with a cadmium reductor and the determination of the resulting nitrite by the formation and extraction of a diazo orange (118).

Reagents

Standard Stock Nitrate Solution. Transfer 0.6068 g analytic-grade sodium nitrate to a 1-liter volumetric flask and dilute to volume with distilled water.

Standard Nitrate Solution I. Transfer 10.00 ml of the standard stock nitrate solution to a 1-liter volumetric flask and dilute to volume with distilled water (1 ml = 1.00 μg nitrate N).

Standard Nitrate Solution II. Transfer 10.00 ml of the standard stock nitrate solution to a 100-ml volumetric flask and dilute to volume with distilled water (1 ml = 10.0 μg nitrate N).

Buffer Solution. Add 20 ml concentrated hydrochloric acid and 50 ml concentrated ammonium hydroxide to 500 ml distilled water and dilute to 1 liter. The pH should be 9.7.

Methanol–Hydrochloric Acid Solution. Add 8.5 ml concentrated hydrochloric acid to 1 liter methanol.

1-Naphthylamine Solution. Dissolve 1.00 g 1-naphthylamine in 50 ml glacial acetic acid and add 150 ml distilled water. If the solution is colored, add 1 g powdered zinc and shake for 3 min. Filter through a medium-porosity filter paper and collect the filtrate in an amber bottle. This reagent solution should be prepared fresh every two to three days and kept under refrigeration.

Reductor Column. Place four zinc rods in 500 ml of a 20% w/v cadmium sulfate solution and allow the reduction reaction to proceed for about 4 hr. Remove the supernatant solution and scrape the cadmium crystals from the zinc rods. Place the cadmium particles in 500 ml distilled water and mix for a few seconds in a Waring Blender. Wash cadmium with distilled water and remove cadmium particles. Sieve the cadmium powder so that particles which pass the #20 sieve are retained by the #40 sieve. Use a column provided with an internal coarse, sintered-glass plate and a capillary delivery tube. Place a wetted glass wool plug above the sintered-glass plate and fill the column with water. Add the cadmium powder and tap the column gently until a 6- to 8-cm column of cadmium is obtained. Attach the reservoir and capillary delivery tube. Add 25 ml 0.10N hydrochloric acid to the reservoir and then add two 25-ml portions of distilled water. Add 25 ml of a 1:10 dilution of the buffer solution as the final wash of the column.

Procedure

Transfer 20.00 ml of the sample solution containing 1 to 20 μg nitrate nitrogen to a 30-ml beaker. Pipet 0, 1, 2, 5, 10, 15, and 20 ml standard nitrate solution I to 30-ml beakers. Pipet 5, 10, 15, and 20 ml standard nitrate solution II to 30-ml beakers.

Add 5 ml of the buffer solution to each solution. Transfer each solution to the reservoir of the cadmium reductor column. As the reservoir empties, add four successive 20-ml portions of distilled water. Collect each eluate in a 200-ml conical flask. Add 2 ml of the 1-naphthylamine reagent and mix. Place the solution in the dark for 2 hr for color development. Transfer each solution to a 250-ml separatory funnel. Rinse the flask with 5 ml chloroform and add rinsings to the separatory funnel. Shake until most of the color is transferred to the chloroform layer and draw off the lower layer into a 50-ml volumetric flask. Extract with additional small volumes of $CHCl_3$ until the washings are colorless. The volume of chloroform used in extracting the color should not exceed 40 ml. Add 10 ml of the methanol—HCl solution and dilute to volume with chloroform. Measure the absorbance of each solution against a blank solution at 555 nm in 1.000-cm cells. Plot absorbance versus μg nitrate nitrogen (per 50 ml final volume).

L. MISCELLANEOUS METHODS

Space considerations forbid the inclusion of every spectrophotometric method for the determination of ammonia, nitrite, and nitrate. Table I lists literature references that describe additional methods and offers the available essential information on the reagent involved, optimum wavelength, and the applicable concentration range of the pertinent constituent.

TABLE I. Miscellaneous General Methods for the Determination of Nitrogen

Constituent	Method or reagent (wavelength; molar absorptivity; optimum concentration range)	References
Ammonia	conversion to trichloramine by hypochlorite; reaction with cadmium iodide –linear starch reagent (615 nm; 10 –300 ppb)	(250)
	mercury (II) methylthymol blue (610 nm; 5.84×10^3)	(128)
	turbidimetric: molybdophosphoric acid (600 nm; 10– 300 ppm)	(93)
	kinetic-catalytic: effect of ligand substitution: mercury(II)–o-cresolphthalein, trans-1,2-diaminocyclohexane-N,N,N′N′-tetraacetic acid (583 nm; 0–0.055 ppm)	(212)
Nitrate	formation of ion-association complex, nitratobis(2,9-dimethyl-1, 10-phenanthroline) copper(I), (MIBK) (456 nm; 0.05–4 ppm)	(247, 248)
	2,4-xylenol	(31, 143)
	2,6-xylenol	(95, 96, 153)
	phenarazenic acid	(177)
	iron(II) sulfate (525 nm; 25–125 ppm)	(210)
	crystal violet	(249)
	orange I	(148)

4-methylumbelliferone (420 nm; 0.7–70 ppm)   (205)
nitrotoluene (toluene) (284 nm; 1–18 ppm)   (41)
nile blue A (o-dichlorobenzene) (650 nm; 0.2–10 ppm)   (178)
indirect: rhenium, 1-furildioxime   (42)
formation of ion-association complex, tetraphenylarsonium   (47)
   nitrate (269 nm; 6–30 ppm)
ultraviolet absorptivity   (35, 80)
automated: reduction, Griess   (101, 216)

Nitrite   modified Griess methods, p-nitroaniline, azulene (515 nm;   (78)
   5.2 x 10^4)
sulfanilamide, N-(1-naphthyl)ethylenediamine hydrochloride,   (141)
   dbl. extrn. with $CHCl_3$, 0.02N HCl (543 nm; 6.2 x 10^6)
8-aminoquinoline (n-heptanol) (465 nm; 0.01–16 ppm)   (73)
3,3'-diaminobenzidine, toluene-3,4-diamine (350 nm;   (138)
   3.5 x 10^3)
sulfanilamide, N-(1-naphthyl)ethylenediamine,   (165)
   dodecylbenzenesulfonate (CCl_4)
2,3-diaminonaphthalene ($C_2 Cl_4$)   (238)
diaminodiphenylsulfone, diphenylamine (450 nm;   (211)
   1.5 x 10^4; 0.1–2.5 ppm)
free-radical chromogens; 1-methyl-2-quinoloneazine   (196)
   (520 nm; 1.27 x 10^6)
dichlorosulfitomercurate, formaldehyde, p-rosaniline   (144)
   (560 nm; 0.2–10 μg)
indirect: thiourea, iron(III)   (108)
brucine   (191)

III. APPLICATIONS

A. AMMONIA NITROGEN IN WATER BY NESSLER METHOD (3)

Ammonia nitrogen occurs in variable concentrations in many surface and ground waters. When encountered in raw surface waters, it often denotes sanitary pollution. Its presence in ground water is usually the result of natural reduction processes. Ammonia is also sometimes introduced in the chloramine disinfection of water.

The ammonia determination should be performed on fresh water samples. When a prompt analysis is impractical, microbiologic activity should be retarded by storing the sample at a low temperature, preferably just above freezing. The addition of 0.8 ml concentrated sulfuric acid to 1 liter sample may also serve to maintain the nitrogen balance of the water for 24 hr. Since free available chlorine progressively reacts with ammonia and albuminoid nitrogen, the residual chlorine should be removed by carefully adding the exact equivalent of one of the following reducing agents: sodium thiosulfate, sodium sulfite, sodium arsenite, and phenylarsene oxide. The removal of residual chlorine should be accomplished without leaving a residual of reductant in the sample.

The quantitative recovery of free ammonia nitrogen is possible when the sample is treated with a phosphate buffer solution in amounts that will maintain a pH near 7.4 during the distillation process (5). The ammonia nitrogen is finally determined by Nesslerization of a suitable aliquot of the distillate. The distillation method is suitable for the determination of trace amounts of ammonia nitrogen, also where it is desirable to run a subsequent albuminoid nitrogen or organic nitrogen determination, and in those instances where interference of any nature may be suspected or encountered.

Reagents

Phosphate Buffer Solution, pH 7.4. Dissolve 14.3 g potassium dihydrogen phosphate, KH_2PO_4, and 68.8 g dipotassium hydrogen phosphate, K_2HPO_4, in ammonia-free water and dilute to 1 liter.

Boric Acid Solution. Dissolve 20 g boric acid, H_3BO_3, in ammonia-free water and dilute to 1 liter.

Nessler Reagent. Dissolve 100 g mercury(II) iodide and 70 g potassium iodide, KI, in 75–100 ml ammonia-free distilled water. Add this mixture slowly, and with stirring, to a cooled solution of 160 g sodium hydroxide pellets dissolved in 700 ml ammonia-free distilled water. Dilute to 1 liter with ammonia-free distilled water. Allow the precipitate to settle, preferably for several days, before using the yellowish reagent. A small amount of sediment continues to deposit upon standing. The formation of an appreciable quantity of precipitate can affect the reagent's sensitivity and colorimetric reproducibility and, therefore, should be regarded with suspicion. Do not shake the reagent or stir the sediment. Use only the clear supernatant solution for color development.

Procedure

Steam out the entire distillation assembly by boiling water in the flask until the distillate shows no trace of ammonia. Empty the distilling flask and add 500 ml of the neutralized (pH ca. 7) sample or a smaller volume (if the ammonia nitrogen exceeds 2 mg per liter) diluted to 500 ml with ammonia-free water. When the ammonia nitrogen level of a water falls below 0.05 mg per liter, or when an albuminoid nitrogen value is necessary, use sample volumes up to 1.0 liter. Add 10 ml phosphate buffer solution. For most water samples, this volume of phosphate solution is sufficient to maintain a pH of 7.4 ± 0.2 during distillation; if not, add another 10 ml or more of the phosphate buffer. Where the calcium content exceeds 250 mg per liter, first add up to 40 ml phosphate buffer solution per liter sample and then adjust the pH to 7.4 with acid or base (197). Add a few glass beads, and distill at the rate of 6–10 ml per min, collecting the distillate in 50-ml portions or larger, until ammonia free. Nessler tubes or volumetric flasks of 50- or 100-ml size are suitable receivers.

Absorb the ammonia distillate below the surface of 10 ml boric acid solution. Neutralize the boric acid with sodium hydroxide. Add 1 ml Nessler reagent to a 50-ml portion of the distillate or a suitable aliquot diluted to 50 ml with ammonia-free distilled water. Mix thoroughly and measure the absorbance after 10 or 30 min in a 1- or 5-cm cell in the wavelength region of 400–425 nm. Use distilled water in the reference cell. Carry through the entire procedure a blank of ammonia-free distilled water and an ammonium sulfate standard in the same nitrogen range as the unknown, for the purpose of applying the pertinent corrections. When maximum accuracy is desired, prepare the calibration plot with ammonia standards that have been distilled and Nesslerized in the same way as the unknown.

B. ALBUMINOID NITROGEN IN WATER BY NESSLER METHOD

After the free ammonia nitrogen has been expelled by distillation, application of a strongly alkaline potassium permanganate solution to a water sample often induces an additional evolution of ammonia. This supplementary release of ammonia represents the so-called albuminoid nitrogen and results in great measure from the action of boiling alkaline potassium permanganate on the unsubstituted amino groups of many amino acids, polypeptides, and proteins. These latter nitrogenous materials are important constituents of organic pollution in a water supply and frequently exert a significant chlorine demand in treatment plants practicing free residual chlorination. The recovery of unsubstituted amino nitrogen by the albuminoid nitrogen determination is estimated at approximately 80% (214).

The same precautions with respect to sample collection and storage apply to the albuminoid nitrogen as to the ammonia nitrogen determination.

Special Reagent

Alkaline Potassium Permanganate Reagent. Place 16 g potassium permanganate in a 3-liter Pyrex beaker and dissolve in ammonia-free distilled water. Add 238 g sodium hydroxide pellets or 404 g potassium hydroxide pellets and enough ammonia-free distilled water to dilute to 2.5 liters. Concentrate to 2 liters on an electric hot plate. Determine the ammonia nitrogen blank in 50 ml of the reagent and use the results as a basis for correction in subsequent determinations. Store the reagent in a Pyrex bottle.

Procedure

The albuminoid nitrogen determination best follows the ammonia nitrogen determination in sequence. Measure an initial volume of 500 to 1000 ml of the neutralized (pH ca. 7) sample into an 800-ml Kjeldahl flask or a 2-liter distilling flask. Choose a sample size not in excess of 1 mg albuminoid nitrogen. Add 10 ml or more of pH 7.4 phosphate buffer solution and several glass beads, and collect 50–200 ml ammonia nitrogen distillate, depending on the ammonia nitrogen concentration. Next, add 50 ml alkaline potassium permanganate reagent and collect 200 or 250 ml albuminoid nitrogen distillate in a volumetric flask or other suitable receiver. Collect the distillate and Nesslerize as described in the ammonia nitrogen determination (Section IIIA).

C. AMMONIA NITROGEN IN WATER BY INDOPHENOL METHOD

This method is applicable to the determination of ammonia nitrogen in water at the 0.01- to 0.5 μg nitrogen per ml level. If the water sample is colored, turbid, very basic, or very acidic, it is advisable to distill and use the distillate instead of the water sample.

Reagents

Phenate Reagent Solution. Dissolve 2.5 g sodium hydroxide and 10 g phenol in 100 ml ammonia-free water. Prepare this reagent each week as it darkens on standing.

Hypochlorous Acid Solution. Add 10 ml of a 5% commercial bleach solution to 40 ml distilled water. Adjust pH to 6.5–7.0 with hydrochloric acid. Prepare this reagent each week.

Manganese(II) Sulfate Solution. Dissolve 50.0 mg $MnSO_4 \cdot H_2O$ in 100 ml distilled water.

Procedure

Transfer a 10.00-ml sample or aliquot of distillate to a 50-ml beaker. Add 0.05 ml of the manganese(II) sulfate solution. Use a magnetic stirrer to stir the solution vigorously. Add 0.50 ml of the hypochlorous acid solution and add immediately dropwise 0.60 ml of the phenate reagent.

Use a 10.00-ml aliquot of ammonia-free water and prepare a blank solution. Measure the absorbance of the sample solution at 630 nm using the reagent blank solution in the reference cell.

Preparation of Calibration Graph. Transfer 0-, 1-, 5-, 10-, 15-, 20-, and 25-ml aliquots of the standard ammonium solution (1 ml = 0.5 μg N) to 25-ml volumetric flasks and dilute to volume with ammonia-free distilled water. Transfer 10.00-ml aliquots to 50-ml beakers and continue with procedure as just described. Plot absorbance versus μg nitrogen per ml.

D. NITRITE IN WATER BY MODIFIED GRIESS METHOD

The presence of nitrite in a raw water supply usually denotes microbiologic activity. A passing stage in oxidation or reduction processes, nitrite seldom occurs in more than trace amounts. When detected in treated water, at the plant or in the distribution system, nitrite may result from the bacterial conversion of ammonium nitrogen which is artificially fed in the chloramine disinfection of the supply.

This modified Griess method is applicable to waters containing no more than 0.5 μg nitrite per ml. The main interferences were discussed previously in respect to the Rider-Mellon modifications (14,19,184).

Reagents

1-Naphthylamine Hydrochloride Solution. Dissolve 0.60 g 1-aminonaphthalene hydrochloride in about 75 ml nitrite-free water containing 1 ml hydrochloric acid in a 100-ml volumetric flask. Dilute to 100 ml.

Sulfanilic Acid Solution. Dissolve 0.60 g 4-aminobenzenesulfonic acid in 70 ml concentrated hydrochloric acid, dilute to 100 ml with water, and mix.

Acetic Acid Solution. Add one volume glacial acetic acid to three volumes water and mix.

Standard Stock Solution of Sodium Nitrite. Dry about 6.2 g sodium nitrite, $NaNO_2$, in a 125-ml tared g.s. conical flask for about 24 hr, cool in a desiccator with anhydrous magnesium perchlorate as desiccant (Dehydrite or Anhydrone) and dry until constant weight is obtained. Adjust the weight of the dried sodium nitrite in the flask to 6.000 g. Add 50 ml nitrite-free water and mix until dissolution is complete. Transfer to a 1-liter volumetric flask and dilute to volume with nitrite-free water. Dry a 1-liter storage bottle for 1 hr in a 175°C oven. Transfer a 50.00-ml aliquot of the stock solution to a 1-liter volumetric flask, dilute to mark with nitrite-free water, and mix thoroughly. Transfer this solution to the dry, sterile storage bottle and add 1 ml chloroform. This solution contains 0.2000 mg of NO_2^- per ml.

Standard Nitrite Solution. Transfer 100 ml of the standard stock solution of sodium nitrite to a 1-liter volumetric flask and dilute to volume with nitrite-free water. Transfer a 50-ml aliquot of this intermediate standard solution to a 1-liter volumetric flask and dilute to volume with nitrite-free water. Transfer this solution to another dry, sterile storage bottle and add 1 ml chloroform. This standard solution contains 1.00 μg NO_2^- per ml and should be prepared fresh each day from the stock solution.

EDTA Solution. Dissolve 0.5 g disodium ethylenediaminetetraacetate dihydrate in water and dilute to 100 ml.

Sodium Acetate Solution. Dissolve 27.5 g sodium acetate trihydrate in water, dilute to 100 ml, and filter if necessary.

Aluminum Hydroxide Gel. Dissolve 125 g potassium aluminum sulfate, $KAl(SO_4)_2 \cdot 12H_2O$, in 1 liter water. Add 55 ml ammonium hydroxide slowly with stirring. Wash the precipitate by decantation with water until the supernatant solution is chloride and ammonia free. (Use phenolphthalein to test for basicity and the nitric acid—silver nitrate test for chloride.) The hydrous aluminum oxide gel is used to remove any color from the water sample.

Procedure

Transfer a 100-ml aliquot of water sample to a 150-ml beaker. (If the water has an appreciable color, 500 ml of the sample should be shaken with 3 ml of the hydrous aluminum oxide gel and the supernatant filtered to give a clear sample. The pH of the water sample should not be above 8.) Add 2.0 ml EDTA solution and 2.0 ml of the sulfanilic acid reagent solution. Mix thoroughly and allow solution to stand for 5—10 min. Add 2.0 ml of the 1-naphthylamine hydrochloride reagent, mix, add 2.0 ml of the sodium acetate solution, and mix thoroughly. The pH of the solution should be checked that it is in the 2—2.5 range. After 20 min, measure the absorbance at 520 nm using 1.000-cm cells with water in the reference cell. If the nitrite concentration is less than 0.05 ppm, the use of 2- or 5-cm absorption cells is recommended.

Preparation of Calibration Graph. Use a buret to transfer 0, 1, 3, 5, 7, 10, 20, 30, 40, and 50 ml of the standard nitrite solution (1.00 μg NO_2^- per ml) to ten 150-ml beakers. Use a buret to add 50, 49, 47, 45, 43, 40, 30, 20, and 10 ml distilled water, respectively, to the first nine beakers. These solutions correspond to 0.00, 0.02, 0.06, 0.10, 0.14, 0.20, 0.40, 0.60, 0.80, and 1.00 ppm nitrite. Add 2 ml EDTA solution and 2.0 ml of the sulfanilic acid reagent solution to each beaker. Mix thoroughly and wait 5 to 10 min before adding 2.0 ml of the 1-naphthylamine hydrochloride solution. Mix each solution after adding 2.0 ml of the sodium acetate solution. After 20 min, measure the absorbance of each solution at 520 nm using 1.000-cm cells, with water in the reference cell. Plot absorbance versus ppm nitrite.

E. NITRITE IN CURED MEATS

Pork and sometimes beef are preserved using sodium nitrate and/or sodium nitrite in addition to salt, sugar, and wood smoke. Because of the toxicity of nitrite, it is often advisable to determine the nitrite nitrogen concentration in cured meats. The procedure given is similar to an A.O.A.C. method (28). The fact that nitrates are converted to nitrites by bacterial reduction in the lower intestine of adults and in the stomach of infants must also be taken into account in considering the potability of waters and the safety of food products. The meat sample should be finely comminuted in a food chopper and thoroughly mixed.

Reagents

Standard Nitrite Solution. Dissolve 0.2750 g silver nitrite, $AgNO_2$, in about 200 ml nitrite-free water in a 250-ml volumetric flask. Add 20 ml of a 0.10M sodium chloride solution, dilute to volume with nitrite-free water, mix, and allow the silver chloride precipitate to settle. Transfer 100.0 ml of the supernatant solution to a 1-liter volumetric flask and dilute to the mark with nitrite-free water. Transfer a 10.00-ml aliquot of this standard stock solution to another 1-liter volumetric flask and dilute to volume with nitrite-free water. This dilute standard solution contains 100 μg nitrogen per ml.

Modified Griess Reagent Solution. Dissolve 0.5 g sulfanilic acid in 150 ml of 3:17 acetic acid solution. Add 0.125 g 1-naphthylamine hydrochloride to 20 ml nitrite-free water and boil until dissolution is complete. Add this hot solution to 150 ml of a 3:17 acetic acid solution. Mix the two solutions. Filter and store in an amber glass bottle.

Procedure

Transfer a 5.0-g sample of a finely ground and homogenized sample to a 50-ml beaker. Add about 50 ml hot (80°C) nitrite-free water and stir vigorously with a glass rod until a fine mixture is obtained. Transfer to a 500-ml Pyrex volumetric flask, using hot water to effect quantitative transfer. Dilute to approximately 350 ml and place on a steam bath for 2 hr. Shake occasionally. Add 5 ml of a saturated mercury(II) chloride solution and mix. After cooling to room temperature, dilute to volume with nitrite-free water and mix.

Filter a portion of the sample solution and transfer 10-, 25-, and 50-ml aliquots to 50-ml volumetric flasks. Dilute the two partially filled flasks to the mark with nitrite-free water and add 50 ml nitrite-free water to a fourth 50-ml volumetric flask. Add 2 ml of the modified Griess reagent to each flask, mix thoroughly, and allow 1 hr for development of color. Use the reagent blank solution in the reference cell and measure absorbance at 520 nm.

Preparation of Calibration Graph. Transfer 0, 1, 5, 10, 25, 35, and 50 ml of the standard nitrite solution (0.1 μg N per ml) to each of seven 50-ml volumetric flasks and dilute the flasks to volume with nitrite-free water. Add 2 ml modified Griess reagent to each flask, mix thoroughly, and wait 1 hr before measuring absorbance. Use 1.000-cm cells and the reagent blank solution in the reference cell and measure the absorbance at 520 nm. Plot absorbance versus μg nitrogen per ml final solution.

F. NITRATE IN PLANT MATERIALS

The nitrate content of colored extracts obtained from plant tissue or soils can be determined by the phenoldisulfonic acid method. The chloride is removed by extracting the sample with an excess of silver sulfate solution. The excess silver ion is then precipitated at pH 6.5 with a sodium dihydrogen phosphate solution. After the sample has been made alkaline with calcium carbonate, the colored organic matter is oxidized with hydrogen peroxide. The final traces of peroxides are eliminated by heating the dry residue, and the sample is flooded with phenoldisulfonic acid reagent (113).

Reagents

Phosphate Solution. Dissolve 138 g sodium dihydrogen phosphate monohydrate in 500 ml distilled water. Adjust the solution to pH 6.5 with concentrated sodium hydroxide and dilute to 1 liter.

Calcium Sulfate Powder. Remove the nitrate by washing the calcium sulfate twice with distilled water, dry at 60–70°C, and grind to a fine powder.

Calcium Carbonate Suspension. Suspend 1 g calcium carbonate powder in 200 ml distilled water.

EDTA Solution. Dissolve 25 g disodium ethylenediaminetetraacetate dihydrate in 75 ml distilled water. Add concentrated ammonium hydroxide and heat if necessary to effect dissolution. Dilute to 100 ml. Prepare a dilute reagent solution by diluting 5 ml of the EDTA solution to 1 liter with distilled water.

Procedure (113)

After the dried material has been ground to pass a 40-mesh screen, weigh 0.100 g of the sample and transfer to a centrifuge tube, bottle, or flask. Add 0.8–1.0 g powdered calcium sulfate to aid in the filtering process. Follow with 25 ml silver sulfate solution (3.5 g per liter), swirl, and add 1 ml phosphate solution. Stopper the flask and agitate the contents with a mechanical shaker for 5 to 10 min. Remove the solids by centrifugation in 50-ml tubes at 1000 x gravity for 10 min or by filtration through folded No. 12 Whatman filter paper. Place an aliquot of the centrifugate or filtrate, containing 5–500 μg nitrate nitrogen, in a 100-ml beaker or evaporating dish. Add 2 ml of the calcium carbonate suspension and evaporate to a volume of 10 ml. The calcium carbonate neutralizes the acids contributed by the reagents and the original plant extract as well as the acids resulting from the digestion of organic matter. The decomposition of the peroxides in the final drying step is also improved by the calcium carbonate.

Add 1.0 ml 30% hydrogen peroxide. Cover the beaker or dish with a watch glass and digest the sample on the steam bath for 2 hr. Withdraw the watch glass and evaporate the sample to dryness, heating the dry residue for an additional 30 min to remove the last traces of peroxides. Flood the cooled residue rapidly with 2.5 ml phenoldisulfonic acid reagent and rub with a glass rod. Occasionally, a small amount of residue resists solution, in which event 70 ml of the dilute EDTA solution should be added within 5 to 10 min of the addition of the phenoldisulfonic acid reagent. Low results and charring may result from extended contact of the phenoldisulfonic acid reagent with the dry residue. Add 15 ml or more 1:1 ammonium hydroxide to develop the maximum yellow color. Should a turbidity appear at this stage, add a sufficient number of drops of the concentrated EDTA solution to dissolve the precipitate. Dilute the sample to a suitable volume and measure the absorbance at 410 nm. Carry a reagent blank as well as nitrate blanks through every step of the procedure, and apply the proper corrections to the results.

Use distilled water alone as the extracting medium for nitrate samples in which the associated chloride content is less than 1% on a dry weight basis. In such instances, omit the application of silver sulfate, phosphate solution, and EDTA solution.

G. NITROGEN DIOXIDE IN AIR

The Griess-Saltzman reaction is used to determine $0.01-10\ \mu g\ NO_2$ per liter air (0.005—5 ppm). Sampling is accomplished using a fritted-glass bubbler with maximum porosity of 60 μm and a flow rate of 0.4 liter per min (21,192). When the concentration is above 5 ppm evacuated bottles or glass syringes can be used for sampling. The nitrogen dioxide is absorbed by the Griess-Saltzman reagent to form a purple azo dye. After about 15 min, the absorbance is measured at 550 nm. An empirical standardization using a standard sodium nitrite solution is employed (192,193,194).

Reagents

N-(1-Naphthyl)ethylenediamine Dihydrochloride. Dissolve 0.10 g N-(1-naphthyl)-ethylenediamine dihydrochloride in 100 ml distilled water.

Nitrite-Free Water. Distill alkaline permanganate—distilled water solution using an all-glass distilling apparatus.

Griess-Saltzman Reagent. Dissolve 5.0 g anhydrous sulfanilic acid in approximately 1 liter distilled water containing 140 ml glacial acetic acid. Add 20 ml of the 0.1% N-(1-naphthyl)ethylenediamine dihydrochloride solution and dilute to 1 liter. Store in an amber bottle in the refrigerator. Use reagent only at room temperature.

Standard Stock Nitrite Solution. Dissolve 2.030 g analytical reagent-grade sodium nitrite in nitrite-free water and dilute to 1 liter.

Standard Nitrite Solution. Transfer 10.00 ml of the standard stock solution to a 1-liter volumetric flask and dilute to volume with nitrite-free distilled water. One ml of the standard nitrite solution gives a color equivalent to 10 microliters NO_2 (10 ppm in 1 liter air at S.T.P.).

Procedure

Assemble a glass-fritted bubbler (21,193) with a rotameter and suction pump positioned after the bubbler. Pipet 10.0 ml of the Griess-Saltzman reagent into the dry bubbler. Start the pump and draw the air sample through the reagent solution at the rate of 0.4 liter per min for 10–20 min. Measure the total volume of air sample, the temperature, and the pressure. Measure the absorbance at 550 nm against unexposed reagent in the reference cell. If the absorbance is very high, dilute equivalent volumes of exposed and unexposed reagent solutions to specific volumes and measure the absorbance. Obtain the corrected absorbance by multiplying the observed absorbance by the appropriate dilution factor.

Standardization. Transfer 0, 0.1, 0.3, 0.5, 0.7, and 1.0 ml of the standard nitrite solution to 25-ml volumetric flasks. Dilute to volume with the Griess-Saltzman reagent and mix thoroughly. After 15 min, measure the absorbance at 550 nm against a solution containing no nitrite in the reference cell. Plot the absorbance of standard colored solutions versus ml of the standard nitrite solution used, and determine the slope S of the line drawn through the origin and the point corresponding to $A = 1.0$. Calculate the standardization constant K, defined as the number of microliters NO_2 required by 1 ml of the Griess-Saltzman reagent to give an absorbance of 1.0. Therefore

$$K = \frac{S \times 10 \text{ ml}}{25 \text{ ml}} = 0.40S$$

Calculate the concentration of nitrogen dioxide in the air sample using the following expression:

$$\text{ppm of } NO_2 = A_{\text{meas}} \times \frac{K}{V}$$

where V equals volume of air sample at S.T.P. in liters per ml of Griess-Saltzman reagent.

H. MISCELLANEOUS APPLICATIONS

Only a sampling of applied methods for the determination of nitrogen has been given in this section. As an aid to those having to determine nitrogen in other materials, representative spectrophotometric methods have been tabulated in Table II.

TABLE II. Spectrophotometric Methods for the Determination of Nitrogen in Various Materials

Constituent	Material	Method or reagent	Reference
N(NH$_3$)	carbon steel, low alloy steel, wrought iron	Nessler	(23)
	steel	bispyrazolone (CCl$_4$)	(117)
	steel	thymol, chloramine-T	(82)
	beryllium	indophenol	(46)
	fertilizers	automated, indophenol	(79)
	organic Si compounds	indophenol	(209)
	blood	indophenol	(110, 135, 215)
	blood	UV derivative spectrometry	(91)
	protein	ninhydrin	(67)
	soil extracts	automated, indophenol	(203)
Organic N	water	Nessler	(202)
NH$_3$	sea water	indophenol (hexanol)	(161)
	sea water	indophenol, thymol, acetone	(188)
	sea water	hypobromite, iodine–starch	(85)
	water	indophenol	(10)
	water	automated, indophenol	(97)
	air	bispyrazolone	(170)
NO$_3^-$	airborne particles	1-aminopyrene	(195)
	cellulose nitrates	phenoldisulfonic acid	(81)
	propellants	iron(II) sulfate	(169)
	water	brucine	(16, 70, 112)
	water	phenoldisulfonic acid	(11, 18, 213)
	water	2,6-xylenol	(25)
	water	reduction, bispyrazolone	(172)
	water	UV absorptivity	(13, 26, 103, 156)
	water	automated: reduction, Griess	(101, 120)
	water	automated: reduction, brucine	(115)
	sea water	reduction: sulfanilic acid, and 1-naphthylamine or N-(1-naphthyl)ethylenediamine	(145, 154, 155)
	sea water	reduction: Griess	(243)
	soil extracts	automated: reduction, Griess	(101)
NO$_2^-$	water	automated: Griess	(101, 120)
	water	p-aminobenzoic acid, N-(1-naphthyl)ethylenediamine dihydrochloride	(61)
	water	nitrosophenol–iron(II)	(174)
	water	modified Griess	(92, 152)
N$_2$H$_4$	water		(20)

REFERENCES

1. Amer. Public Health Assoc., *Standard Methods for the Examination of Water and Wastewater,* 13th ed., Washington, D.C., 1971, p. 221.

2. *Ibid.,* pp. 123—124.

3. *Ibid.,* pp. 222—230.

4. *Ibid.,* p. 223.

5. *Ibid.,* p. 224.

6. *Ibid.,* p. 225.

7. *Ibid.,* pp. 226—228.

8. *Ibid.,* p. 227.

9. *Ibid.,* pp. 227—228.

10. *Ibid.,* pp. 232—233.

11. *Ibid.,* pp. 233—237.

12. *Ibid.,* p. 235.

13. *Ibid.,* pp. 237—239.

14. *Ibid.,* pp. 240—243.

15. *Ibid.,* p. 241.

16. *Ibid.,* pp. 461—464.

17. *Ibid.,* pp. 465—467.

18. Amer. Soc. for Testing and Materials, Philadelphia, *1973 Annual ASTM Standards, Part 23*, ASTM Designation: D992-71, p. 363.

19. *Ibid.,* ASTM Designation: D1254-67, p. 366.

20. *Ibid.,* ASTM Designation: D1385-67, p. 318.

21. *Ibid.,* ASTM Designation: D1607-69, p. 874.

22. Amer. Soc. for Testing and Materials, Philadelphia, *1974 Annual ASTM Standards, Part 12*, pp. 254—256.

23. *Ibid.,* p. 567.

24. Allerton, F. W., *Analyst, 72*, 349 (1947).

25. Andrews, D. W. W., *Analyst., 89*, 730 (1964).

26. Armstrong, F. A. J., *Anal. Chem., 35*, 1292 (1963).

27. Ashton, F. L., *J. Soc. Chem. Ind.* (London), *54*, 389T (1935).

28. Association of Official Analytical Chemists, *Official Methods of Analysis of A.O.A.C.,* 11th ed., Washington, D.C., 1970, p. 393.

29. Baker, P. R. W., *Talanta, 8*, 57 (1961).

30. Barker, A. V., and R. J. Volk, *Anal. Chem., 36*, 439 (1964).

31. Barnes, H., *Analyst, 75*, 388 (1950).

32. Barnes, H., and A. R. Folkard, *Analyst, 76*, 599 (1951).

33. Bartow, E., and J. S. Rogers, *Univ. Illinois Bull., Water Survey Ser., 7*, 14 (1909).

34. Basargin, N. N., and E. A. Chernova, *Zh. Anal. Khim.*, **23**, 102 (1968).

35. Bastian, R., R. Wiberling, and F. Palilla, *Anal. Chem.*, **29**, 1795 (1957).

36. Batey, T., M. S. Cresser, and I. R. Willett, *Anal. Chim. Acta*, **69**, 484 (1974).

37. Batten, J. J., *Anal. Chem.*, **36**, 939 (1964).

38. Beatty, R. L., L. B. Berger, and H. H. Schrenk, U.S. Bur. Mines, *Rept. Invest. No. 3687*, Washington, D.C., 1943.

39. Beeghly, H. F., *Anal. Chem.*, **24**, 1095 (1952).

40. Berthelot, M., *Rép. Chim. Appl.* 1, 284 (1859).

41. Bhatty, M. K., and A. Townshend, *Anal. Chim. Acta*, **56**, 55 (1971).

42. Bloomfield, R. A., J. C. Guyon, and R. K. Murmann, *Anal. Chem.*, **37**, 248 (1965).

43. Bolleter, W. T., C. J. Bushman, and P. W. Tidwell, *ibid.*, **33**, 592 (1961).

44. Bremner, J. M., and D. R. Keeney, *Anal. Chim. Acta*, **32**, 485 (1965).

45. Brown, R. H., G. D. Duda, S. Korkes, and P. Handler, *Arch. Biochem. Biophys.* **66**, 301 (1957).

46. Bundy, J. K., and G. C. Goode, *Anal. Chim. Acta*, **37**, 394 (1967).

47. Burns, D. T., A. C. Fogg, and A. Willcov, *Mikrochim. Acta*, **1971**, 205.

48. Burstrom, H., *Svensk. Kem. Tidskr.*, **54**, 139 (1942); *Chem. Zentr.*, I, 1196 (1943).

49. Chamot, E. M., and D. S. Pratt, *J. Amer. Chem. Soc.*, **31**, 922 (1909).

50. Chamot, E. M., and D. S. Pratt, *ibid.*, **32**, 630 (1910).

51. Chamot, E. M., D. S. Pratt, and H. W. Redfield, *ibid.*, **33**, 366 (1911).

52. Chaney, A. L., and E. P. Marbach, *Clin. Chem.*, **8**, 130 (1962).

53. Chester Beatty Research Institute, London, *Precautions for Workers Who Handle Carcinogenic Aromatic Amines*, Institute of Cancer Research, Royal Cancer Hospital, London, 1966.

54. Chow, T. J., and M. Johnstone, *Anal. Chim. Acta*, **27**, 441 (1962).

55. Clarkson, T. W., and L. Ferraio, *Clin. Chem.*, **15**, 433 (1969).

56. Conway, E. J., *Micro-Diffusion Analysis and Volumetric Error*, 2nd ed., Crosby-Lockwood, London, 1947.

57. Conway, E. J., and A. Byrne, *Biochem. J.* (London), **27**, 419 (1933).

58. Cotton, R. H., *Ind. Eng. Chem., Anal. Ed.*, **17**, 734 (1945).

59. Crosby, N. T., *Analyst*, **93**, 406 (1968).

60. Day, H. F., E. Bernstorf, and R. T. Hill, *Anal. Chem.*, **21**, 1290 (1949).

61. Dey, A. P., *J. Proc. Inst. Chemists* (India), **37**, 249 (1965).

62. Dienst, S. G., and B. Morris, *J. Lab. Clin. Med.*, **64**, 495 (1964).

63. Dillon, P. L., M. J. Caldwell, and A. J. Gehrt, *J. Assoc. Offic. Anal. Chem.*, **55**, 101 (1972).

64. Edwards, G. P, *J. Water Pollution Control Fed.*, **34**, 1112 (1962).

65. Emmet, R. T., *Anal. Chem.*, **41** 1648 (1969).

66. Fadrus, H., and J. Maly, *Z. Anal. Chem.*, **246**, 239 (1969).

67. Fels, G., and R. Veatch, *Anal. Chem.*, **31**, 451 (1959).

68. Fenton, J. C. B., *Clin. Chim. Acta*, **7**, 163 (1962).

69. Fish, V. B., *Anal. Chem.*, **24**, 760 (1952).

70. Fisher, F. L., E. R. Ibert, and H. F. Beckman, *Anal. Chem.*, **30**, 1972 (1958).

71. Folin, O., and H. Wu, *J. Biol. Chem.*, **38**, 81 (1919).

72. Follett, M. J., and P. W. Ratcliff, *J. Sci. Food Agr.*, **14**, 138 (1963).

73. Foris, A., and T. R. Sweet, *Anal. Chem.*, **37**, 701 (1965).

74. Foxwell, G. E., *Gas World*, **64**, 10 (1916).

75. Frankenburg, W. G., A. M. Gottscho, S. Kissinger, D. Bender, and M. Ehrlech, *Anal. Chem.*, **25**, 1784 (1953).

76. Friedrich, A., E. Kühaas, and R. Schürch, *Z. Physiol. Chem.*, **216**, 68 (1933).

77. Fujinuma, H., Y. Shimada, and S. Hirano, *Bunseki Kagaku*, **20**, 131 (1971).

78. Garcia, E. E., *Anal. Chem.*, **39**, 1605 (1967).

79. Gehrke, C. W., W. L. Baker, G. F. Krause, and C. H. Russell, *J. Assoc. Offic. Anal. Chem.*, **50**, 382 (1967).

80. Goldman, E., and R. J. Jacobs, *J. Amer. Water Works Assoc.*, **53**, 187 (1961).

81. Gordon, L. L., and B. Leopold, *Anal. Chem.*, **30**, 2057 (1958).

82. Goto, H., Y. Kakita, and I. Atsuga, *Bunseki Kagaku*, **12**, 727 (1963).

83. Goto, H., Y. Kakita, and I. Atsuga, *Sci. Rep. Res. Inst., Tahaku Univ., Ser. A.*, **19**, 50 (1967).

84. Grandval, A., and H. Lajoux, *Compt. Rend.*, **101**, 62 (1885); *J. Chem. Soc.*, **48**, 1093 (1885).

85. Grasshoff, K., *Z. Anal. Chem.*, **234**, 13 (1968).

86. Greenberg, A. E., J. R. Rossum, N. Moskowitz, and P. A. Villarruz, *J. Amer. Water Works Assoc.*, **50**, 821 (1958).

87. Griess, P., *Ber.*, **12**, 426 (1879).

88. Grunbaum, B. W., P. L. Kirk, L. G. Green, and C. W. Koch, *Anal. Chem.*, **27**, 384 (1955).

89. Gunning, J. W., *Z. Anal. Chem.*, **28**, 188 (1889).

90. Haase, L. W., *Chemiker-Ztg.*, **50**, 372 (1926).

91. Hager, R. N., Jr., D. R. Clarkson, and J. Savory, *Anal. Chem.*, **42**, 1813 (1970).

92. Hagino, K., *Bunseki Kagaku*, **11**, 237 (1962).

93. Hargis, L. G., *Anal. Lett.*, **1**, 471 (1968).

94. Harrison, G. A. F., *Talanta*, **9**, 533 (1961).

95. Hartley, A. M., and R. I. Asai, *Anal. Chem.*, **35**, 1207 (1963).

96. Hartley, A. M., and R. I. Asai, *J. Amer. Water Works Assoc.*, **52**, 255 (1960).

97. Harwood, J. E., and D. J. Huyser, *Water Res.*, **4**, 501 (1970).

98. Hashitani, H., and H. Yoshida, *Bunseki Kagaku,* **19**, 355, 1081, 1564 (1970).

99. Hawk, P. B., B. L. Oser, and W. H. Summerson, *Practical Physiological Chemistry,* 13th ed., Blakiston, New York, 1954, p. 1329.

100. Heisler, E. G., J. Siciliano, S. Krulick, W. L. Porter, and J. W. White, Jr., *J. Agr. Food Chem.*, **21**, 971 (1973).

101. Hendriksen, A., and A. R. Selmer-Olsen, *Analyst,* **95**, 514 (1970).

102. Hiller, A., J. Plazin, and D. D. Van Slyke, *J. Biol. Chem.*, **176**, 1401 (1948).

103. Hoather, R. C., and R. F. Rackam, *Analyst,* **84**, 549 (1959).

104. Hofmann, M., *Z. Med. Lab. Tech.*, **10**, 86 (1969).

105. Holler, A. C., and R. V. Huch, *Anal. Chem.*, **21**, 1385 (1949).

106. Horn, D. B., and C. R. Squire, *Clin. Chim. Acta,* **17**, 99 (1967).

107. Howell, J. A. and D. F. Boltz, *Anal. Chem.*, **36**, 1799 (1964).

108. Hutchinson, K., and D. F. Boltz, *ibid.*, **30**, 54 (1958).

109. Ilosvay, M. L., *Bull. Soc. Chim. Fr.*, **2**, 388 (1889).

110. Imler, M., A. Frick, A. Stahl, B. Peter, and J. Stahl, *Clin. Chim. Acta,* **37**, 245 (1972).

111. James, W. E., F. A. Slesinski, and H. B. Pierce, *J. Lab. Clin. Med.*, **27**, 113 (1941).

112. Jenkins, D., and L. L. Medsker, *Anal. Chem.*, **36**, 610 (1964).

113. Johnson, C. M., and A. Ulrich, *ibid.*, **22**, 1526 (1950).

114. Jonnard, R., *Ind. Eng. Chem., Anal. Ed.*, **17**, 246, 746 (1945).

115. Kahn, L., and F. T. Brezenski, *Environ. Sci. Technol.*, **1**, 488, 492 (1967).

116. Kallmann, S., E. W. Hobart, H. K. Oberthin, and W. C. Brienza, Jr., *Anal. Chem.*, **40**, 332 (1968).

117 Kamada, H., and K. Sato, *Bunseki Kagaku,* **6**, 150 (1957).

118. Kamm, L., G. G. McKeown, and D. M. Smith, *J. Assoc. Offic. Anal. Chem.*, **48**, 892 (1965).

119. Kammori, O., Y. Hiyama, and W. Hotta, *Nippon-Kinzoku Gakkaishi,* **29**, 126 (1965).

120. Kamphake, L. J., S. Hannah, and J. M. Cohen, *Water Res.*, **1**, 205 (1967).

121. Keay, J., and P. M. A. Menage, *Analyst,* **95**, 379 (1970).

122. Kirk, P. L., *Anal. Chem.*, **22**, 354 (1950).

123. Kirk, P. L., *ibid.*, p. 611.

124. Kjeldahl, J., *Z. Anal. Chem.*, **22**, 366 (1883).

125. Koch, F. C., and T. L. McMeekin, *J. Amer. Chem. Soc.*, **46**, 2066 (1924).

126. Kolthoff, I. M., E. B. Sandell, E. J. Meehan, and S. Bruckenstein, *Quantitative Chemical Analysis*, 4th ed., Macmillan, New York, 1969, pp. 834, 846.

127. Komatsu, S., and K. Hagino, *Nippon Kagaku Zasshi*, **88**, 1157 (1967).

128. Komatsu, S., T. Nomura, T. Nakamura, and H. Suzuki, *ibid.*, **91**, 865 (1970).

129. Kruse, J. M., and M. G. Mellon, *Anal. Chem.*, **25**, 1188 (1953).

130. Kruse, J. M., and M. G. Mellon, *Sewage Ind. Wastes*, **24**, 1098 (1952).

131. Lake, G. R., et al., *Anal. Chem.*, **24**, 1806 (1952).

132. Lambert, J. L., and F. Zitomer, *ibid.*, **32**, 1684 (1960).

133. Lambert, R. S., and R. J. Dubois, *ibid.*, **43**, 955 (1971).

134. Lear, J. B., and M. G. Mellon, *ibid.*, **29**, 293 (1957).

135. Leffler, H. H., *Amer. J. Clin. Pathol.*, **48**, 233 (1967).

136. Levin, B., V. G. Oberholzer, and T. P. Whitehead, *Analyst*, **75**, 561 (1950).

137. Liebhafsky, H. A., and L. B. Bronk, *Anal. Chem.*, **20**, 588 (1948).

138. Lin, E., and K. L. Cheng, *Mikrochim. Acta*, **1970**, 652.

139. Lubochinsky, B., and J. P. Zolta, *Bull. Sté. Chim. Biol.*, **36**, 1363 (1954).

140. Lyubimov, V. I., N. P. L'vov, and B. Kirsteine, *Prikl. Biokhim. Mikrobiol.*, **4**, 120 (1968).

141. Macchi, G. R., and B. S. Ceson, *Anal. Chem.*, **42**, 1809 (1970).

142. McCutchan, P., and W. F. Roth, *ibid.*, **24**, 369 (1952).

143. MacDonald, A. M. G., *Ind. Chem. Mfr.*, **31**, 568 (1968).

144. Máchová, I., and J. Dokládalová, *Chem. Tech. Berl.*, **19**, 767 (1967).

145. Matsunaga, K., and M. Nishimura, *Anal. Chim. Acta*, **45**, 350 (1969).

146. Matsunaga, K., T. Oyama, and M. Nishimura, *ibid.*, **58**, 228 (1971).

147. Matsunaga, K., K. Sugino, and M. Nishimura, *Bunseki Kagaku*, **19**, 1255 (1970).

148. Middleton, K. R., *Chem. Ind.* (London), **1957**, 1147.

149. Miller, G. L., and E. E. Miller, *Anal. Chem.*, **20**, 481 (1948).

150. Miller, G. E., and J. D. Rice, *Amer. J. Clin. Pathol.*, **39**, 97 (1963).

151. Moeller, G., *Z. Anal. Chem.*, **245**, 155 (1969).

152. Montgomery, H. A. C., and J. F. Dymock, *Analyst*, **86**, 414 (1961).

153. Montgomery, H. A. C., and J. F. Dymock, *ibid.*, **87**, 374 (1962).

154. Morris, A. W., and J. P. Riley, *Anal. Chim. Acta*, **29**, 272 (1963).

155. Mullin, J. B., and J. P. Riley, *ibid.*, **12**, 464 (1955).

156. Navone, R., *J. Amer. Water Works Assoc.*, **56**, 781 (1964).

157. Nelson, E. L., *J. Assoc. Offic. Anal. Chem.*, **43**, 468 (1960).

158. Nelson, G. H., M. Levine, and J. H. Buchanan, *Ind. Eng. Chem., Anal. Ed.*, **4**, 56 (1932).

159. Nelson, J. L., L. T. Kurtz, and R. H. Bray, *Anal. Chem.*, **26**, 1081 (1954).

160. Nessler, J., *Chem. Zentr.*, **27**, N.F.I., 529 (1856).

161. Newell, B. S., *J. Mar. Biol. Assoc. U.K.*, **47**, 271 (1967).

162. Nichols, M. L., and C. O. Willits, *J. Amer. Chem. Soc.*, **56**, 769 (1934).

163. Niedermaier, T., *Z. Anal. Chem.*, **223**, 336 (1966).

164. Nishimura, M., and K. Matsunaga, *Bunseki Kagaku,* **18**, 154 (1969).

165. Nishimura, M., K. Matsunaga, and K. Matsuda, *ibid.,* **19**, 1096 (1970).

166. Noble, E. D. *Anal. Chem.*, **27**, 1413 (1955).

167. Noll, C. A., *Ind. Eng. Chem., Anal. Ed.,* **3**, 311 (1931).

168. Noll, C. A., *ibid.,* **17**, 426 (1945).

169. Norwitz, G., *Anal. Chem.*, **34**, 227 (1962).

170. Okita, T., and S. Kanamori, *Atmosph. Environ.*, **5**, 621 (1971).

171. O'Neal, J. M. and K. G. Clark, *J. Assoc. Offic. Anal. Chem.*, **47**, 1054 (1964).

172. Pappenhagen, J. M., *Anal. Chem.*, **30**, 282 (1958).

173. Pappenhagen, J. M., and M. G. Mellon, *ibid.,* **25**, 341 (1953).

174. Peach, S. M., *Analyst,* **86**, 757 (1961).

175. Pepkowitz, L. P., *Anal. Chem.*, **24**, 900 (1952).

176. Perez, M., *Acta Cient. Compostelano,* **5**, 159 (1968).

177. Pietsch, R., *Mikrochim. Acta,* **1956**, 1490, 1672.

178. Pokorny, G., and W. Likussar, *Anal. Chim. Acta,* **42**, 253 (1968).

179. Polley, J. R., *Anal. Chem.*, **26**, 1523 (1954).

180. Prochazkova, L., *ibid.,* **36**, 865 (1964).

181. Purs, J., *Chem. Listy,* **65**, 1295 (1971.

182. Reiner, M., Ed., *Standard Method of Clinical Chemistry,* Vol. I, American Association of Clinical Chemists, Academic Press, New York, 1953, pp. 118–121.

183. Reitmeier, R. F., *Ind. Eng. Chem., Anal. Ed.,* **15**, 393 (1943).

184. Rider, B. F., and M. G. Mellon, *ibid.,* **18**, 96 (1946).

185. Riley, J. P., *Anal. Chim. Acta,* **9**, 575 (1953).

186. Roller, E. M., and N. McKaig, *Soil Science,* **47**, 397 (1939).

187. Rose, E. L., and H. Ziliotto, *Ind. Eng. Chem., Anal. Ed.,* **17**, 211 (1945).

188. Roskam, R. I., and D. Langen, *Anal. Chim. Acta,* **30**, 56 (1964).

189. Rommers, P. J., and J. Visser, *Analyst,* **94**, 653 (1969).

190. Russell, J. A., *J. Biol. Chem.*, **156**, 457 (1944).

191. Saito, G., K. Sugimoto, and K. Hagino, *Bunseki Kagaku,* **20**, 542 (1971).

192. Saltzman, B. E., *Anal. Chem.*, **26**, 1949 (1954).

193. Saltzman, B. E., *ibid.,* **33**, 1100 (1961).

194. Saltzman, B. E., and A. F. Wartburgh, Jr., *ibid.,* **37**, 1261 (1965).

195. Sawicki, E., H. Johnson, and T. W. Stanley, *ibid.*, **35**, 1934 (1963).
196. Sawicki, E., T. W. Stanley, J. Pfaff, and A. Johnson, *ibid.*, **35**, 2183 (1963).
197. Sawyer, C. N., *ibid.*, **25**, 816 (1953).
198. Schall, E. D., and D. W. Hatcher, *J. Assoc. Offic. Anal. Chem.*, **51**, 763 (1968).
199. Scheurer, P. G., and F. Smith, *Anal. Chem.*, **27**, 1616 (1955).
200. Schimpff, G. W., and R. E. Pottinger, *Ind. Eng. Chem., Anal. Ed.*, **13**, 337 (1941).
201. Schlenker, F. S., *ibid.*, **19**, 471 (1947).
202. Schmid, M., and Z. Schweiz, *Hydrology*, **30**, 244 (1968).
203. Selmer-Olsen, A. R., *Analyst*, **96**, 565 (1971).
204. Shuey, P. M., *Ind. Eng. Chem., Anal. Ed.*, **19**, 882 (1947).
205. Skujins, J. J., *Anal. Chem.*, **36**, 240 (1964).
206. Sobel, A. E., A. Hirschman, and L. Besman, *ibid.*, **19**, 927 (1947).
207. Stender, J. H., *Fed. Regist.*, **38**, 10929 (1973).
208. Steyermark, A., H. K. Alber, V. A. Aluise, W. D. Huffman, J. A. Kuck, J. J. Moran, and C. O. Willits, *Anal. Chem.*, **23**, 523 (1951).
209. Strukova, M. P., and G. I. Veslora, *Zh. Anal. Khim.*, **26**, 1731 (1971).
210. Swann, M. H., and M. L. Adams, *Anal. Chem.*, **28**, 1630 (1956).
211. Szekely, E., *Talanta*, **14**, 941 (1967).
212. Tabata, M., S. Funabashi, and M. Tanaka, *Anal. Chim. Acta*, **62**, 289 (1972).
213. Taras, M. J., *Anal. Chem.*, **22**, 1020 (1950).
214. Taras, M. J., *J. Amer. Water Works Assoc.*, **45**, 56 (1953).
215. Ternberg, J. L., and F. B. Hershey, *J. Lab. Clin. Med.*, **56**, 766 (1960).
216. Terrey, D. R., *Anal. Chim. Acta*, **34**, 41 (1966).
217. Tetlow, J. A., and A. L. Wilson, *Analyst*, **89**, 453 (1964).
218. Thaler, H., and W. Sturm, *Z. Anal. Chem.*, **250**, 120 (1970).
219. Thomas, P., *Bull, Soc. Chim.*, **11**, 706 (1912); *ibid.*, **13**, 398 (1913).
220. Thompson, J. F., and G. R. Morrison, *Anal. Chem.*, **23**, 1153 (1951).
221. Vanetten, C. H., and M. B. Wiele, *ibid.*, **23**, 1338 (1951).
222. Van Slyke, D. D., and G. E. Cullen, *J. Biol. Chem.*, **19**, 211 (1914).
223. Van Slyke, D. D., and A. Hiller, *ibid.*, **102**, 499 (1933).
224. Wada, E., and A. Hattori, *Anal. Chim. Acta*, **56**, 233 (1971).
225. Walker, R. I., and W. H. Shipman, *J. Chromatagr.*, **50**, 157 (1970).
226. Warington, R., *J. Chem. Soc.*, **39**, 229 (1881).
227. Weatherburn, M. W., *Anal. Chem.*, **39**, 971 (1967).
228. Weichselbaum T. E., J. C. Hagerty, and H. B. Mark, Jr., *ibid.*, **41**, 848, 1015 (1964).

229. Weiss, K. B., and D. F. Boltz, *Anal. Chim. Acta,* **55**, 77 (1971).

230. Welcher, F. J., *Organic Analytical Reagents,* Vol. I, Van Nostrand, New York, 1947, p. 254.

231. Welcher, F. J., *ibid.,* Vol. II, p. 495.

232. Welcher, F. J., *ibid.,* Vol. IV, p. 188.

233. West, P. W., and G. L. Lyles, *Anal. Chim. Acta,* **23**, 227 (1960).

234. West, P. W., and T. P. Ramachandran, *ibid.,* **35**, 317 (1966).

235. West, P. W., and P. Sarma, *Mikrochim. Acta,* **1957**, 506.

236. Weston, R. S., *Proc. Amer. Chem. Soc.,* **27**, 281 (1905).

237. White, L. M., and M. C. Long, *Anal. Chem.,* **23**, 363 (1951).

238. Wiersma, J. H., *Anal. Lett.,* **3**, 123 (1970).

239. Willits, C. O., and C. L. Ogg, *J. Assoc. Offic. Agr. Chem.,* **33**, 179 (1950); **34**, 607 (1951).

240. Winkler, L. N., *Chem. Ztg.,* **23**, 454 (1899); *ibid.,* **25**, 586 (1901).

241. Wolf, B., *Ind. Eng. Chem., Anal. Ed.,* **16**, 446 (1944).

242. Wolf, B., *ibid.,* **19**, 334 (1947).

243. Wood, E. D., F. A. J. Armstrong, and F. A. Richards, *J. Marine Biol. Assoc. U.K.,* **47**, 23 (1967).

244. Woodward, P., *Analyst,* **78**, 727 (1953).

245. Yamaguchi, R., and T. Machida, *Yakugaku Zasshi,* **88**, 1383 (1968).

246. Yamaguchi, R., T. Machida, and M. Ueki, *ibid.,* **89**, 804 (1969).

247. Yamamoto, Y., N. Okamoto, and E. Tao, *Anal. Chim. Acta,* **47**, 127 (1969).

248. Yamamoto, Y., N. Okamoto, and E. Tao, *Nippon Kagaku Zasshi,* **89**, 399 (1968).

249. Yamamoto, Y., S. Uchikawa, and K. Akabori, *Bull. Chem. Soc. Japan,* **37**, 1718 (1964).

250. Zitomer, F., and J. L. Lambert, *Anal. Chem.,* **34**, 1738 (1962).

CHAPTER

8

OXYGEN

GORDON A. PARKER

University of Toledo, Toledo, Ohio

The essential presence of oxygen in life processes necessitates accurate and precise knowledge of its presence in the atmosphere and, as dissolved oxygen, in natural waters. Numerous methods are available for the determination of oxygen; and among these, colorimetric methods serve an important role. Other oxidants are also present in our environment: ozone, hydrogen peroxide, and organic peroxides. These, in contrast to oxygen, contribute to the pollution of our environment. An accurate and precise estimation of their presence is also important. In addition to environmental studies, it is often necessary to know the concentration of oxygen or ozone or hydrogen peroxide in other systems. Colorimetric methods for determination of these components in a variety of systems are available. As a closely related topic, the colorimetric determination of water, both as a vapor and as a component in nonaqueous media, is considered. Existing colorimetric procedures for these substances have, generally, tried to utilize a specific reagent to react with the component of interest. Direct colorimetric measurement of either the reaction product or a substance formed from the reaction product and another reagent serve for quantitative determination.

OXYGEN

I. METHODS OF DETERMINATION

A. WINKLER METHOD

The Winkler method (316) is perhaps the oldest and best-known method for determination of oxygen. It uses an alkaline iodide solution added to an aqueous sample containing dissolved oxygen. A manganese(II) solution is then added, and the freshly precipitated manganese(II) hydroxide formed reacts with the oxygen present. The manganese(II) is oxidized to a higher valence state (50). Upon acidification, the oxidized manganese reacts with the iodide present, oxidizing it to free iodine. Iodine, equivalent to the amount of oxygen present, is then determined by any of several available methods. An alternate approach to this procedure allows the oxidized manganese species itself to react with a color-forming reagent, and this species is used for determining the equivalent amount of oxygen. Numerous precautions are necessary before the Winkler method can be relied upon to give accurate results (208,272). These include proper sampling procedure to avoid introduction of additional oxygen into the sample and removal of, or correction for, the amount of dissolved oxygen in the various reagent solutions used in the determination. In addition, the presence of other oxidizing species which attack manganese(II) hydroxide or the presence of reducing species which react with the oxidized manganese must either be eliminated or appropriate corrections made to account for their interference. The volatility of the iodine formed upon acidification must also be considered.

1. Methods Involving Iodine Determination

a. Iodine–Starch Complex

The various parameters affecting the determination of oxygen by measurement of the iodine–starch complex are known (21,146,317). These include the stability of the iodine–starch color, optimum reagent concentration, minimum amount of iodine necessary for color formation, optimum wavelength for colorimetric measurement, and temperature effects. A reagent blank is necessary as a reference solution and is prepared in a manner analogous to that of the sample solution but with the order of reagent addition changed. For the blank, an identical sample solution containing alkaline iodide is acidified before addition of manganese(II) solution. Oxygen does not attack manganese(II) in acid solution, but, hopefully, all other substances present that interfere will oxidize manganese(II) to a higher valence state in the acid medium. The absorbance of the iodine–starch complex in the blank effectively serves to cancel the similar effect produced by the same interfering substances in the sample solution. In one method to correct for dissolved oxygen in the reagents used for determination, four identical samples are analyzed (234). Two of these samples are treated in the usual way, that is, one as a blank solution with reagents added in reverse order and the other, as the sample solution. To the second pair of samples, twice the amount of each reagent is added, and a sample and blank solution are prepared as before.

Any difference between the values obtained from these pairs of samples is attributed to dissolved oxygen in the reagent solutions. The observed oxygen content from the first pair of samples is corrected by the amount of this difference. Another method of correction for dissolved oxygen in the reagent solutions assumes a known solubility of oxygen for all solutions used at the temperature and pressure of the experiment. This solubility is subtracted from the experimentally measured value. To assure sufficient iodine for adequate color intensity of iodine—starch complex, a known amount of iodine is sometimes added to both sample and blank. Being present in both solutions, its absorbance is canceled, and only the color present from the iodine equivalent to the oxygen present is recorded.

The following procedure is for determination of dissolved oxygen in concentrations ranging from 0.005 to 0.400 mg oxygen per liter. Values from duplicate samples did not differ by more than 0.002 mg oxygen per liter (231).

Solutions

Manganese(II) Sulfate. Dissolve 450 g $MnSO_4 \cdot 4H_2O$ in distilled water and dilute to 1 liter.

Potassium Sulfate. Dissolve 100 g K_2SO_4 in distilled water and dilute to 1 liter.

Iodine, 0.01N. Dissolve 40 g KI in 25 ml distilled water. Dissolve 1.27 g reagent-grade iodine in this solution and dilute to 1 liter in a volumetric flask. Store in an amber bottle. Standardize this solution daily.

Iodine, 0.0001N. Dilute exactly 10.0 ml of the above iodine solution to 1 liter in a volumetric flask. The concentration of this solution is exactly 0.01 times that of the more concentrated stock solution.

Potassium Iodide. Dissolve 140 g KI and 600 g NaOH in distilled water and dilute to 1 liter.

Sulfuric acid, 18N.

Starch, 1%. Disperse 1 g soluble potato starch in 10 ml distilled water and pour into 90 ml boiling water. Prepare fresh each week.

Procedure

An elongated gas collection bulb of 60-ml capacity with openings at both ends having stopcocks and with fine-bore glass tubing extending beyond the stopcocks is used as the reaction vessel. Two marks on the tubing at each end indicate a volume of 0.2 ml. Flush the reaction vessel several times with sample solution. Close the lower stopcock and allow the vessel to fill. Withdraw excess sample from the upper tube by suction until the liquid level is at the lower marking nearest the stopcock. Add 0.2 ml KI solution. Open the lower stopcock and then the upper stopcock to allow the KI solution to fall into the sample solution. Close both stopcocks and mix thoroughly. Invert the sample tube and add the remaining reagents from the other end to avoid any contact with traces of KI solution which might adhere to the tubing walls. Now add 0.2 ml $MnSO_4$ solution. Mix thoroughly, and after 2 min add 0.2 ml sulfuric acid. Mix until the precipitate is completely dissolved. Transfer the sample to a 100-ml volumetric flask and add 25 ml K_2SO_4 solution, exactly 5.00 ml dilute iodine solution, and 5 ml starch solution. Dilute to volume. Measure the absorbance at 580 nm within 30 min against a reagent blank prepared in a similar manner but with the reagents added in reverse order, that is KI followed by H_2SO_4 followed by $MnSO_4$. Use 4.0-cm cells. The color is unstable above 30°C. A calibration curve is prepared by carrying out the above procedure upon solutions of known iodine concentration. A correction of 0.005 mg oxygen per liter is substracted from the experimental value obtained to correct for dissolved oxygen in the reagent solutions.

b. Iodine Extraction

Instead of adding starch to the iodine liberated by the Winkler method, some prefer direct measurement of the iodine color. This is reported utilizing a series of potassium dichromate color standards for comparison (289) and by direct spectrophotometric measurement at 450 nm (295). An alternate approach is to extract the iodine formed into an organic solvent before colorimetric measurement. Various solvents are proposed, including carbon tetrachloride (9), toluene (228), *o*-xylene (273), and chloroform (94,176). For a procedure utilizing *o*-xylene extraction for the determination of oxygen in gaseous samples, see Section IIA.

c. Iodine as Triiodide Ion

Direct ultraviolet spectrophotometric measurement of triiodide ion, formed upon acidification in the presence of excess iodide, is also used for determination of oxygen in the Winkler procedure (51,52,223). This modification is faster than modifications employing addition of starch or extraction of iodine into an organic phase. Here, measurements are made directly upon the sample solution.

The following procedure is for determination of dissolved oxygen concentrations below 0.05 mg oxygen per liter. Ten repeated determinations on identical samples gave a mean value of 0.023 mg oxygen per liter with a standard deviation of ±0.0002 mg oxygen per liter (42).

Solutions

Manganese(II) Sulfate. Dissolve 25 g $MnSO_4 \cdot 4H_2O$ or 20 g $MnSO_4 \cdot H_2O$ in 500 ml distilled water.

Potassium Iodide. Dissolve 60 g KI and 14 g KOH in 200 ml distilled water. This solution discolors upon standing and should be prepared fresh as needed.

Phosphoric Acid. To 400 ml distilled water, add 100 ml 85% phosphoric acid.

Potassium Hydrogen Biiodate, $3.333 \times 10^{-4} M$. Dissolve exactly 0.1300 g dry reagent-grade $KH(IO_3)_2$ in distilled water and dilute to 1 liter in a volumetric flask.

Potassium Hydrogen Biiodate, $8.333 \times 10^{-6} M$. Dilute exactly 25 ml of the above biiodate solution to 1 liter in a volumetric flask. This solution is equivalent to 0.025 mg oxygen per liter.

Iodine, $1.25 \times 10^{-3} M$. Dissolve 25 g KI and 0.1 g KOH in 200 ml distilled water. Dissolve 0.32 g reagent-grade iodine in this solution and dilute to 1 liter in a volumetric flask. This solution should be basic to retard air oxidation of iodide ion. Store in an amber bottle. The concentration of this solution increases slowly as iodide is oxidized to iodine. Standardize this solution daily.

Procedure

A syringe fitted with a large-bore (16 gauge) cannula is flushed several times with the sample solution. Draw exactly 25 ml of the sample into the syringe. Immediately, add 0.5 ml manganese(II) sulfate solution through the tip of the sample syringe from a 1-ml syringe fitted with a small-bore (20 gauge) cannula. This is followed, at once, by 0.5 ml alkaline iodide solution added in the same manner. Cover the tip of the sample syringe and mix thoroughly. After 30 sec, add 0.5 ml phosphoric acid and mix again. Avoid excessive sunlight which might cause photochemical reduction of iodine. Measure the absorbance of triiodide ion at 352 nm in a 1-cm silica cell against a reagent blank prepared by adding 0.5 ml phosphoric acid and 0.5 ml alkaline iodide to a syringe containing 25 ml distilled water. Shake well and add 0.5 ml manganese(II) sulfate. To eliminate dissolved oxygen in the reagents, they are stored in aspirator bottles fitted with rubber stoppers and deaerated with nitrogen for 30 min before use. Standardization is achieved by preparing solutions of known iodine concentration using dilute solutions of potassium hydrogen biiodate and treating them as described above, except that 0.5 ml phosphoric acid is added before the alkaline iodide and manganese(II) sulfate to avoid possible oxygen interferences. If other interfering ions are thought to be present, a 25-ml sample of the unknown oxygen solution is first acidified with 0.5 ml phosphoric acid solution. This is followed by 0.5 ml alkaline iodide solution and exactly 0.5 ml standard iodine solution accurately measured from a microliter syringe fitted with a small-bore cannula. Deviations from expected oxygen concentration, in terms of known iodine equivalents, are attributed to interferences, and an appropriate correction is made on the value obtained from the sample solutions measured with reagents added in their normal sequence and without additional iodine present.

d. Iodine with 3,3'-Dimethylnaphthidine

3,3'-Dimethylnaphthidine is oxidized by iodine liberated from the Winkler method forming a red-purple oxidation product (24,25). The absorbance of this species obeys Beer's law for dissolved oxygen concentrations below 0.06 mg oxygen per liter.

The following procedure for dissolved oxygen produced a standard deviation of ±0.0013 mg oxygen per liter from measurement of 45 known solutions varying in oxygen concentration from 0.001 to 0.038 mg oxygen per liter.

Solutions

Manganese(II) Sulfate. Dissolve 400 g $MnSO_4 \cdot 4H_2O$ in distilled water and dilute to 1 liter.

Potassium Iodide. Dissolve 150 g KI and 700 g KOH in distilled water, cool, and dilute to 1 liter.

Sulfuric Acid, 9N.

3,3'-Dimethylnaphthidine. Dissolve 0.025 g 3,3'-dimethylnaphthidine in 50 ml glacial acetic acid. Warm to assist solution. Cool before use. Prepare fresh each week.

Iodine Solution, 0.005N. Dissolve 25 g KI in 20 ml distilled water. Dissolve 0.64 g reagent-grade iodine in this solution and dilute to 1 liter in a volumetric flask. Store in an amber bottle. Standardize this solution daily.

Procedure

Samples are collected in narrow-necked 250-ml bottles. The exact volume of each bottle used is accurately known. The sample bottle is placed in a container sufficiently deep to catch overflow from the sample bottle, as it is flushed with sample solution. The sample bottle is eventually immersed inside the larger container. This procedure avoids contact with atmospheric oxygen. Stopper the filled sample bottle while immersed with a cup-type rubber septum that has also been washed with sample solution. Leave the septum with its soft lip upturned and filled with distilled water to form an air-tight seal. The distilled water is changed each time reagent is added by syringe through the seal. Add in order, using separate syringes for each reagent, 0.1 ml of each of the following solutions: alkaline iodide, manganese(II) sulfate, sulfuric acid. A platinum—iridium needle for the dilute sulfuric acid is less subject to attack from acid solution than is a stainless steel needle. Mix the sample flask thoroughly after each of the reagents is added.

After the hydroxide precipitate dissolves, add exactly 0.100 ml iodine solution from a micrometer syringe. Remove the rubber seal and add 1.0 ml 3,3'-dimethylnaphthidine solution. Measure the absorbance in a Spekker absorptiometer using an Ilford No. 605 filter and a 4-cm cell. A reagent blank is prepared by adding, in order, to a sample bottle filled with distilled water, 0.1-ml portions of alkaline iodide solution, dilute surfuric acid, and manganese(II) sulfate. Next, add exactly 0.100 ml iodine solution. Follow this with 1.0 ml 3,3'-dimethylnaphthidine. The presence of a known amount of iodine in sample and blank is to correct for possible interference reactions. A calibration curve is prepared by measuring solutions of known oxygen content or from solutions of known iodine concentration. A correction of 0.0012 mg oxygen per liter is substracted from the experimental value obtained to correct for dissolved oxygen in the reagent solutions (237).

2. Methods Involving Manganese Determination

a. o-Tolidine Reaction

If hydrochloric acid is used in place of sulfuric acid for the acidification step in the Winkler determination, the free chlorine, formed by reaction of chloride with the oxidized manganese species, is equivalent to the amount of oxygen originally present and can be determined by reaction with o-tolidine to form a yellow-colored o-tolidine oxidation product (123,197). It was later shown that the oxidized manganese species itself will react with o-tolidine to produce a yellow coloration satisfactory for the quantitative determination of oxygen (103,136). Iron(II) and other reducing agents, if present in an aqueous oxygen sample, cause serious interference with the o-tolidine and other Winkler methods. Procedures have been developed to correct for iron(II) interference by determining the amount of iron(III) formed from reaction between any iron(II) present and oxygen prior to the addition of o-tolidine. Iron(III) is determined by reaction with thiocyanate and colorimetric measurement. The amount of iron(III) present is converted to equivalents of oxygen and added to the amount of oxygen found (309). An alternate procedure uses potassium permanganate to oxidize interfering reducing substances prior to oxygen determination (102). Passage through a mixed ion exchange resin before oxygen analysis is also suggested as a means of eliminating cationic and anionic interferences (225,234). Removal of interference from sulfite, dithionite, and hydrazine by addition of bromine is suggested. Excess bromine is destroyed with sulfosalicylic acid before oxygen determination (243,311,312).

The following procedure is for determination of gaseous samples containing oxygen by reacting the sample with an alkaline suspension of manganese(II) hydroxide in the presence of a foaming agent. o-Tolidine is added to produce a yellow coloration. The procedure is applicable for gaseous samples containing from 2 to 50 microliters oxygen. A series of 27 measurements resulted in a precision of ±1.0 microliter of oxygen for samples in this concentration range (179).

Solutions

Manganese(II) Sulfate. Dissolve 400 g $MnSO_4 \cdot 2H_2O$ in distilled water and dilute to 1 liter.

Sodium Hydroxide. Dissolve 400 g NaOH in distilled water and dilute to 1 liter.

o-**Tolidine.** Dissolve sufficient o-tolidine in 80% concentrated sulfuric acid to produce a saturated solution.

Dodecylbenzenesulfonate, 1%. Dissolve 1 g of sodium dodecylbenzenesulfonate in distilled water and dilute to 100 ml.

Potassium Permanganate. Dilute exactly 17.80 ml $0.100N$ $KMnO_4$ solution to 1 liter with distilled water in a volumetric flask. One milliliter of this solution is equivalent to 10.0 microliters oxygen at 760 torr and $0°C$.

Procedure

The reaction is carried out in an elongated tube, approximately 95 cm in length and 3.5 cm in diameter. The tube is attached to a vacuum line and, in addition, has a three-way vacuum stopcock. One setting of this stopcock connects to the sample gas container and the other leads to a cup reservoir. It is from this reservoir that various reagent solutions are added to the reaction vessel. Connect the reaction tube to vacuum and evacuate. Fill the tube with purified, oxygen-free nitrogen through the sample entrance using a flow rate of approximately 125 ml/min. Repeat the vacuum purge. Flush the tube several times with nitrogen purging after each filling and finally evacuate the reaction tube. Fill the reagent reservoir with 15 ml water, 0.1 ml NaOH solution, and 1 ml dodecylbenzenesulfonate solution. Purge this mixture of dissolved oxygen by passing purified nitrogen through the solution for 2 to 3 min. Now, carefully, open the three-way stopcock and allow this mixture to be sucked into the reaction tube. Close the stopcock while a small amount of liquid remains in the reagent reservoir in order to avoid the entrance of atmospheric oxygen.

Again evacuate the reaction tube for about 10 min to assure the absence of all gaseous components. Place 1 ml water and 0.1 ml $MnSO_4$ solution into the reagent reservoir. Purge this solution as before to remove dissolved oxygen. Allow this solution to pass into the reaction tube through the three-way stopcock. Avoid the entrance of any atmospheric gases. Close the reaction tube to the vacuum line. Attach the sample gas container to the reaction vessel through a connecting tube containing a vacuum gauge. The amount of sample gas is calculated from the known pressure, temperature, and reaction vessel volume. Close off and remove the sample container. Vigorously shake the reaction tube in a direction parallel to its longitudinal axis. Fifteen minutes of shaking at a rate of two to three strokes per second are sufficient.

Place 1 ml water and 0.3 ml o-tolidine solution in the reagent cup. Purge this mixture of oxygen and add it to the reaction tube, avoiding the entrance of any atmospheric gases. Mix the contents of the reaction tube. A yellow coloration is observed. Rinse the contents of the reaction tube into a 100-ml volumetric flask with distilled water and dilute to the mark with additional distilled water. Transfer a portion of this solution to a 1-cm cuvette and measure the absorbance at 430 nm. For absorbance readings greater than 0.5, dilute 20.0 ml of the solution to 100 ml in a second volumetric flask and record the absorbance of this, more dilute, solution. A reagent blank solution is prepared by following the same complete procedure but evacuating the reaction tube instead of introducing any sample gas.

A standard curve is prepared from solutions of varying potassium permanganate concentration. Place in each of a series of 100-ml volumetric flasks 75 ml water and 0.3 ml o-tolidine solution. Add varying amounts of potassium permanganate solution. Zero to 1.00 ml $KMnO_4$ stock solution corresponds to oxygen concentrations from zero to 10.0 microliters oxygen. For samples containing greater amounts of oxygen, 1.00 to 5.00 ml $KMnO_4$ solution is reacted with o-tolidine in 500-ml volumetric flasks. These correspond to oxygen concentrations between 10.0 and 50.0 microliters oxygen. The blank solution for the standards contains only 0.3 ml o-tolidine diluted to the appropriate volume.

b. Manganese Complexation

The red-orange complex formed by oxidation of reduced manganese-formaldoxime complex in alkaline solution by dissolved oxygen is used as a method for oxygen determination (290). The oxidized form of the complex is reduced by reaction with metallic zinc. In the reduced form, contact with dissolved oxygen restores the complex to its colored form. Absorbance values are measured at 475 nm, and the measured absorbance values are related to the amount of oxygen present by preparation of a suitable calibration curve.

If ethylenediaminetetraacetic acid, EDTA, is present with manganese(II) in a dissolved oxygen sample, a red Mn(III)–EDTA complex forms upon addition of base. This complex is used for the spectrophotometric determination of oxygen (74).

The following procedure uses 1,2-cyclohexanediaminetetraacetic acid (CyDTA) in place of EDTA to form a similar complex with manganese(III) for the determination of aqueous oxygen solutions. A calibration plot of over 20 data points determined by this procedure, when compared with similar samples determined by a potentiometric modification of the Winkler method, produced, for samples containing less than 11 mg oxygen per liter, slopes having a relative standard deviation of 0.83%. A molar absorptivity for Mn(III)–CyDTA of 339 liters $\cdot$ mole^{-1} $\cdot$ cm^{-1} is reported (257).

Solutions

Manganese(II) Chloride, 3M. Dissolve 594 g MnCl$_2$ $\cdot$ 4H$_2$O in distilled water and dilute to 1 liter.

Sodium Hydroxide, 8M. Dissolve 320 g NaOH in distilled water and dilute to 1 liter.

Sulfuric Acid, 5M.

1,2-Cyclohexanediaminetetraacetic Acid, Solid.

Procedure

Samples are collected in 300-ml biologic oxygen demand bottles, each of accurately known volume. Add to the sample from a syringe 2 ml 3M MnCl$_2$. The syringe needle is dipped into the sample during addition of reagent. From a separate syringe, add 2 ml 8M sodium hydroxide solution, again dipping the needle into the sample solution to avoid contact with atmospheric oxygen. Stopper the sample bottle and mix thoroughly. Allow the precipitate to settle. Add 0.5 g solid CyDTA and mix thoroughly. Now add 1.4 ml 5M sulfuric acid. The pH of the solution is now 3.0 ± 0.2. Mix thoroughly. The red-purple color of Mn(III)–CyDTA appears. Filter the sample through a medium-porosity sintered glass filter to remove undissolved CyDTA. Measure the absorbance of the sample solution at 500 nm in a 1-cm cell. The absorbance decreases approximately 3% per hour at room temperature. A reagent blank in the reference cell of the spectrophotometer is prepared in the same manner but from a sample that has been dearated with nitrogen for 1 hr prior to analysis. A calibration curve is prepared by analyzing samples of known oxygen content. Cu(II) at the 10-ppm level does not interfere. Fe(III) at the 10-ppm level causes slight interference. Ca(II) at the 10-ppm level interferes.

B. INDIGO CARMINE METHOD

Oxidation of the reduced, leuco, form of indigo carmine by oxygen is used for determination of oxygen in aqueous solutions and in gaseous mixtures (90,189). Various reducing agents are used to form the yellow leuco base of indigo carmine. Originally, glucose in potassium carbonate solution was employed. Since then, sodium dithionite (54,55,245) and ammoniacal zinc amalgam (15,16,313) have been used. The leuco base of indigo carmine, when oxidized, first forms a red semiquinone oxidation product. Continued oxidation results in a blue, completely oxidized form. Procedures are reported which, because of a large excess of indigo carmine, ensure only partial oxidation to the red form (185,186). Other procedures arrange conditions to ensure complete oxidation to the blue form (202,229). Still other modifications, using glycerol in addition to glucose (or dextrose) in potassium hydroxide solution, report varying observed colors for the oxidized indigo carmine solution depending upon the extent of oxidation. The extent of oxidation depends on the amount of oxygen present. Numerous color standards, prepared by mixing appropriate portions of cobalt(II) chloride, iron(III) chloride, and copper(II) sulfate solutions and ranging from yellow-green through orange, red, purple, blue, and blue-green, are associated with oxygen concentrations up to 0.17 mg oxygen per liter solution. Visual matching of the color from the sample solution under study with the appropriate color standard immediately gives the oxygen content of that sample (11,46).

To avoid the question of complete oxidation of the leuco base, a procedure is reported which measures the decrease in absorbance of reduced indigo carmine rather than the increase in absorbance of either the partially oxidized or completely oxidized form produced by reaction with oxygen. This procedure is given in Section IIB.

C. COPPER–AMMINE METHOD

The oxidation of ammoniacal copper(I) by oxygen and colorimetric measurement of the copper(II)–ammine complex formed is used for determination of oxygen in various gases and liquids (29,131,133,144, 182,212,232). The sample is passed through a solution of copper(I)–ammine or over metallic copper immersed in a solution containing ammonia and ammonium chloride (79,188). In the latter case, oxygen is reduced on the copper surface forming a mixture of copper(I) and copper(II) oxides. Earlier procedures using metallic copper either applied a numerical correction to account for incomplete formation of copper(II) oxide or, after the sample had reacted, allowed the mixed oxides to remain in contact with the metallic copper. This converts all oxidized copper to copper(I), which can be determined by potentiometric means (77,298).

An alternate procedure allows air to pass over the mixed oxides after their reaction with sample and after removal from contact with metallic copper. In this way, all the oxidized copper is converted to copper(II), which can be determined colorimetrically (68,239). Contact with hydrogen peroxide is also used to achieve oxidation (140). The copper–ammine procedure is applied to determination of oxygen in gaseous samples. Dissolved oxygen is also determined by sweeping the oxygen from a liquid sample with nitrogen and following the procedure as developed for gaseous samples. Removal of interfering gases is possible to some extent by passing the gaseous oxygen sample through an appropriate scrubbing solution prior to passing through the copper containing solution. A silver nitrate scrubber is, for example, used to remove hydrogen sulfide and mercaptans from gaseous petroleum fractions prior to oxygen determination.

Procedures similar to those using copper(II)–ammine are reported using diethyldithiocarbamate for spectrophotometric determination of copper(II) formed from reaction between oxygen and ammoniacal copper (44,191).

The following procedure for oxygen is based upon colorimetric measurement of the copper(II)–ammine complex formed from oxygen in nonaqueous solvents. It is applied to the determination of oxygen in metal alkyl solutions and in solvents to be used with aluminum alkyl catalysts or Grignard reagents. Data are presented for oxygen concentrations up to approximately 100 mg oxygen per liter. One can expect both accuracy and precision to be about ±5% (213).

Solutions

Reduced Copper Solution. Dissolve 5 g copper(II) sulfate and 75 g ammonium chloride in 500 ml 5% aqueous ammonium hydroxide. Place the solution in a container closed with a serum-type rubber stopper and containing copper wire. Reduction is complete when the solution loses its dark-blue color. It is not necessary that the solution be completely colorless.

Benzene.

***n*-Butanol.**

Ammonium Hydroxide, 15*M*.

Procedure

The reaction vessel consists of a cuvette permanently attached to the bottom of a glass-stoppered bulb of about 200-ml capacity (127). It may be necessary to modify the sample compartment of the measuring instrument to accommodate a vessel of this size. A calibration mark on the cuvette indicates a volume of 3.0 ml. Place 100 ml benzene and 15 ml *n*-butanol in the reaction vessel. Flush the air space above the liquid with nitrogen and cover the opening of the reaction vessel with a serum-type rubber stopper. Add from a syringe 10 ml reduced copper solution and 10 ml 15M ammonium hydroxide. Mix thoroughly and withdraw the aqueous phase with a syringe. Again add 10 ml reduced copper solution and 10 ml 15M ammonium hydroxide. Mix thoroughly and withdraw all but 3.0 ml of the aqueous phase. This is the reference solution. Record its absorbance at 640 nm. Now add to the reference solution 10.0 ml of the sample. Shake vigorously for 3 min and read the increased absorbance from the copper(II)–ammine which has now formed in the reaction vessel. Correct this absorbance for the blank reading. The oxygen content is found from a calibration curve prepared previously by recording absorbance measurements of samples containing known amounts of oxygen. Those substances that either oxidize copper(I) to copper(II) or reduce copper(II) formed from reaction with oxygen interfere with this determination and should be removed prior to analysis.

D. IRON–THIOCYANATE METHOD

Iron(II) hydroxide is used in place of Mn(II) hydroxide for determination of oxygen by a procedure analogous to the Winkler method. Freshly precipitated iron(II) hydroxide is allowed to come in contact with the oxygen-containing sample. Iron(II) is oxidized to iron(III). After sufficient time to ensure complete reaction, the solution is acidified and the iron(III) present determined by appropriate means. Colorimetric determination as the thiocyanate complex is the most common method (310). Sulfosalicylic acid is also reported for colorimetric determination of iron(III) in oxygen analysis (98).

The following procedure is for determination of oxygen in coke-oven gas mixtures. It is applicable in the range of 0.010 to 1.00% oxygen by volume. One can expect an oxygen value whose accuracy is within ±2% of the true oxygen content (266).

Solutions

Iron(II) Ammonium Sulfate. Dissolve 15 g $FeSO_4(NH_4)_2SO_4 \cdot 6H_2O$ and 8.7 ml 12M hydrochloric acid in distilled water. Dilute to 1 liter. Prepare this solution fresh every second day.

Potassium Thiocyanate. Dissolve 10 g KSCN in distilled water and dilute to 1 liter.

Iron(III) Ammonium Sulfate. Dissolve 0.860 g $Fe(NH_4)(SO_4)_2 \cdot 12H_2O$ in distilled water and dilute to 100 ml in a volumetric flask. Dilute 10.0 ml of this solution and 1.0 ml $12M$ HCl with distilled water to 1 liter in a volumetric flask. One milliliter of this solution is equivalent to 0.001 ml dry oxygen gas at standard temperature and pressure.

Sulfuric Acid. To 800 ml distilled water, add 200 ml concentrated ($18M$) sulfuric acid.

Potassium Hydroxide. Dissolve 300 g KOH in 700 ml distilled water.

Sodium Hydroxide. Dissolve 500 g NaOH in 500 ml distilled water.

Hydrochloric acid, $12M$.

Hydrochloric acid, $1M$.

Procedure

The reaction vessel for this determination consists of a glass bottle of approximately 175-ml capacity provided with a recessed ground-glass stopper notched so that upon rotation reagents can be added to the flask without removing the stopper and upon further rotation again form a closed seal (265). A gas inlet tube reaches to the bottom of the flask and a gas outlet tube is located near the top of the vessel. Both inlet and outlet tubes are provided with glass stopcocks. The exact volume of the flask is known. Two test tube scrubbers, the first containing sulfuric acid solution and the second containing potassium hydroxide solution, are placed between the sample source and the reaction vessel inlet tube. These serve to remove volatile bases, hydrogen sulfide, and hydrogen cyanide that might be present with oxygen in a coke-oven gas sample. If not removed, these substances interfere with the oxygen determination. Before attaching the reaction vessel, pass at least 5–6 liters sample gas through the scrubbers at a flow rate between 30 and 50 liters per hr. Now, while gas is still flowing, attach the reaction vessel, containing 15 ml iron(II) solution, to the scrubber train. Attach the reaction vessel inlet tube to the scrubber. The outlet stopcock is open and the ground-glass stopper closed. Pass an additional 14 liters or more sample gas through the reaction vessel. Close the outlet and then the inlet stopcocks. Adjust the pressure within the flask by momentarily opening the outlet stopcock. Record temperature and pressure. Sample volume is determined from the volume of the reaction flask, from the temperature, and from the pressure.

Add 25 drops sodium hydroxide solution to the funnel neck of the recessed ground-glass stopper. Cool the reaction vessel with running water. Rotate the glass stopper and allow the sodium hydroxide to flow slowly into the flask. Wash the residual caustic into the flask with two portions of distilled water of 1 or 2 ml each. No air should enter the flask during these additions. Place the flask on a mechanical shaker and agitate vigorously for 1 hr. After shaking, chill the flask in cold water or, if unsaturated hydrocarbons are thought to be present, evacuate to a partial vacuum. Place 10.0 ml $12M$ hydrochloric acid in the funnel neck of the recessed glass stopper. Cool the reaction flask with running water. Rotate the glass stopper and allow the acid to enter the flask. Wash residual acid into the flask with two portions of distilled water of 1 to 2 ml each. Mix until all of the iron(II) hydroxide dissolves. It may be necessary to warm the flask in hot water, with the glass stopper open, provided no partial vacuum exists within the flask. The acidified solution is safe from interference from atmospheric oxygen. Quantitatively transfer the contents of the reaction flask to a volumetric flask. Add sufficient additional hydrochloric acid to make the final solution $1M$ in HCl. If a 100-ml volumetric flask is used, this should require about 0.7 ml additional $12M$ HCl. Now, add distilled water to dilute the sample to the mark of the volumetric flask.

Various means are available for determining iron(III) with thiocyanate. To continue with the procedure presented, visual comparison using Nessler tubes is employed. Into each of two 50-ml Nessler tubes, place 10 ml potassium thiocyanate solution. Place an accurately measured aliquot of sample solution into one tube. The aliquot should, upon color development, produce a color comparable to one produced by no more than 5 ml iron(III) standard solution. Add, if necessary, additional $1M$ HCl to produce a total of 2.5 ml $1M$ HCl in the tube, and add sufficient $1M$ HCl to equal the amount in the sample-containing tube. Add an accurately measured amount of iron(III) standard until the color intensity of the two tubes is about equal. Dilute to 20.0 ml with distilled water and add sufficient additional iron(III) standard until the colors match in both tubes. The sum of these volumes is the total amount of iron(III) in the standard tube and is equivalent to the amount of iron(III) in the sample tube. A blank correction is necessary and is determined in the following way. Place 25 drops sodium hydroxide solution and 15–20 ml distilled water in a 50-ml volumetric flask. Add 10.0 ml $12M$ hydrochloric acid to the flask. Cool, add 15 ml iron(II) solution, and dilute to the mark with distilled water. To each of two Nessler tubes, add 10 ml potassium thiocyanate solution. To the sample tube, add 5.0 ml blank solution and dilute to 25.0 ml with distilled water. To the remaining Nessler tube, add 9.2 ml $1M$ HCl and sufficient iron(III) standard, accurately measured, to nearly match the color of the first tube. Dilute to 25.0 ml with distilled water and add sufficient additional iron(III) standard until the colors match in both tubes. Ten times the total amount of iron standard in the blank is subtracted from the amount of iron necessary to match the oxygen-containing sample solution. The oxygen concentration of the sample gas is calculated from the volume of sample and the equivalent amount of iron(III).

E. PYROGALLOL METHOD

The color of an alkaline pyrogallol solution in the presence of oxygen is used for oxygen determination in gaseous samples. An early procedure employed visual comparison of unknown samples against iodine solutions of known concentration (7).

The following procedure is for the determination of oxygen in inert gases. A linear relation between oxygen content and absorbance is observed for oxygen in amounts up to 800 μg (314).

Reagents

Pyrogallol, Solid.

Potassium Hydroxide, Solid.

Procedure

The reaction vessel is a round-bottom 1-liter flask, of accurately known volume, with a center opening and adjacent side opening. Each opening is provided with a stopcock and with a ball joint for attaching other flasks and/or a vacuum line. Opposite the center opening, a cuvette is permanently attached to the bottom of the flask. It may be necessary to modify the cell compartment of the measuring instrument to accommodate this assembly. Solid potassium hydroxide, 2.5 g, is placed in the flask through the center opening. The center opening is closed and a vacuum source connected to the side opening. The flask is evacuated and, in addition, gently flaming is used to remove adsorbed gases. The side opening is closed. A small round-bottom flask with two openings, one at the top and the other at the bottom, is now attached to the center connection of the reaction vessel. Into this smaller flask place 25 ml distilled water and 1 g pyrogallol. Connect to a water aspirator and degas the pyrogallol solution. After all dissolved gases are removed, open the center stopcock and slowly allow all but 0.5 ml of the pyrogallol solution to drain into the reaction vessel Close the stopcock and remove the smaller flask. The alkaline pyrogallol solution should be nearly colorless. Set the reaction vessel upright, allowing the liquid to fill the cuvette. Place the flask in a water bath at 70°C. Remove and record the absorbance value at 15-min intervals until successive values correspond. This absorbance is the blank correction for the oxygen determination. After obtaining a satisfactory blank reading, remove the flask from the instrument and attach a two-opening connector to the side opening of the reaction vessel. One opening of this connector is attached to a vacuum source and the other, to the sample source. Apply vacuum and purge the connector fitting of trapped air. Shut off the vacuum and open the stopcock leading to the sample gas. Allow sufficient sample into the reaction flask to give an oxygen content within the range of this method. The exact amount of sample gas is calculated from the known volume of the reaction flask, from the temperature, and from the pressure read from a gauge in the sample gas line.

Close the stopcock and remove the connecting joint assembly. Immerse the reaction vessel into the 70°C water bath. Remove and take absorbance readings at regular intervals until successive readings give consistent values. Three or four readings at 15-min intervals may be necessary for consistent readings. Initial readings generally give higher absorbance values but do not produce a linear relation with oxygen concentration. They are, therefore, unsatisfactory. The final absorbance reading, after correction for blank, is used to determine oxygen content of sample gas. The amount of oxygen is found from a calibration curve prepared by analyzing samples of known oxygen content using this same procedure.

F. METHYL VIOLOGEN METHOD

A major difficulty in the determination of oxygen is elimination of possible contaminations from atmospheric oxygen and from dissolved oxygen in reagent solutions. Elaborate precautions are proposed to remove these sources of error. They result, however, in complicated procedures and the need for specially designed apparatus. One attempt to overcome these difficulties proposes the formation of the reagent for oxygen analysis in situ by photochemical change. The photochemical reduction of methyl viologen (1,1'-dimethyl-4,4'-bipyridinium dichloride) in the presence of proflavine (3,6-diaminoacridine), EDTA, and a buffer at pH 6.5 produces a characteristic deep-blue radical cation of methyl viologen. Reaction of this reduced radical cation species with oxygen returns it to the colorless oxidized form. These changes are followed spectrophotometrically and used for the determination of gaseous oxygen (287). Traces of oxygen in the generating solution are corrected by a suitable blank before sample addition. No further addition of reagents is necessary after the sample is placed in the reaction vessel. Photochemical reduction is achieved by exposing the reagent mixture to a high-intensity light. After generation, normal indirect laboratory lighting does not produce further reduction, and one can safely proceed with the analysis. Oxidizing gases other than oxygen do, however, interfere.

The method is unique in two respects. Methyl viologen is not consumed in the reaction. This allows repetitive analyses to be conducted using the same methyl viologen solution over and over again. EDTA, which serves as the electron donor for the methyl viologen reduction, is consumed; but, if the initial reagent mixture contains a relatively large amount of this reagent, many determinations can be carried out before it becomes necessary to change the reagent mixture. The second feature of this procedure is applied to samples containing more oxygen than necessary to completely oxidize the reduced methyl viologen present. The oxidation is allowed to proceed until the absorbance has decreased to some convenient value. Then, additional reduced methyl viologen is generated and allowed to continue reacting with oxygen present in the sample. If necessary, the regeneration is repeated several times during the course of a single determination.

The following procedure is for the determination of oxygen in nitrogen, carbon monoxide, hydrogen, and ethylene. With a 300-ml reaction vessel, it is applicable for oxygen concentrations up to 35 ppm. With repetitive generation of reduced methyl viologen, oxygen concentrations up to 200 ppm are determined.

Solutions

Methyl Viologen, 0.01M. Dissolve 2.60 g methyl viologen in distilled water and dilute to 1 liter.

Proflavin Dihydrochloride, 0.001M. Dissolve 0.32 g proflavin dihydrochloride in distilled water and dilute to 1 liter.

Phosphate Buffer, pH 6.5. Mix appropriate amounts of $0.2M$ $Na_2HPO_4 \cdot 7H_2O$ and $0.2M$ NaH_2PO_4 to produce a solution of pH 6.5. Measure the pH with a pH meter.

EDTA, 0.25M. Dissolve 93 g $Na_2H_2EDTA \cdot 2H_2O$ in 800 ml distilled water. Adjust to pH 6.5 with phosphate buffer and dilute to 1 liter with additional distilled water.

Proflavin–Methyl Viologen–EDTA Solution. Combine 18 ml $0.001M$ proflavin, 45 ml $0.01M$ methyl viologen, 36 ml $0.25M$ EDTA, and 90 ml phosphate buffer with 261 ml distilled water. Store in an amber bottle in the dark or in a refrigerator. The solution is stable for several months.

Iron(III) Sulfate, 0.1M. Dissolve 40 g $Fe_2(SO_4)_3$ in $0.36N$ H_2SO_4. Dilute to 1 liter with additional $0.36N$ H_2SO_4. Standardize this solution to accurately determine the iron(III) concentration.

Iron(III) Sulfate, 0.0020M. Using a 1-liter volumetric flask, dilute an appropriate amount of standardized iron(III) sulfate with 100 ml $0.25M$ EDTA solution and sufficient distilled water to give a final solution exactly $0.0020M$.

Procedure

The reaction vessel consists of a gas collection tube with a Teflon stopcock at each end. A cuvette is permanently attached at a right angle to the length of the tube at a point midway between the stopcocks. A tube of about 300-ml capacity serves for the analysis of samples containing up to 35 ppm oxygen. Reaction tubes ranging from 50 to 500 ml capacity are suggested depending upon the amount of oxygen in the sample gas. An alternate approach with samples containing large amounts of oxygen is to use the repetitive generation technique described above. It is necessary to know the exact volume of the reaction vessel. It may be necessary to modify the cell compartment of the measuring instrument to accommodate larger reaction vessels. Place 15 ml proflavin–methyl viologen–EDTA solution in the reaction vessel. Allow the solution to drain into the cuvette. Deaerate the vessel and stock solution by passing sample gas through the tube for several minutes at a flow rate of 300 to 1000 cm^3 per min. With gas flowing, expose the solution in the cuvette to the bright light of a high-intensity light source (120 to 200 foot-candle output at 12 in.). A deep-blue color is observed. Continue exposing the solution until sufficient reduced methyl viologen radical cation is formed to give an absorbance reading between 0.7 and 0.9 absorbance units. Measurement is made at 580 nm. As the absorbance approaches the desired value, mix the solution and sample gas in the reaction tube by tipping the tube. Return to an upright position (cuvette down) and continue generating reduced methyl viologen reagent until the desired absorbance is obtained.

Shut off the light source. This absorbance value represents the absorbance of the reduced reagent. It is important that no additional direct light strike the reaction tube at this time. Indirect laboratory light is allowable. Thirty to 60 sec after the initial absorbance reading is obtained, close the stopcocks stopping the flow of sample gas. Disconnect the gas source. Momentarily, open one of the stopcocks to adjust the gas pressure within the tube to atmospheric pressure. Sample size is determined from the volume of the reaction vessel, from the temperature, and from the pressure.

Now tip the reaction vessel and allow the reduced methyl viologen reagent to come into contact with the sample gas. Shake the tube for about 20 sec, allow the liquid to drain back into the cuvette, and record the decreased absorbance value at 580 nm. Repeat the shaking and again record the absorbance after a 20-sec interval. Repeat this procedure until successive identical absorbance readings are obtained. This usually occurs after five or six readings. Record the difference in absorbance values between that of the unreacted methyl viologen radical cation and that of the methyl viologen radical cation present after reaction.

Iron(III) sulfate is used to calibrate this procedure. As a separate determination, sweep the reaction vessel with nitrogen. Generate reduced methyl viologen reagent as described above until sufficient reagent is present to give an absorbance of between 0.7 and 0.9 absorbance units, Add exactly 0.2 ml dilute iron(III) sulfate standard solution with a hypodermic syringe. Do not allow any of the iron(III) solution to drain into the cuvette containing the reduced methyl viologen. Again flush the reaction tube with nitrogen for 1 to 2 min. Record the initial absorbance of the reduced reagent. Now tip the reaction tube and allow the iron(III) solution to come into contact with the reduced reagent. Return the tube to an upright position and allow the reagent solution to drain back into the cuvette. Repeat this mixing a second time. After the second mixing, read the absorbance of the remaining reduced methyl viologen. A blank solution containing exactly 0.2 ml distilled water in place of dilute iron(III) solution is also reacted in a similar manner.

The absorbance difference for the iron(III) solution is corrected for the difference obtained from the water blank. The corrected absorbance difference is converted to the corresponding equivalent oxygen concentration by taking into consideration the volume of the reaction vessel, the temperature, the sample pressure, the vapor pressure of water, and the stoichiometry of the reactions betwen iron(III) and oxygen with reduced methyl viologen radical cation. One mole iron(III) reacts with one mole reduced methyl viologen reagent. One mole molecular oxygen reacts with two moles reduced methyl viologen reagent at pH 6.5 to yield one mole hydrogen peroxide and two moles oxidized methyl viologen. The calibration factor obtained (amount of oxygen per unit of absorbance difference), when multiplied by the absorbance difference for an oxygen-containing sample, gives the amount of oxygen present in that sample.

G. MISCELLANEOUS METHODS

Amidol (2,4-diaminophenol dihydrochloride) is suggested as a colorimetric reagent for dissolved oxygen (143). Contact with dissolved oxygen produces a red-colored oxidation product in solutions buffered at pH 5.1 with citrate buffer. The intensity of the color produced is compared with standards prepared by mixing appropriate amounts of cobalt chloride and potassium dichromate (110). An alternate procedure measures the absorbance of the colored solutions spectrophotometrically at 490 nm (95). Oxygen is determined at concentrations up to 5 mg per liter, with the best results obtained for oxygen in the concentration range of 0.2 to 2.5 mg per liter. Thirty min is required for maximum color formation. The color begins to fade after 60 min. Reducing substances, if present in the sample, cause interference. These include nitrite, sulfite, hydrogen sulfide, and iron(II). Chlorine in excess of 1.5 mg per liter and iron(III) in excess of 5 mg per liter also interfere. Results from this procedure and the Winkler method agreed within 5%.

The reaction of reduced, red anthraquinone-β-sulfonate in alkaline solution to its oxidized colorless form is used for determination of trace oxygen in nitrogen, helium, argon, and various hydrocarbon gases (38,65,274,282). The sample gas is swept through a solution of reduced anthraquinone-β-sulfonate by oxygen-free carrier gas. The analysis apparatus is designed to allow the solution, after contact with sample gas, to flow through an absorption cell. Absorbance is measured and the solution is degassed, reduced, and recycled for further contact with sample gas. The decrease in absorbance at 530 nm is recorded at known flow rates. Electrolytic reduction of aqueous sulfuric acid in a separate part of the apparatus serves as a source of known amounts of oxygen. This generated oxygen passes through the apparatus, and calibration curves are prepared from the observed absorbance values and the known oxygen concentrations. Several modifications of the original apparatus design are proposed (152,158,238, 264,280,301). The procedure is applied to gaseous oxygen in the concentration range from 0.01 to 15 ppm, with a standard deviation varying from ±0.004 at lower oxygen concentrations to ±0.2 at higher concentrations.

Large amounts of acetylene and hydrogen do not interfere with this method. Carbon monoxide in amounts less than 10 times that of oxygen also does not interfere. Carbon dioxide, in large amounts, interferes. Zinc amalgam is generally used to prepare the reduced anthraquinone-β-sulfonate although photochemical reduction is also suggested.

Photochemically reduced anthraquinone-β-sulfonate is monitored spectro-photometrically at wavelengths in the ultraviolet region of the spectrum. A photonmetric titration procedure, analogous to a coulometric titration but utilizing the number of generated photons (light intensity x time) in place of the number of generated coulombs (current x time) to determine equivalents of reacting species, is reported (174). A one-to-one correspondence exists between the micromoles of generated photons and the micromoles of reacting oxygen.

Formation of blue, oxidized methylene blue from a solution of the colorless leuco form is proposed for estimation of gaseous oxygen (73,163,164,246). The leuco form is prepared by mixing a solution of methylene blue with a small amount of a reducing sugar such as lactose in the presence of sodium hydroxide. An oxygen-containing gas passed through this solution oxidizes the leuco form back to its blue color. The intensity of the color measured at 660 nm is related to the amount of oxygen in the sample by comparing with absorbance values obtained from gaseous mixtures of known oxygen content reacted in a similar manner. The method is applicable to samples containing less than 0.1 mg oxygen per liter.

A 1:1 water-ethanol solution containing 3% ammonia and 0.1% safranine T (3,7-diamino-2,8-dimethyl-5-phenylphazinium chloride) when passed through a zinc amalgam-filled column forms the colorless, leuco form of safranine T. Contact between this reduced species and a solution containing dissolved oxygen causes oxidation of the leuco form to the red, oxidized form (1,2,145).

Colorimetric measurement of the oxidized form at 520 nm results in an absorbance curve which obeys Beer's law for solutions containing up to 0.03 mg oxygen per liter solution. Color formation is complete after 3 min at room temperature. The reaction is stoichiometric, requiring two moles of safranine T for each mole of oxygen. Hydrazine does not interfere.

Oxygen is known to react with pentacyanocobaltate(II) ion to form a peroxy-bridged complex. Pentacyanocobaltate(II) ion also reacts with halopentamminecobalt(III) ions to form inert halopentacyanocobaltate(III) ions and labile aquopentamminecobalt(II) ions. This latter ammine complex also reacts with EDTA forming the cobalt(II)–EDTA complex. With this series of reactions in mind, it is shown that trace amounts of any species that consume EDTA affect the catalytic enhancement of EDTA upon the rate of exchange between copper(II)–EDTA and nickel(II)–trien (trien = triethylenetetraamine) (285). The rate of this exchange is followed spectrophotometrically at 550 nm, the absorbance maximum of copper(II)-trien being formed in the exchange reaction. This sequence of reactions is utilized for determination of trace amounts of oxygen (192). An excess of pentacyanocobaltate(II) ions is brought into contact with an oxygen-containing sample solution. After reaction, the excess pentacyanocobaltate(II) determines the amount of aquopentamminecobalt(II) available for reaction with EDTA. The amount of remaining EDTA determines the rate of copper(II)–EDTA and nickel(II)–trien exchange. The amount of oxygen present bears a one-to-one relationship with the amount of EDTA consumed. Dissolved oxygen is determined in the concentration range of 0.08 to 0.60 mg oxygen per liter, with an accuracy of about 0.01 mg oxygen per liter and a standard deviation of about ±0.05 mg oxygen per liter.

II. APPLICATIONS

The determination of oxygen is, for all methods presented, dependent upon the availability of acceptable standards. Measurements for unknown concentrations of oxygen are compared with similar measurements obtained from standard solutions. Standards can be accurately known solutions of any suitable oxidizing agent. For example, in the modified Winkler procedure presented earlier, utilizing triiodide ion formation and direct absorbance measurement of the ultraviolet absorbance maximum, primary standard potassium hydrogen biodate serves as a source of triiodide for preparation of a calibration curve. Other procedures use saturated solutions of dissolved oxygen as the standard from which a calibration curve is prepared. Appropriate dilutions of saturated oxygen solution with oxygen-free solvent provide a range of oxygen standards. The oxygen concentration of a saturated solution, at known temperature and pressure, is obtained from solubility tables (96,165,284,296). These values for oxygen are the subject of considerable discussion, and there is disagreement among various researchers about the validity of some of the tabular values. These disagreements center about errors in the procedures used for determination of oxygen, about improper accounting for possible interferences, and about errors in achieving oxygen saturation without supersaturation. These considerations have resulted in several revised tables of oxygen solubilities in both pure and natural waters (27,50,53,208). A study of oxygen solubilities in Winkler reagents is also available, and corrections are given for dissolved oxygen values determined when using air-saturated solutions of these reagents (214).

Various techniques are proposed to achieve a saturated solution of air, or oxygen, in water. These include, for example, placing 2 liters water in a 5-liter flask and stirring vigorously for 6 hr at 30°C. No supersaturation is reported. An alternate procedure consists of bubbling air through water with intermittent shaking. No supersaturation occurs at high temperatures. Another procedure provides for a shaking motion to equilibrate air in contact with water in a rectangular box. Three hr of shaking is satisfactory to achieve equilibrium. A countercurrent technique in which water and air saturated with water vapor are brought into contact from opposite directions is also suggested (114).

To calibrate procedures for the analysis of gaseous oxygen mixtures, gas samples containing known percentages of oxygen are used. These are available commercially, prepared by electrolytic generation (181) or by a steady flow technique. This later technique is used to prepare oxygen in helium over the concentration range of 10 to 10,000 ppm oxygen (8).

A. OXYGEN IN CERTAIN GASES

Suggested procedures and apparatus for collection of gaseous samples are available (11). These include filling an evacuated container, repeated displacement of air from a container by sample gas, absorption of sample gas in a suitable absorbent, and condensation of gas sample in a suitable cold trap.

Several of the procedures presented in Section I apply to the determination of oxygen in gaseous samples. The iron—thiocyanate procedure is for determination of oxygen in coke-oven gases. The methyl viologen procedure is for oxygen mixed with nitrogen, hydrogen, carbon monoxide, or ethylene. The methylene blue procedure is applied to determination of atmospheric oxygen. Pyraogallol, anthraquinone-β-sulfonate, and a modified Winkler method using extracted iodine are all used for oxygen determination in inert atmospheres.

The following procedure is for the determination of oxygen in argon, carbon dioxide, helium, hydrogen, or nitrogen. It is based upon a modification of the Winkler method and employs o-xylene as the organic phase for extraction of iodine formed during reaction of gaseous oxygen with Winkler reagent solutions. The method is applicable for oxygen in the concentration range of 1 to 30 ppm oxygen , with a standard deviation of ±0.33 ppm oxygen (273).

Solutions

Manganese(II) Chloride, 0.1M. Dissolve 10 g $MnCl_2 \cdot 4H_2O$ in 500 ml distilled water that has been slightly acidified with hydrochloric acid. Boil the solution for 5 to 10 min while purging with carbon dioxide. Continue purging and cool to room temperature. Dilute to 500 ml.

Potassium Iodide, 0.3M. Boil 100 ml distilled water in a 200-ml flask while purging with carbon dioxide. Continue passing carbon dioxide through the water as it cools. After cooling, add 5 g KI. Dissolve the KI and cool the solution to room temperature while continuing to purge with carbon dioxide. Prepare fresh daily.

Sodium Hydroxide, 1.5M. Boil 100 ml distilled water in a 200-ml flask while purging with helium. Continue passing helium through the water as it cools. After cooling, weigh and immediately add 6 g NaOH pellets to the water. Dissolve the NaOH and cool the solution to room temperature while continuing to purge with helium. Prepare fresh daily.

Hydrochloric Acid, 12M.

o-**Xylene.**

Potassium Dichromate, 0.00100N. Weigh exactly 0.294 g reagent-grade $K_2Cr_2O_7$ and dissolve in distilled water. Dilute to 1 liter in a volumetric flask.

Procedure

The reaction vessel is a 1-liter round-bottom flask fitted at the top with a Y-connection. Each arm of the Y terminates with a vacuum stopcock. Opposite this opening is a standard taper joint to which a test tube, also with standard taper joint, can be attached. To one of the upper two connecting arms, a vacuum gauge and the sample-containing flask are attached. To the other connection, a three-way stopcock is attached. One setting of this stopcock connects a reagent preparation container of about 30-ml capacity to the reaction vessel. The other setting connects the reaction vessel to a vacuum train consisting of a moisture trap, a cold trap, a vacuum gauge, and the vacuum source.

Assemble the vacuum train. Close the reaction vessel stopcock leading to the sample vessel while, at the same time, opening the reaction vessel stopcock leading to the reagent preparation flask and vacuum train. Set the three-way stopcock to connect the reaction vessel to the vacuum line while closing off the reagent preparation vessel. Place 10.0 ml 1.5M NaOH and 5.0 ml 0.3M KI into the test tube side vessel. Attach this to the reaction flask. Evacuate the system to a pressure less than 0.003 torr. Place 25.0 ml 0.1M $MnCl_2$ solution into the reagent preparation flask. This flask is enclosed within a heating mantle. By appropriate control of the heating mantle, boil the $MnCl_2$ solution within the flask. At the same time, purge the solution with carbon dioxide. These precautions are necessary to ensure minimum contamination of oxygen from the reagent solutions.

After boiling, cool the solution to room temperature while continuing to purge with carbon dioxide. When the solution is at room temperature, open the three-way stopcock and allow the $MnCl_2$ solution to be sucked into the evacuated reaction vessel. Close the three-way stopcock while a small amount of solution is left in the preparation vessel. This ensures that no air will be drawn into the system. Check the vacuum pressure to assure that it is still at its previous setting. Close the stopcock on the reaction vessel leading to the reagent preparation flask and open the stopcock leading to the sample container. The sample is conveniently contained in a 1-liter round-bottom flask with a three-way stopcock. One setting of this stopcock allows the flask to be filled with sample gas, and the alternate setting allows sample gas to flow into the reaction vessel. A vacuum gauge is connected between these containers. Sample size is calculated from the known reaction vessel volume, the temperature, and the pressure.

After the reaction vessel is filled with sample gas, shut off and disconnect the sample container. Remove the reaction vessel from the vacuum train and place it on an appropriate mechanical mixer. A mixer rotation speed of about 5 rpm is satisfactory. Rotate the reaction vessel for 2 hr. At the end of this time, reinsert the reaction vessel in the vacuum system but invert the connections, that is, the stopcock that was formerly attached to the sample bulb is now attached to the three-way stopcock leading to the reagent preparation vessel. With the reaction vessel stopcock closed, evacuate the vacuum train to a pressure less than 0.003 torr. Open the stopcock joining the reaction vessel and the vacuum train. Check the pressure to see that the proper vacuum has been maintained. Into the reagent preparation flask, which is closed from the rest of the system, place 20.0 ml boiled distilled water which has been purged with carbon dioxide. Add to this 5.0 ml $12M$ HCl. Heat this mixture to boiling while, at the same time, purging with carbon dioxide. Continue purging while cooling the acid solution to room temperature. When cool, open the three-way stopcock and allow the acid solution to be sucked into the evacuated reaction vessel. Close this stopcock while a small amount of acid solution remains in the preparation vessel. Now add 10.0 ml o-xylene to the reaction vessel by way of the reagent preparation vessel. Be careful to exclude the entry of any atmospheric air. For samples containing appreciable amounts of oxygen, it is desirable to use 25.0 ml o-xylene extractant. A white precipitate of manganese(II) hydroxide in the reaction vessel indicates low oxygen-containing samples, and 10.0 ml o-xylene is sufficient. A brown precipitate indicates higher oxygen-containing samples, and 25.0 ml o-xylene should be used. Close the stopcock on the reaction vessel. Remove the reaction vessel from the vacuum train and gently mix the contents of the flask. The iodine formed is extracted from the acidified solution into the o-xylene layer. When extraction is complete, open the flask to the atmosphere and separate the two liquid layers. Filter the o-xylene solution through paper into a 10-ml volumetric flask or, if the larger amount of o-xylene was used, into a 25-ml volumetric flask. Dilute to volume with additional o-xylene. The iodine-containing o-xylene solution is stable up to 4 hr if stored in the dark. Transfer a portion of this solution to a 1-cm cell and measure the absorbance at 495 nm against a reagent blank prepared in an analogous manner but having no contact with any sample. A calibration curve is prepared by measuring the absorbance of solutions containing known quantities of iodine extracted into the same volume of o-xylene as used for the sample extraction. Known iodine solutions are prepared by reacting known amounts of standard potassium dichromate solution with an excess of potassium iodide.

B. DISSOLVED OXYGEN IN WATER

Suggested procedures and apparatus for collecting water samples are available (11,235). Containers for sampling from large bodies of water range from gallon-size bottles weighted with lead shot (132) to a series of syringes mounted on a 2.5-meter board which, after immersion, are filled simultaneously by a spring-activated mechanism (41). In general, bottles for dipping into larger bodies of water have an inlet tube reaching to the bottom of the vessel and an outlet tube near the top. This arrangement allows the entering liquid to displace air present in the bottle (137). One procedure recommends filling collection bottles with nitrogen prior to sample collection. This avoids possible contamination from air. In addition, horizontal placement of collection bottles in water rather than vertical placement is recommended (207). Collection vessels which also serve as reaction containers are preferred (255). This avoids transfer of sample solutions. Data reporting a decrease in precision for dissolved oxygen determination with increased sample transfer are available (115).

The use of rubber tubing, not only for sampling but in the analysis apparatus, is not recommended. Oxygen is permeable to rubber tubing, and one runs the danger of both loss of oxygen from sample and contamination of sample from atmospheric oxygen. Some plastic tubing is also unsatisfactory. Poly(vinyl chloride) tubing is not recommended. Polyethylene tubing is acceptable below 44°C, and nylon tubing is acceptable below 90°C (210,236).

Some of the procedures presented in Section I apply to the determination of dissolved oxygen. Iodine—starch complex formation and the manganese—tolidine reaction are both used to determine oxygen in boiler feed waters and in natural waters. Colorimetric measurement of oxygen concentrations with triiodide ion formation and by reaction of iodine with 3,3′-dimethylnaphthidine are used in the determination of aqueous oxygen solutions. Methods using Mn–CyDTA complexation and safranine T reduction are also applicable to aqueous oxygen determination. The procedure using copper–ammine complexation is applied to oxygen dissolved in metal alkyl solutions. A procedure for determination of oxygen in blood uses the color formed from the reaction of iron with sulfosalicylic acid (98).

The following procedure utilizes the reaction of dissolved oxygen with the reduced form of indigo carmine for oxygen determination. The decrease in absorbance of reduced indigo carmine at pH 11.5 is directly related to the amount of oxygen present in a sample. As presented below, the procedure is applicable for oxygen up to a concentration equaling 50% of its maximum solubility at the temperature and pressure of measurement. Repetitive measurements agree within ±0.5%. The method is fast and easily performed with a minimum amount of apparatus and reagent solutions (286).

Solutions

Indigo Carmine, 0.045%. Dissolve 22.5 mg indigo carmine, 4.0 g glucose, and 5.0 g potassium carbonate in distilled water. Dilute to 500 ml with additional distilled water. Place the solution in a 1-liter bottle and close with a rubber septum. Flush the air space above the solution with nitrogen by inserting two number-22 hypodermic needles through the rubber septum. Attach a source of oxygen-free nitrogen to one of the needles. After purging, leave a slight positive nitrogen pressure over the solution in the bottle. Reduction of indigo carmine is complete in about two days at 25°C. Do not heat, as reduction at higher temperatures gives turbid solutions. Allow this solution to stand for one to two weeks before use. Solutions aged for shorter periods of time give less precise absorbance readings.

Oxygen-Free Water. Prepare as needed by passing prepurified nitrogen through a sintered-glass sparger immersed in distilled water in a 500-ml three-necked flask. The sparger is sealed into one opening. The second opening is closed with a rubber septum. It is through this opening that oxygen-free water is withdrawn by syringe as needed. The third opening contains a valve to allow for release of nitrogen. The water is purged vigorously for 10 min prior to use and gently thereafter.

Oxygen-Saturated Water. Shake vigorously a flask partially filled with distilled water for several minutes. Allow the solution to stand for an additional several minutes before use. Record temperature and pressure and consult an appropriate solubility table for the concentration of dissolved oxygen under these experimental conditions.

Procedure

The reaction between indigo carmine solution and sample takes place in a 1-ml gas-tight syringe. A small steel ball is placed within the bore of the syringe to facilitate mixing of solutions. The syringe also serves as the spectrophotometric cell, that is, it is placed directly into the cell compartment of the measuring instrument. This technique is both expedient and eliminates contact with atmospheric oxygen which could occur during solution transfer. It may be necessary to adapt the measuring instrument being used to accommodate the syringe as a sample cell. Rinse the syringe several times with oxygen-free water. Fill the syringe with exactly 0.5 ml oxygen-free water and 0.5 ml indigo carmine reagent solution. Shake for 30 sec to mix the solutions.

Place the syringe in the sample compartment and rotate the barrel until the absorbance reading is at a minimum. This rotation is necessary each time in order to reproducibly align the syringe in the optical path. Adjust the absorbance reading to some appropriate setting, for example, 0.5 absorbance units. Remove the syringe, discard the blank and again rinse several times with oxygen-free water. Place 0.5 ml sample solution in the syringe followed by 0.5 ml indigo carmine reagent solution. Shake the syringe for 30 sec. Place the sample into the cell compartment and record the absorbance exactly 1 min after the initial mixing. The time interval is critical and must be accurately measured. Absorbance readings are made at 410 nm. The amount of oxygen in the sample is read from a calibration curve prepared with appropriately diluted samples of oxygen-saturated water and oxygen-free water. The temperature of sample and standard solutions should be identical, otherwise a correction is necessary to account for the difference in oxygen solubility under different conditions. The calibration curve is repeated every four to five days for accurate results.

Adjusting the portions of sample and reagent solution allows a suitable calibration curve to be prepared over different ranges of oxygen concentration. For example, 0.75 ml sample and 0.25 ml indigo carmine solution are suggested for preparation of a working curve over the concentration range of 0 to 20% oxygen saturation.

Nitrate, sulfate, chloride, and carbonate do not interfere with this procedure. Nitrite does not interfere in concentrations below 1000 ppm. Iron(II) and iron(III) interfere if present in amounts greater than 5 ppm and 700 ppm, respectively. Calcium and magnesium may precipitate as carbonates and interfere. This is avoided by using a monohydrogen phosphate—hydroxide buffer at pH 11.5.

OZONE

I. METHODS OF DETERMINATION

A. IODINE METHOD

Determination of iodine formed from oxidation of iodide ion by ozone is one of the oldest methods of ozone analysis. The reaction occurs over a wide pH range but with different stoichiometry at different pH values (3,148,291,292). In determining ozone, some suggest an alkaline iodide medium, while others recommend a solution containing only aqueous potassium iodide. Still others prefer a neutral buffered iodide solution. The addition of buffer prevents an increase in pH from release of hydroxide ion formed, with oxygen and iodine, during reaction in neutral solution. Aluminum chloride—ammonium chloride or boric acid is sometimes used as buffer, although a phosphate buffer is most often chosen. Reaction between ozone and iodide in acid medium yields hydrogen peroxide in addition to oxygen and iodine. For samples of high ozone content, employing iodometric titration of liberated iodine with thiosulfate and starch indicator, the initial pH is shown to be of no consequence between limits of 2.3 and 12.3, provided the sample is acidified after contact with ozone and before thiosulfate titration (31,33).

However, conflicting reports appear in the literature regarding the merits of various procedures employing different pH values and different buffering media for trace ozone determination with iodide. Certain studies show that relatively more iodine is produced for an ozone—iodide reaction occurring in acid medium than for the same reaction occurring in neutral medium when trace amounts of ozone are reacted. Furthermore, the relative amount of iodine decreases below the stoichiometric amount for trace ozone—iodide reaction which takes place in alkaline medium. The reaction solution in alkaline medium is acidified prior to measurement of iodine content (5,89). An experimentally determined multiplication factor of 1.54 is suggested for correcting ozone values in the 0.01- to 30-ppm (v/v) range when ozone—iodide reaction occurs in alkaline solution (49). The correction is reported to be necessary in order to bring ozone values obtained in alkaline medium into agreement with those obtained for identical samples reacting in a neutral buffered solution. Further studies point out the complex nature of the ozone—iodide reaction (37,253). From numerous pH values and buffer media tested, a 1% KI solution dissolved in $0.1M$ KH_2PO_4 and $0.1M$ Na_2HPO_4 gave iodine values most closely corresponding to a simple stoichiometric reaction for ozone concentrations in the 2- to 80-ppm (v/v), range.

Other reports present additional procedures for trace ozone—iodide reaction occurring in alkaline medium (11,72,251) and in acetic acid—sodium acetate buffer (80). This latter report contains data indicating satisfactory agreement between ozone values from either acid- or phosphate-buffered neutral iodide solution.

Regardless of the procedure selected, certain factors are important in any ozone analysis. The first of these concerns the preparation of accurately known standards for trace ozone concentrations. Secondly, it is necessary to remember that with many ozone samples collected in the field, other oxidants present cause iodine formation. Often, the result of a determination represents total oxidant content of a sample from all sources. With air samples, this can include contributions from nitrogen oxides, hydrogen peroxide, acyl peroxides, alkyl hydroperoxides, peracids, and peroxyacyl nitrates. Recent studies using kinetic spectrophotometry show that various oxidants react at different rates with iodide ion and other reagents. The individual identities of these oxidants are assessed by this technique (62,240).

Reducing agents, too, interfere in ozone determination by reducing the iodine formed from ozone—iodide reaction. Notable among these interferences are sulfur dioxide and hydrogen sulfide. A chromium(III) oxide—sulfuric acid scrubber is proposed to remove sulfur dioxide from air samples prior to ozone determination (254). This treatment, however, converts any nitrous oxide, which might also be present in the sample, to nitrogen dioxide. The latter, itself, reacts with iodide ion to produce free iodine. Silica gel saturated with potassium dichromate—sulfuric acid is also reported for removal of reducing gases prior to ozone—iodide reaction (268). The volatility of iodine in acid solution, possible oxidation of iodide ion by oxygen, decomposition of ozone in basic solution, and interfering side reactions must also be considered in ozone determinations.

Ozone determination is reported from reaction in alkaline potassium iodide solution followed by acidification with either phosphoric or acetic acid prior to measuring the amount of iodine present. Ozone determination using potassium iodide in neutral dihydrogen phosphate—monohydrogen phosphate buffer is also described (139,250). Finally, the procedure cited above, in which ozone and potassium iodide react in acetic acid—sodium acetate buffer is available. Most procedures, regardless of choice of reacting conditions, utilize spectrophotometric measurement of triiodide ion absorption for determination of iodine formed from ozone reaction. A few procedures recommend addition of starch and subsequent measurement of the blue iodine—starch complex (241). Details of an iodine—starch method for ozone in neutral buffered potassium iodide are presented in Section IIA.

A novel indicating technique consists of the addition of phenol red or other suitable alkaline indicating acid—base indicator to neutral unbuffered potassium iodide solution. Hydroxide ion formed with iodine upon reaction of ozone and iodide ion increases pH to the point where the indicator changes color. The amount of sample brought into contact with the solution at this instant, determined from the product of gas flow rate and contact time, serves as a convenient relative measure of ozone content for various samples (40).

The following procedure is for the determination of ozone and other oxidants in air by reaction with basic potassium iodide solution. Iodine formed is measured spectrophotometrically as triiodide ion at 352 nm (277). A linear calibration curve in terms of iodine equivalents was obtained for solutions containing up to 10 ppm aqueous iodine. Assuming a stoichiometric reaction, the equivalent weight of ozone is 24.0.

Solutions

Potassium Iodide, 1%. Dissolve 10 g KI and 4 g NaOH in distilled water and dilute to 1 liter.

Potassium Iodate, 0.100N. Dissolve 3.57 g KIO_3 in distilled water and dilute to 1 liter in a volumetric flask.

Potassium Iodate, $1.0 \times 10^{-4} N$. Transfer 1.00 ml of the above solution to a 1-liter volumetric flask and dilute to volume with distilled water.

Phosphoric Acid, 36%.

Procedure

Place 25.0 ml KI solution into a 125-ml gas-washing bottle with coarse fritted disk. Draw ozone-containing air sample into the bottle at a rate of 0.5 liters per min for 10 min. Sample volume is calculated from the product of sample rate and sampling time. Temperature and pressure corrections are necessary for samples measured under varying experimental conditions. A plug of glass wool in the sample line before contact with the alkaline iodide solution serves to remove particulate matter. Shut off the sample inlet and thoroughly agitate the contents of the flask. Transfer a 10.0-ml portion of the solution to a 25-ml volumetric flask. Add 2.0 ml 36% phosphoric acid, mix, cool, and let stand for about 5 min. Dilute to the mark with distilled water. For samples known to contain peroxide interference, boil the iodide aliquot briefly prior to acidification. The iodine absorbance of the solution is measured at 352 nm in a 1-cm silica cell. A reagent blank is treated in the same manner but with ozone-free air drawn through the alkaline iodide solution.

Consistent time intervals between acidification and measurement are necessary for all solutions. A calibration curve is prepared by adding appropriate aliquots of dilute potassium iodate standard solution to 25.0 ml alkaline potassium iodide. This solution is then treated as described above. Iodine produced upon acidification is measured spectrophotometrically, and the calibration curve is prepared and labeled in terms of absorbance versus iodine concentration. Ozone concentration is calculated from the equivalent iodine value. Neither a fivefold difference in potassium iodide concentration nor a fourfold difference in sampling rate seriously affects the precision of ozone values obtained by this procedure.

The following procedure is for determination of ozone and other oxidants in air by reaction with acidic potassium iodide solution. Iodine formed is measured spectrophotometrically as triiodide ion at 355 nm (80). The method applies for air samples containing between 0.01 and 0.70 ppm ozone at 1 atmosphere pressure and 25°C. Sulfur dioxide and hydrogen sulfide do not interfere.

Solutions

Potassium Iodide, 1.5%. Dissolve 15 g KI and 4 g NaOH in distilled water and dilute to 1 liter.

Acetic Acid. Dilute 50 ml glacial acetic acid to 300 ml with distilled water.

Potassium Iodate, 0.1%. Dissolve exactly 1.000 g KIO_3 in distilled water and dilute to 1 liter in a volumetric flask.

Potassium Iodate, 0.001%. Transfer 10.00 ml of the above solution to a 1-liter volumetric flask and dilute to volume with distilled water. One ml of this solution is equivalent to 35.4 μg iodine when added to the acidified KI solution.

Procedure

Place 10 ml alkaline KI solution and 3 ml acetic acid solution into a midget impinger flask. The pH of this solution is approximately 3.8. Attach the inlet tube of the impinger to an automatic air sampler. A piece of glass wool in the connecting line serves to remove particulate matter. Ozone-containing air is passed into the reagents at a flow rate of 14.1 liters per min for 30 min. Sample size is calculated from the product of gas flow rate and contact time. Temperature and pressure corrections are necessary for samples measured under varying experimental conditions. A 15-min sampling time is recommended for samples containing over 0.4 ppm ozone. Avoid contact with direct sunlight during sampling. Disconnect the automatic sampler. Mix the contents of the impinger flask and transfer the solution quantitatively to a 25-ml volumetric flask. Dilute to the mark with distilled water. The iodine absorbance is measured at 355 nm in a 1-cm silica cell against a reagent blank prepared at the time of sampling. The same procedure is employed for the blank, except that no air is drawn into the impinger containing the blank solution.

A calibration curve relating iodine concentration to absorbance is prepared by placing from 0.07 to 6.32 ml dilute potassium iodate solution in separate 25-ml volumetric flasks. These amounts correspond to iodine values ranging from 25 to 225 μg iodine per 25 ml sample. Add 10 ml alkaline KI solution to each flask. Mix thoroughly and add 3 ml acetic acid solution to each flask. Mix again, cool, and dilute each flask to the mark with distilled water. These solutions are measured against a reagent blank prepared in the same manner but containing no potassium iodate. Ozone concentration is calculated from the equivalent iodine value for each sample. Possible iodine loss from volatility in acid solution during sampling is insignificant, and temperature variations from 25 to 35°C have no effect upon results.

B. IRON–THIOCYANATE METHOD

Oxidation of iron(II) to iron(III) by ozone and other oxidants followed by colorimetric determination of iron(III) with thiocyanate is reported (91,293,302). One rapid method of determination uses paper impregnated with iron salts and potassium thiocyanate. Ozone-containing sample passing through the paper causes reddening. Knowledge of sample flow rate relative to the area of the colored spot serves as an estimate of the amount of ozone presents (75,76). An accuracy of ±10% is reported for trace amounts of ozone by this method. Other oxidants present in the sample interfere.

The following procedure is for determination of ozone and other oxidants in air by reaction with iron(II) solution. Subsequent addition of thiocyanate produces a red iron(III)–thiocyanate complex. The method is best for ozone concentrations greater than 2.0 ppm (v/v). For ozone concentrations less than this, the molar absorptivity of the iron complex appears to increase with decreasing amounts of ozone. Variations of sample bubbler frit size and sample collection rate also affect results. The method is unaffected by a substantial delay between the time of sample collection and colorimetric measurement (60,62).

Solutions

Iron(II) Ammonium Sulfate. Dissolve 0.1 g $FeSO_4(NH_4)_2SO_4 \cdot 6H_2O$ and 1.0 ml $6N$ H_2SO_4 in distilled water. Dilute to 200 ml.

Ammonium Thiocyanate. Dissolve 50 g NH_4SCN in distilled water and dilute to 100 ml. Store this solution in the dark.

Ozone. Prepare ozone by exposing ultrapure tank air to a generator consisting of several 4-watt mercury vapor lamps. This mixture is diluted with varying additional amounts of air to produce samples containing a range of ozone concentrations. Standardize these samples by an appropriate procedure.

Procedure

Place 10.0 ml iron(II) solution in a gas washing bottle containing a fritted bubbler of 50-micron pore size. Pore size is critical and should be identical for both sample and standards. Draw ozone-containing sample into the bottle at a rate of 0.2 liters per min for 10 min. Sample volume is calculated from the product of sample rate and sampling time. Temperature and pressure corrections are necessary for samples measured under varying experimental conditions. Sampling rate is also critical and should be identical for both sample and standards. Shut off the sample inlet and thoroughly mix the contents of the bottle. Add 2.0 ml ammonium thiocyanate solution. Mix thoroughly and quantitatively transfer the contents of the flask to a 25-ml volumetric flask. Dilute to the mark with distilled water. For samples of high ozone content, a suitable aliquot of the reaction solution is diluted to 25 ml in a second volumetric flask. The absorbance of the iron(III)—thiocyanate solution is measured at 480 nm in a 1-cm cell against a reagent blank which has undergone the same treatment except that no ozone-containing sample is allowed contact with the blank solution. A calibration curve is prepared by using the above procedure with samples of known ozone content.

C. DIPHENYLAMINESULFONATE METHOD

A procedure for ozone determination is reported in which an acidic solution of diphenylaminesulfonate is reacted with ozone (36). A blue oxidization product is formed with maximum absorbance at 593 nm. Optimum pH is about 3. Other oxidants, too, attack diphenylaminesulfonate; but because they apparently form different oxidation products, their presence can be ascertained. Nitrogen dioxide produces a yellow-green coloration; chlorine and hydrogen peroxide, violet solutions; and sulfur dioxide, a bright yellow solution. Interference from nitrogen dioxide, in particular, is less than exhibited with either alkaline or neutral buffered iodide procedures. Results obtained on identical ozone samples agreed with those found by the neutral buffered iodide method.

The following procedure is for determination of ozone in air by reaction with sodium diphenylaminesulfonate in perchloric acid solution. Data are reported for samples containing up to 40 μg ozone. Ten repeated determinations on identical samples produced average deviations of 0.45 and 0.77% when 1.0- and 7.5-cm cells, respectively, were used for abosrbance measurements.

Solutions

Sodium Diphenylaminesulfonate, 1%. Dissolve 1 g sodium diphenylamine-sulfonate in 0.02% perchloric acid. Dilute to 100 ml with additional perchloric acid.

Ozone. Prepare ozone by exposing air to an ultraviolet lamp. The amount of ozone in various mixtures is controlled by regulating the current passing through the lamp or by using pure oxygen with air. Standardize these samples by an appropriate independent procedure.

Procedure

Place 10.0 ml diphenylaminesulfonate solution in a midget impinger. Pass ozone containing air through the solution at a rate of 2.83 liters per min for 10.0 min. Collection time is varied depending upon the amount of ozone in a particular sample. Sample volume is calculated from the product of sample rate and sampling time. Temperature and pressure corrections are necessary for samples measured under varying experimental conditions. Shut off the sample inlet and mix the contents of the flask thoroughly. Add additional distilled water to the solution to replace any loss due to evaporation. Mix thoroughly. A portion of this solution is transferred to 1-cm cell and the absorbance is measured at 593 nm against a reagent blank that has undergone similar treatment but that has not had contact with ozone-containing sample.

The above procedure is repeated with samples of known ozone content to prepare a proper calibration curve. For samples of low ozone content, cells of longer optical path are recommended. No interference is found from oxygen in air samples. The color produced by ozone reaction upon diphenylaminesulfonate is stable for 2 to 3 hr after formation and can be prevented from changing for even longer periods if kept refrigerated near 0°C.

D. 4,4′-DIMETHOXYSTILBENE METHOD

A specific colorimetric method for determination of ozone is reported utilizing a reaction in which ozone is used to cleave an ethylenic double bond (39). One of the cleavage products, an aldehyde, is reacted further to yield a colored product. Absorbance values for this product are related to initial ozone concentration. 4,4′-Dimethoxystilbene is the reagent cleaved by ozone. Anisaldehyde formed is determined colorimetrically by an established procedure (261). Neither nitrogen dioxide, peroxyacetic acid, peroxyacetyl nitrate, nor methyl hydroperoxide at a concentration of 1 mg per liter gave a detectable color by this procedure. One mg per liter sulfur dioxide, when present with ozone in a sample, caused no interference. Hydrogen sulfide did, however, interfere. Precautions are necessary with this procedure because of the corrosive nature of trifluoroacetic acid and trifluoroacetic anhydride used in the determination.

The following procedure is for determination of gaseous ozone mixtures. A minimum concentration of 0.01 mg ozone per ml can be detected.

Solutions

4,4′-Dimethoxystilbene. Dissolve 5 mg 4,4′-dimethoxystilbene in *sym*-tetra-chloroethane. Dilute to 100 ml with additional solvent. Prepare fresh every week. The tetrachloroethane prior to use is passed through a short column containing sodium bicarbonate.

Fluoranthene. Dissolve 5 g fluoranthene in chloroform. Dilute to 100 ml with additional solvent. Prepare fresh every week.

Anisaldehyde. Dissolve 3.0 mg anisaldehyde in *sym*-tetrachloroethane in a 100-ml volumetric flask. Dilute to the mark with additional solvent.

Trifluoroacetic Acid.

Trifluoroacetic Anhydride.

Procedure

An inverted bubbler is used as the sample collecting flask. It consists of two concentric tubes. The inner tube terminates in a glass frit and, at the opposite end, has a removable adaptor which is connected to a vacuum source. The outer tube has a sidearm for attaching the sample source inlet. Reagent solution is placed in the inner tube. The tubes are placed one inside the other, and a sample line is connected to the sidearm of the outer tube. When vacuum is applied, sample is drawn into the outer tube through the frit and forced to mix with reagent solution in the inner tube. Place 3.0 ml 4,4'-dimethoxystilbene solution into the inner tube of each of two sample collectors. Attach the outer tubes and connect the flasks in series. Attach the sample source to the sidearm of the first collector and vacuum to the inner tube of the second collector. Pass ozone containing sample through the reagent solution at an accurately known flow rate somewhere between 0.10 and 0.15 liter per min. Collect sample for a period varying from 15 min to 1 hr depending upon the amount of ozone in the sample gas. Sample volume is calculated from the produce of sample flow rate and sampling time. Temperature and pressure corrections are necessary for samples measured under varying experimental conditions.

Shut off the sample source and vacuum systems. Transfer 1.0 ml solution from each collector to separate 10-ml volumetric flasks. Add to each flask 1.0 ml fluoranthene solution and 0.8 ml trifluoroacetic anhydride. Mix thoroughly and allow to stand for 5 min. Dilute the contents of each flask to volume with trifluoroacetic acid. If a blue color appears in the solution from the second bubbler, it, too, is measured for ozone reaction, and the total ozone equivalent is determined from the sum of that found in each bubbler. If no color appears, assume that all the ozone has reacted in the first bubbler and discard the second solution. Absorbance of the solution in the 10-ml volumetric flask is measured in a 1-cm cell at 610 nm against a reagent blank prepared in a similar manner but having no contact with sample.

Ozone values are obtained from a calibration curve. Solutions containing appropriate known amounts of anisaldehyde serve for preparation of this curve. Alternately gaseous samples containing known amounts of ozone can provide a suitable calibration curve. Values of identical ozone samples taken from curves prepared by both procedures showed that the ozonolysis step proceeded with 98% yield of anisaldehyde from 4,4′-dimethoxystilbene.

E. 1,2-DI(4-PYRIDYL)ETHYLENE METHOD

Cleavage of the ethylenic double bond of 1,2-di(4-pyridyl)ethylene by ozone and subsequent measurement of the aldehyde produced is reported as a method for determination of atmospheric ozone (124). The method is simple and fairly specific for ozone. Pyridine 4-aldehyde formed by the reaction is determined colorimetrically by modification of an established procedure (260). Hydrogen peroxide and some alkyl hydroperoxides interfere with this procedure (62). However, most other substances present in concentrations normally found in atmospheric samples do not interfere when 30-min sampling times are employed (125). These include sulfur dioxide, hydrogen sulfide, nitrogen dioxide, acrolein, formaldehyde, peracetic acid, 1-hexene, and peroxyacetyl nitrate. The latter two substances do interfere when 24-hr samples are collected. All of the above cause interference if present in concentrations larger than those normally found in atmospheric samples.

Prolonged sampling times cause no interference in the absence of 1-hexene and peroxyacetyl nitrate provided sufficient reagent solution is present, allowing for evaporation, to cover the absorber frit. The use of a midget impinger in place of an absorber frit is not satisfactory. Frit porosity and sample collection rate are not critical. Collection temperature is not critical, although for samples collected below 16°C, water must be added to the reagent solution to prevent freezing. Addition of water to reagent solution results in a decrease in sensitivity for ozone by this method. The reagent solution, after exposure to atmospheric sample, is stable up to four days. Pyridine 4-aldehyde formed from ozone reaction with reagent is combined with 3-methyl-2-benzothiazolinone-hydrazone hydrochloride (3-MBTH) at elevated temperature. Heating for 20 min in a boiling water bath is sufficient. Shorter heating times give lower absorbance readings. Longer heating, up to 40 min, produces no further change in absorbance from that observed after a 20-min heating period. The colored product is stable up to 1 hr. This method is more advantageous than that employing 4,4′-dimethoxystilbene because it avoids using trifluoroacetic anhydride and trifluoroacetic acid, both of which are corrosive.

The following procedure is for determination of atmospheric ozone. A linear calibration curve is obtained for ozone concentrations up to 3.65 mg ozone per liter sample. One mole ozone is empirically found to generate 1.24 mole pyridine 4-aldehyde by this procedure. Repeated determinations upon 10 samples gave values varying by ±5% of the mean value.

Solutions

1,2-Di(4-pyridyl)ethylene, 0.5%. Dissolve 0.5 g 1,2-di(4-pyridyl)ethylene in glacial acetic acid. Dilute to 100 ml with additional glacial acetic acid. Prepare fresh every two weeks.

3-Methyl-2-benzothiazolinonehydrazone Hydrochloride, 0.2%. Dissolve 0.2 g 3-methyl-2-benzothiazolinonehydrazone hydrochloride monohydrate in distilled water. Dilute to 100 ml with additional distilled water.

Pyridine 4-aldehyde. Dissolve 1.00 g redistilled pyridine 4-aldehyde in glacial acetic acid. Dilute to 1 liter in a volumetric flask with additional glacial acetic acid.

Pyridine 4-Aldehyde, Dilute. Transfer 20.0 ml of the above solution to a 1-liter volumetric flask. Dilute to volume with glacial acetic acid. One ml of this solution contains 20 μg reagent. This corresponds to 7.3 μg ozone per ml.

Procedure

Samples are collected in a gas washing bottle having an extracoarse fritted bubbler. Place 15.0 ml 1,2-di(4-pyridyl)ethylene solution in this container. Attach to the sample source and draw sample into the flask at a rate of 0.5 liter per min for 30 min. Collection time is increased for samples containing very small amounts of ozone. Sample volume is calculated from the product of sample rate and sampling time. Temperature and pressure corrections are necessary for samples measured under varying experimental conditions. A water trap between the sample-collecting bottle and the flow gauge prevents acetic acid from contacting the flow gauge. Disconnect the sample source. Add sufficient glacial acetic acid to restore the liquid level to 15.0 ml. This is necessary because some evaporation occurs during sample collection. Transfer 10.0 ml of the reacted 1,2-di(4-pyridyl)ethylene solution to a test tube. Add 1.0 ml 3-MBTH solution. Heat the mixture in a boiling water bath for 20 min. Remove from the bath and cool under running water. Transfer a portion of this solution to a 1-cm cell and measure the absorbance at 442 nm. A reagent blank is prepared in the same way, but no sample is drawn through the blank solution. Instead, 15.0 ml 1,2-di(4-pyridyl)ethylene solution is allowed to stand in an open container for the time duration that a like solution is collecting sample. Both solutions are treated identically from this point. The amount of ozone is obtained from a calibration curve value and the known total volume of sample.

A calibration curve is prepared by placing from 1.0 to 5.0 ml diluted pyridine 4-aldehyde in separate 10-ml volumetric flasks. Add 1.0 ml 3-MBTH solution and dilute to the mark with glacial acetic acid. Heat these solutions as described above. After cooling, measure the absorbance of each solution against a reagent blank containing only 1.0 ml 3-MBTH diluted to 10 ml with glacial acetic acid in a volumetric flask. These standards, 0.0 to 10.0 μg pyridine 4-aldehyde per ml sample, correspond to ozone concentrations of 0.0 to 3.65 mg ozone per liter.

F. DIACETYLDIHYDROLUTIDINE METHOD

Oxidation of diacetyldihydrolutidine by ozone is used for determination of atmospheric ozone (217). The reaction proceeds rapidly in aqueous solution thus avoiding use of objectionable reagents necessary with ozonolysis reactions. A decrease in absorbance of reduced lutidine is taken as a measure of the amount of reacting ozone. The reaction is not stoichiometric since diacetyl-lutidine, the expected product, is not observed. Nevertheless, under suitable conditions, one obtains quantitative estimation of ozone. The collection efficiency of reagent solutions containing less than 100 micromoles diacetyldi-hydrolutidine per ml is less than unity. In this case, it is necessary to use more than one bubbler in series to trap all of the ozone present in a sample. Quantitative results occur for measurements made over the pH range of 5 to 9. A buffered solution is recommended. Solutions exposed to ozone are stable for at least 1 hr.

Oxygen in air causes negligible interference. Hydrogen peroxide and t-butyl hydroperoxide at concentrations normally found in atmospheric samples also cause negligible interference. If present in higher concentrations, they do interfere. Sulfur dioxide in amounts comparable to ozone causes only slight interference with the buffered reagent. A scrubbing solution containing hydrogen peroxide eliminates sulfur dioxide interference. A mixture of nitrogen oxides in amounts comparable to ozone causes approximately 5% error in ozone values.

The following procedure is for determination of atmospheric ozone. As presented, it is applicable to mixtures containing less than 0.36 mg ozone per liter. An experimental error of ±3% was reported. Values obtained are comparable with those found by the neutral buffered iodide procedure.

Solutions

Diacetyldihydrolutidine. The reagent is prepared by dissolving 150 g ammonium acetate in 1 liter water. Add 3 ml glacial acetic acid and 2 ml acetylacetone. Stir until dissolved and allow to stand for 1 hr. Add exactly 0.5 ml 40% formaldehyde. Allow the solution to stand overnight. Filter the yellow crystals through a coarse-fluted filter paper. Wash with water and air dry. Store in an amber bottle. The reagent solution is prepared by placing a few crystals of diacetyldihydrolutidine in a few milliliters distilled water and mixing. Filter the undissolved crystals. The entire solution is transferred to a 1-cm cell and additional distilled water is added to produce an absorbance of 0.8 absorbance units at 412 nm. Distilled water serves as a suitable reagent blank. This solution contains approximately 100 micromoles reagent per ml. The solution is removed from the cell, and sufficient solid disodium hydrogen phosphate is added to produce a solution $0.01M$ in this salt. The buffered diacetyldihydrolutidine solution is stored in an amber bottle away from sunlight. Prepare fresh reagent every third day.

Ozone, Prepare ozone by exposing oxygen to an ultraviolet lamp. Additional varying amounts of oxygen result in a series of gaseous mixtures of varying ozone concentration. Standardize these samples by an appropriate independent procedure.

Procedure

Sample gas containing ozone is drawn through 4.0 ml diacetyldihydrolutidine solution in a midget impinger at a flow rate of 1.0 liter per min for 10 min. If the total amount of ozone collected exceeds 0.36 mg, the amount of diacetyldihydrolutindine is insufficient and a second impinger in series with the first is necessary to collect all the ozone present. An alternate procedure is to use a shorter collection time. Sample volume is calculated from the product of sample rate and sampling time. Temperature and pressure corrections are necessary for samples measured under varying experimental conditions. Disconnect the sample source. Add additional distilled water to restore liquid volume to 4.0 ml. This is necessary because some evaporation occurs during sample collection. Transfer a portion of this solution to the 1-cm cell and measure the absorbance at 412 nm against a distilled water blank. Ozone concentration is read from a calibration curve prepared by using gaseous mixtures of known ozone content by the procedure described above. The decrease in absorbance is proportioned to increasing ozone concentration.

G. MANGANESE–TOLIDINE METHOD

Ozone produces a yellow-colored reaction product with *o*-tolidine. The reaction is applied to ozone determination (241). In a related procedure, ozone is first allowed to react with manganese(II) ion. The oxidized manganese(III) species then reacts with *o*-tolidine forming a yellow-colored product,(320). Substitution of sodium arsenite for manganese(II) in the procedure is suggested as a means of removing interfering effects from other oxidants. One sample is mixed with *o*-tolidine prior to addition of arsenite, and a second identical sample is added to a solution containing both *o*-tolidine and arsenite. The difference in the values obtained from these samples is taken as a measure of ozone concentrations (284).

The following procedure is for determination of ozone dissolved in water solution. It employs manganese(II) as a phosphate complex for reaction with ozone. Subsequent addition of *o*-tolidine produces the expected yellow coloration (135). The method is applicable for ozone concentrations greater than 0.01 mg ozone per liter. Ten repeated determinations of a 0.51-mg ozone per liter solution resulted in a standard deviation of ±3.1%.

Solutions

Sodium Pyrophosphate, 0.75*M*. Dissolve 199 g $Na_4P_2O_7$ and 208.0 ml 3.6*M* H_2SO_4 in distilled water and dilute to 1 liter.

Manganese(II) Sulfate, 3*M*. Dissolve 507 g $MnSO_4 \cdot H_2O$ in distilled water and dilute to 1 liter.

Manganese(II) Pyrophosphate, in 6*M* H_2SO_4. Mix together 60 ml 0.75*M* sodium pyrophosphate, 35 ml 18*M* sulfuric acid, and 5 ml 3*M* manganese(II) sulfate solution. Cool.

Manganese(II) Pyrophosphate, in 0.45*M* H_2SO_4. Mix together 60 ml 0.75*M* sodium pyrophosphate, 12.5 ml 3.6*M* H_2SO_4, and 5 ml 3*M* manganese(II) sulfate solution. Dilute to 100 ml with distilled water.

***o*-Tolidine, 0.135%.** Dissolve 1.35 g *o*-tolidine in 500 ml boiled distilled water. Add an additional 350 ml distilled water and 150 ml 12*M* HCl.

Manganese(III) Pyrophosphate, 0.0005*M*. Combine 10 ml manganese(II) pyrophosphate in 0.45*M* H_2SO_4 and 1.00 ml 0.0500*N* $K_2Cr_2O_7$. Allow the solution to stand for 15 min and then dilute to 100 ml in a volumetric flask.

Procedure

To 100.0 ml sample solution, add 5.0 ml manganese(II) pyrophosphate in 6*M* sulfuric acid. Mix thoroughly. After 5 min, add 5.0 ml *o*-tolidine solution and again mix thoroughly. After an additional 5 min, transfer a portion of this solution to a 2-cm cell and measure the absorbance using an appropriate colored filter, for example, Hilger & Watts No. 601. The absorbance maximum occurs at about 440 nm. A reagent blank consists of a second 100.0-ml sample to which is added 5.0 ml 6*M* H_2SO_4 solution. A calibration curve is prepared by diluting accurately measured amounts of manganese(III) pyrophosphate, between 0 and 5.0 ml, to 100.0 ml with distilled water. These solutions correspond to ozone concentrations between 0.0 and 0.60 mg ozone per liter. Add to each diluted standard 5.0 ml *o*-tolidine solution and an additional 5.0 ml distilled water. Mix and allow the solution to stand for 5 min. Measure the absorbance of each of these solutions against a distilled water blank.

Chlorine or chlorine dioxide, if present in the water sample, can be measured by deaerating the sample after addition of manganese(II) pyrophosphate but before addition of *o*-tolidine. As an alternate procedure, the ozone-containing sample can be deaerated directly. The dissolved gases are collected in manganese(II) pyrophosphate and acidified with $0.45M$ sulfuric acid, using a series of fritted-glass wash bottles. With this adaptation, the addition of 1 ml $0.1M$ chromium(III) sulfate to the manganese(II) solution is necessary to catalyze the oxidation of manganese(II). This adaptation has the advantage of avoiding interferences from heavily polluted waters.

H. MISCELLANEOUS METHODS

The reaction of ozone with the yellow, reduced form of indigo carmine is used for ozone determination (82). Ozone-containing gas is passed through a $5 \times 10^{-5}M$ solution of indigo carmine buffered at pH 6.85 with phosphate buffer. The absorbance change measured at 578 nm is applicable for ozone concentrations within the range of 1 to 5×10^{-6} vol-%. An alternate technique employs indigo carmine and magnesium sulfate heptahydrate impregnated on silica gel (205). Ozone-containing air passed over this mixture causes discoloration. Comparison of color intensity obtained with those from passing samples of known ozone content provides a measure of the concentration of ozone in the sample.

The reduced form of phenolphthalein, prepared by refluxing an aqueous basic solution of phenolphthalein with zinc dust, is used for determination of organic oxidants, including ozone (118,288). The red-purple oxidation product is measured colorimetrically.

Leuco malachite green (N,N,N′,N′-tetramethyl-*p,p*′-diaminotriphenyl-methane) or Arnold's qualitative reagent (N,N,N′,N′-tetramethyl-*p,p*′-diaminodiphenylmethane) react quantitatively with aqueous ozone solutions (104,168). The colored oxidation product is extracted into chloroform and absorbance measurements are made at 492 nm. Prior addition of potassium permanganate prevents interference from manganese dioxide, chlorine, and hydrogen peroxide. The reaction occurs at pH 2 in a phosphoric acid—dihydrogen phosphate buffer. The procedure applies for ozone concentrations varying from 0.02 to 1.00 mg ozone per liter solution.

Atmospheric samples containing ozone produce a colored reaction product when passed through a solution of N-phenyl-2-naphthylamine dissolved in *o*-dichlorobenzene (78,109). Absorbance values measured at 435 nm are dependent upon amount of ozone and sample flow rate. The reaction is sensitive to samples containing a minimum of 1.8 μg ozone. Oxygen and nitrogen oxides in amounts normally found in atmospheric samples do not interfere.

An indirect method of ozone determination involves addition of gaseous nitrogen oxide to an ozone-containing sample. Nitrogen dioxide produced is reacted with a solution containing sulfanilic acid, N-(1-naphthyl)ethylenediamine dihydrochloride, and acetic acid. The red-colored reaction product is measured colorimetrically at 550 to 560 nm (11,249,252).

Other procedures for ozone determination include direct absorbance measurements in both the ultraviolet and visible spectral regions (141,161,184,283). Other suggested color-forming reagents for reaction with ozone include *p*-aminodimethylaniline (224), copper(II) in the presence of periodate (304), and chromium(III) chloride (162).

II. APPLICATIONS

Trace ozone standards, necessary for calibration of any colorimetric procedure for ozone determination, are not readily available. Assuming stoichiometric reaction, equivalent amounts of other oxidants are commonly substituted for ozone in the preparation of calibration curves. Potassium iodate, for example, is used in the iodometric methods presented earlier as a source of equivalent iodine. A potassium dichromate standard is used in the manganese–tolidine method for preparation of a calibration curve. In some of the ozonolysis reactions, known amounts of aldehydes serve as standards. It is possible to generate ozone directly from either air or oxygen. Dilution of the ozone by additional air or oxygen produces mixtures of varying ozone concentration. Generally, these mixtures are analyzed by independent means, usually involving neutral buffered potassium iodide and either a colorimetric or potentiometric iodine determination. Oxygen is recommended over air for ozone generation because with air, nitrogen oxides and/or various peroxides can be produced during ozone formation. Commerical ozone generators are available, or a laboratory assembly can be used. One type of laboratory generator employs a stream of oxygen passing over an ultraviolet lamp (36,276). Generation using electrostatic discharge is also reported (30). Aqueous ozone solutions can also be prepared, and solubility values for aqueous ozone solutions are available (242).

A. ATMOSPHERIC OZONE

Consideration of factors affecting collection of representative atmospheric samples is necessary for accurate ozone determination, and statistical evaluation of each possible effect leads to the most representative method of sample selection (11,155). Wind speed and direction, temperature, type and size of particulate matter, topography, time of day, and presence of adjacent industrial sites are some of the factors to be considered in choosing a sample collection procedure. Commercial sequential air sampling units are available. Portable laboratory units are also described (270). Choice of collection apparatus depends in part upon the subsequent method of determination. With on-site measurement, the atmospheric sample can be collected directly into the reagent solution.

This procedure is also applicable if the reaction products formed from contact between sample and solution are stable over the time interval necessary to transport the solutions back to the laboratory for measurement. Collection of ozone in evacuated containers is not recommended (49). Ozone collection in a liquid oxygen trap is suggested (184). Temporary sample storage in large Mylar bags is proposed (63). For ozone storage, it is necessary initially to expose the bags for 24 hr to ozone concentrations three or four times greater than those of the samples under study. Subsequent conditioning is not necessary. Impinger-type flasks, in which the sample is drawn through a fine capillary and blown into the reagent solution, are satisfactory collection vessels. Fritted-type bubblers are not recommended, and it has been shown that consistantly low results for ozone are obtained when samples are analyzed by a neutral potassium iodide procedure using this type of collector in place of the impinger type (271). The error is attributed to accelerated ozone decomposition by the glass frit. An Allihn-type absorber consisting of concentric perforated bell-shaped caps over the tip of an impinger is proposed as a most effective device for achieving contact between sample and reagent solution (292). An inverted collector having atmospheric sample drawn into a tube filled with reagent solution is described earlier in this chapter in connection with the ozone procedure using 4,4'-dimethoxystilbene. Other collection arrangements are also proposed (11).

More than one collection flask in series may be necessary to assure complete reaction of ozone. Solutions from each absorber are combined prior to subsequent determination steps. Teflon or glass connections are recommended for all apparatus used in ozone determination (6). Even with glass, preliminary exposure to ozone is suggested to assure proper conditioning of the glass surface. Aluminum, stainless steel, and poly(vinyl chloride) tubing require varying exposure to ozone before negligible loss of ozone occurs, and it is necessary to repeat the conditioning procedure after a period of time. For this reason, these materials are not recommended for use in ozone determinations. The above effects are enhanced when low flow rates, less than 1 liter per min, are employed. Polyethylene tubing and rubber tubing are not satisfactory for ozone samples.

Various scrubbers are suggested for removal of interfering substances prior to ozone determination. Acidified chromium(III) oxide or hydrogen peroxide are, for example, recommended for removal of sulfur dioxide from atmospheric ozone samples. One must be certain when employing scrubbing solutions that part of the ozone is not removed. A 15% loss of ozone at 0.8-mg ozone per liter concentration is reported when a chromic acid—potassium permanganate scrubber was used to remove interfering substances prior to ozone determination (128).

The following procedure is for determination of atmospheric ozone. It provides for the interfering effect of nitrogen dioxide which might also be present by measuring the total oxidant concentration in terms of ozone equivalents. Ozone is then removed from a sample and the procedure repeated to ascertain the amount of interfering nitrogen dioxide present. Ozone concentration is obtained from the difference of these results. Sample gas is passed into each of two impingers in parallel, equal volumes of sample going into each impinger. A cotton plug trap is placed before one of the impingers to remove ozone. Both impingers contain neutral, buffered potassium iodide. After sample collection, starch is added and the color intensity observed compared with those of standards (59). Ozone concentrations of up to 0.4 ppm (v/v) in the absence of nitrogen dioxide or up to 0.05 ppm (v/v) in samples containing up to 5 ppm nitrogen dioxide are determined. Repeated determinations of 24 identical samples gave a mean value of 0.090 ppm (v/v) with a standard deviation of ±0.014 when visual color comparison of the iodine—starch color was observed.

Solutions

Potassium Iodide, 2%. Dissolve 20 g KI, 14.2 g Na_2HPO_4, and 13.6 g KH_2PO_4 in distilled water and dilute to 1 liter.

Starch. Dissolve 0.25 g soluble starch in 70 ml boiling water. Cool and dilute to 100 ml. Prepare fresh daily.

Cotton Absorbent. Wash a quantity of cotton with distilled water. Dry at 105°C. Cool in a desiccator. Pack 1 g dried cotton into a glass tube forming a plug approximately 7 cm in length. Store in a desiccator.

Sulfuric Acid, 0.001M.

Potassium Iodate, 0.0167M. Dissolve 3.574 g KIO_3 in distilled water and dilute to 1 liter in a volumetric flask.

Potassium Iodate—Potassium Iodide. Dissolve 45 g KI in distilled water and add exactly 100 ml 0.0167M KIO_3 solution. Dilute to 1 liter in a volumetric flask. Prepare fresh daily. Test this solution by adding to a 10-ml portion 0.02 ml 0.001M sulfuric acid and 5 ml starch. If any blue coloration forms, reject this solution and prepare fresh using new bottles of reagents.

Procedure

Place 10.0 ml buffered iodide solution into each of two impinger bottles. The bottles are connected in parallel with the cotton-filled absorbing tube placed between one of the bottles and the sample source. The sample stream is split and simultaneously passed through both impinger bottles. Flow controls in both branches are adjusted to give equal flow rates of 2.0 liters per min. Collect sample for 20 min. Sample volume is calculated from the product of sample rate and sampling time. Temperature and pressure corrections are necessary for samples measured under varying experimental conditions. Shut off the sample source. Add 5.0 ml starch solution to each sample flask. Mix thoroughly and transfer portions of each solution to cuvettes for colorimetric measurements. Absorbance of the blue iodine—starch complex can be measured colorimetrically at 540 nm; however, as developed by the authors, this procedure is intended for field use. To achieve this end, a color comparator is employed to visually match the colors obtained from the sample solutions with those of known iodine concentration prepared at the sample collection site.

Appropriate standards are prepared by adding exactly known amounts of $0.001M$ sulfuric acid to the standard iodate—iodide solution. This is followed by the addition of starch. Ozone concentration is ascertained assuming a stoichiometric reaction between ozone and excess iodide to form an equivalent amount of iodine. To each of nine test tubes, add 10.0 ml iodate—iodide solution. Add the following amounts of $0.001M$ sulfuric acid: 0.08, 0.14, 0.20, 0.30, 0.41, 0.52, 0.63, 0.74, and 0.85 ml. Now add 5.0 ml starch solution to each test tube. Mix thoroughly. These amounts of acid were empirically found to produce sufficient iodine—starch complex of the proper color to correspond to ozone concentrations of 0.05, 0.08, 0.10, 0.15, 0.20, 0.25, 0.30, 0.35 and 0.40 ppm (v/v) ozone, respectively. Match the unknown and standards using an appropriate color comparator. Actual ozone concentration is obtained from the difference of the concentrations measured from each of the two sample solutions. Sample solution color should be measured within 15 min of formation. Standards are color stable for up to 30 min.

Slightly low results are obtained if only one impinger collection flask is used in each branch of the sample train. For precise work, two impingers in series in each branch are recommended. Sample concentrations with and without ozone are found from the sum of those measured for each flask. This method is useful for removing the interference of nitrogen dioxide. Interferences from other oxidants were not reported, although they, too, would be eliminated provided they did not react with the cotton plug. Sulfur dioxide, although it reacts with iodine, should not interfere assuming it reacts to the same extent with iodine in both branches of the sample train. The procedure is not affected by normal humidities, however, some nitrogen dioxide is removed by the cotton if it becomes damp. Iron fumes, present perhaps at sites where welding equipment is employed, interfere with this procedure.

B. OZONE IN WATER

Although most of the procedures described above apply to the determination of ozone in atmospheric samples, methods for measurement of ozone content in treated water samples are also available.

The following procedure is for determination of ozone dissolved in water. It utilizes the oxidation of manganese(II) to manganese(III) in the presence of pyrophosphate ions and is applied to residual ozone in drinking and industrial waters that have been treated with ozone (134). The method is applicable for ozone concentrations greater than 1.0 mg ozone per liter. A related procedure described in Section IG and using o-tolidine in addition to the above reagents allows for ozone determination at concentrations greater than 0.01 mg ozone per liter. Ten repeated determinations on a sample containing 4.30 mg ozone per liter resulted in a standard deviation of ±12.42%. Chlorine, chlorine dioxide, hypochlorite, chlorite, chlorate, and perchlorate in amounts normally found in purified water do not interfere.

Solutions

Manganese(II) Pyrophosphate. Mix 60 ml $0.75M$ $Na_2H_2P_2O_7$ and 12.5 ml $3.6M$ H_2SO_4 with 15 ml $1M$ $MnSO_4$ and dilute to 1 liter with distilled water.

Sulfuric Acid, $0.72M$.

Potassium Dichromate, $0.00500N$. Dissolve 1.471 g dried potassium dichromate in distilled water and dilute to 1 liter in a volumetric flask.

Procedure

To 100.0 ml sample, add 5.0 ml manganese(II) pyrophosphate solution. Mix thoroughly and allow the mixture to stand 5 min. After standing, transfer a portion of the mixture to a 3.5-cm cell and measure the absorbance using a yellow-green filter, for example, Hilger & Watts No. 605. The absorbance maximum for the manganese(III) pyrophosphate species occurs in the region of 518 to 524 nm. A reagent blank consists of a second 100.0-ml sample to which is added 5.0 ml dilute sulfuric acid solution. A calibration curve is prepared by placing 0.00, 1.00, 2.00, 3.00, 6.00, and 9.00 ml dichromate solution into separate 100-ml volumetric flasks. Add 5.0 ml manganese(II) solution to each flask. Mix and allow to stand for 10 min. After standing, dilute to the mark with distilled water. Add an additional 5.00 ml distilled water to each flask. Mix thoroughly and measure the absorbance of each of these solutions against a distilled water blank using 20.0-cm cells. The contents of these flasks correspond to ozone concentrations of 0.00, 1.20, 2.40, 3.60, 7.20, and 10.80 mg ozone per liter, respectively.

I. METHODS OF DETERMINATION

A. TITANIUM METHOD

The reaction of hydrogen peroxide with titanium(IV) in acid solution producing yellow peroxytitanic acid is well known for colorimetric determination of titanium (256). It is also used for determination of hydrogen peroxide (4,35,92,138). The yellow color of peroxytitanic acid forms immediately and is stable for several hours. The reaction is specific for hydrogen peroxide, although hydroperoxides capable of forming hydrogen peroxide by hydrolysis under the reaction conditions employed interfere. This interference is used to advantage for determination of hydroperoxides in hydrocarbon mixtures by intentionally converting them to their hydrogen peroxide equivalent prior to addition of titanium reagent solution (233). It is reported that only secondary and tertiary hydroperoxides undergo acid hydrolysis to form hydrogen peroxides (62). Inorganic peroxides capable of forming hydrogen peroxide also interfere (244). Formaldehyde, if present with hydrogen peroxide, interferes. This interference is removed, however, by acidifying the sample to a hydrogen ion concentration of $3N$ and allowing it to stand for 2 hr before addition of titanium reagent solution (258). It is reported that Cu(II), Fe(II), Zn(II), Mn(II), Mg(II), nitrate, chlorate, iodate, and peroxydisulfate do not interfere, while phosphate in concentrations greater than $0.005M$, Al(III), Ni(II), and borate interfere (149). Chromate interferes but is removed by an anion exchange technique prior to titanium addition (247). Atmospheric oxygen does not interfere (91).

Various titanium salts are used to prepare the titanium(IV) reagent. These include $TiOSO_4$, $TiCl_4$, and $K_2TiO(C_2O_4)_2$. The wavelength of maximum absorbance for peroxytitanic acid is reported to vary between 400 and 420 nm depending on the experimental conditions and the choice of titanium salt (319).

A novel adaptation of the peroxytitanic acid procedure for atmospheric hydrogen peroxide employs a packed tube of silica gel impregnated with alcoholic titanium(IV) sulfate. By recording the time necessary to observe development of a yellow color upon passage of samples containing known amounts of hydrogen peroxide vapor at known flow rates, a calibration curve of hydrogen peroxide concentration versus sample volume provides a convenient reference for subsequent determination of unknown hydrogen peroxide-containing samples (206).

To increase the sensitivity of the peroxytitanic acid method for hydrogen peroxide, formation of a ternary complex is suggested. A ligand capable of combining with peroxytitanic acid is used. The absorptivity of the resulting species is greater than that of peroxytitanic acid itself. Xylene orange (219) and 8-hydroxyquinoline (61) are possible auxiliary ligands. A specific procedure employing the later reagent is presented in Section IIA.

The following procedure is for determination of aqueous hydrogen peroxide solution using titanium(IV) sulfate reagent (62). A molar absorptivity for peroxytitanic acid of 770 liters per mole-cm at 407 nm is reported.

Solutions

Titanium(IV) Sulfate. Carefully add 4.6 g $TiOSO_4$ and 20 g $(NH_4)_2SO_4$ to 100 ml $18M$ H_2SO_4 in a beaker. Heat until only a small residue remains. Cool and carefully pour into 350 ml distilled water. Cool and transfer to a 500-ml volumetric flask. Dilute to volume with additional distilled water. If the solution is turbid, filter through a fine sintered-glass crucible before use.

Sulfuric Acid, $9M$.

Hydrogen Peroxide. Transfer 1.0000 g 30% H_2O_2 solution to a 500-ml volumetric flask. Dilute to volume with distilled water. Standardize this solution by an appropriate volumetric means. The solution should contain approximately 600 μg hydrogen peroxide per ml.

Hydrogen Peroxide, Dilute. Transfer exactly 50.0 ml of the above solution to a 500-ml volumetric flask. Dilute to volume with distilled water.

Procedure

Transfer an accurately known volume of sample, 1 to 3 ml, depending upon the hydrogen peroxide concentration, to a 10-ml volumetric flask. Add 1.0 ml titanium(IV) solution and 2.5 ml $9M$ sulfuric acid. Mix thoroughly. Place the flask in a water bath at 60°C for 10 min. At the end of this time, remove the flask, cool and dilute to volume with distilled water. Transfer a portion of this solution to a 1-cm cell and measure the absorbance at 407 nm against a reagent blank prepared in the same manner but with an equal volume of distilled water in place of hydrogen peroxide-containing solution. A calibration curve is prepared by using between 0 and 3.00 ml dilute hydrogen peroxide standard solution in place of the unknown sample. Follow the same procedure. Interference from t-butyl hydroperoxide is reported.

The following procedure is for determination of aqueous hydrogen peroxide solution using titanium(IV) chloride reagent (319). Beer's law is followed for hydrogen peroxide solutions containing from 1 to 30 mg H_2O_2 per liter.

Solutions

Titanium(IV) Chloride. Cool in separate beakers, each surrounded with crushed ice, 10 ml $6N$ HCl and 10 ml water-white titanium(IV) chloride. Add the chilled titanium(IV) chloride dropwise to the chilled HCl solution. Allow the mixture to stand at ice temperature until the solids that form dissolve. Dilute to 1 liter with additional 6N HCl. The solution is stable at room temperature and contains approximately 4 mg Ti(IV) per ml.

Hydrogen Peroxide. Transfer exactly 1 g 30% H_2O_2 solution to a 500-ml volumetric flask. Dilute to volume with distilled water. Standardize this solution by an appropriate volumetric means. The solution should contain approximately 600 μg hydrogen peroxide per ml.

Hydrogen Peroxide, Dilute. Transfer exactly 50.0 ml of the above solution to a 500-ml volumetric flask. Dilute to volume with distilled water.

Procedure

Transfer an accurately known aliquot of sample solution, 1 to 5 ml, to a 10-ml volumetric flask. Dilute to volume with titanium(IV) chloride solution. Mix and transfer a portion of the solution to a 1-cm cell. Measure the absorbance at 415 nm against a reagent blank containing only titanium(IV) reagent solution. A calibration curve is prepared by placing accurately known amounts, 1.00 to 5.00 ml, dilute hydrogen peroxide standard solution into separate 10-ml volumetric flasks. Dilute each to volume with titanium(IV) reagent solution, mix, and measure the absorbance at 415 nm against a titanium(IV) reagent blank. Interferences from t-butyl hydroperoxide and 1-ethoxyethyl hydroperoxide are reported.

B. IODINE METHOD

Triiodide ion formed from reaction of iodide ion and hydrogen peroxide is used for hydrogen peroxide determination. Colorimetric measurement of blue iodine—starch complex upon addition of starch solution (259) or direct ultraviolet measurement of triiodide ion itself (226) is used for this determination. The rate of hydrogen peroxide reaction with iodide ion increases with increasing acidity, and kinetic procedures based upon the rate of triiodide formation are suggested for distinguishing between hydrogen peroxide and other organic peroxides which also react but at different rates (5,253). In general, the longer the alkyl group of the organic peroxide, the more slowly it reacts with triiodide ions (240). Addition of ammonium molybdate to neutral, phosphate-buffered iodide solution after the addition of hydrogen peroxide sample catalyses their reaction and results in rapid conversion of iodide ion to triiodide (62). A procedure for determination of a variety of organic peroxide types in organic solvents using iodide ion reaction is presented in Section IIC.

The following procedure is for determination of aqueous hydrogen peroxide solutions. Beer's law is followed for solutions containing up to a maximum of 12 μg hydrogen peroxide in a final volume of 10 ml (222).

Solutions

Potassium Iodide, 0.2M. Dissolve 23.2 g KI in distilled water and dilute to 1 liter. Store in the dark and prepare fresh every several days.

Ammonium Molybdate. Dissolve 5 g $(NH_4)_6Mo_7O_{24} \cdot 4H_2O$ in distilled water and dilute to 1 liter.

Hydrogen Peroxide. Dilute 1.0 ml 30% H_2O_2 with distilled water in a 1-liter volumetric flask. Dilute to volume with additional distilled water. Standardize this solution against a standard solution of potassium permanganate.

Hydrogen Peroxide, Dilute. Transfer exactly 10.0 ml of the above solution to a 1-liter volumetric flask and dilute to volume with distilled water.

Procedure

Transfer an exactly known amount of neutral hydrogen peroxide-containing solution, 1.00 to 4.00 ml, to a 10-ml volumetric flask. Add 5.00 ml potassium iodide solution and 0.1 ml ammonium molybdate solution. Mix and dilute to volume with distilled water. After 5 min, measure the absorbance of triiodide ion produced at 353 nm against a reagent blank containing only iodide, molybdate, and distilled water. Use 1-cm silica cells. A calibration curve is prepared by treating exactly known amounts of diluted standard hydrogen peroxide solution, 1.00 to 4.00 ml, by the same procedure.

C. IRON–THIOCYANATE METHOD

Hydrogen peroxide and other organic peroxides oxidize iron(II) to iron(III). Iron(III) is subsequently determined by formation of the red thiocyanate complex and colorimetric measurement (91,227). The reaction is utilized for determination of a variety of organic peroxides in both liquid (302) and atmospheric (293) samples. Kinetic data are reported which distinguish between hydrogen peroxide and various other oxidants (62,240). The procedure presented in Section IB of the ozone section applies equally well to hydrogen peroxide determination.

Other color-forming agents are also suggested for reaction with iron(III) produced from hydrogen peroxide oxidation. These are sulfosalicylic acid (56) and 7-iodo-8-quinolinol-5-sulfonic acid (215). An alternate approach to hydrogen peroxide determination by oxidation of iron(II) to iron(III) uses 1,10-phenanthroline to form the colored iron(II)–1,10-phenanthroline complex with excess iron(II) remaining after peroxide reaction. Details of this procedure are presented in Section IIB.

D. BENZOHYDROXAMIC ACID METHOD

Destruction of the colored complex formed with benzohydroxamic acid and vanadium(V) in 1-hexanol is reported as an undesirable interference in the colorimetric determination of vanadium (200). This interfering reaction is now used for determination of hydrogen peroxide (201). Formation of a complex between vanadium(V) and benzohydroxamic acid in 1-hexanol is prevented if hydrogen peroxide is present because of preferential formation of peroxyvanadic acid. Decreasing absorbance values at 450 nm and pH 3 for the vanadium(V)–benzohydroxamic acid species with increasing amounts of hydrogen peroxide are a quantitative measure of hydrogen peroxide concentration. Hydroperoxides and dialkyl peroxides present with hydrogen peroxide in a test sample do not appreciably interfere if a water extraction of hydrogen peroxide is employed. Strong oxidizing and reducing agents that attack either hydrogen peroxide or vanadium(V) interfere. Destruction of the uranium(VI) complex of benzohydroxamic acid also serves to determine hydrogen peroxide by this method. With the uranium complex, however, measurements are made at 380 nm and pH 6.

The following procedure is for determination of hydrogen peroxide in water-immiscible organic liquids. It is also applicable to water-miscible liquids by omitting the preliminary water extraction step. The method is sensitive to 1 micromole hydrogen peroxide.

Solutions

Ammonium Vanadate, 0.001M. Dissolve 0.117 g NH_4VO_3 in distilled water and dilute to 1 liter in a volumetric flask.

Potassium Benzohydroxamate, 0.100M. Recrystallize commercially available potassium benzohydroxamate from a solution containing 20% distilled water and 80% methanol. Dissolve 17.5 g recrystallized salt in distilled water. Adjust the solution to pH 5 and dilute to 1 liter with additional distilled water. The solution is stable for approximately one month (318).

Potassium Benzohydroxamate, 0.010M. Dilute 100 ml of the above solution with distilled water to 1 liter in a volumetric flask.

1-Hexanol. Commercially available 1-hexanol is purified of peroxide contamination by distilling over potassium hydroxide. Further peroxide formation is prevented by saturating the alcohol with distilled water and storing in a metal container.

Vanadium(V)–Benzohydroxamic Acid. Add 10.0 ml 0.010M benzohydroxamic acid solution to 20.0 ml vanadium(V) solution. Adjust the pH of the solution to 3.0 using 6M sulfuric acid and a pH meter. Add 20.0 ml peroxide-free 1-hexanol and mix thoroughly. Centrifuge and remove the organic layer containing the desired complex. The complex is best stored in a metal container, although a darkened glass bottle is satisfactory. The complex is stable for several months when stored in a metal container.

Hydrogen Peroxide. Transfer exactly 1 g 30% H_2O_2 to a 500-ml volumetric flask. Dilute to volume with distilled water. Standardize this solution by an appropriate means. The solution should contain approximately 600 μg hydrogen peroxide per ml.

Hydrogen Peroxide, Dilute. Transfer exactly 50.0 ml of the above solution to a 500-ml volumetric flask. Dilute to volume with distilled water.

Procedure

Place an exactly known amount of water-immiscible sample into a separatory funnel. Add 15 to 20 ml distilled water which has been adjusted to pH 2.0 with sulfuric acid. Use a pH meter. Mix thoroughly, allow the layers to separate, and transfer the aqueous layer to a second separatory funnel. Add 10.0 ml vanadium(V)–benzohydroxamic acid solution. Mix thoroughly for 10 min, allow the layers to separate, and transfer a portion of the organic layer to a 1-cm cell. Measure the absorbance at 450 nm against a blank solution of 1-hexanol. For samples miscible with water, vanadium(V)–benzohydroxamic acid solution is added directly to a known amount of sample. A calibration curve is prepared by carrying out the above procedure using between 1.00 and 10.0 ml diluted standard hydrogen peroxide solution. A curve of decreasing absorbance versus increasing hydrogen peroxide concentration is prepared.

E. 1,2-DI(4-PYRIDYL)ETHYLENE METHOD

Formation of pyridine 4-aldehyde from reaction of 1,2-di(4-pyridyl)ethylene with subsequent colorimetric determination by an established procedure is applied to the colorimetric determination of ozone (see ozone, Section IE). It is also applied to the determination of hydrogen peroxide (126).

The following procedure is for determination of aqueous solutions containing hydrogen peroxide. A linear calibration curve is obtained for hydrogen peroxide concentrations ranging from 0.30 to 15.0 mg H_2O_2 per liter. One mole hydrogen peroxide is empirically found to generate 1.68 mole pyridine 4-aldehyde by this procedure. Duplicate values of samples gave a precision range of ±2%.

Solutions

1,2-Di(4-pyridyl)ethylene, 0.5%. Dissolve 0.5 g 1,2-di(4-pyridyl)ethylene in glacial acetic acid. Dilute to 100 ml with additional glacial acetic acid. Prepare fresh daily.

3-Methyl-2-benzothiazolinonehydrazone Hydrochloride, 0.25%. Dissolve 0.25 g 3-methyl-2-benzothiazolinonehydrazone hydrochloride monohydrate in distilled water. Dilute to 100 ml with additional distilled water.

Hydrogen Peroxide. Dilute 1.0 ml 30% H_2O_2 with distilled water to 1 liter in a volumetric flask. Standardize this solution by an appropriate iodometric method.

Procedure

Place 5.0 ml 1,2-di(4-pyridyl)ethylene solution and 1.00 ml aqueous sample into a test tube. Mix and place the test tube in a boiling water bath for 15 min. Remove the test tube and cool to room temperature under running water. Add 1.0 ml 3-MBTH solution. Mix and again heat in a boiling water bath for 1 to 1.5 min. Remove the test tube and cool to room temperature under running water. Transfer a portion of this solution to a 1-cm cell and measure the absorbance at 442 nm. A reagent blank is prepared in the same manner but with 1.00 ml distilled water substituted for the sample solution. A standard curve is prepared by placing known amounts of hydrogen peroxide solution, between 0.10 and 3.50 ml, into separate 10-ml volumetric flasks. Dilute each flask to volume with distilled water. Exactly 1.00 ml of each standard solution is carried through the above procedure. The micrograms hydrogen peroxide corresponding to each test solution is calculated from the known hydrogen peroxide concentration and the appropriate dilution factor. A calibration curve relating absorbance to micrograms hydrogen peroxide per ml test solution is prepared.

F. *p*-DIMETHYLAMINOBENZALDEHYDE METHOD

p-Dimethylaminobenzaldehyde is slowly oxidized by aqueous hydrogen peroxide at pH 5.8 to produce a colored oxidation product. Colorimetric measurement at 525 nm yields absorbance values proportional to the amount of hydrogen peroxide present (69,70). The presence of a variety of inorganic salts, sugars, amino acids, starch, and proteins generally resulted in only minor absorbance changes from those observed when these materials were absent. Changes in absorbance values are, however, observed with changing temperature, pH, and time of color development.

The following procedure is for determination of aqueous hydrogen peroxide. A linear calibration curve results for hydrogen peroxide within the limits of 85 and 1000 μg H_2O_2. Ten repeated determinations on the same hydrogen peroxide solution produced a mean absorbance value of 0.377, with a standard deviation of ±0.0010 absorbance unit.

Solutions

p-Dimethylaminobenzaldehyde. Dissolve 20.0 g *p*-dimethylaminobenzaldehyde in 95% ethanol. Dilute to 1 liter with additional ethanol. Store this solution in the dark and use within 2 hr of preparation.

Citric Acid–Sodium Citrate Buffer. Mix together 29.5 ml 0.1M citric acid and 200 ml 0.1M sodium citrate. Adjust to pH 5.8 ± 0.02 with additional strong acid or strong base. Use a pH meter.

Hydrogen Peroxide. Dilute between 0.3 and 3.0 ml 30% hydrogen peroxide to 1 liter in separate volumetric flasks. Standardize each solution against standard potassium permanganate solution.

Procedure

To 4.8 ml citrate buffer in a screw-capped test tube, add 0.4 ml *p*-dimethyl-aminobenzaldehyde solution and 1.00 ml hydrogen peroxide sample. Cap and mix thoroughly for 5 sec. Transfer the tube to a constant-temperature bath at 45°C. Allow the sample to remain in the bath for 5 hr. At the end of this time, remove the sample and rapidly, within 1 min, cool the sample tube by first placing it in an ice bath for 15 sec and then in a 25°C water bath for 45 sec. The solution temperature for colorimetric measurement should be 25 ± 2°C. Transfer a portion of the sample to a 1-cm cell and record the absorbance at 525 nm against a reagent blank containing buffer and *p*-dimethylaminobenzaldehyde but not hydrogen peroxide. Only a slight difference in absorbance value is observed if water is used as the reagent blank. A calibration curve is prepared by allowing to react exactly 1.00 ml of each of the hydrogen peroxide standards using this procedure.

G. OXIDASE METHOD

Enzymatic reaction with hydrogen peroxide in the presence of an oxidizable chromatogen is used for hydrogen peroxide determination. Colorimetric measurement of the oxidized species formed serves as a measure of the amount of hydrogen peroxide originally present in the sample. Horseradish peroxidase is the enzyme most often used, although lactoperoxidase is employed for determination of hydrogen peroxide in milk and cream (177). Various oxidizable organic species are suggested for color formation. These include

o-tolidine (190,248), benzidine (12,221), and o-dianisidine (199). It is necessary to buffer the sample for optimum color formation. The exact pH chosen depends upon the choice of organic reagent. Oxidase reaction is used for hydrogen peroxide determination in a variety of biologic materials (166) and indirectly for determination of glucose by forming from glucose an equivalent amount of hydrogen peroxide through reaction with glucose oxidase (196).

The following procedure is for determination of aqueous hydrogen peroxide using horseradish peroxidase and leuco crystal violet (211). The method applies to samples containing between 0.06 and 0.50 mg H_2O_2 per liter. Eight repeated determinations upon a solution containing 0.34 mg hydrogen peroxide per liter resulted in a deviation from the mean value of less than 2%. Adenine, valine, tyrosine, and tryptophan in concentrations of $8.0 \times 10^{-4} M$ do not interfere. Cerium(IV) and periodate, because they also oxidize crystal violet, interfere.

Solutions

Leuco Crystal Violet. Commercially available leuco crystal violet is purified by first dissolving it in 0.5% (v/v) aqueous hydrochloric acid and extracting with benzene. Evaporate the benzene extract and recrystallize the residue three times from benzene. Dissolve 50 mg purified leuco crystal violet in about 80 ml 0.5% (v/v) hydrochloric acid. Dilute to 100 ml with additional acid solution. Store in a refrigerator.

Horseradish Peroxidase. Dissolve 10 mg commercially available material in 10 ml distilled water. Prepare fresh as needed. Store in a refrigerator.

Acetic Acid–Sodium Acetate Buffer. Mix equal volumes of $2M$ sodium acetate and $2M$ acetic acid. Adjust to pH 4.5 using additional acetic acid and a pH meter.

Hydrogen Peroxide, 0.1M. Dilute 11.3 ml 30% H_2O_2 to 1 liter in a volumetric flask with distilled water. Determine the exact concentration of this solution using an iodometric procedure. Store in a refrigerator.

Hydrogen Peroxide, Dilute. Transfer exactly 0.50 ml of the above solution to a 1-liter volumetric flask and dilute to volume with distilled water. Store in a refrigerator.

Procedure

Place a known amount of sample solution, between 1.00 and 5.00 ml, in a 10-ml volumetric flask. Add 1.0 ml leuco crystal violet, followed by 0.5 ml horseradish peroxidase. Mix and add 4.0 ml buffer. Dilute to volume with distilled water and mix thoroughly. Color development is almost instantaneous.

Transfer a portion of solution to a 1-cm cell and measure absorbance at 596 nm against a reagent blank prepared in the same manner but containing no hydrogen peroxide. A calibration curve is prepared by carrying out the above reaction sequence on exactly known amounts, between 0.30 and 3.00 ml, of diluted standard hydrogen peroxide solution.

H. MISCELLANEOUS METHODS

Various peroxy acid species, formed from allowing to react hydrogen peroxide with transition metals other than titanium, are used in colorimetric procedures for hydrogen peroxide determination. An acidic solution of sodium vanadate forms a peroxy acid upon reaction with hydrogen peroxide. Absorbance measurements of this species near 450 nm serve to determine hydrogen peroxide in the ppm range (4,119). The method is applied to both biologic materials (305) and atmospheric samples (47). Reactions with ammonium molybdate (142), ammonium molybdate-impregnated paper (99), and uranyl nitrate (32) are also reported.

Determination of excess chromium(VI) after its reaction with a variety of substances, including hydrogen peroxide, is employed for colorimetric determination. Direct absorbance measurement of remaining chromium(VI) in the dichromate form (106) and reaction of excess chromium(VI) in acid solution with o-dianisidine followed by colorimetric measurement at 470 nm (48) are suggested.

The decrease in absorbance of alkaline potassium hexacyanoferrate(III) at 418 nm because of reduction to potassium hexacyanoferrate(II) upon contact with hydrogen peroxide is another procedure for determination of hydrogen peroxide (14). In a related method, the hexacyanoferrate(II) formed is converted at pH 3.5 in the presence of a mercury catalyst to iron(II), and this is subsequently determined colorimetrically with 1,10-phenanthioline (153,154).

Other inorganic reactions employed for hydrogen peroxide determination include oxidation of the cobalt(II) complex of nitrilotriacetic acid to the corresponding cobalt(III) complex (57). Reaction occurs at pH 11. Spectrophotometric titration of hydrogen peroxide with aqueous xenon trioxide at pH 1, followed at 200 nm in the ultraviolet absorption region, is also reported. Both reacting species absorb at this wavelength. The procedure is sensitive to a lower limit of 20 mg hydrogen peroxide per liter (170).

Reactions with a variety of other organic reagents are also reported for detection and quantitative determination of hydrogen peroxide and organic peroxides. These include reaction with phenolphthalein (195), malachite green (111), indigo carmine (169), p-phenylenediamine (129), substituted phenylenediamines (85,120,122), and the glucoside phlorizin (178).

II. APPLICATIONS

Standard solutions of hydrogen peroxide are prepared by diluting concentrated commercially available hydrogen peroxide to the desired concentration and standardizing by an established analytic procedure (262). Titration techniques are commonly used with standard solutions of either potassium permanganate or cerium(IV) sulfate. Reaction of hydrogen peroxide with iodide ion-liberating iodine and subsequent titration with standard sodium thiosulfate solution is an alternate procedure. Gasometric or polarographic determinations are also employed. Aqueous solutions of hydrogen peroxide are subject to decomposition and hence should be used soon after they are prepared and standardized. Elevated temperatures, trace metallic impurities which catalyze decomposition, and ultraviolet radiation all contribute to decreasing hydrogen peroxide stability.

A. ATMOSPHERIC HYDROGEN PEROXIDE

Hydrogen peroxide and a variety of organic peroxides are found in the atmosphere. Procedures for their determination generally result in a total peroxide content. Kinetic differences do, however, exist and are discussed above in connection with the titanium, iodide ion, and iron(III)—thiocyanate methods. Specific procedures for atmospheric peroxide determination are available using titanium, iron(III)—thiocyanate, vanadium, and phenolphthalein.

The following procedure is for determination of atmospheric samples containing hydrogen peroxide vapor. It is based upon colorimetric measurement of the peroxytitanic acid–8-hydroxyquinoline complex (61). Hydrogen peroxide in the ppm (v/v) range is determined. The molar absorptivity of the complex is reported as 3.06×10^3 liters per mole-cm.

Solutions

Titanium(IV) Sulfate. Into 100 ml $18M$ sulfuric acid, place 4.6 g $TiOSO_4$ and 20 g $(NH_4)_2SO_4$. Heat gently for several minutes to dissolve all soluble material. Cool and carefully pour the mixture into 350 ml distilled water. Cool and filter through a sintered-glass crucible to remove all undissolved matter. Dilute the solution to 500 ml with additional distilled water.

Titanium(IV) Sulfate, Dilute. Dilute 10 ml of the above solution to 500 ml with additional distilled water. This solution contains approximately 50 mg Ti(IV) per ml.

8-Hydroxyquinoline, 0.1%. Dissolve 0.1 g 8-hydroxyquinoline in chloroform. Dilute to 100 ml with additional chloroform. Store this solution in a cool, dark place.

Sodium Acetate, 5%. Dissolve 5 g NaOAc in distilled water. Dilute to 100 ml with additional distilled water.

Sodium Sulfate, Anhydrous.

Hydrogen Peroxide, 30%.

Procedure

Place 5.0 ml dilute titanium(IV) solution into a gas collection flask fitted with a fritted bubbler. Pass sample gas through the bubbler at a rate of 0.5 liter per min for 10 min. Sample volume is calculated from the product of sample rate and sampling time. Temperature and pressure corrections are necessary for samples measured under varying experimental conditions. Transfer the peroxytitanic acid solution to a separatory funnel. Add sufficient distilled water to adjust the volume to 15 ml. Add sufficient sodium acetate solution to adjust the pH to 4.2 ± 0.2. Use a pH meter. Now add 10 ml 8-hydroxyquinoline solution and agitate the funnel continuously for 5 min. Allow the layers to separate. Remove the chloroform layer and dry it briefly by placing it in contact with anhydrous sodium sulfate. Transfer a portion of this solution to a 1-cm cell and measure the absorbance at 450 nm. A reagent blank is prepared by the same procedure, except for omitting any contact with hydrogen peroxide. A calibration curve is prepared by injecting microliter quantities of 30% hydrogen peroxide into a stream of metered air. The suspension is passed into a bag made of fluorinated ethylene–propylene copolymer (FEP Teflon, du Pont). The hydrogen peroxide concentration of this standard is determined by an independent colorimetric method. The iodine method is recommended. Known volumes of sample from the storage bag, containing a known concentration of hydrogen peroxide, are passed into titanium(IV) solutions as described and carried through the colorimetric procedure. A calibration curve of absorbance reading versus hydrogen peroxide concentration is prepared.

B. HYDROGEN PEROXIDE IN AQUEOUS SOLUTION

The majority of procedures presented above are for aqueous solutions. Specific utilization of these procedures for problems of special interest are reported. These include use of titanium (203) and phenolphthalein (86) for determination of hydrogen peroxide formed during radiolysis studies of aqueous solutions. Another application involves determining hydrogen peroxide in the presence of various peroxy acids of sulfur, using titanium (67) or spectrophotometric titration with cerium(IV). Absorbance of the cerium species is monitored in the ultraviolet region of the spectrum (193). A procedure for hydrogen peroxide in protein-containing biologic solutions using the ternary complex peroxytitanic acid–xylenol orange is available (219). Determination of hydrogen peroxide in rainwater using 7-iodo-8-quinolinol-5-sulfonic acid is reported (130).

The following procedure is for determination of aqueous solutions of hydrogen peroxide based upon the ability of hydrogen peroxide to oxidize iron(II) to iron(III). The residual iron(II) after reaction is determined by the well-known 1,10-phenanthroline method (20). Other species present in the solution that react with iron(II) and/or hydrogen peroxide interfere. The procedure is applicable for hydrogen peroxide concentrations within the range of 0.1 to 2.5 mg H_2O_2 per liter. Ten repeated determinations upon a solution containing 0.66 mg H_2O_2 per liter resulted in a mean value of 0.67 mg H_2O_2 per liter, with a standard deviation of 1.8%. If bathophenanthroline (4,7-dimethyl-1,10-phenanthroline) is substituted for 1,10-phenanthroline in the reaction procedure, the sensitivity is increased. Hydrogen peroxide within the range of 0.03 to 0.10 mg H_2O_2 per liter can now be determined. With this more sensitive chromatogen, ten repeated determinations of a solution containing 0.066 mg H_2O_2 per liter resulted in a mean value of 0.065 mg H_2O_2 per liter, with a standard deviation of 2.8%.

Solutions

Iron(II)   Ammonium   Sulfate.   Dissolve   exactly   0.2100 g $FeSO_4$ $(NH_4)_2SO_4 \cdot 6H_2O$ in distilled water. Add 5 ml 72% perchloric acid and dilute to 1 liter in a volumetric flask. The solution resists oxidation for up to 10 weeks if, after dilution, 5 g of $\frac{1}{2}$-in. by $\frac{1}{16}$-in. pieces of aluminum rod is placed in the storage container.

1,10-Phenanthroline. Dissolve 50 mg 1,10-phenanthroline monohydrate in 100 ml distilled water. Heat until dissolved.

Bathophenanthroline. Dissolve 20 mg 4,7-dimethyl-1,10-phenanthroline in 50 ml 95% ethanol.

Sodium Acetate, 0.5M. Dissolve 68.0 g NaOAc $\cdot$ $3H_2O$ in distilled water and dilute to 1 liter.

Perchloric Acid, 72%.

Hydrogen Peroxide. Dilute 20 ml 30% hydrogen peroxide to 1 liter in a volumetric flask with boiled redistilled water. Add 20 mg acetanilide to prevent decomposition. Standardize this solution by an appropriate titrimetric or colorimetric procedure.

Hydrogen Peroxide, Dilute. Transfer exactly 0.50 ml the above solution to a 1-liter volumetric flask and dilute to the mark with boiled redistilled water.

Procedure

To avoid standardization of the iron(II) solution used in this procedure, the following preliminary steps are taken. Place an accurately known aliquot of hydrogen peroxide-containing sample, perhaps 10.0 ml, and 10 ml 1,10-phenanthroline solution, 0.1 ml 72% perchloric acid, and 10 ml sodium acetate solution into a 50-ml volumetric flask. Add iron(II) solution to the flask from a microburet and mix thoroughly to assure uniform color distribution. Continue to add iron(II) solution until the color intensity produced reaches some suitable value, perhaps 1.0 absorbance unit, when measured against a water blank. Record the exact volume of iron(II) solution necessary to achieve this condition and then discard the preliminary solution. The exact volume of iron(II) solution ascertained above is now added to each of two separate 50-ml volumetric flasks. To one of these flasks, add the accurately known volume of hydrogen peroxide-containing sample and to the other, an equal amount of distilled water. Now add to each flask 0.1 ml 72% perchloric acid, 10 ml sodium acetate solution, and 5 ml 1,10-phenanthroline solution. Dilute both flasks to their final volumes with distilled water and mix thoroughly. Allow the flasks to stand for 15 min to assure full color development. Transfer portions from each flask to separate matched 1-cm cells. The cell containing solution without hydrogen peroxide sample is placed in the sample compartment of a double-beam spectrophotometer. Place the cell containing solution with hydrogen peroxide into the reference compartment. Record the absorbance difference between these two solutions at 510 nm. This difference is proportional to the concentration of hydrogen peroxide in the sample.

A calibration curve is prepared by repeating the above procedure on various aliquots, between 3.00 and 30.0 ml, of dilute hydrogen peroxide standard solution. For this series, the amount of iron(II) solution added to each flask is first determined using the sample containing the largest test aliquot. Once determined, this amount of iron(II) solution is used for all calibration samples. If the more sensitive bathophenanthroline is used in place of 1,10-phenanthroline, the same procedure is followed except that 95% ethanol is used for final dilution and absorbance readings are made at 533 nm. The standard hydrogen peroxide aliquots must also be adjusted to be within the proper range for this more sensitive reagent.

C. HYDROGEN PEROXIDE IN ORGANIC SOLVENTS

Various peroxide species are often found in organic solvents. Their content is important not only because possible undesirable side reactions can occur, but also because they constitute an extreme safety hazard. Peroxide-containing solvents possess powerful explosive capability (71). Procedures for total peroxide content are based on reaction with one of several reagents. A titanium procedure is available similar to that in Section IA above (319). Procedures using benzoyl leuco methylene blue (93) and N,N-dimethyl-p-phenylenediamine sulfate (84) are also reported. Most common, however, is an iodometric procedure (303). A mixture of potassium iodide, acetic acid, and chloroform constitutes the necessary reagent solution. Upon contact with peroxide-containing sample, iodine is liberated. Early methods called for titrimetric determination using standard sodium thiosulfate solutions. Colorimetric procedures are now available. To decrease the interference of atmospheric oxygen in this iodometric method, various other solvent combinations are suggested. These include substitution of 2-propanol for the solvent medium (167), use of acetic anhydride and sodium iodide in place of acetic acid and potassium iodide (220), and use of citric acid in a mixed t-butanol and carbon tetrachloride solvent (121).

The following procedure is for determination of hydrogen peroxide and a variety of organic peroxides in organic solvents using iodide–iodine oxidation (23). Results were obtained for 19 various organic peroxides representative of the following peroxide groups: hydroperoxide, diacyl, diaroyl, perester, and ketone peroxide. Of the peroxides tested, only t-butyl hydroperoxide and dicumyl peroxide failed to react quantitatively. Solvents studied were methanol, 2-propanol, pentane, hexane, chloroform, benzene, toluene, ethyl ether, acetone ethyl acetate, and vinyl acetate. Spectrophotometric measurements of iodine are made at 470 nm instead of 362 nm to avoid possible solvent interference. The method is applicable for solutions containing up to 400 µg active oxygen. Active oxygen is defined as the amount of oxygen released when an amount of hydrogen peroxide equivalent to the total peroxide content of a sample is converted to water and oxygen.

Solutions

Potassium Iodide. Dissolve 50 g KI in distilled water. Dilute to 100 ml with additional distilled water. Prepare fresh as needed.

Acetic Acid–Chloroform. Combine 600 ml glacial acetic acid and 300 ml chloroform.

Iodine. Dissolve 0.1270 g iodine in acetic acid–chloroform solvent. Dilute to 100 ml in a volumetric flask with additional solvent mixture. This solution contains 1.27 mg iodine per ml and is equivalent to 80.0 µg active oxygen per ml.

Procedure

Pipet 5.00 ml peroxide-containing sample into a 25-ml volumetric flask and dilute to volume with acetic acid–chloroform solvent. Insert a capillary tube into the bottom of the flask and purge with a fine stream of nitrogen for 1.5 min. Add 1.00 ml KI solution and continue to purge for 1 additional min.

Remove the capillary, stopper, mix, and set aside in the dark for 1 hr. At the end of this time, quickly transfer a portion of this solution to a 1-cm cell and immediately measure the absorbance at 470 nm against a reagent blank solution. Prolonged exposure to the atmosphere during transfer allows for extensive air oxidation and erroneous results. A calibration curve is prepared by transferring to separate 25-ml volumetric flasks exactly known amounts, up to 5.00 ml, standard iodine solution. Treat each sample by the procedure described above. Record absorbance versus μg active oxygen for each sample. Alternately, a calibration curve can be prepared using appropriate amounts of diluted standardized hydrogen peroxide solution. For samples containing up to 40 μg active oxygen, it is suggested that the entire sequence of reactions be performed in an absorption cell fitted with a purging capillary tube. This avoids possible contact with atmospheric oxygen by eliminating the transfer step. The solvent mixture is modified for this more sensitive range by addition of 4% water. Five drops of deaerated 50% potassium iodide solution are used in place of the 1 ml recommended above. Measurements at 410 nm exhibit a larger absorbance value than do those at 470 nm and are recommended for this lower concentration range. A calibration curve for this concentration range is prepared from an iodine solution containing 63.4 mg iodine per liter acetic acid–chloroform solvent mixture.

WATER

I. METHODS OF DETERMINATION

A. COBALT CHLORIDE METHOD

Color changes exhibited by cobalt(II) salts dissolved in various organic solvents upon addition of increasing amounts of water have been studied extensively (17,19,156,157). For cobalt(II) chloride, di-, tri-, tetra-, and hexahydrates are reported varying in color from violet to red brown. The color change from blue to pink exhibited by Co(II) chloride dissolved in an organic solvent containing water serves both as a qualitative test and quantitatively for determining small amounts of water. In addition to cobalt(II) chloride (13), the bromide (229) and iodide (18) are also used for quantitative determination. Cobalt(II) salts are reported for estimation of water in acetone (28) and other organic liquids (183). A procedure for the study of evaporation rates from water surfaces is also reported (216).

The following procedure is for determination of adsorbed moisture on samples insoluble in ethanol. The water is extracted into ethanol and cobalt(II) chloride is added. The procedure is applicable for extracts containing from 2 to

10 mg water per ml (100). A reagent solution containing 300 mg cobalt(II) ion produces a nearly linear calibration curve for this range of water content. Decreasing absorbance values with increasing water content are recorded at 671 nm. The absorbance value, although constant for several days, is temperature dependent, and thermostated cells are necessary for most reliable results. Close agreement between values obtained by this method and the Karl Fisher titration method (204) is reported for various materials.

Solutions

Cobalt(II) Chloride. Dissolve 4.00 g $CoCl_2 \cdot 6H_2O$ in absolute ethanol. Dilute to 1 liter with additional absolute ethanol.

Water. Weigh exactly 2.50 g water into a 100-ml volumetric flask. Dilute to volume with absolute ethanol. This solution contains 25 mg water per ml.

Procedure

Place an accurately weighed amount of solid material, between 1.00 and 5.00 g, depending upon the amount of water present, into a glass-stoppered flask. Add exactly 50.0 ml absolute ethanol, stopper, and mix vigorously for 5 to 10 min. Allow the solid material to settle. Centrifuge if necessary. Transfer 25.0 ml of the solution to a 50-ml volumetric flask. Add 15.0 ml cobalt(II) chloride solution. Mix and place the flask in a water bath at $25 \pm 0.1°C$. After temperature equilibrium is reached, fill to the mark with additional absolute ethanol also at $25 \pm 0.1°C$. Mix the contents of the flask to achieve uniform distribution and transfer a portion of solution to a 1-cm cell. Measure the absorbance at 671 nm against a reagent blank. It is recommended that the instrument cell compartment be thermostated at $25 \pm 0.1°C$. An increase in absorbance is observed with increasing temperature. A calibration curve is prepared by placing accurately known amounts of water solution, between 4.0 and 20.0 ml, into separate 50-ml volumetric flasks and carrying out the above procedure.

B. LITHIUM–COPPER CHLORIDE METHOD

Lithium chloride dissolved in acetone forms upon addition of an acetone solution of copper(II) chloride the red-orange complex lithium trichloro-cuprate(II) (281). Absorbance measurements at 366 nm are less if water is present in the reaction medium, and this absorbance decrease is now used for determination of water in acetone and other organic solvents (147). The decrease is linear for solutions containing up to 5% (v/v) water. Although direct absorbance measurement is possible, a photometric titration technique is recommended for yielding more precise data.

The following procedure is for determination of up to 5% (v/v) water in acetone. Ten repeated determinations on a solution containing 2.0% water resulted in a mean value of 2.02%, with a variance of 1.6%. Values for water in ketone, ester, and ether solutions exhibited similar accuracy and precision.

Solutions

Lithium Chloride, 0.0100M. Place 0.0424 g LiCl and 50 ml dried distilled acetone in a flask which has been dried at 120°C. Shake for 1 hr or reflux for 20 min. Transfer the solution to a dry 100-ml volumetric flask and dilute to volume with additional acetone.

Copper(II) Perchlorate, 0.00100M. Dissolve 0.0371 g $Cu(ClO_4)_2 \cdot 6H_2O$ in dried, distilled acetone. Transfer to a 100-ml volumetric flask and dilute to volume with additional acetone.

Water. Place exactly 0.00, 0.10, 0.20, 0.50, 1.00, and 2.00 ml distilled water in separate 50-ml volumetric flasks. Fill each to volume with anhydrous acetone. These solutions contain, respectively, 0, 0.2, 0.4, 1.0, 2.0, and 4.0% (v/v) water.

Procedure

Place 25.0 ml moist acetone sample into a photometric titration cell. The cell should have provision for stirring the solution during titration. Add exactly 3.00 ml 0.0100M lithium chloride solution. Place the cell in the instrument cell compartment and start the stirrer at low speed. Add copper(II) perchlorate solution slowly and continuously from a buret. Observe the absorbance change at 366 nm against an anhydrous acetone reference solution as titrant is added. Record the maximum aborbance value. The percentage water in the sample is read from a calibration curve of maximum absorbance value versus increasing water content. To prepare the calibration curve, proceed as directed above with each of the standard water solutions. Plot maximum absorbance for each solution against water content of that sample. For determination of water in other solvents, also proceed as described above. It is suggested, however, that standard water samples for each solvent be prepared by appropriate dilution of a water saturated sample of the solvent under study. Water content for each saturated solvent at the experimental temperature is obtained from an appropriate reference source. Aldehydes do not interfere with water determination in ketones, ester, and ethers. Lower alcohols interfere. Ethanol interferes to a lesser extent than methanol, and further decrease in interference is observed with increasing alcohol carbon content.

C. DIETHYLALUMINUM HYDRIDE–2-ISOQUINOLINE METHOD

Diethylaluminum hydride forms a red complex with isoquinoline in hydrocarbon solvents. If water is present in the solvent, it reacts preferentially with the diethylaluminum hydride, resulting in a corresponding decrease in the amount of isoquinoline complex formed. The decreased absorbance observed for the diethylaluminum hydride—2-isoquinoline species is taken as a direct measure of the amount of water present (213). The presence of other organic compounds that contain functional groups capable of reacting with diethylaluminum hydride interfere. Oxygen, too, interferes and must be rigorously excluded from the reaction vessel and reagent solutions. An alternate approach is to determine the amount of oxygen present and substract this value as water equivalents from the experimentally obtained results.

The following procedure is for determination of water in hydrocarbon and aromatic solvents. It is applied to determination of water in these solvents prior to their use in preparation of Grignard-type reagents. A decrease of 1.00 absorbance unit is reported for a sample containing 1.15 mg water. Most satisfactory results were obtained when absorbance readings fell within the range 1.5 to 0.8 absorbance unit. One can expect both accuracy and precision to be about ±5%.

Solutions

Benzene. Dry benzene is prepared by distilling a solution of reagent-grade benzene containing 1% triethylaluminum.

Isoquinoline, 10%. Dissolve 10 ml freshly distilled isoquinoline in dry benzene. Dilute to 100 ml with additional dry benzene.

Diethylaluminum Hydride. The reagent should have a minimum purity of at least 50%.

Water. Dry benzene is freed from dissolved oxygen by bubbling with oxygen-free nitrogen for several hours at a flow rate of 50 ml per min. Moisture is introduced into the benzene by diverting the nitrogen stream through a water scrubber prior to contact with the deoxygenated benzene. Varying amounts of water are introduced by regulating the duration of contact between nitrogen and water. Standardize each of the water-containing benzene solutions by Karl Fisher titration. These solutions are stored in septum-covered bottles and kept in a water- and oxygen-free dry box. They should be used within 2 hr of preparation.

Procedure

Reaction and colorimetric measurement are carried out in a 200-ml glass-stoppered bulb with a cuvette permanently attached opposite the glass-stoppered opening (127). It may be necessary to modify the sample compartment of the spectrophotometer to accommodate a vessel of this size. The dry container is covered with a septum and flushed with oxygen-free nitrogen prior to use. Introduce, with a hypodermic syringe, 20 ml dry benzene and 5 ml isoquinoline solution into the flask. Add diethylaluminum hydride until the red complex formed exhibits an absorbance value of between 1.0 and 1.5 units. Measurements are made at 460 nm against a solvent blank. It is necessary to thoroughly mix the contents of the flask prior to measurement to assure that any traces of oxygen and/or moisture present have reacted. Inject a known amount of sample into the reaction vessel, mix thoroughly, and record the decreased absorbance value. A calibration curve is prepared by placing known amounts of each of the known standard water-containing benzene samples into a reaction flask containing diethylaluminum hydride—2-isoquinoline complex. The change in absorbance is plotted versus the water content of each standard. It is necessary to correct the observed absorbance reading for varying volumes of sample or standard added.

D. CHLORANILIC ACID METHOD

A procedure using chloranilic acid (2,5-dichloro-3,6-dihydroxy-*p*-benzoquinone) for water in organic solvents over the concentration range of 0 to 100% water is given in Section IIB.

E. MISCELLANEOUS METHODS

Titration by the Karl Fisher method is probably the most common procedure for determining water (87). A variety of techniques are employed for endpoint detection with this titration. Included among these are photometric titration procedures (64,263,278,308).

Various inorganic species are utilized for color formation with water-containing samples. The blue color of copper(II) sulfate pentahydrate formed from contact between anhydrous copper(II) sulfate and water is reported for trace water determination in kerosene (209) and other organic solvents (160). The effect of water on copper(II)—chloro complexes is also reported for determination of water in acetone (107). The solubility of chromium(VI) as dichromate increases in water—ethanol mixtures with increasing amounts of water. Samples containing a small amount of acetic acid, to prevent chromate formation, are saturated with potassium dichromate. From 5 to 30% water by volume is determined upon measuring the absorbance at 520 nm and comparing values from a previously prepared calibration curve (198). Potassium dichromate and concentrated sulfuric acid when added to samples of foodstuffs produce a color related to the moisture content of the sample. A colorimetric procedure for estimating the moisture content of fruit pulp using this procedure is reported (269).

Nessler's reagent is used for colorimetric determination of ammonia formed from reaction of water-containing samples and magnesium nitride. Colorimetric measurement at 470 nm is related to the amount of water present (275). Molar ratios of 3:1 and 3:2 for water to ammonia are observed from this reaction depending upon the reaction temperature (112). Colorimetric measurement of iodine released from iodine—EDTA complex proportional to the water content is reported for various alcoholic tinctures (22). A vanadium-impregnated silica gel is reported for colorimetric determination of atmospheric moisture (279).

Numerous organic reagents are suggested for determination of water. Fuchsin is used for ascertaining the moisture content of sugar. Powdered reagent is thoroughly mixed with a sugar sample and the color obtained compared with that of color standards (187). Hexanitrohydrazobenzene forms a colored product with various solvents containing water. Colorimetric measurement at 533 nm is related to water content varying from 0.2 to 5.0% by volume (58). Adduct formation between halides and azo dyes is affected by the presence of water in the reaction medium. Decreasing absorbance values measured at 460 nm for the complex formed from boron trifluoride and p-methoxyazobenzene in acetonitrile with increasing water content are proportional to the amount of water present. Values obtained from adding known increments of water to the adduct solution are extrapolated to determine the amount of water originally present (117).

Solvatochromism, the change in wavelength of maximum absorption and intensity of absorption with changing polarity of the solvent medium, is utilized for determination of water. Various reagents exhibit the solvatochromic effect, and methods for determining water in pyridine (43), alcohols (230), and a variety of other organic solvents (81) are reported. One recent study recommends a series of betaine dyes for measuring the water content of various solvents (172). Absorbance values recorded at a fixed optimum wavelength are related to the amount of water present. Nonlinear calibration curves are obtained because the method is not based upon Beer's law. For water in ethanol, 2,4,6-triphenyl-N-(3,5-diphenyl-4-hydroxyphenyl)pyridinium betaine, with measurements at 570 nm, exhibits greatest sensitivity. It and 1-methyl-6-hydroxyquinolinium betaine are also sensitive for water in isopropanol, measurements being made at 610 nm and 490 nm, respectively. These and other betaine dyes are also recommended for water in acetonitrile, pyridine, dioxane, and acetone.

A photometric titration of water in glacial acetic acid is available, based upon the sulfuric acid-catalyzed hydrolysis of acetic anhydride titrant. Absorbance values are monitored for acetic anhydride at its ultraviolet absorbance maximum of 256 nm (45). In other procedures, direct measurement of water absorption bands in the near infrared are reported for determination of water in fuming nitric acid (307), hydrazines (66), and a variety of organic solvents (113,116, 159,300,315). Turbidimetric procedures for water in various media are also described based upon measurements with colored solutions (101,173,194) and upon measurements of suspended hydrolysis products (171,175).

II. APPLICATIONS

Standards containing known amounts of water in an anhydrous medium are necessary for accurate calibration of procedures. The ability of various drying agents to remove moisture from gas streams is reported (297). Likewise, procedures for preparation of anhydrous methanol and pyridine (88), acetone (151), and other organic solvents (172) are available. Techniques for the preparation of accurately known concentrations of water vapor in gaseous media are reported. One approach passes nitrogen containing known trace amounts of hydrogen and oxygen over heated platinum in the presence of a catalyst to convert them to an equivalent amount of water (105). Another procedure, suitable for low ppm concentrations of moisture in nitrogen, employs dilution of known amounts of water vapor in evacuated vessels with dry carrier gas (108). Methanol standards containing known amounts of water are prepared by diluting weighed samples of water with anhydrous solvent to a known volume (306). Ethanol—water standards are prepared by adding appropriate amounts of water and measuring the density of the resulting solution. Exact composition is obtained from appropriate tabulated values (180). Known amounts of water in dioxane also serve as standard solutions (34). Titration with the Karl Fisher procedure serves to standardize a variety of water-containing solutions. Karl Fisher reagent is itself standardized by using one of the standards discussed above or by reaction with a hydrated salt of known composition. Sodium tartrate dihydrate (218), calcium acid malate hexahydrate (267), and various other hydrates (150) are recommended. In addition to standard solutions, careful control of experimental conditions is essential for accurate water determination. Reagent purity, atmospheric humidity, and the dryness of apparatus are all important factors (294).

A. WATER IN GASES

The presence of atmospheric moisture is, of course, closely related to weather forecasting, and instruments for determining relative humidity and associated quantities are readily available (97).

The following procedure is for determination of water in cylinders of oxygen (10). Using visual indication, it is simply and rapidly performed. Filter paper strips soaked in cobalt(II) chloride solution serve as the indicating medium. An alternate approach uses cobalt(II) chloride-impregnated silica gel. With the latter, the distance of a colored band within the packed adsorption tube indicates the water content of gaseous sample flowing over the silica gel. Data for the filter paper method agreed favorably with those obtained by a standard gravimetric method for determining water using phosphorous pentoxide.

Solutions

Cobalt(II) Chloride Indicating Paper. Dip filter paper into a solution containing 30 g $CoCl_2 \cdot 6H_2O$ per liter aqueous solution. After impregnating with cobalt(II) chloride, hang the paper and allow it to air dry. When dry, cut off the bottom cm of the paper. The remaining portion should have a uniform distribution of cobalt(II) chloride. Cut the paper into 2- x 14-cm strips and store in a desiccator over anhydrous calcium chloride.

Diethyl Ether. Diethyl ether, purified and peroxide free, is dried over anhydrous sodium sulfate. The dried ether is stored in a container with sodium sulfate wrapped in filter paper.

Water. Prepare solutions containing 0.5, 1.0, 2.0, 5.0, and 10.0 mg water per 100 ml dry ether by diluting a solution of 50.0 mg water in 100 ml dry ether. Use immediately after preparation.

Procedure

Six 200-ml gas bottles containing fritted-glass bubblers are assembled. Five of the bottles contain the ether—water standards and the sixth, an equal amount of dry ether. A cobalt(II) chloride indicating strip is added to each bubbler and held in place with a stopper. Pass sample gas through the bubbler containing the dry ether at a flow rate of 1 liter per min for 10 min. Allow all bubbler flasks to stand for 15 min with occasional shaking. Now match the color of the indicating strip in the unknown sample flask with that of the nearest corresponding standard. Optimum results are obtained for samples releasing between 1 and 10 mg water per 100 ml ether—water standard.

B. WATER IN ORGANIC SOLVENTS

Applications regarding various water determinations under a variety of conditions are thoroughly reviewed (83).

The following procedure is for determination of water in ethanol and other solvents based upon reaction with chloranilic acid (26). Pyridine and other nitrogen-containing solvents interfere because they, too, react with this reagent. The method is applicable for solutions containing from 2 to 100% water. A series of repeated measurements resulted in a relative average deviation of 1.5%.

Solutions

Chloranilic Acid, 0.002M. Dissolve 0.418 g 2,5-dichloro-3,6-dihydroxy-*p*-benzoquinone in absolute ethanol and dilute to 1 liter in a volumetric flask with additional absolute ethanol. Store in a dark bottle in a refrigerator.

Water. Transfer exactly known amounts of water, between 0.20 and 10.0 ml, to separate 10-ml volumetric flasks. Dilute each flask to volume with absolute ethanol.

Procedure

Place 1.00 ml water-containing solvent and 9.00 ml absolute ethanol in a flask. Transfer 5.00 ml of this solution to a second flask and add 5.00 ml chloranilic acid solution. Measure the absorbance of this solution in a 1-cm cell at 520 nm against a reagent blank containing 5.00 ml absolute ethanol and 5.00 ml chloranilic acid solution. A calibration curve is prepared by substituting 1.00 ml of each of the standard water solutions for the sample and carrying out the above procedure.

REFERENCES

1. Aleskovskii, V. B., V. A. Koval'tsov, I. N. Fedorov, and G. P. Tsyplyatnikov, *Zavod. Lab.*, **28**, 1440 (1962).
2. Aleskovskii, V. B., V. A. Koval'tsov, I. N. Fedorov, and G. P. Tsyplyatnikov, *Zavod. Lab.*, **30**, 105 (1964).
3. Allen, N., *Ind. Eng. Chem., Anal. Ed.*, **2**, 55 (1930).
4. Allsopp, C. B., *Analyst* (London), **66**, 371 (1941).
5. Altshuller, A. P., C. M. Schwab, and M. Bare, *Anal. Chem.*, **31**, 1987 (1959).
6. Altshuller, A. P., and A. F. Wartburg, *Int. J. Air Water Pollut.*, **4**, 70 (1961).
7. Ambler, H. R., *Analyst* (London), **59**, 14 (1934).
8. Angely, L., E. Levart, G. Guiochon, and G. Peslerbe, *Anal. Chem.*, **41**, 1446 (1969).
9. Arnott, J., and J. McPheat, *Engineering*, **176**, 103 (1953).
10. Asami, T., *Anal. Chem.*, **40**, 648 (1968).
11. ASTM, *1968 Book of ASTM Standards, Part 23: Water; Atmospheric Analysis*, American Society for Testing and Materials, Philadelphia, 1968.
12. Avi-dor, Y., E. Cutolo, and K. G. Paul, *Acta Physiol. Scand.*, **32**, 314 (1954).
13. Ayres, G. H., and B. V. Glanville, *Anal. Chem.*, **21**, 930 (1949).
14. Aziz, F., and G. A. Mirza, *Talanta*, **11**, 889 (1964).
15. Babkin, R. L., *Elek. Stantsii*, **25**, 16 (1954); *Chem. Abstr.*, **49**, 9199d (1955).
16. Babkin, R. L., and K. P. Epeikina, *Teploenergetika*, **9** (2), 48 (1962); *Chem. Abstr.*, **57**, 5725c (1962).
17. Babko, A. K., and L. L. Shvechenko, *Dopov. Akad. Nauk Ukr. RSR*, **1958**, 970; *Chem. Abstr.*, **53**, 6737d (1959).

18. Babko, A. K., and L. L. Shvechenko, *ibid.*, **1958**, 1212; *Chem. Abstr.*, **53**, 9890i (1959).

19. Babko, A. K., and M. M. Tananaiko, *Ukr. Khim. Zh.*, **24**, 298 (1958); *Chem. Abstr.*, **52**, 19666b (1958).

20. Bailey, R., and D. F. Boltz, *Anal. Chem.*, **31**, 117 (1959).

21. Bairstow, S., J. Francis, and G. H. Watt, *Analyst* (London), **72**, 340 (1947).

22. Balatre, P., A. M. Guyot-Hermann, J. C. Guyot, R. Khairzada, and M. Traisnel, *Bull. Soc. Pharm. Lille*, **2**, 91 (1968); *Chem. Abstr.*, **70**, 60892x (1969).

23. Banerjee, D. K., and C. C. Budke, *Anal. Chem.*, **36**, 792 (1964).

24. Banks, J., *Analyst* (London), **79**, 170 (1954).

25. Banks, J., *ibid.*, **84**, 700 (1959).

26. Barreto, R. C. R., and H. S. R. Barreto, *Anal. Chim. Acta*, **26**, 494 (1962).

27. Battino, R., and H. L. Clever, *Chem. Rev.*, **66**, 395 (1966).

28. Bender, A. E., M. Burnham, and D. S. Miller, *Chem. Ind.* (London), **1953**, 293.

29. Berezina, Yu. I., M. M. Korosteleva, and V. V. Broit, *Zh. Anal. Khim.*, **20**, 262 (1965).

30. Beroza, M., and B. A. Bierl, *Mikrochim. Acta*, **1969**, 720.

31. Birdsall, C. M., A. C. Jenkins, and E. Spadinger, *Anal. Chem.*, **24**, 662 (1952).

32. Blanquet, P., and G. Dalmai, *Bull. Soc. Chim. Fr.*, **1957**, 419.

33. Boelter, E. D., G. L. Putnam, and E. I. Lash, *Anal. Chem.*, **22**, 1533 (1950).

34. Bonauguri, E., and G. Seniga, *Ann. Chim.* (Paris), **45**, 805 (1955).

35. Bonet-Maury, P., *C. R. Acad. Sci., Paris*, **218**, 117 (1944).

36. Bovee, H. H., and R. J. Robinson, *Anal. Chem.*, **33**, 1115 (1961).

37. Boyd, A. W., C. Willis, and R. Cyr., *Anal. Chem.*, **42**, 670 (1970).

38. Brady, L. J., *Anal. Chem.*, **20**, 1033 (1948).

39. Bravo, H. A., and J. P. Lodge, *Anal. Chem.*, **36**, 671 (1964).

40. Britaev, A. S., *Tr. Tsentr. Aerolog. Observ.*, **37**, 13 (1960); *Chem, Abstr.*, **56**, 6663b (1962).

41. Broenkow, W. W., *Limnol. Oceanogr.*, **14**, 288 (1969).

42. Broenkow, W. W., and J. D. Cline, *Limnol. Oceanogr.*, **14**, 450 (1969).

43. Brooker, L. G. S., G. H. Keyes, and D. W. Heseltine, *J. Amer. Chem. Soc.*, **73**, 5350 (1951).

44. Brooks, F. R., M. Dimbat, R. S. Treseder, and L. Lykken, *Anal. Chem.*, **24**, 520 (1952).

45. Bruckenstein, S., *Anal. Chem.*, **31**, 1757 (1959).

46. Buchoff, L. S., N. M. Ingber, and J. H. Brady, *Anal. Chem.*, **27**, 1401 (1955).

47. Bukolov, I. E., and N. I. Nechiporenko, *Gig. Sanit.*, **32** (6), 68 (1967); *Chem. Abstr.*, **67**, 67362w (1967).

48. Buscarons, F., J. Artigas, and C. Rodriguez-Roda, *Anal. Chim. Acta*, **23**, 214 (1960).

49. Byers, D. H., and B. E. Saltzman, *Amer. Ind. Hyg. Assoc. J.*, **19**, 251 (1958).

50. Carritt, D. E., *J. Marine Res.*, **24**, 286 (1966).

51. Carpenter, J. H., *Limnol. Oceanogr.*, **10**, 135 (1965).

52. Carpenter, J. H., *ibid.*, **10**, 141 (1965).

53. Carpenter, J. H., *ibid.*, **11**, 264 (1966).

54. Cauwenberge, H. van, and J. de Clerck, *Bull. Assoc. Anciens Etud. Brass. Univ. Louvain*, **51**(3), 97 (1955); *Chem. Abstr.*, **50**, 3708c (1956).

55. Cauwenberge, H. van, and J. de Clerck, *Int. Tijdschr. Brouw. en Mout.*, **16**, 66 (1956–57); *Chem. Abstr.*, **51**, 3364i (1957).

56. Celechovsky, J., E. Krejci, and V. Krejci, *Cesk. Farm.*, **6**, 103 (1957); *Chem. Abstr.*, **53**, 20695b (1959).

57. Cheng, K. L., *Anal. Chem.*, **30**, 1035 (1958).

58. Cherkesov, A. I., and L. V. Cherkesova, *Materialy XXII [Dvadtsat Vtoroi] Nauchn. Konf. Saratovsk. Gos. Ped. Inst., Fak. Estestvozh., Fiz., Matem., Saratov, Sb.*, **1961**, 100; *Chem. Abstr.*, **59**, 26c (1963).

59. Cohen, I. C., A. F. Smith, and R. Wood, *Analyst* (London), **93**, 507 (1968).

60. Cohen, I. R., and J. J. Bufalini, *Environ. Sci. Technol.*, **1**, 1014 (1967).

61. Cohen, I. R., and T. C. Purcell, *Anal. Chem.*, **39**, 131 (1967).

62. Cohen, I. R., T. C. Purcell, and A. P. Altshuller, *Environ. Sci. Technol.*, **1**, 247 (1967).

63. Conner, W. D., and J. S. Mader, *Amer. Ind. Hyg. Assoc. J.*, **25**, 291 (1964).

64. Connors, K. A., and T. Higuchi, *Chemist-Analyst*, **48**, 91 (1959).

65. Corcoran, J. T., *Anal. Chem.*, **27**, 1018 (1955).

66. Cordes, H. F., and C. W. Tait, *Anal. Chem.*, **29**, 485 (1957).

67. Csanyi, L. J., *Anal. Chem.*, **42**, 680 (1970).

68. Dammers-deKlerk, A., and B. Boots-Meurs, *Anal. Chim. Acta*, **16**, 297 (1957).

69. DasGupta, B. R., *Anal. Chem.*, **42**, 659 (1970).

70. DasGupta, B. R., and D. A. Boroff, *Anal. Chem.*, **40**, 2060 (1968).

71. Davis, A. G., *Organic Peroxides*, Butterworth, London, 1961.

72. Davis, I., U.S. *Dept. Comm., Office Tech. Serv., AD 267,251*, Washington, D.C., 1961; 6 pp.

73. Davis, P. S., *Metallurgia*, **62**, 49 (1960).

74. De Carvalho, A. H., J. G. Calado, and M. Legrand de Moura, *Rev. Port. Quim.*, **5**, 15 (1963).

75. Deckert, W., *Fresenius' Z. Anal. Chem.,* **150**, 421 (1956).

76. Deckert, W., *ibid.,* **153**, 189 (1956).

77. Deinum, H. W., and J. W. Dam, *Anal. Chim. Acta,* **3**, 353 (1949).

78. Delman, A. D., A. E, Ruff, B. B. Simms, and A. R. Allison, in *Ozone Chemistry and Technology,* Advances in Chemistry Series No. 21, American Chemical Society, Washington, D.C., 1959, p. 119.

79. Dement'eva, M. I., and E. V. Skvortsova, *Tr. Vses. Nauchn.-Issled. Inst. Khim. Pererabotki Gazov,* **6**, 270 (1951); *Chem. Abstr.,* **49**, 9439d (1955).

80. Deutsch., S., *J. Air Pollut. Control Assoc.,* **18**, 78 (1968).

81. Dimroth, K., and C. Reichardt, *Fresenius' Z. Anal. Chem.,* **215**, 344 (1966).

82. Dorta-Schaeppi, Y., and W. D. Treadwell, *Helv. Chim. Acta,* **32**, 356 (1949).

83. Drozdov, V. A., A. P. Kreshkov, and S. I. Petrov, *Usp. Khim.,* **38**, 113 (1969); *Chem. Abstr.,* **70**, 83848b (1969).

84. Dugan, P. R., *Anal. Chem.,* **33**, 1630 (1961).

85. Dugan, P. R., and R. D. O'Neill, *Anal. Chem.,* **35**, 414 (1963).

86. Dukes, E. K., and M. L. Hyder, *Anal. Chem.,* **36**, 1689 (1964).

87. Eberius, E., *Wasserbestimmung mit Karl-Fisher-Lösung,* Verlag Chemie, Gmbh, Weinheim, 1958.

88. Eberius, E., and W. Kowalski, *Chem.-Ztg., Chem. App.,* **81**, 75 (1957).

89. Effenberger, E., *Fresenius' Z. Anal. Chem.,* **134**, 106 (1951).

90. Efimoff, W. W., *Biochem. Z.,* **155**, 371 (1925).

91. Egerton, A. C., A. J. Everett, G. J. Minkoff, S. Rudrakanchana, and K. C. Salooja, *Anal. Chim. Acta,* **10**, 422 (1954).

92. Eisenberg, G. M., *Ind. Eng. Chem., Anal. Ed.,* **15**, 327 (1943).

93. Eiss, M. I., and P. Giesecke, *Anal. Chem.,* **31**, 1558 (1959).

94. Elliott, J. W., *Progressive Fish Culturist,* **25**, 42 (1963).

95. Ellis, M. M., and M. D. Ellis, *Sewage Works J.,* **15**, 1115 (1943).

96. Elmore, H. L., and T. W. Hayes, *J. Sanit. Eng. Div., Proc. Amer. Soc. Civil Eng.,* **86**, SA4, 41 (1960).

97. Ewing, G. W., *J. Chem. Educ.,* **45**, A377 (1968).

98. Exton, W. G., F. Schattner, S. Korman, and A. R. Rose, *J. Lab. Clin. Med.,* **30**, 84 (1945).

99. Fedotov, V. P., *Gig. Sanit.,* **30**(9), 61 (1965); *Chem. Abstr.,* **63**, 17026c (1965).

100. Ferguson, B. C., and N. M. Coulter, *Proc. Indiana Acad. Sci.,* **63**, 124 (1953).

101. Fikhtengol'ts, V. S., and R. V. Zolotareva, *Vestn. Tekhn. i Ekon. Inform. Nauchn.-Issled. Inst. Tekhn.-Ekon. Issled. Gos. Kom. Sov. Min. SSSR po Khim.,* **1961**, 15; *Chem. Abstr.,* **57**, 13184g (1962).

328 OXYGEN

102. Freier, R., *Chem.-Ztg., Chem. App.*, **76**, 844 (1952).

103. Freier, R., *Vom Wasser*, **19** 124 (1952).

104. Fuhrmann, H., *Staub*, **25**, 266 (1965).

105. Gamache, L. D., *Anal. Instrum.*, **1965** (Pub. 1966), 167.

106. Garcia, F. C., and M. L. Garrido, *Anales Real Soc. Espan. Fis. y Quim.*, **52B**, 237 (1956); *Chem. Abstr.*, **50**, 12737h (1956)

107. Gazo, J., and J. Truchly, *Chem. Zvesti*, **18**, 655 (1964); *Chem. Abstr.*, **62**, 28f (1965).

108. Gelezunas, V. L., *Anal. Chem.*, **41**, 1400 (1969).

109. German, A., J. Panouse-Perrin, and A. M. Quero, *Ann. Pharm. Fr.*, **25**, 115 (1967); *Chem. Abstr.*, **67**, 57032z (1967).

110. Gilcreas, F. W., *J. Amer. Water Works Assoc.*, **27**, 1166 (1935).

111. Glavind, J., and S. Hartmann, *Acta Chem. Scand.*, **3** 1021 (1949).

112. Gol'dinov, A. L., V. I. Lukhovitskii, and G. Ya. Mal'kova, *Zh. Anal. Khim.*, **16**, 724 (1961).

113. Goulden, J. D. S., and D. J. Manning, *Analyst* (London), **95**, 308 (1970).

114. Grasshoff, K., *Kiel Meeresforsch*, **20**, 143 (1964); *Chem. Abstr.*, **62**, 8361a (1965).

115. Green, E. J., and D. E. Carritt, *Analyst*(London), **91**, 207 (1966).

116. Greinacher, E., W. Luttke, and R. Mecke, *Z. Elektrochem.*, **59**, 23 (1955).

117. Gutmann, V., and A. Steininger, *Allg. Prakt. Chem.*, **18**, 282 (1967).

118. Haagen-Smit, A. J., and M. Brunelle, *Int. J. Air Water Pollut.*, **1**, 51 (1958).

119. Hartkamp, H., *Angew. Chem.*, **71**, 651 (1959).

120. Hartman, L., C. N. Hooker, and H. E. Watt, *New Zealand J. Sci. Technol.*, **35B**, 307 (1954).

121. Hartman, L., and M. D. L. White, *Anal. Chem.*, **24**, 527 (1952).

122. Hartmann, S., and J. Glavind, *Acta Chem. Scand.*, **3**, 954 (1949).

123. Haslam, J., and G. Moses, *J. Soc., Chem. Ind.*, **57**, 344 (1938).

124. Hauser, T. R., and D. W. Bradley, *Anal. Chem.*, **38**, 1529 (1966).

125. Hauser, T. R., and D. W. Bradley, *ibid.*, **39**, 1184 (1967).

126. Hauser, T. R., and M. A. Kolar, *Anal. Chem.*, **40**, 231 (1968).

127. Henderson, S. R., and L. J. Synder, *Anal. Chem.*, **31**, 2113 (1959).

128. Hendricks, R. H., and L. B. Larsen, *Amer. Ind. Hyg. Assoc. J.*, **27**, 80 (1966).

129. Herold, G., and G. Fuchs, *Lebensmittelchem. Gerichtl. Chem.*, **19**, 121 (1965); *Chem. Abstr.*, **66**, 18099m (1967).

130. Higashino, T., T. Doi, and S. Musha, *J. Chem. Soc. Japan, Pure Chem. Sect.*, **73**, 363 (1952).

131. Hirano, S., and M. Kitahara, *J. Chem. Soc. Japan, Ind. Chem. Sect.*, **56**, 325 (1953).

132. Hirsch, A. A., *Water Sewage Works,* **110**, 147 (1963).

133. Hoffmann, E., *Mikrochemie,* **25**, 82 (1938).

134. Hofmann, P., and P. Stern, *Anal. Chim. Acta,* **45**, 149 (1969).

135. Hofmann, P., and P. Stern, *ibid.,* **47**, 113 (1969).

136. Holland, P., *Chem. Ind.*(London), **1959**, 218.

137. Houlihan, J. E., and P. E. L. Farina, *Analyst*(London), **81**, 377 (1956).

138. Humpoletz, J. E., *Austr. J. Sci.,* **12**, 111 (1949).

139. Hunold, G. A., and W. Pietrulla, *Fresenius' Z. Anal. Chem.,* **178**, 271 (1961).

140. Imai, H., S. Chaki, and Y. Tanaka, *Bunseki Kagaku,* **10**, 191 (1961).

141. Inn, E. C. Y., and Y. Tanaka, in *Ozone Chemistry and Technology,* Advances in Chemistry Series No. 21, American Chemical Society, Washington, D.C., 1959, p. 263.

142. Isaacs, M. L., *J. Amer. Chem. Soc.,* **44**, 1662 (1922).

143. Isaacs, M. L., *Sewage Works J.,* **7**, 435 (1935).

144. Ito, T., and Y. Hoshino, *Kogyo Kagaku Zasshi,* **59**, 500 (1956); *Chem. Abstr.,* **52**, 2648d (1958).

145. Ivanitskaya, A. S., and A. A. Mostofin, *Elek. Stantsii,* **35**, 79 (1964); *Chem. Abstr.,* **61**, 11755f (1964).

146. Ivanoff, A., *C.R. Acad. Sci., Paris,* **254**, 4493 (1962).

147. Jackwerth, E., and H. Speaker, *Fresenius' Z. Anal. Chem.,* **171**, 270 (1959).

148. Jacobs, M. B., *The Chemical Analysis of Air Pollutants,* Interscience, New York, 1960.

149. Janicek, G., and J. Pokorny, *Chem. Listy,* **49**, 1315 (1955).

150. Johnson, C. A. in *Advances in Pharmaceutical Sciences,* Vol. II, Bean, H. S., A. H. Beckett, and J. E. Carless, Eds., Academic Press, New York, 1967, p. 224.

151. Jordan, K., and W. R. Fischer, *Fresenius' Z. Anal. Chem.,* **168**, 182 (1959).

152. Karasek, F. W., R. J. Loyd, D. E. Lupfer, and E. A. Houser, *Anal. Chem.,* **28**, 233 (1956).

153. Karas-Gasparec, V., *Acta Pharm. Jugoslav.,* **16**, 85 (1966); *Chem. Abstr.,* **66**, 82151z (1967).

154. Karas-Gasparec, V., and T. Pinter, *Croat. Chem. Acta,* **34**, 131 (1962).

155. Katz, M., *Proc. Int. Clean Air Conf. London,* **1959** (Publ. 1960), 172.

156. Katzin, L. I., and J. R. Ferraro, *J. Amer. Chem. Soc.,* **74**, 2752 (1952).

157. Katzin, L. I., and E. Gebert, *J. Amer. Chem. Soc.,* **72**, 5464 (1950).

158. Kavan, I., and J. Base, *Chem. Prum.,* **12**, 252 (1962); *Chem. Abstr.,* **57**, 7899a (1962).

159. Keyworth, D. A., *Talanta,* **8**, 461 (1961).

160. Khoroshaya, E. S., A. A. Avilov, G. I. Kovrigina, and Z. A. Koroleva, Zavod. Lab., **21**, 542 (1955).

161. Kiffer, A. D., and L. G. Dowell, *Anal. Chem.*, **24**, 1796 (1957).

162. Kikuchi, S., S. Suzuki, and K. Koyanagi, *Seisan-Kenkyu*, **19**, 119 (1967); *Chem. Abstr.*, **68**, 71918v (1968).

163. Kling, A., *Bull. Acad. Med.*, **119**, 178 (1938); *Chem. Abstr.*, **32**, 8299[3] (1938).

164. Kling, A., and M. Claraz, *C. R. Acad. Sci., Paris*, **203**, 319 (1936).

165. Klots, C.,E., and B. B. Benson, *J. Marine Res.*, **21**, 48 (1963).

166. Kminkova, M., M. Gottwaldova, and J. Hanus, *Chem. Ind.*(London), **1969**, 519.

167. Kokatnur, V. R., and M. Jelling, *J. Amer. Chem. Soc.*, **63**, 1432 (1941).

168. Koppe, P., and A. Muhle, *Fresenius' Z. Anal. Chem.*, **210**, 241 (1965).

169. Krause, A., and J. Slawek, *Monatsh. Chem.*, **99**, 1494 (1968).

170. Krueger, R. H., J. P. Warringer, and B. Jaselskis, *Talanta*, **15**, 741 (1968).

171. Kubinova, M., O. Vilim, and V. Svoboda, *Collect. Czech. Chem. Commun.*, **26**, 1320 (1961).

172. Kumoi, S., K. Oyama, T. Yano, and H. Kobayashi, *Talanta*, **17**, 319 (1970).

173. Kusakov, M. M., A. Yu. Koshevnik, and A. E. Mikirov, *Inzh.-Fiz. Zh.*, **3**, (11), 11 (1960); *Chem. Abstr.*, **55**, 18069i (1961).

174. Kuwana, T., *Anal. Chem.*, **35**, 1398 (1963).

175. Kwiatkowski, E., and J. Szychlinski, *Chem. Anal.*(Warsaw), 6, 541 (1961).

176. LaMont, B. D., and N. E. Gordon, *U.S. At. Energy Comm.*, WCAP-917, Washington, D.C., 1958, 13 pp.

177. Lechner, E., and F. Kiermerer, *Z. Lebensm.-Unters. Forsch.*, **133**, 372 (1967); *Chem. Abstr.*, **67**, 98974b (1967).

178. Lehongre, G., J. Neumann, and J. Lavollay, *Bull. Soc. Chim. Biol.*, **32**, 1023 (1950).

179. Leithe, W., and A. Hofer, *Mikrochim. Acta*, **1968**, 1066.

180. Levy, G. B., J. J. Murtaugh, and M. Rosenblatt, *Ind. Eng. Chem., Anal. Ed.*, **17**, 193 (1945).

181. Lewartowicz, E., *J. Electroanal. Chem.*, 6, 11 (1963).

182. Lin, K., *Hua Hsueh*, **1959**, 166; *Chem. Abstr.*, **54**, 2083a (1960).

183. Line, R. A., and H. Hoftiezer, *U.S. Pat. 2,761,312* (1956).

184. Littman, F. E., and C. W. Marynowski, *Anal. Chem.*, **28**, 819 (1956).

185. Loomis, W. F., *Anal. Chem.*, **26**, 402 (1954).

186. Loomis, W. F., *ibid.*, **28**, 1347 (1956).

187. Lysyanskii, E. B., *Sakh. Prom.*, **35**(12), 40 (1961); *Chem. Abstr.*, **56**, 13149c (1962).

188. MacHattie, I. J. W., and J. E. Maconachie, *Ind. Eng. Chem., Anal. Ed.,* **9**, 364 (1937).

189. Macura, H., and G. Werner, *Die Chemie,* **56**, 90 (1943).

190. Main, E. R., and L. E. Shinn, *J. Biol. Chem.,* **128**, 417 (1939).

191. Malkova, E. M., T. L. Radovskaya, M. P. Belozerova, and Z. T. Berestneva, *Izmeritel. Tekhn.,* **1963**, 54; *Chem. Abstr.,* **59**, 10744a (1963).

192. Margerum, D. W., and R. H. Stehl, *Anal. Chem.,* **39**, 1351 (1967).

193. Mariano, M. H., *Anal. Chem.,* **40**, 1662 (1968).

194. Matsuyama, G., *Anal. Chem.,* **29**, 196 (1957).

195. McCabe, L. C., *Ind. Eng. Chem.,* **45**, 111A (Sept. 1953).

196. McComb, R. B., and W. D. Yushok, *J. Franklin Inst.,* **265**, 417 (1958).

197. McCrumb, F. C., and W. R. Kenny, *J. Amer. Water Works Assoc.,* **21**, 400 (1929).

198. Meditsch, J. O., *Chemist-Analyst,* **45**, 49 (1956).

199. Meerov, G. I., *Zh. Anal. Khim.,* **21**, 128 (1966).

200. Meloan, C. E., and W. W. Brandt, *Anal. Chem.,* **33**, 102 (1961).

201. Meloan, C. E., M. Mauck, and C. Huffman, *Anal. Chem.,* **33**, 104 (1961).

202. Meyling, A. H., and G. H. Frank, *Analyst*(London), **87**, 60 (1962).

203. Miller, N., and E. MacPherson, *Natl. Research Council Canada, At. Energy Project Div. Research CRC 352* (NRC No. 1617), Ottawa; 1947, 2 pp.

204. Mitchell, J., and D. M. Smith, *Aquametry,* Interscience, New York, 1948.

205. Mokhov, L. A., and V. P. Dzedzichek, *Zavod. Lab.,* **25**, 1304 (1959).

206. Mokhov, L. A., and N. S. Mareeva, *Zh. Prikl. Khim.,* **35**, 2573 (1962); *Chem. Abstr.,* **58**, 8351f (1963).

207. Montgomery, H. A. C., and A. Cockburn, *Analyst*(London), **89**, 679 (1964).

208. Montgomery, H. A. C., N. S. Thom., and A. Cockburn, *J. Appl. Chem.*(London), **14**, 280 (1964).

209. Morekhin, M. G., S. I. Ageev, O. E. Matyash, and T. G. Chechina, *Zavod. Lab.,* **28**, 670 (1962).

210. Morgan, R., *Proc. Roy. Aust. Chem. Inst.,* **35**, 82 (1968).

211. Mottola, H. A., B. E. Simpson, and G. Gorin, *Anal. Chem.,* **42**, 410 (1970).

212. Mugdan, M., and J. Sixt, *Angew. Chem.,* **46**, 90 (1933).

213. Mungall, T. G., and J. H. Mitchen, *Anal. Chem.,* **33**, 1330 (1961).

214. Murray, C. N., J. P. Riley, and T. R. S. Wilson, *Deep-Sea Res. Oceanogr. Abstr.,* **15**, 237 (1968); *Chem. Abstr.,* **69**, 40982c (1968).

215. Musha, S., T. Higashino, and T. Doi, *J. Chem. Soc. Japan, Pure Chem. Sect.,* **72**, 995 (1951).

216. Mysels, K. J., *Science,* **129**, 96 (1959).

217. Nash, T., *Atmos. Environ.,* **1**, 679 (1967).

218. Neuss, J. D., M. G. O'Brien, and H. A. Frediani, *Anal. Chem.*, **23**, 1332 (1951).

219. Nordschow, C. D., and A. R. Tammes, *Anal. Chem.*, **40**, 465 (1968).

220. Nozaki, K., *Ind. Eng. Chem., Anal. Ed.*, **18**, 583 (1946).

221. Okolov, F. S., *Gig. Sanit.*, **12**(8), 30 (1947); *Chem. Abstr.*, **43**, 1119i (1949).

222. Ovenston, T. C. J., and W. T. Rees, *Analyst*(London), **75**, 204 (1950).

223. Ovenston, T. C. J., and J. H. E. Watson, *Analyst*(London), **79**, 383 (1954).

224. Palin, A. T., *Water and Water Engineering,* **57**, 271 (1953).

225. Parkhouse, D., *Chem. Ind.*(London), **1955**, 588

226. Patrick, W. A., and H. B. Wagner, *Anal. Chem.*, **21**, 1279 (1949).

227. Patti, F., and P. Bonet-Maury, *C.R. Acad. Sci., Paris*, **239**, 976 (1954).

228. Pepkowitz, L. P., and E. L. Shirley, *Anal. Chem.*, **25**, 1718 (1953).

229. Pfister, R. J., and D. J. Kerley, *ASTM Bull.*, **127**, 17 (1944).

230. Phillips, J. P., and R. W. Keown, *J. Amer. Chem. Soc.*, **73**, 5483 (1951).

231. Pieters, H. A. J., and W. J. Hanssen, *Anal. Chim. Acta*, **2**, 712 (1948).

232. Pilarczyk, H., G. Miller, and N. Paterok, *Chem. Anal.*(Warsaw), **12**, 899 (1967).

233. Pobiner, H., *Anal. Chem.*, **33**, 1423 (1961).

234. Potter, E. C., *J. Appl. Chem.*(London), **7**, 285 (1957).

235. Potter, E. C., *ibid.*, **7**, 297 (1957).

236. Potter, E. C., and G. E. Everitt, *J. Appl. Chem.*(London), **10**, 48 (1960).

237. Potter, E. C., and J. F. White, *J. Appl. Chem.*(London), **7**, 459 (1957).

238. Pour, V., and J. Muller, *Chem. Prum.*, **9**, 630 (1959); *Chem. Abstr.*, **54**, 8166h (1960).

239. Powell, J. S., and P. C. Jay, *Anal. Chem.*, **21**, 296 (1949).

240. Purcell, T. C., and I. R. Cohen, *Environ. Sci. Technol.*, **1**, 431 (1967).

241. Rawson, A. E., *Water Water Eng.*, **57**, 56 (1953).

242. Rawson, A. E., *Water Water Eng.*, **57**, 102 (1953).

243. Rebsdorf, A., *Int. Ver. Theor. Angew. Limnol., Verh.*, **16**, 459 (1965); *Chem. Abstr.*, **68**, 16013w (1968).

244. Reichert, J. S., S. A. McNeight, and H. W. Rudel, *Ind. Eng. Chem., Anal. Ed.*, **11**, 194 (1939).

245. Rothchild, H., and I. M. Stone, *J. Inst. Brew.*, **44**, 425 (1938).

246. Rothstein, V. P., and V. N. Shemyakin, *Teploenergetika*, **9**(2), 54 (1962); *Chem. Abstr.*, **57**, 3214i (1962).

247. Rynasiewicz, J., *Anal. Chem.*, **26**, 355 (1954).

248. Salomon, L. L., and J. E. Johnson, *Anal. Chem.*, **31**, 453 (1959).

249. Saltzman, B. E., *Anal. Chem.*, **26**, 1949 (1954).

250. Saltzman, B. E., *U.S. Public Health Service 999-AP-11, D-1-D-5,* Washington, D.C., 1965.

251. Saltzman, B. E., *ibid., E-1-E-6,* **1965.**

252. Saltzman, B. E., and N. Gilbert, *Amer. Ind. Hyg. Assoc., J.,* **20,** 379 (1959).

253. Saltzman, B. E., and N. Gilbert, *Anal. Chem.,* **31,** 1914 (1959).

254. Saltzman, B. E., and A. F. Wartburg, *Anal. Chem.,* **37,** 779 (1965).

255. Samsoni, Z., *Energia Atomtech.,* **15,** 251 (1962); *Chem. Abstr.,* **57,** 13177b (1962).

256. Sandell, E. B., *Colorimetric Determination of Traces of Metals,* Interscience, New York, 1959.

257. Sastry, G. S., R. E. Hamm, and K. H. Pool, *Anal. Chem.,* **41,** 857 (1969).

258. Satterfield, C. N., and A. H. Bonnell, *Anal. Chem.,* **27,** 1174 (1955).

259. Savage, D. J., *Analyst*(London), **76,** 224 (1951).

260. Sawicki, E., T. R. Hauser, T. W. Stanley, and W. Elbert, *Anal. Chem.,* **33,** 93 (1961).

261. Sawicki, E., T. Stanley, and T. Hauser, *Chemist-Analyst,* **47,** 31 (1958).

262. Schumb, W. C., C. N. Satterfield, and R. L. Wentworth, *Hydrogen Peroxide,* Reinhold, New York, 1955.

263. Scopp, H. A., and C. P. Evans, *Anal. Chem.,* **28,** 143 (1956).

264. Seris, G., Ph. Vernotte, Mme. Klein, and A. M. Clave, *Chim. Anal.*(Paris), **42,** 200 (1960).

265. Shaw, J. A., *Ind. Eng. Chem., Anal. Ed.,* **12,** 668 (1940).

266. Shaw, J. A., *ibid.,* **14,** 891 (1942).

267. Shead, A. C., *Anal. Chem.,* **24,** 1451 (1952).

268. Sianu, E., and C. Radulian, *Igiena,* **15,** 561 (1966); *Chem. Abstr.,* **67,** 5513n (1967).

269. Siddappa, G. S., and D. P. Das, *Curr. Sci.,* **23,** 157 (1954).

270. Silber, G., *Plant Disease Reporter,* **45,** 310 (1961).

271. Silber, G., *ibid.,* **46,** 137 (1962).

272. Sillars, I. M., and R. S. Silver, *J. Soc. Chem. Ind.,* **63,** 177 (1944).

273. Silverman, L., and W. Bradshaw, *Anal. Chim. Acta,* **12,** 526 (1955).

274. Silverman, L., and W. Bradshaw, *ibid.,* **14,** 514 (1956).

275. Singliar, M., and J. Zubak, *Chem. Prum.,* **6,** 426 (1956); *Anal. Abstr.,* **4,** 3556 (1957).

276. Skare, I., *Air Water Pollut.,* **9,** 601 (1965).

277. Smith, R. G., and P. Diamon, *Amer. Ind. Hyg. Assoc. Quart.,* **13,** 235 (1952).

278. Sobel, H., *Anal. Chem.,* **25,** 1756 (1953).

279. Societé Anon. des Manufactures des Glaces et Produits Chimiques de Saint-Gobain, Chauny & Cirey, Fr. Pat. 1,005,351 (1952).

280. Sourek, R., and A. Recka, *Chem. Prum.*, **9**, 71 (1959); *Chem. Abstr.*, **53**, 13681d (1959).

281. Specker, H., H. Hartkamp, and E. Jackwerth, *Fresenius' Z. Anal. Chem.*, **163**, 111 (1958).

282. Stafford, C., J. E. Puckett, M. D. Grimes, and B. J. Heinrich, *Anal. Chem.*, **27**, 2012 (1955).

283. Stair, R., in *Ozone Chemistry and Technology*, Advances in Chemistry Series No. 21, American Chemical Society, Washington, D.C., 1959, p. 269.

284. *Standard Methods for The Examination of Water and Waste Water*, 12th ed., American Public Health Association, New York, 1965.

285. Stehl, R. H., D. W. Margerum, and J. J. Latterell, Anal. Chem., **39**, 1346 (1967).

286. St. John, P. A., J. D. Winefordner, and W. S. Silver, *Anal. Chim. Acta*, **30**, 49 (1964).

287. Sweetser, P. B., *Anal. Chem.*, **39**, 979 (1967).

288. Tada, O., and K. Nakaaki, *Rodo Kagaku*, **41**, 294 (1965); *Chem. Abstr.*, **64**, 20507d (1966).

289. Takahashi, K., and Y. Hashimoto, *Suisan Zoshok*, **8**, 7 (1960); *Chem. Abstr.*, **62**, 22b (1965).

290. Tanaka, M., *Mikrochim. Acta*, **1955**, 1048.

291. Thorp, C. E., *Ind. Eng. Chem., Anal. Ed.*, **12**, 209 (1940).

292. Thorp, C. E., *Bibliography* of *Ozone Technology*, Armour Research Foundation of Illinois Institute of Technology, Chicago, 1954.

293. Todd, G. W., *Anal. Chem.*, **27** 1490 (1955).

294. Tranchant, J., *Bull. Soc. Chim. Fr.*, **1968**, 2216.

295. Trotti, L., and D. Sacks, *Arch. Oceanogr. Limnol.*(Venice), **12**, 257 (1962); *Chem. Abstr.*, **58**, 12295d (1963).

296. Truesdale, G. A., A. L. Downing, and G. F. Lowden, *J. Appl. Chem.*(London), **5**, 53 (1955).

297. Trusell, F., and H. Diehl, *Anal. Chem.*, **35**, 674 (1963).

298. Uhrig, K., F. M. Roberts, and H. Levin, *Ind. Eng. Chem., Anal. Ed.*, **17**, 31 (1945).

299. Ungureanu, C., *Energetica* (Bucharest), **12**, 182 (1964); *Chem. Abstr.*, **61**, 11755g (1964).

300. Vendt, V. P., *Dokl. Akad. Nauk SSSR*, **73**, 689 (1950); *Chem. Abstr.*, **44**, 10596i (1950).

301. Waclawik, J., and S. Waszak, *Chem. Anal.*(Warsaw), **8**, 633 (1963).

302. Wagner, C. D., H. L. Clever, and E. D. Peters, *Ind. Eng. Chem., Anal. Ed.*, **19**, 980 (1947).

303. Wagner, C. D., R. H. Smith, and E. D. Peters, *Ind. Eng. Chem., Anal. Ed.*, **19**, 976 (1947).

304. Wagnerova, D. M., K. Eckschlager, and J. Veprek-Siska, *Collect. Czech. Chem. Commun.*, **32**, 4032 (1967).

305. Warburg, O., and G. Krippahl, *Z. Naturforsch. B*, **18**, 304 (1963).

306. Wernimont, G., and F. J. Hopkinson, *Ind. Eng. Chem., Anal. Ed.*, **15**, 272 (1943).

307. White, L., and W. J. Barrett, *Anal. Chem.*, **28**, 1538 (1956).

308. Whittum, J. B., *Anal. Chem.*, **23**, 209 (1951).

309. Wickert, K., *Werkst. Korros.*, **2**, 209 (1951).

310. Wickert, K., and E. Ipach, *Vom Wasser*, **18**, 337 (1950).

311. Wickert, K., and E. Ipach, *Fresenius' Z. Anal. Chem.*, **139**, 181 (1953).

312. Wickert, K., and E. Ipach, *ibid.*, **140**, 350 (1953).

313. Wickert, K., and E. Jaap, *Fresenius' Z. Anal. Chem.*, **145**, 338 (1955).

314. Williams, D. D., C. H. Blachly, and R. R. Miller, *Anal. Chem.*, **24**, 1819 (1952).

315. Willis, H. A., and R. G. J. Miller, *Spectrochim. Acta*, **14**, 119 (1959).

316. Winkler, L. W., *Chem. Ber.*, **21**, 2843 (1888).

317. Winslow, E. H., and H. A. Liebhofsky, *Ind. Eng. Chem., Anal. Ed.*, **18**, 565 (1946).

318. Wise, W. M., and W. W. Brandt, *Anal. Chem.*, **27**, 1392 (1955).

319. Wolfe, W. C., *Anal. Chem.*, **34** 1328 (1962).

320. Zehender, F., and W. Stumm, *Mitt. Geb. Lebensmittelunters. Hyg.*, **44**, 206 (1953); *Chem. Abstr.*, **47**, 8946b (1953).

PHOSPHORUS

D. F. BOLTZ, CHARLES H. LUECK[1] and ROBERT J. JAKUBIEC[2]

Department of Chemistry, Wayne State University, Detroit, Michigan

[1] E. I. du Pont de Nemours & Co., Inc., Textile Fibers Department, Old Hickory, Tennessee
[2] President, Enviro-Test, Inc., Westmont, Illinois

The spectrophotometric determination of traces of orthophosphate ions is one of the most widely performed determinations in applied analysis, despite the fact that phosphorus occurs only to the extent of about 0.12% in the lithosphere.

Phosphorus is essential for the formation of bone resulting in the forming of a tricalcium phosphate complex. Calcium monohydrogen phosphate, $CaHPO_4$, is often added to foods as a mineral supplement. Phosphorus compounds, such as creatine phosphate, are presumably involved in muscle contraction and in the utilization of carbohydrates. Phosphates also play an important role in buffering the blood at the proper pH value. The determination of small amounts of phosphorus in blood serum, tissues, and other biologic samples is often performed in clinical analysis.

Phosphate compounds, particularly "superphosphate," $Ca(H_2PO_4)_2$, are used as fertilizers. The determination of phosphate in soils, in plants, in foods, and in feeds is important in studying plant growth, nutrition, and the fertility of soils.

Phosphates are often added to boiler waters to prevent scale formation and the embrittlement of boilers. The phosphate precipitates calcium ions as hydroxyapatite, $3Ca_3(PO_4)_2 \cdot Ca(OH)_2$, which does not form boiler scale. The concentration of phosphate seldom exceeds 30 ppm phosphorus, or 100 ppm phosphate. Hard water can also be softened by treatment with sodium hexametaphosphate, $Na_6P_6O_{18}$, or $(NaPO_3)_6$, which sequesters the calcium by the formation of a chelate complex, thereby preventing the consumption of soap by the calcium ions. The extensive use of phosphates in combination with soaps or detergents has led to serious problems in water pollution. The spectro-photometric determination of the total phosphate content of water is, therefore, a highly significant determination in detecting contamination and in controlling water treatments and industrial waste disposal systems.

A number of organic phosphates are utilized commercially. For example, triphenyl phosphate, $(C_6H_5)_3PO_4$, is a plasticizer, and tricresyl phosphate is an additive for gasoline. Many organophosphorus pesticides such as parathion, malathion, ethion, and thimet are used quite extensively.

Phosphorus is found in ferrous metallurgical products because of the occurrence of phosphates in iron ore and in the limestone used as a flux. The phosphorus is usually in the form of iron phosphide, Fe_3P, which is soluble in the ferrite and decreases the solubility of carbon in the iron. High concentrations of phosphorus in steel cause embrittlement, the effect being more pronounced with increasing percentage of carbon in the steel. The phosphorus content of most steels is less than 0.01% and is often detrimental if over 0.03%. In dissolving steel, it is necessary to convert the phosphide to phosphate with an appropriate oxidizing reagent. Cast iron, which is used in parts requiring machining to very close tolerances, for example, crankshafts, may have a phosphorus content of 0.5%. Phosphorus is also found to some extent in nonferrous metallurgical materials, such as the phosphor bronzes which are used in bearings.

Industrial electroplating uses sodium orthophosphates, tripolyphosphates, and pyrophosphates in cleaning operations and zinc and iron phosphates as rust inhibitors.

This chapter is devoted to a presentation of suitable methods of effecting preliminary analytical separations and spectrophotometric methods for determining phosphorus in the form of orthophosphate and the application of these methods to the determination of small amounts of phosphorus in specific materials. For alternative methods of determining phosphorus in a variety of materials, the treatise by Halman should be consulted (21).

Inasmuch as the methods of separation and determination are based on the presence of orthophosphates, the following brief guide is given for the conversion of other forms of phosphorus to orthophosphate.

Phosphide. Dissolution of the sample in aqua regia, followed by fuming with perchloric acid, is recommended for most metallurgical materials. Certain procedures recommend the use of permanganate as an oxidant and hydrogen peroxide to remove the excess reagent.

Pyrophosphate. Pyrophosphate ($P_2O_7{}^{4-}$) is not appreciably converted to orthophosphate in neutral or basic solutions. However, in acidic solutions, pyrophosphate undergoes a hydrolytic conversion to orthophosphoric acid according to the following equation:

$$H_4P_2O_7 + H_2O \longrightarrow 2H_3PO_4$$

Sufficient concentrated nitric acid should be added to result in approximately a $5N$ solution and the solution heated for 2 hr to completely convert pyrophosphate to orthophosphate.

Phosphite. Phosphite (HPO_3^{2-}) in acidic solution, or phosphorous acid, H_3PO_3, is oxidized by bromine as shown in the following equation to give orthophosphoric acid.

$$H_3PO_3 + Br_2 + H_2O \longrightarrow H_3PO_4 + 2H^+ + 2Br^-$$

Concentrated nitric acid or concentrated sulfuric acid will also oxidize phosphorous acid when the solutions are evaporated almost to dryness at low temperatures ($100-150°C$).

Hypophosphite. Hypophosphite salts in acidic solution give hypophosphorous acid, H_3PO_2, which can be oxidized by hot, concentrated nitric acid. Small amounts ($5-50\,\mu g$ P) of hypophosphorous acid have been converted to orthophosphoric acid by evaporating almost to dryness a sulfuric acid solution containing ammonium peroxydisulfate (22).

Tripolyphosphate. Sodium triphosphate, $Na_5P_3O_{10}$, is water soluble, and the triphosphate hydrolyzes very slowly at ordinary temperatures to give orthophosphate. Conversion is accomplished by heating in a boiling water bath with nitric acid for about 1 hr.

Metaphosphates. Sodium salts of the metaphosphates, sodium hexametaphosphate or Graham's salt, undergo hydrolysis in sodium hydroxide solution ($2M$) to give both orthophosphate and pyrophosphate. Hence, after prolonged heating in basic solution, the concentrated nitric acid–evaporation treatment must follow to assure conversion of pyrophosphate to orthophosphate.

I. SEPARATIONS

The principal separations involve the partition of phosphate from arsenate, silicate, and germanate ions, which also form heteropoly complexes. In certain colorimetric methods, the removal of copper(II), iron(III), nickel(II), and chromium(III or VI) is desirable because of the color of these ions.

A. PRECIPITATION METHOD

Traces of phosphate can be coprecipitated as aluminum phosphate with aluminum hydroxide, which serves as a carrier from a solution buffered at pH of about 5.5. Copper(II), nickel(II), zinc(II), chromium(VI), cobalt(II), and manganese(II) ions will remain in solution. It has been recommended that the precipitate be separated by centrifuging, inasmuch as filter paper ash often contains traces of phosphate. The phosphate in seawater has been concentrated in this manner prior to its colorimetric determination (42). Hydrous zirconium oxide has also been recommended as a collector. Vanadium(V) will coprecipitate when hydrous aluminum oxide is used.

Germanium and arsenic can be precipitated as sulfides. Arsenic, in either the tervalent or quinquevalent state, is precipitated from solution $6N$ in hydrochloric acid or $6N$ in sulfuric acid. Only the tervalent arsenic is completely precipitated from the solutions $6N$ in acid, but usually it requires a one- or two-day digestion period for the precipitate to settle out. Care is also required in washing the precipitate in order to prevent peptization of the precipitate.

The separation of silicate from phosphate is effected by fuming with perchloric acid to dehydrate the silicic acid and render it insoluble so that it can be removed by filtration or centrifugation. The preliminary removal of the silica using the perchloric acid has been used in the colorimetric determination of phosphorus in iron ore (13), limestone (11), and biological materials (39).

In the analysis of steels, the carbides, nitrides, and hydrolysis products of niobium, tantalum, and tungsten are removed by filtration before the perchloric acid fuming treatment.

B. DISTILLATION METHOD

Arsenic can be removed from phosphate solutions by volatilizing as arsenious bromide from a hydrobromic–perchloric acid solution or from a hydrobromic–sulfuric acid solution (44). If arsenious bromide is to be volatilized, the reduction of quinquevalent arsenic to a tervalent state can be accomplished by sulfurous acid or hydroiodic acid. Arsenious bromide is easily volatilized at a temperature of 200–250°C.

Arsenic and antimony, after a preliminary reduction to the tervalent state by hydrazine sulfate in sulfuric acid solution, can be separated by distillation of arsenic(III) chloride and antimony(III) chloride. Tin(IV) chloride is also removed by distillation (25).

The separation of germanate and phosphate is conveniently performed by volatilization of germanium bromide from hydrobromic acid–perchloric acid or by hydrobromic acid–sulfuric acid solutions. The use of a gas sweep expedites removal of the volatile halides of germanium and arsenic.

Tin, either in the divalent or quadrivalent state, can be volatilized as bromide from hydrobromic acid–perchloric acid or hydrobromic acid–sulfuric acid solutions. Chlorides should be absent.

Silicon can be removed by the volatilization of silicon tetrafluoride from sulfuric acid solutions. Chromium can be removed by volatilizing as chromyl chloride (61). Osmium and ruthenium are volatilized as tetroxides.

C. EXTRACTION METHOD

Molybdophosphoric acid is extractable from aqueous solutions by oxygenated organic solvents such as alcohols, ethers, esters, and ketones. Wadelin and Mellon (62) investigated the liquid–liquid extraction of heteropoly acids and found that a 20% by volume solution of 1-butanol in chloroform selectively extracted the molybdophosphoric acid in the presence of arsenate, silicate, and germanate ions. Jakubiec and Boltz (32) used propylene carbonate: chloroform to extract molybdophosphoric acid. Approximately 5 ppm arsenic as arsenate and 20 ppm silicon as silicate did not interfere. Stoll (60) extracted with ethyl acetate to separate molybdophosphoric acid from molybdosilicic acid. Paul (55) recommended isobutyl acetate to extract molybdophosphoric acid from molybdosilicic and molybdoarsenic acids from a highly acidic solution containing perchloric acid. Clabaugh and Jackson (14), after forming molybdophosphoric acid, molybdosilicic acid, and molybdoarsenic acid at pH 1.8, extracted molbydophosphoric acid with ether after adding sufficient hydrochloric acid to give an acidity corresponding to about a $1.2M$ concentration. Hurford and Boltz (29) recommended a solution with an acidity of $1M$ hydrochloric acid to extract molybdophosphoric acid with diethyl ether prior to ultraviolet spectrophotometric or atomic absorption spectrometric determination.

Molybdovanadophosphoric acid was extracted with 1-pentanol by Jakubiec and Boltz (31), while Maksimova and Kozlovskii (49) used diethyl ether and 3-methylbutanol and Kinnunen and Wennerstrand (34) used 1-butanol and 1-butylacetate.

D. ION EXCHANGE METHOD

A strong cation exchange resin will remove many cations such as iron(III), calcium, magnesium, and aluminum from solutions 0.3–0.4M in hydrochloric acid. The column is washed with a dilute 0.01–0.02M hydrochloric acid solution. Large amounts of iron cannot be removed satisfactorily because of the tendency for positively charged iron(III) phosphate complexes to be formed and retained by the ion exchange resin (27).

A weakly basic anion exchanger in chloride form can be used to separate phosphate and silicate, provided the phosphorus concentration is at least twice that of silicon. The silicate is eluted prior to the phosphate. A solution of 0.05M in hydrochloric acid and 0.1M in potassium chloride and a 0.5M potassium chloride solution were used as elutriants (41).

E. ELECTRODEPOSITION METHOD

Lead and antimony, which interfere in certain colorimetric determinations because a turbidity results when color-forming reagent is added, particularly in sulfuric acid medium, can be removed from an acidic solution by electro-deposition on a platinum cathode. Copper likewise can be removed. Lead can also be deposited as lead dioxide at a platinum anode from nitric acid solutions.

Phosphate can be separated from many reducible metal ions by employing an electrolytic separation with a mercury cathode. The following ions are removed at the mercury cathode: iron(II), chromium(III), bismuth(III), zinc(II), cadmium(II), copper(II), tin(II), nickel(II), cobalt(II), silver(I), gold(III), molybdenum(VI), and germanium(IV). Aluminum(III), titanium(IV), vanadium(V), uranium(VI) and zirconium(IV) remain in solution with the phosphate.

II. METHODS OF DETERMINATION

The most extensively used spectrophotometric methods for the deter-mination of phosphorus are based on the initial formation of either molybdo-phosphoric acid or molybdovanadophosphoric acid.

A. MOLYBDOPHOSPHORIC ACID METHOD

A yellow color develops when excess molybdate solution is added to an acidic solution of orthophosphate ions (9). The resultant color is due to the formation of molybdophosphoric acid, $H_3PMo_{12}O_{40}$. The optimum conditions for the development of this heteropoly complex consist of having a final molybdate concentration of approximately $0.04M$ and a final acidity of about $0.257M$ in respect to nitric or perchloric acid. The optimum concentration range is 1–15 ppm phosphorus when absorbance measurements are made at 380 nm. The system shows conformity to Beer's law. The most serious interferring ions are silicate, arsenate, tungstate, vanadate, iron(III), tin(II), and bismuth. Nickel(II), copper(II), and fluoride, if in excess of 40, 100, and 25 ppm, respectively, also interfere.

Standard Phosphate Solution. Dissolve 0.4395 g potassium dihydrogen phosphate in distilled water and dilute to 1 liter. Each ml contains 0.1 mg phosphorus.

Molybdate Reagent. Dissolve 100 g sodium molybdate, $Na_2MoO_4 \cdot 2H_2O$, in distilled water, filter, and dilute to 1 liter with distilled water.

Procedure

The sample solution should contain not more than 1 mg phosphorus as orthophosphate per 25 ml solution and should be neutral to litmus. Transfer a 25-ml aliquot to a 50-ml volumetric flask. Add 5 ml of a $2.5N$ nitric acid (or $2.5N$ perchloric acid) and slightly less than 15 ml distilled water. Add 5 ml of the molybdate solution, dilute to graduation mark with distilled water, and mix thoroughly. Use either distilled water or a reagent blank solution and measure the transmittance or absorbance at 380 nm using 1-cm cells.

B. EXTRACTION MOLYBDOPHOSPHORIC ACID METHOD

The extraction of molybdophosphoric acid eliminates the interference due to silicate, arsenate, and germanate. A 20% by volume solution of 1-butanol in chloroform has been recommended as an appropriate extractant to isolate the molybdophosphoric acid not only from appreciable amounts of silicate, arsenate, and germanate but also from the excess molybdate, which exhibits ultraviolet absorption (62). Thus, spectrophotometric measurement of the extract at 310 nm increases the inherent sensitivity of the molybdophosphoric acid method. However, it is necessary to control very carefully the concentrations of the reagents.

The following ions do not interfere: acetate, ammonium, barium, beryllium, borate, bromide, cadmium, calcium, chloride, chromium(III), cobalt, copper(II), iodate, iodide, lithium, magnesium, manganese(II), mercury(II), nickel, nitrate, potassium, selenium(IV), sodium, strontium, and tartrate. Gold(III), bismuth(III), dichromate, lead, nitrate, thiocyanate, thiosulfate, thorium, uranyl, and zirconyl ions must be absent. Up to 1 mg fluoride, periodate, permanganate, vanadate, and zinc can be tolerated. The amount of aluminum, iron(III), and tungstate should not exceed 10 mg.

Special Solutions

Standard Phosphate Solution. Dissolve 0.1098 g potassium dihydrogen phosphate in distilled water and dilute to 1 liter. Store in polyethylene bottle. Each ml of this solution contains 0.025 mg phosphorus.

Molybdate Reagent. Dissolve 7.5 g sodium molybdate dihydrate, $Na_2MoO_4 \cdot 2H_2O$, in 200 ml water, add 100 ml concentrated hydrochloric acid, and dilute to 500 ml. Store in a polyethylene bottle.

Extractant. Mix 100 ml 1-butanol with 400 ml chloroform.

Procedure

The sample solution should contain not more than 0.04 mg phosphorus per 25 ml should have a pH of 5–9. Transfer a 15-ml aliquot to a 125-ml separatory funnel and use a pipet to add 10 ml of the molybdate reagent solution. Shake the funnel for 1 min, allow the layers to separate, and drain the lower layer into a 25-ml volumetric flask. Add another 10 ml of the extractant and repeat the extraction step. Add the lower layer to the volumetric flask. Dilute to the graduation mark with the extractant. Measure the transmittance or absorbance at 310 nm using 1-cm silica cells with the extractant in the reference cell.

Alternate Extraction Procedure Using Propylene Carbonate (37)

Special Solutions

Acidic Molybdate Solution. Dissolve 25 g reagent-grade sodium molybdate, $Na_2MoO_4 \cdot 2H_2O$, in 10N sulfuric acid and dilute to 1 liter with 10N sulfuric acid.

Sulfuric Acid Solution. 1.0N.

Procedure

Transfer an amount of sample containing not more than 7.5 μg soluble phosphorus(V) as phosphate to a 50-ml volumetric flask. Add 5.0 ml acidic molybdate solution, then dilute to the mark with distilled water. Transfer solution to a 125-ml separatory funnel with the aid of 10 ml of the 1.0N sulfuric acid solution to rinse the flask. Add 5.0 ml of propylene carbonate to the separatory funnel and shake the funnel gently for 5 sec. Add 1.0 ml chloroform to the funnel and shake the funnel vigorously with the mechanical shaker for 60 sec. Allow the phases to separate and then drain the lower layer into a 5-ml volumetric flask.

Add an additional 1.0 ml chloroform to the separatory funnel and again shake the funnel vigorously for 60 sec with the mechanical shaker. Allow the phases to separate and drain the lower layer into the same 5-ml volumetric flask. Add to each flask 0.5 ml propylene carbonate and dilute to volume with chloroform. Measure the absorbance at 308 nm using chloroform in the reference cell. A reagent blank prepared in a similar manner except for addition of sample solution should also be measured against the chloroform in the reference cell.

C. INDIRECT MOLYBDOPHOSPHORIC ACID METHOD

A very sensitive, indirect ultraviolet spectrophotometric method is based on (a) a preliminary extraction of the molybdophosphoric acid with an immiscible solvent, (b) a retrograde extraction with a basic buffer solution which extracts the molybdate originally complexed with the phosphate, and (c) the measurement of the absorbance of molybdate solution at 230 nm (45). The final acidity of the solution should be 0.9–1.6N in perchloric acid. The optimum concentration range is 0.1–0.5 ppm phosphorus, and the system shows conformity to Beer's law.

The following ions cause no interference: acetate, aluminum, ammonium, bromide, calcium, chloride, chromium(III), cobalt, copper(II), dichromate, fluoride, iron(III), lead(II), manganese(II), molybdate, nickel, oxalate, perchlorate, permanganate, potassium, silver, sodium, sulfate, vanadate, and zinc. Silicate, arsenate, arsenite, germanate, and nitrite interfere and must be absent or removed prior to the initial extraction. A maximum of 200 ppm nitrate and 20 ppm tungstate can be tolerated.

Special Solutions

Standard Phosphate Solution. Dissolve 0.4395 g potassium dihydrogen phosphate in water and dilute to 1 liter. Dilute a 100-ml aliquot of this solution to 1 liter. This dilute solution contains 0.01 mg phosphorus per ml. Store in a polyethylene bottle.

Molybdate Reagent. Dissolve 50 g sodium molybdate dihydrate, $Na_2MoO_4 \cdot 2H_2O$, in water and dilute to 500 ml. Store in a polyethylene bottle.

Buffer Solution. Dissolve 53.5 g ammonium chloride and 70 ml concentrated ammonium hydroxide in water and dilute to 1 liter. Store in a polyethylene bottle.

Extractant. Mix 500 ml diethyl ether and 100 ml isobutyl alcohol.

Procedure

The sample solution should contain not more than 0.065 mg phosphorus per 25 ml and should be slightly acidic to litmus. Transfer a 25-ml aliquot to a 125-ml separatory funnel and add 5 ml 72% perchloric acid. Add 10 ml water and 5 ml of the molybdate reagent and mix thoroughly. Extract with 40 ml of the ether—isobutyl alcohol extractant, shaking the funnel for 1 min. Allow the layers to separate and discard the lower aqueous layer. Swirl the funnel to collect any water droplets and discard the additional aqueous layer. Wash twice with 25-ml portions of a 1:10 perchloric acid solution, shaking the funnel at least 30 sec each time. Swirl the funnel after each washing and discard the remaining aqueous phase. Wash the funnel with a jet of water to remove any remaining trace of molybdate solution.

Add 30 ml of the basic buffer solution. Shake the funnel for 30 sec and drain the lower aqueous phase into a 100-ml volumetric flask. Add another 15 ml of the basic buffer solution to the funnel and repeat the extraction procedure. Add the aqueous layer to the volumetric flask containing the first extract. Swirl the funnel to remove the final traces of aqueous phase, which should also be added to the flask. Dilute to the graduation mark with water and mix well. Measure the transmittance or absorbance at 230 nm using 1-cm silica cells with a reagent blank solution in the reference cell.

D. HETEROPOLY BLUE METHOD

Orthophosphate and molybdate ions combine in acidic solution to form molybdophosphoric acid which, upon selective reduction, produces a heteropoly blue complex. If the acidity at the time of reduction corresponds to about $1N$ in sulfuric or perchloric acid and if the appropriate reductant is used, the resulting heteropoly blue complex exhibits maximum absorbance in the 820–830 nm wavelength region. At lower acidities, the blue compound exhibits maximum absorbance in the 650–700 nm region. Boltz and Mellon (8) have used the term "heteropoly blue" to designate the reduction product having an absorbance maximum at 830 nm and the term "molybdenum blue" to designate the blue reduction product exhibiting absorbance in the 650–700 nm region. Although many reductants have been used under a variety of conditions, hydrazine sulfate (52) and chlorostannous acid (17) are the most often recommended reductants. In using hydrazine sulfate, the reduction should be carried out with

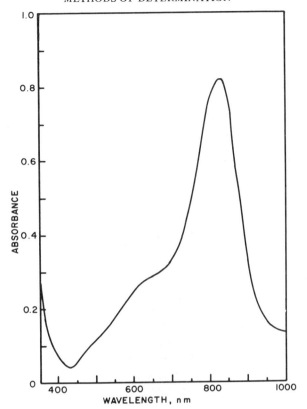

Fig. 1. Absorption spectrum for the heteropoly blue of phosphorus (1 ppm P, 1-cm cells, reagent blank solution used in reference cell).

solutions having a final acidity of about $1N$ in respect to perchloric or sulfuric acid and should be maintained for 10 min at a temperature of about $100°C$. This heteropoly blue system shows conformity to Beer's law at 650 and 830 nm (Fig. 1). The optimum concentration range is 0.1–0.2 ppm phosphorus.

The following ions do not interfere: aluminum, ammonium, cadmium, chromium(III), copper(II), cobalt(II), calcium, iron(II), magnesium, manganese(II), nickel, zinc, chloride, bromide, acetate, citrate, silicate, fluoride, vanadate, and borate. The following ions must be absent: tin(II), nitrate, and arsenate. The iron(III) concentration should not exceed 200 ppm. Only 10 ppm tungstate can be tolerated. Lead(II), bismuth(III), barium, and antimony(III), interfere because of the formation of precipitates or a turbidity in the presence of sulfuric acid.

Special Solutions

Standard Phosphate Solution. Dissolve 0.2197 g potassium dihydrogen phosphate in redistilled water and dilute to 1 liter. Each ml is equivalent to 0.05 mg phosphorus.

Molybdate Solution. Dissolve 25 g sodium molybdate, $Na_2MoO_4 \cdot 2H_2O$, in 10N sulfuric acid and dilute to 1 liter with 10N sulfuric acid.

Hydrazine Sulfate Solution. Dissolve 1.5 g hydrazine sulfate, $N_2H_6SO_4$, in redistilled water and dilute to 1 liter.

Procedure

The sample solution should contain not more than 0.06 mg phosphorus as the orthophosphate per 25 ml solution and should be neutral to litmus. Transfer a 25-ml aliquot of the prepared solution to a 50-ml Pyrex volumetric flask. Add 5 ml of the molybdate solution and 2 ml of the hydrazine sulfate solution. Dilute to the graduation mark with distilled water and mix thoroughly. Immerse the flask in a boiling water bath for 10 min, remove, and cool rapidly. Again shake the flask and adjust the meniscus to the mark with a few drops of water, if necessary. Use 1-cm cells and measure the transmittance or absorbance at 830 nm with either distilled water or a reagent blank. The Beer's law plot goes through the origin only if a reagent blank solution is used in the reference cell. The absorption cells should be examined to make sure that gas bubbles have not collected at the glass—solution interface. A gentle tapping of the cells will result in the release of the gas. Spectrophotometric measurements can also be made at 650 nm although the heteropoly blue method is only about 40% as sensitive when measurements are made at this wavelength.

E. EXTRACTION HETEROPOLY BLUE METHOD

Instead of reducing molybdophosphoric acid in aqueous solution, the molybdophosphoric acid can be extracted with an immiscible solvent and then reduced to the heteropoly blue by shaking the extract with a reductant (5,43,45,54,61). The extraction of the reduced molybdophosphoric acid from aqueous solutions by organic solvents has also been studied (38).

The acidity at the time of the formation of the molybdophosphoric acid may be varied from 0.5 to 1.5M in respect to perchloric acid, with the higher acidity being recommended if an appreciable quantity of iron(III) is present. A large excess of the molybdate solution is also necessary. A final molarity of 0.04M molybdate is sufficient.

The acidity of the chlorostannous acid should be about 2N in hydrochloric acid. A much higher acidity results in a decrease in the intensity of the resulting heteropoly blue color, whereas a very low acidity results in no reduction to the heteropoly blue.

Butanol, isobutyl alcohol, octanol, and isobutyl acetate have been used as extractants. When isobutyl alcohol is used as the extractant, the absorption spectrum shows absorbance maxima at 625 and 725 nm. The optimum concentration range is 0.2–1.5 ppm phosphorus.

The following ions do not interfere: acetate, bromide, carbonate, chloride, citrate, dichromate, fluoride, iodate, nitrate, nitrite, oxalate, permanganate, sulfate, ammonium, aluminum, barium, bismuth(III), cadmium, calcium, chromium(III), cobalt(II), copper(II), iron(II), iron(III), lead(II), lithium, magnesium, manganese(II), nickel(II), potassium, silver, sodium, thorium(IV), uranyl, and zinc. Arsenic(III), iodide, and thiocyanate should not exceed 50 ppm and not more than 25 ppm silicate or tin(IV) are permissible. Arsenic(V), germanium(IV), gold(III), thiosulfate, tungstate, vanadium(V), and tin(II) must be absent. Both germanium and arsenic can be removed by volatilizing as the bromide (44).

Special Solutions

Standard Phosphate Solution. Dissolve 0.1098 g potassium dihydrogen phosphate in distilled water and dilute to 1 liter. Each ml contains 0.025 mg phosphorus.

Molybdate Solution. Dissolve 25 g sodium molybdate, $Na_2MoO_4 \cdot 2H_2O$, in distilled water and dilute to 250 ml. Filter if solution is turbid.

Chlorostannous Acid Solution. Dissolve 2.38 g stannous chloride, $SnCl_2 \cdot 2H_2O$, in 170 ml concentrated hydrochloric acid and dilute to 1 liter with distilled water. Add several pellets of metallic tin.

Procedure

The sample solution should contain not more than 0.06 mg phosphorus per 25 ml. Transfer 25 ml of the slightly acidic sample solution to a 100-ml beaker and add 5 ml perchloric acid. Add 10 ml distilled water and 5 ml of the molybdate reagent. Mix and allow the solution to stand for several minutes. Transfer the solution to a 125-ml separatory funnel. Rinse the beaker with a small amount of distilled water and add the washings to the separatory funnel. Extract with 40 ml isobutyl alcohol. Shake for 1 min. Drain and discard the lower aqueous layer. Shake the isobutyl alcohol extract with two successive 25-ml portions of distilled water. Discard the lower aqueous layer. Swirl the solution in the funnel to collect the droplets of water into one globule and discard this water.

Add 25 ml of the chlorostannous acid reagent and shake for at least 15 sec.

Discard the lower aqueous layer and drain the alcohol phase into a 50-ml volumetric flask. Wash the funnel with 10 ml isobutyl alcohol and add the washings to the 50-ml volumetric flask. Dilute to the graduation mark with isobutyl alcohol. After mixing thoroughly, measure the transmittance or absorbance in 1-cm cells at 725 nm using a reagent blank in the reference cell.

F. MOLYBDOVANADOPHOSPHORIC ACID METHOD

A mixed heteropoly acid is formed when a molybdate solution is added to an acidic solution containing orthophosphate and vanadate ions (51). Kitson and Mellon (37) studied this method thoroughly and made recommendations concerning the optimum concentration of reagents. The optimum acidity is about $0.5N$ in nitric acid. Sulfuric, perchloric, and hydrochloric acids can be used instead of nitric acid. The final vanadate concentration should be about $0.002M$, and the final molybdate concentration should be about $0.01M$. When the acidity is less than $0.2N$, a yellow-orange color develops in the absence of orthophosphate ions. An acidity of higher than about $0.75N$ results in the slow development of the desired color. It is essential to adjust the acidity and then add successively the vanadate and molybdate reagents. The optimum concentration range is 5–40 ppm phosphorus using 1-cm cells and measuring the absorbance at 460 nm.

One advantage of the molybdovanadophosphoric acid method, despite its much lower sensitivity than that of the heteropoly blue, is its relative freedom from the interferences caused by silicate and arsenate. The following do not interfere: aluminum, ammonium, barium, beryllium, manganese(II), mercury(II), potassium, silver, sodium, strontium, tin(II), uranyl, zinc, zirconyl, acetate, arsenate, benzoate, bromide, carbonate, chlorate, citrate, cyanide, formate, iodate, lactate, molybdate, nitrate, nitrite, oxalate, perchlorate, periodate, pyrophosphate, salicylate, selenate, silicate, sulfate, sulfite, tartrate, and tetraborate. Copper(II) and nickel ions change the color of the solution, but if absorbance measurements are made at 460 nm, as much as 1000 ppm of these ions can be tolerated. Iron(II), sulfide, thiosulfate, and thiocyanate, if present in excess of about 100 ppm, interfere by either reducing the molybdovanadophosphoric acid or the excess molybdate to molybdenum blue.

Bismuth, thorium, arsenate, chloride, and fluoride inhibit development of the color. If present in more than small amounts, a longer time, perhaps 30 min, should be allowed for the maximum development of color. The arsenate concentration should not exceed 100 ppm; the chloride, 500 ppm; and the fluoride, 10 ppm. A bismuth concentration of 400 ppm is permissible. Cobalt(II) should not exceed 100 ppm, and chromium(III), about 10 ppm. Iodide, dichromate, and permanganate should be absent. Cerium(IV), tin(IV), and silver interfere because of the formation of a precipitate or a turbidity. When large amounts of silicate, chloride, or iron(III) are present in the sample solution, a preliminary evaporation to incipient dryness with perchloric acid is recommended. This treatment removes the chloride ions and dehydrates the silica, which can be removed by filtration. Iron(III) does not have so high an absorptivity at 460 nm in perchloric acid solution as in hydrochloric acid solution. Also, it is possible to compensate externally for large amounts of iron(III) by using a properly diluted aliquot to the sample solution in the reference cell.

Michelson (50) modified the molybdovanadophosphoric method by using very dilute reagents and measuring the absorbance at 315 nm. Nitric acid is not used in this modified method. A final acidity of $0.07N$ hydrochloric acid was used. The recommended concentration range is 0.2–1.5 ppm phosphorus.

Special Solutions

Standard Phosphate Solution. Dissolve 0.4395 g potassium dihydrogen phosphate in distilled water and dilute to 1 liter. Store in a polyethylene bottle. Each ml of this solution contains 0.1 mg phosphorus.

Vanadate Reagent. Dissolve 2.5 g ammonium vanadate, NH_4VO_3, in 500 ml boiling water. Cool the solution and add 20 ml concentrated nitric acid. Cool and dilute to 1 liter. Store in a polyethylene bottle.

Molybdate Reagent. Dissolve 50 g ammonium molybdate, $(NH_4)_6 Mo_7O_{24} \cdot 4H_2O$, in 1 liter warm water ($50°C$). If sodium molybdate, $Na_2MoO_4 \cdot 2H_2O$, is used, 65 g of the salt should be dissolved. Store in a polyethylene bottle.

Procedure

The sample solution should not contain more than 2.0 mg phosphorus per 25 ml and should be slightly acidic to litmus. Transfer 25 ml of the sample solution to a 50-ml volumetric flask and add 5 ml of a 1:2 nitric acid solution and 5 ml of the vanadate solution. Add 5 ml of the molybdate solution, dilute to the mark with water, and mix thoroughly. After 5 min, measure the transmittance or the absorbance at 460 nm using 1-cm cells. A reagent blank solution should be used in the reference cell unless the external compensation technique is being employed.

G. EXTRACTION MOLYBDOVANADOPHOSPHORIC ACID METHOD

The extraction of molybdovanadophosphoric acid by an immiscible or organic solvent is the basis of another ultraviolet spectrophotometric method for the determination of phosphorus (31). The recommended procedure utilizes a mixed reagent, similar in composition to that recommended by Michelson (50),

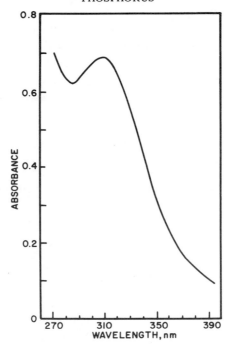

Fig. 2. Absorption spectrum for molybdovanadophosphoric acid extracted into 1-pentanol (0.84 ppm P, 1-cm cells, reagent blank solution extracted into 1-pentanol used in reference cell).

and 1-pentanol as the extractant (Fig. 2). Absorbance measurements are made at 308 nm. The optimum concentration range is 0.2–1.1 ppm phosphorus.

Special Solutions

Standard Phosphate Solution (3.0 μg Phosphorus per ml). Dissolve 0.2203 g reagent–grade potassium dihydrogen phosphate in distilled water and dilute to 1 liter with distilled water. Transfer a 60-ml aliquot of this solution to 1-liter volumetric flask and dilute to the mark with distilled water. Store in a polyethylene bottle.

Stock Vanadate Solution. Dissolve 2.34 g reagent-grade ammonium vanadate in 500 ml distilled water, add 28 ml concentrated hydrochloric acid, and dilute to 1 liter with distilled water. Store in a polyethylene bottle.

Stock Molybdate Solution. Dissolve 3.23 g reagent-grade ammonium molybdate in 100 ml distilled water. Store in a polyethylene bottle.

Stock Hydrochloric Acid Solution. Prepare a $2.5N$ solution. Store in a polyethylene bottle.

Mixed Reagent. Combine 6.75 ml stock vanadium solution with 12.50 ml stock hydrochloric ácid solution. Add 12.50 ml stock molybdate solution and dilute to 1 liter with distilled water. Store in a polyethylene bottle. Prepare a new solution after about two weeks.

Dilute Hydrochloric Acid Solution. Prepare a $1.2N$ hydrochloric acid solution. Store in a polyethylene bottle.

Extractant. 1-Pentanol, reagent grade.

Procedure

Transfer by volume a solution containing not more than 15 µg phosphorus, as soluble phosphate, to a 125-ml separatory funnel containing 5.0 ml of the mixed reagent. Add 10.0 ml 1-pentanol and 35 ml of the dilute hydrochloric acid wash solution. Shake vigorously for 30 sec with the mechanical shaker. Allow the phases to separate for about 10 min. Drain off and discard the aqueous layer.

Add an additional 35 ml of the dilute acid solution to the funnel and shake vigorously for 30 sec using the mechanical shaker. Allow phases to separate. Drain off and discard aqueous layer. Filter through dry cotton into a clean, dry spectrophotometer cell after discarding the first ml of the organic layer. Measure the absorbance at 308 nm against the reference solution prepared in a similar manner, except for the addition of the sample solution.

H. INDIRECT 2-AMINO-4-CHLOROBENZENETHIOL METHOD

This method, developed by Djurkin, Kirkbright, and West (15), is based on the formation of a green molybdenum—2-amino-4-chlorobenzenethiol complex (35), with the molybdenum being furnished by the molybdate initially incorporated in an extract of molybdophosphoric acid. Fiftyfold excesses of silicon, germanium, arsenic, and antimony do not interfere. A thirtyfold excess of tungsten does not interfere provided an excess of 1,2-dihydroxybenzene-3,5-disulfonic acid is added prior to extraction. The method is very sensitive, with a molar absorptivity of 9.69×10^4 when the absorbance is measured at 710 nm.

Special Solutions

2-Amino-4-chlorobenzenethiol Hydrochloride Solution. Dissolve 1 g 2-amino-4 chlorobenzenethiol hydrochloride (Eastman No. 3279) in 95% ethanol and dilute to 25 ml. A fresh reagent solution should be prepared each day.

Molybdate Solution. Dissolve 7.5 g sodium molybdate dihydrate in 200 ml distilled water, add 100 ml hydrochloric acid, and dilute to 500 ml with distilled water.

Standard Phosphate Solution. Dissolve 0.1098 g potassium dihydrogen phosphate in distilled water and dilute to 1 liter. One ml contains 25 μg phosphorus. Dilute a 25.00-ml aliquot of this standard solution to 250 ml with water in a volumetric flask to obtain a standard solution containing 1.00 μg phosphorus per ml.

Extractant. Prepare a 1:4 mixture of 1-butanol and chloroform mixture.

Buffer Solution. Prepare a buffer solution of pH 2.2 by mixing 33.5 ml 0.2M hydrochloric acid and 250 ml 0.2M potassium chloride solution and diluting to 1 liter with distilled water.

Hydrochloric Acid Solution. 2M Hydrochloric acid.

Ammonia Solution. 2M Ammonium hydroxide.

Procedure

Transfer 5 ml sample solution containing 0.2–5 μg phosphorus as the orthophosphate to a 50-ml separatory funnel. Add 10 ml of the molybdate reagent and dilute to 25 ml with distilled water. Extract with 10 ml of the butanol–chloroform solution using a shaking time of 2 min. Transfer organic phase to a dry 50-ml separatory funnel and repeat the extraction step with another 10 ml extractant. Combine the butanol–chloroform extracts and wash by contacting with 10 ml 2N hydrochloric acid for 1 min. Discard the aqueous phase. Add 10 ml of a 2M ammonia solution to the organic extract and again shake for 1 min.

Transfer the basic aqueous solution to a 50-ml beaker and rinse funnel with 8 ml 2M hydrochloric acid and 2–5 ml distilled water. After combining washings with basic aqueous solution, adjust the pH to approximately 2.2 by adding 2N hydrochloric acid. Transfer solution to a 100-ml volumetric flask, add 3 ml of the buffer solution, and dilute to mark with distilled water.

Transfer a 25-ml aliquot of this solution to a 50-ml separatory funnel, add 0.25 ml of the 2-amino-4-chlorobenzenethiol reagent, mix, and allow to stand for 15 min. Add 10.0 ml chloroform and shake funnel for 1 min. Filter chloroform extract through filter paper into 1.000-cm absorption cells. Measure the absorbance immediately at 710 nm against a reagent blank solution.

I. INDIRECT CHLORANILATE METHOD (23)

This method, the only method presented in this chapter which does not involve the formation of heteropoly acids, is based on the addition of insoluble lanthanum chloranilate to an orthophosphate solution with the liberation of chloranilate equivalent to the phosphate and the precipitation of lanthanum phosphate:

$$2PO_4^{3-} + La_2(C_6Cl_2O_4)_3 + 3H^+ \longrightarrow 2LaPO_4 + 3HC_6Cl_2O_4^-$$

The optimum concentration range for this method is 3–300 ppm phosphate. Nitrate and chloride up to 400 ppm do not interfere. Fluoride must be absent. Sulfate interferes but the addition of 800 ppm sulfate to both standard and sample solutions compensates for sulfate present in sample.

The solution is adjusted to pH 7 with a succinate buffer solution and diluted with an equal volume of 95% ethanol prior to precipitation of the lanthanum phosphate. The absorbance is measured at 530 nm, although more sensitivity can be realized by measuring in the 300–335 nm region (6). Lead(II) chloranilate has also been suggested as precipitant, and various chloranilate salts have been used to determine polymetaphosphates, orthophosphates, and pyrophosphates in mixtures (26).

Special Solutions

Standard Phosphate Solution. Dissolve 0.7165 g potassium dihydrogen phosphate in distilled water and dilute to 1 liter. Each ml contains 0.500 mg PO_4^{3-}.

Buffer Solution. Prepare a 0.03M sodium succinate solution and adjust to pH 7.

Lanthanum Chloranilate. Prepared from chloranilic acid and lanthanum oxide according to the following procedure (23): Dissolve 20 g chloranilic acid in 4.6 liters distilled water and heat the solution. Dissolve 17 g lanthanum oxide (99.9%) in 27 ml nitric acid (sp. gr. 1.42) and 200 ml water. Add the hot chloranilic acid solution to the lanthanam(III) nitrate solution and mix thoroughly. Allow solution to stand for at least 12 hr and filter to obtain a fine, dense precipitate of lanthanum chloranilate. Wash with distilled water until the filtrate is nitrate free. Dry the dark-gray product under an infrared heating lamp and store in a dark bottle.

Procedure

Transfer 10 ml sample or standard solution containing 0.15–5 mg phosphate to a 50-ml volumetric flask. Add 5 ml of the 0.03M sodium succinate (pH 7) buffer solution and 25 ml 95% ethanol. Add 0.1 g of the lanthanum chloranilate and dilute to 50 ml with distilled water. Shake the flask vigorously for 10 min on a mechanical shaker. Filter the solution through filter paper and discard the first 5–10 ml filtrate. Measure the absorbance at 530 nm against a reagent blank solution. A calibration graph should be prepared by following this same procedure using 0.5–10 ml standard phosphate solution.

J. INDIRECT CARMINIC ACID METHOD (40)

This method is based on the formation of 12-molybdophosphoric acid, the extraction of the molybdophosphoric acid into an immiscible organic solvent mixture, a retrograde extraction of the molybdate into a basic aqueous solution, and the determination of the equivalent molybdenum by the carminic acid method. The method is suitable for the determination of 0.03–0.18 ppm phosphorus.

Special Solutions

Standard Phosphate Solution. Dissolve 2.198 g potassium dihydrogen phosphate in double-distilled water and dilute to 1 liter. Transfer 20.00 ml of this stock phosphate solution to a 1-liter volumetric flask and dilute to volume with distilled water. This dilute standard solution contains 10 μg phosphorus as orthophosphate per ml.

Molybdate Reagent Solution. Dissolve 25.0 g ammonium molybdate, $(NH_4)_6Mo_7O_{24} \cdot 4H_2O$, in double-distilled water and dilute to 250 ml in a volumetric flask.

Carminic Acid Solution. Dissolve 0.2462 g carminic acid in 10.0 ml 0.10N sodium hydroxide. Add 10 ml 0.20M acetic acid, mix well, and dilute to 100 ml with double-distilled water.

Acidic Washing Solution. Add 50.0 ml concentrated hydrochloric acid to distilled water in a 500-ml volumetric flask and dilute to volume.

Buffer Solution. Dissolve 53.5 g ammonium chloride in distilled water containing 70 ml concentrated ammonium hydroxide and dilute to 1 liter with distilled water.

Extractant. Spectrograde diethyl ether.

Procedure

Transfer by microburet 0.50–2.00 ml of the standard phosphate solution to a 125-ml separatory funnel. (In the case of an unknown phosphate solution, transfer an aliquot containing no more than 20 μg phosphorus, as orthophosphate, in a 25-ml aliquot.) Add 1.0 ml of a 1:2 hydrochloric acid solution and dilute to about 40 ml with double-distilled water. Add 4.0 ml of the 10% ammonium molybdate reagent, mix well, and allow to stand for 10 min. The pH at this stage should be approximately 1.3. Add 5.0 ml concentrated hydrochloric acid, mix thoroughly, and allow solution to stand for another 5 min.

Add 45 ml diethyl ether to the solution in the separatory funnel and shake the mixture for 3 min. Rinse the separatory funnel stopper with 2 ml diethyl ether using a capillary pipet and collect rinsings in the separatory funnel. After separation of the layers, remove carefully the lower aqueous layer and rinse the stem of the funnel with a jet stream of distilled water and wipe dry. Add 10 ml of acidic hydrochloric acid washing solution to the ether extract of molybdophosphoric acid and shake for 20 sec to remove excess molybdate. Discard aqueous wash layer. Again rinse the tip of funnel stem with distilled water and wipe dry. Repeat this washing procedure three times in order to ensure complete removal of excess molybdate.

Add 30 ml of the ammonium chloride—ammonia buffer to the ether extract and shake for 60 sec. Transfer the lower aqueous layer containing the equivalent molybdate and phosphate to an 80-ml beaker. Add 15 ml of the basic buffer to the ether layer and shake thoroughly for 30 sec. Withdraw this aqueous layer and combine with the molybdate solution in the 80-ml beaker. Add glacial acetic acid dropwise with stirring as the pH is monitored by a pH meter until a pH of about 4.75 is obtained. Transfer this solution carefully to a 100-ml volumetric flask. Add 10.0 ml of the $5 \times 10^{-3} M$ carminic acid reagent, mix well, and dilute to 100 ml with distilled water.

Measure the absorbance in 1.000-cm silica cells using a reagent blank solution, prepared in a similar manner except the extraction stages were omitted, in the reference cell. The absorbance at 565 nm is used to prepare calibration plots for phosphorus.

III. APPLICATIONS

A. TOTAL PHOSPHATE IN WATER

The eutrophication of many natural waters has been attributed to the presence of appreciable concentrations of phosphate, so that many spectrophotometric determinations are performed to detect contamination or to control treatment of industrial effluents. Soluble orthophosphate and polyphosphates, in addition to organic-bound phosphates, may be found in water samples. The heteropoly blue method is recommended for determining phosphate in waters because of the small amounts of phosphate usually present and its tolerance to significant amounts of silicate. Iron(III), chromium(VI), arsenic(V), nitrate, and tannin should be absent. Tannin can be removed by treatment with a hypochlorite solution during initial acidic digestion of sample (1). The addition of sulfamic acid has been recommended when nitrites are present in the samples (19). In the case of organic-bound phosphorus, a preliminary oxidation with a small amount of perchloric acid has been recommended (56). Special care must be exercised to clean all glassware with hot 1:3 hydrochloric acid and to rinse thoroughly with distilled water.

Special Solutions

Sulfuric–Nitric Acid Solution. Add slowly 300 ml concentrated sulfuric acid to approximately 600 ml distilled water. Cool, add 4.0 ml concentrated nitric acid, and dilute to 1 liter.

Ammonium Molybdate Reagent. Dissolve 25 g ammonium molybdate, $(NH_4)_6Mo_7O_{24} \cdot 4H_2O$, in about 200 ml distilled water. Add slowly with care 280 ml concentrated sulfuric acid to 400 ml distilled water and cool to room temperature. Add the molybdate solution to the sulfuric acid solution, mix, and dilute to 1 liter.

Tin(II) Chloride Solution. Dissolve 1.25 g tin(II) chloride, $SnCl_2 \cdot 2H_2O$, in 100 ml glycerol. Heat in a water bath and stir until dissolution is complete.

Standard Phosphate Solution. Dissolve 0.7165 g potassium dihydrogen phosphate, KH_2PO_4, in distilled water and dilute to 1 liter. Transfer exactly 100.0 ml of this standard stock solution to a 1-liter volumetric flask and dilute to mark with distilled water. Each ml of this solution contains 50.0 μg PO_4^{3-}.

Procedure

Transfer a 100-ml aliquot of water sample to a 250-ml conical flask and add one drop of a 0.5% phenolphthalein indicator. If indicator gives color, add the sulfuric-nitric acid mixture until the color is discharged. Add 1 ml of this acidic solution. Boil for approximately 2 hr, adding distilled water to maintain a volume of about 50 ml. After cooling the solution, neutralize with dilute sodium hydroxide. Transfer to a 100-ml volumetric flask and dilute to mark with distilled water.

Pipet 50.0 ml of this sample solution, which should not contain more than 30 μg PO_4 per ml, to a dry 125-ml conical flask. Add 2.0 ml of the molybdate reagent and mix thoroughly. Add 0.5 ml of the tin(II) chloride solution and mix thoroughly. After exactly 10 min, measure the absorbance or transmittance at either 650 nm or 820 nm using 1.000-cm cells. Compare with a calibration graph prepared in an analogous manner.

B. ORTHOPHOSPHATE IN WATER

The molybdovanadophosphoric acid method is applicable to the routine determination of orthophosphate in industrial waters, for example, boiler water (1). Tannin should be absent, and iron(II) should not exceed about 10 ppm.

Special Solutions

Molybdate–Vanadate Reagent. Dissolve 40 g ammonium molybdate, $(NH_4)_6Mo_7O_{24} \cdot 4H_2O$, in 400 ml water. Dissolve 1.0 g ammonium meta-vanadate, NH_4VO_3, in 500 ml of a 2:3 nitric acid solution. Add the molybdate solution to the acidic vanadate solution, mix thoroughly, and dilute to 1 liter with distilled water.

Standard Phosphate Solution. Dissolve 0.1433 g potassium dihydrogen phosphate, KH_2PO_4, in distilled water and dilute to 1 liter. Each ml contains 0.100 mg PO_4^{3-}

Procedure

Transfer a 50.0-ml aliquot of a filtered sample containing 0.1–1.2 mg PO_4^{3-} to a 125-ml conical flask. Add 25.0 ml of the molybdate vanadate reagent solution and mix thoroughly. After 5 min, measure the absorbance at 400 nm using 2.000-cm cells.

Calculate the ppm phosphate by reference to a calibration curve prepared using 50-ml aliquots of standard phosphate solutions prepared by diluting appropriate volumes (1–25 ml) of the standard phosphate solution to 100 ml.

C. PHOSPHORUS IN PLAIN CARBON STEELS AND CAST IRON

Phosphorus in steel has been determined by the heteropoly blue method (2,18,20,33,47), the extraction–heteropoly blue method (27,30,43,44,48,54), the extraction–molybdophosphoric acid method (62), the molybdovanado-phosphoric acid method (3,7,12,16,24,37,53,58), and the extraction–molybdovanadophosphoric acid method (6,59). Although the molybdovanadophosphoric acid method is rapid and often suitable for the determination of phosphorus in carbon steels containing no vanadium (24), the heteropoly blue methods are usually the methods selected because of their inherent sensitivity and tolerance to traces of silicon.

Special Solutions

Ammonium Molybdate Solution. Dissolve 20 g ammonium molybdate, $(NH_4)_6Mo_7O_{24} \cdot 4H_2O$, in 10$N$ sulfuric acid and dilute to 1 liter with 10N sulfuric acid.

Hydrazine Sulfate Solution. Dissolve 1.5 g hydrazine sulfate, $N_2H_6SO_4$, in water and dilute to 1 liter.

Standard Phosphate Solution. Dissolve 0.2292 g sodium monohydrogen phosphate, Na_2HPO_4, in about 500 ml water. Add 10 ml perchloric acid, dilute to 1 liter, and mix thoroughly (1 ml = 0.05 mg P).

Sodium sulfite Solution. Dissolve 100 g sodium sulfite, Na_2SO_3, in water and dilute to 1 liter.

Aqua Regia Reagent. Mix one volume of concentrated nitric acid with three volumes concentrated hydrochloric acid.

Ammonium Molybdate–Hydrazine Sulfate Reagent. This solution is unstable and should be prepared just prior to being used. Dilute 125 ml of the ammonium molybdate solution to about 400 ml. Add 50 ml of the hydrazine sulfate solution and dilute to 500 ml.

Procedure

Preparation of Calibration Graph. Transfer 1, 2, 3, 5, and 10 ml standard phosphate solution (1 ml = 0.05 mg P) to six 100-ml volumetric flasks, and have a seventh flask for the blank solution. Add 20 ml perchloric acid to each of the seven flasks, dilute to volume with distilled water, and mix.

Pipet 10-ml aliquots of each solution and transfer to 100-ml borosilicate glass volumetric flasks. Add 15 ml of the sulfite solution. Heat carefully each solution to boiling and boil gently for about 1 min. Add 50 ml of the ammonium molybdate–hydrazine sulfate reagent. Dilute almost to the mark with distilled water and mix thoroughly. Place each flask in a boiling water bath for about 15 min. Cool rapidly and dilute to the mark with distilled water. Measure the absorbance of each solution against water at 650 nm and 830 nm, using 1.000-cm cells. Subtract the absorbance of the blank solution from other absorbance values before constructing the calibration graph. The absorbance readings at 830 nm are applicable to samples containing 0.002–0.05% phosphorus, while the absorbance measurements at 650 nm are more applicable to samples in the 0.05–0.3% phosphorus range.

Procedure for Sample. Transfer a 1.000-g sample to a 250-ml conical flask. A reagent blank solution is also prepared using this same general procedure. Add slowly 15 ml aqua regia. When the reaction is complete, add 10 ml perchloric acid and evaporate to fumes of perchloric acid. Remove the flask at once, cool, and add 20 ml 1:4 hydrobromic acid. Evaporate the solution until the dense, white fumes are expelled from top of flask for about 1 min. Cool this solution and add 50 ml 1:4 perchloric acid solution. When metal salts have dissolved completely, transfer the solution to a 100-ml volumetric flask. Cool, dilute to the mark with distilled water, and mix thoroughly. Transfer two 10-ml aliquots of this sample solution to 100-ml volumetric flasks and a 10-ml aliquot of reagent blank solution to another 100-ml volumetric flask.

To the solution that is to be used as a background compensation reference solution, add 15 ml of the sulfite solution. Heat carefully to boiling and boil gently for about 1 min. Add 50 ml of a 1:12 sulfuric acid solution, cool, dilute to mark with distilled water, and mix well. This solution will be used in the reference cell when the absorbance of the sample solution is measured.

Add 15 ml of the sulfite solution to the sample solution and to the reagent blank solution and heat each solution to boiling. Add 50 ml of the ammonium molybdate–hydrazine sulfate reagent. Dilute almost to the mark with distilled water. Place both flasks in a boiling water bath for about 15 min. Cool each solution rapidly, adjust the volume to exactly 100 ml, and mix. Measure the absorbance of the sample solution at 650 nm and/or at 830 nm, using 1.000-cm cells and the background compensation reference solution in the reference cell. Measure the absorbance of the blank solution using water in the reference cell. Subtract the absorbance of the reagent blank solution from the absorbance obtained from the sample solution before referring to the calibration graph to determine phosphorus in 100 ml of the final solution being measured.

D. PHOSPHORUS IN ALLOY STEELS (61)

Special Solutions

Ammonium Molybdate Solution. Dissolve 50 g ammonium molybdate tetra-hydrate, $(NH_4)_6Mo_7O_{24} \cdot 4H_2O$, in 400 ml water. Add 115 ml concentrated sulfuric acid, cool, and dilute to 1 liter.

Stock Tin(II) Chloride Solution. Dissolve 10 g tin(II) chloride dihydrate, $SnCl_2 \cdot 2H_2O$, in 25 ml concentrated hydrochloric acid. Prepare fresh daily.

Tin(II) Chloride Reagent. Add 1 ml stock tin(II) chloride solution to 20 ml $1N$ sulfuric acid. Prepare this reagent just prior to use.

Ascorbic Acid Solution. Dissolve 100 g l-ascorbic acid in 400 ml water and dilute to 500 ml. Prepare fresh daily.

Standard Phosphate Solution. Dissolve 0.0439 g KH_2PO_4 in distilled water and dilute to 2 liters (1 ml = 5 μg P).

Procedure

Transfer a 0.5-g sample to a 125-ml conical flask. Add 15 ml 1:2 nitric acid–hydrochloric acid mixture and heat gently. After dissolution, add 15 ml perchloric acid and evaporate until perchloric acid fumes are evolved for 1 min.

If the alloy contains more than 0.5% chromium, the following additional treatment is necessary to remove chromium as chromyl chloride: Add 5 ml hydrochloric acid to the residue and evaporate just to fumes of perchloric acid. Repeat the hydrochloric acid treatment several times to remove chromium. Cool and proceed with the hydrobromic acid treatment.

If niobium, tantalum, and tungsten are present with more than 0.5% chromium, the following treatment must be utilized: cool the residue from the perchloric acid fuming and add 30 ml water. Remove carbides, nitrides, and hydrolyzed niobium, tantalum, and tungsten by filtering with a fine-porosity filter paper. Wash this precipitate with 1:20 perchloric acid and evaporate to fumes. Add 5 ml hydrochloric acid and evaporate to volatilize chromyl chloride.

Dissolve the residue in 10 ml water. (If any tungstic acid precipitates, remove by filtration.) Add 5 ml hydrobromic acid (48%) and evaporate until fumes of perchloric acid are evolved and a syrupy consistency is obtained. Add 15 ml of the ascorbic acid solution to complex titanium and allow the solution to stand 30 min. Transfer this solution to a 125-ml separatory funnel and add rinsings of 1:50 perchloric acid and dilute to 50 ml with the 1:50 perchloric acid.

Add 10 ml of the ammonium molybdate solution to the separatory funnel and swirl. After allowing to stand 15 min, add 20 ml isobutyl alcohol and shake for 1 min. Allow the phases to separate and discard the aqueous phase. After about 3 min, discard the residual aqueous phase that has formed. Add 15 ml of the tin(II) chloride reagent to the separatory funnel and shake for 30 sec. Allow the phases to separate and discard the aqueous phase. Transfer the isobutyl alcohol extract to a dry 50-ml volumetric flask. Use methanol to rinse the separatory funnel and dilute the solution to volume. After 1 hr, measure the absorbance at 725 nm.

Preparation of Calibration Graph. Use 0.5 g samples of phosphorus-free iron and follow the same procedure as used for the alloy. Spike samples after initial dissolution with 0, 1, 2, 3, 4, 5, 6, 7, and 8 ml of the standard phosphorus solution.

E. PHOSPHORUS IN COPPER-BASE ALLOYS

The molybdovanadophosphoric acid method is suitable for the determination of 0.01–1.2% phosphorus in copper-base alloys, provided the silicon and the arsenic contents do not exceed 1% (46,47).

Special Solutions

Standard Phosphate Solution. Dissolve 0.4394 g potassium dihydrogen phosphate in distilled water and dilute to 1 liter. Each ml of this solution contains 0.10 mg phosphorus.

Mixed Acid Reagent. Add 320 ml concentrated nitric acid and 120 ml hydrochloric acid to about 500 ml water and dilute to 1 liter.

Hydrogen Peroxide, 3%. Dilute 10 ml 30% hydrogen peroxide to 100 ml and store in an amber bottle.

Vanadate Reagent. Dissolve 2.5 g ammonium vanadate (NH_4VO_3) in 500 ml hot water. When dissolution is complete, add 20 ml 1:1 nitric acid, cool, and dilute to 1 liter.

Molybdate Reagent. Dissolve 100 g ammonium molybdate, $(NH_4)_6Mo_7O_{24} \cdot 4H_2O$, in 600 ml water at $50°C$ and dilute to 1 liter. Filter if necessary.

Pure Copper. Less than 0.002% P.

Procedure

Weigh 0.5–1 g sample, depending upon the phosphorus content, and transfer to a 150-ml beaker. Weigh an equivalent-size sample of pure copper and transfer to a 150-ml beaker. Use a buret to add 20.0 ml of the mixed acid reagent and a few silicon carbide particles to serve as boiling chips to each beaker. Heat the covered beakers until dissolution is complete. Add 1 ml 3% hydrogen peroxide, heat, and boil for only 3 min. After cooling the solutions, transfer to 100-ml volumetric flasks and add 10 ml of the vanadate reagent. Add 10 ml of the molybdate reagent, dilute to mark with distilled water, and mix. After 5 min, measure the absorbance at 470 nm, using 1-cm cells and the reagent blank solution in the reference cell.

Preparation of Calibration Graph. Transfer 0.5-g samples of the pure copper to nine 150-ml beakers. Use a buret to transfer 1, 2, 3, 4, 5, 8, 10, and 15 ml of the standard phosphate solution (1 ml = 0.10 mg P) to eight of the beakers. (If the phosphorus content of sample is about 1%, a standard phosphate solution three times as concentrated should be used.) Use a buret to add 15.0 ml of the mixed acid reagent to the nine beakers and add a few silicon carbide particles. Heat the covered beakers until dissolution is complete. Add 1 ml hydrogen peroxide and heat to boiling for only 3 min. Transfer each solution to a 50-ml volumetric flask. Add 5 ml of the vanadate reagent and 5 ml of the molybdate reagent and dilute to volume with distilled water. Mix and allow to stand 5 min. Measure the absorbance of the spiked solution at 470 nm, using 1.000-cm cells and the reagent blank solution in the reference cell. Plot the absorbance values versus μg phosphorus per ml final solution.

F. INORGANIC PHOSPHORUS IN SERUM

The heteropoly blue method is recommended because of its inherent sensitivity.

Special Solutions

Trichloroacetic Acid, 10%. Dissolve 100 g trichloroacetic acid to water and dilute to 1 liter.

Molybdate Reagent. Dissolve 25 g sodium molybdate dihydrate in $10N$ sulfuric acid and dilute to 1 liter with $10N$ sulfuric acid.

Hydrazine Sulfate Reagent, 0.075%. Dissolve 0.750 g hydrazine sulfate in water and dilute to 1 liter.

Standard Phosphate Solution. Dissolve 0.4395 g potassium dihydrogen phosphate in distilled water and dilute to 1 liter. Store in a polyethylene bottle. Each ml of this solution contains 0.1 mg phosphorus per ml.

Procedure

Pipet 9 ml of the 10% trichloroacetic acid into a small centrifuge tube. Add 1 ml serum and mix thoroughly. Centrifuge for 5 min. Pipet 5 ml of the supernatant solution and transfer to a 10-ml volumetric flask. Add 1 ml of the molybdate solution and 1 ml of the 0.075% hydrazine sulfate solution. Add water to the mark and mix thoroughly. Immerse the flask in a boiling water bath for 10 min. Remove the flask and cool rapidly. Adjust the meniscus with a few drops of water, if necessary, and mix well. Transfer the sample to a 1-cm absorption cell and measure the transmittance at 830 nm against a reagent blank. The transmittance can be measured at 650 nm, but the method is not as sensitive when measurements are made at this wavelength. The blank is prepared in an identical manner except that 1 ml water should be used instead of the serum. The mg-% phosphorus can be obtained from a calibration graph prepared by using 5 ml of the following solutions instead of the 5 ml of the supernatant solution used in the regular procedure.

Preparation of Calibration Graph. Transfer 90 ml 10% trichloroacetic acid to each of seven 100-ml volumetric flasks. Add 0, 1, 2, 4, 6, 8, and 10 ml of the standard phosphate solution (1 ml = 0.1 mg P) to the volumetric flasks. Dilute each flask to the mark with water and mix thoroughly. The solutions in these flasks correspond to 0.0, 1.0, 2.0, 4.0, 6.0, 8.0, and 10 mg-% phosphorus, respectively, when 5 ml of each solution is used in the procedure previously outlined.

G. TOTAL PHOSPHORUS IN BIOLOGICAL MATERIALS (4)

Special Solutions

Calcium Nitric Acid Reagent. Dissolve 50 ml calcium carbonate in 1 liter nitric acid.

Ascorbic Acid–Trichloracetic Acid Reagent. Dissolve 1.0 g ascorbic acid in 100 ml 10% trichloroacetic acid.

Ammonium Molybdate Solution. Dissolve 1.0 g ammonium molybdate tetra-hydrate, $(NH_4)_6Mo_7O_{24} \cdot 4H_2O$, in 100 ml distilled water.

Arsenite–Citrate Reagent. Dissolve 2.0 g sodium citrate dihydrate and 2.0 g sodium arsenite in 100 ml 2% acetic acid.

Procedure

Transfer 0.1 ml sample solution to a 25 × 150 mm borosilicate glass test tube. Add 2.0 ml of the calcium–nitric acid reagent and two boiling stones[1] and heat in a micro-Kjeldahl unit until all yellow fumes have disappeared. Cool and add 1.0 ml ascorbic acid–trichloroacetic acid reagent. Add 0.5 ml of the ammonium molybdate reagent and then 1.0 ml of the arsenite–citrate reagent. After 15 min, measure the absorbance at approximately 849 nm, using a reagent blank solution in the reference cell. A calibration graph should be prepared using 0.2–0.5 μg phosphorus in the sample solution taken (0.1 ml).

H. PHOSPHORUS IN ORGANIC COMPOUNDS (36,57)

Special Solutions

Special Digestion Reagent. Prepare phosphorus-free HI by adding 25 g iodine to 125 ml 47% HI containing 1.5% H_3PO_2 as preservative. Distill and collect the first 120 ml distillate. Add 50 ml of this distilled HI solution, 0.6 g $Ca(OH)_2$, 50 ml water, and 500 g phenol to a 1-liter volumetric flask. Dilute to volume with acetic acid.

Ammonium Molybdate Solution. Dissolve 2.5 g $(NH_4)_6Mo_7O_{24} \cdot 4H_2O$ in 1.25N H_2SO_4 and dilute to 1 liter with the 1.25N acid.

Hydrazine Sulfate Solution. Dissolve 0.6 g $N_2H_6SO_4$ in 1 liter water.

---

[1] Boiling stones: British Drug Houses antibumping granules are precleaned by boiling in nitric acid, rinsed with distilled water, and air dried.

Procedure

Transfer a sample containing 1–8 μg phosphorus to a 50-ml beaker and add 2 ml special digestion reagent. Cover the beaker and digest the solution on a steam bath until dissolution is complete, usually for 10–30 min. Remove the cover and expel excess solvents and reagents by volatilization on a hot plate. Complete the removal of carbonaceous residue by combustion for 5–10 min on a furnace maintained at approximately 700°C. After cooling, add 4 ml of the ammonium molybdate solution and 1 ml of the hydrazine sulfate reagent. Heat on a steam bath for 10 min, cool, and measure the absorbance at 830 nm, using a reagent blank solution in the reference cell.

I. TOTAL PHOSPHORUS (P_2O_5) IN FERTILIZERS

The molybdovanadophosphoric acid method is recommended for the determination of total phosphorus in most fertilizers (28). A compensation technique analogous to that used in the official A.O.A.C. method is employed in the following procedure.

Special Solutions

Standard Phosphate Solution. Dissolve 0.7669 g potassium dihydrogen phosphate, KH_2PO_4, in distilled water and dilute to 1 liter. Each ml corresponds to 0.40 mg P_2O_5.

Molybdovanadic Acid Reagent. Dissolve 20 g ammonium molybdate, $(NH_4)_6Mo_7O_{24} \cdot 4H_2O$, in 200 ml hot water and cool. Dissolve 1.0 g ammonium metavanadate, NH_4VO_3, in 125 ml hot water, cool, and add carefully 225 ml perchloric acid. Add slowly with stirring the molybdate solution to the vanadate solution. Dilute to 1 liter with distilled water.

Procedure

Transfer a 1.0-g sample of fertilizer to a 100-ml Kjeldahl flask, add 25 ml nitric acid, and boil for at least 30 min. Cool, add 15 ml perchloric acid, and boil gently until the solution becomes almost colorless and copious fumes are evolved. Do not evaporate to complete dryness. Cool, add 50 ml distilled water, and boil again for about 5 min. Transfer to a 250-ml volumetric flask and dilute to volume with distilled water. (If the sample contains more than 3% phosphorus, further dilution will be necessary.)

Transfer a 5.0-ml aliquot to a 100-ml volumetric flask and add 5.00 ml of the standard phosphate solution (1 ml = 0.4 mg P_2O_5). To another 100-ml volumetric flask, add 5.00 ml of the standard phosphate solution. Dilute to about 50 ml with distilled water and use a buret to add 20.0 ml of the molybdovanadic acid reagent to each flask. Dilute to volume with distilled water, mix, and allow to stand 10 min. Use the solution containing only the standard phosphate solution spike in adjusting the 0 absorbance or 100% transmittance setting at 400 nm. Measure absorbance of the analytical sample against the same reference solution.

Preparation of Calibration Graph. Use a buret to transfer 5, 8, 10, 13, 15, and 17 ml of the standard phosphate to six 100-ml volumetric flasks. Dilute to 50 ml with water. Use a buret to add 20.0 ml of the molybdovanadic acid reagent to each flask. Dilute to volume with distilled water, mix, and allow to stand 5 min. Use the solution containing 5 ml standard phosphate solution in sample and reference absorption cells to adjust the 0 absorbance reading at 400 nm. Measure the absorbance of remaining standard solutions against this same solution in the reference cell. Plot the absorbance versus mg P_2O_5 per ml final solution after correcting for the 2.0 mg P_2O_5 in the reference by subtracting from the "total P_2O_5" added to each solution.

REFERENCES

1. American Public Health Assn., *Standard Methods for Examination of Water and Wastewater*, 13th ed., Washington, D.C., 1971, pp. 518–534.

2. American Society for Testing and Materials, *1974 Annual Book of ASTM Standards*, Part 12, Philadelphia, pp. 558–560.

3. Baghurst, H. C., and V. J. Norman, *Anal. Chem.*, **27**, 1070 (1955).

4. Baginski, E. S., P. F. Foa, and B. Zak, *Clin. Chem.*, **13**, 326 (1967).

5. Berenblum, I., and E. Chain, *Biochem. J.* (London), **32**, 286 (1938).

6. Bertolacini, R. J., and J. E. Barney, *Anal. Chem.*, **30**, 202 (1958).

7. Blazejak-Ditges, D., *Mikrochim. Acta*, **1972**, 65.

8. Boltz, D. F., and M. G. Mellon, *Ind. Eng. Chem., Anal. Ed.*, **19**, 873 (1947).

9. Boltz, D. F., and M. G. Mellon, *Anal. Chem.*, **29**, 749 (1948).

10. Boyer, W. J., *Amer. Soc. Testing Materials, Proc.*, **44**, 774 (1944).

11. Brabson, J. A., J. H. Karchmer, and M. S. Katz, *Ind. Eng. Chem., Anal. Ed.*, **16**, 553 (1944).

12. British Iron and Steel Research Assoc., *J. Iron Steel Inst.*, London, **209**, 364 (1971).

13. Center, E. J., and H. H. Willard, *Ind. Eng. Chem., Anal. Ed.*, **14**, 287 (1942).

14. Clabaugh, W. S. and A. Jackson, *J. Res. Nat. Bur. Stand.*, **62**, 201 (1959).

15. Djurkin, V., G. F. Kirkbright, and T. S. West, *Analyst*, **91**, 89 (1961).

16. Elwell, W. T., and H. N. Wilson, *ibid.*, **81**, 139 (1956).

17. Fontaine, T. D., *Ind. Eng. Chem., Anal. Ed.*, **14**, 77 (1942).

18. Gates, O. R., *Anal. Chem.*, **26**, 730 (1954).

19. Greenberg, A. E., L. W. Weinberger and C. N. Sawyer, *ibid.*, **22**, 499 (1950).

20. Hague, J. L. and H. A. Bright, *J. Res. Nat. Bur. Stand.*, **26**, 405 (1941).

21. Halman, M., Ed. *Analytical Chemistry of Phosphorous Compounds,* Wiley—Interscience, New York, 1972.

22. Harrach, G., and D. F. Boltz, private communication.

23. Hayashi, K., T. Danzuka, and K. Ueno, *Talanta* **4**, 244 (1960).

24. Hill, U. T., *Ind. Eng. Chem., Anal. Ed.*, **19**, 318 (1947).

25. Hillebrand, W. F., G. E. F. Lundell, H. A. Bright, and J. L. Hoffman, *Applied Inorganic Analysis*, 2nd ed., Wiley, New York, 1953, p. 70.

26. Hoffman, E., and A. Saracz, *Z. Anal. Chem.*, **190**, 326 (1962).

27. Holroyd, A., and J. E. Salmon, *J. Chem. Soc.*, **1957**, 159.

28. Horowitz, W., Ed., *Official Methods of Analysis of the Association of Official Analytical Chemists,* 12th ed., 1975, pp. 11—12.

29. Hurford, T. R., and D. F. Boltz, *Anal. Chem.*, **40**, 379 (1968).

30. Inokuma, S., T. Tsuchiga, and S. Isomura, *Bunseki Kagaku*, **15**, 1378 (1966).

31. Jakubiec, R. J., and D. F. Boltz, *Mikrochim. Acta,* **1969**, 181.

32. Jakubiec, R. J., and D. F. Boltz, *ibid.,* **1970**, 1199.

33. Katz, H. L., and K. L. Proctor, *Anal. Chem.*, **19**, 612 (1947).

34. Kinnunen, J., and B. Wennerstrand, *Chemist-Analyst*, **40**, 35 (1951).

35. Kirkbright, G. F., and J. H. Yoe, *Talanta*, **11**, 415 (1964).

36. Kirsten, W. J., *Microchem. J.,* **12**, 307 (1967).

37. Kitson, R. E., and M. G. Mellon, *Ind. Eng. Chem., Anal. Ed.*, **16**, 379 (1944).

38. Klitina, V. I., F. P. Sukakov, and I. R. Alimarin, *Zh. Anal. Khim.*, **20**, 1145 (1965).

39. Koenig, R. E., and C. R. Johnson, *Ind. Eng. Chem., Anal. Ed.*, **14**, 155 (1942).

40. Lee, A., and D. F. Boltz, *Microchem. J.*, **17**, 380 (1972).

41. Lenz, W. R., Jr., and D. F. Boltz, private communication.

42. Levine, H., J. J. Rowe, and F. S. Grimaldi, *Anal. Chem.*, **27**, 258 (1955).

43. Lounamaa, N., and W. Fugmann, *Z. Anal. Chem.*, **199**, 352 (1963).

44. Lueck, C. H., and D. F. Boltz, *Anal. Chem.*, **28**, 1168 (1956).

45. Lueck, C. H., and D. F. Boltz, *ibid.*, **30**, 183 (1958).

46. Lutwak, H. K., *Analyst*, **78**, 661 (1953).

47. Maekawa, S., and K. Kato, *Bunseki Kagaku*, **15**, 957 (1966).

48. Maekawa, S., and K. Kato, *ibid.*, **17**, 597 (1968).

49. Maksimova, N. V., and M. T. Kozlovskii, *Zh. Anal. Khim.*, **2**, 353 (1947).

50. Michelson, O. B., *Anal. Chem.*, **29**, 60 (1957).

51. Mission, G., *Chemiker-Ztg.*, **32**, 633 (1908).

52. Morris, H. J., and H. O. Calvary, *Ind. Eng. Chem., Anal. Ed.*, **9**, 447 (1937).

53. Murray, W. M., Jr., and S. E. Q. Ashley, *Ind. Eng. Chem., Anal. Ed.*, **10**, 1 (1938).

54. Pakalns, P., *Anal. Chim. Acta*, **40**, 1 (1968).

55. Paul, J., *Mikrochim. Acta*, **1965**, 830.

56. Robinson, R. J., *Ind. Eng. Chem., Anal. Ed.*, **13**, 465 (1941).

57. Saliman, P. M., *Anal. Chem.*, **36**, 112 (1964).

58. Schwarz, H., *Mikrochim. Acta*, **1969**, 677.

59. Shimanuki, T., *Bunseki Kagaku*, **16**, 1136 (1967); *ibid.*, **17**, 928 (1968).

60. Stoll, K., *Z. Anal. Chem.*, **112**, 81 (1938).

61. Theakston, H. M., and W. R. Bandi, *Anal. Chem.*, **38**, 1764 (1966).

62. Wadelin, C., and M. G. Mellon, *ibid.*, **25**, 1668 (1953).

CHAPTER

10

SELENIUM AND TELLURIUM

K. L. CHENG

Department of Chemistry, University of Missouri-Kansas City, Kansas City, Missouri

and

RALPH A. JOHNSON

Shell Development Co., Houston, Texas

PHYSICAL AND CHEMICAL PROPERTIES OF SELENIUM AND TELLURIUM

The most important practical methods for determining small quantities of selenium and tellurium are the photometric methods. The chemistry of selenium and tellurium has been neglected in the past, so that relatively few analytic methods for selenium and tellurium are known.

On the basis of reaction modes, all the photometric methods for selenium and tellurium may be classified into four groups: (1) the sol methods, based on the reduction of selenium and tellurium to their elemental states; (2) the complex methods, based on the complex formation with inorganic or organic ligands; (3) the redox methods, based on the formation of colored products of dyes by redox reactions; (4) the miscellaneous methods, including the catalytic reactions, precipitation reactions, etc.

Significant developments include photometric methods for (1) the determination of selenium based on the use of the ortho-substituted aromatic diamines (aliphatic ortho diamines do not form colored complexes with selenium) and (2) the determination of tellurium using bismuthiol II. In both cases, ethylenediaminetetraacetic acid (EDTA) or diethylenetriaminepentaacetic acid (DTPA) has been used to complex interfering metal ions.

In this chapter, the important methods belonging to these four groups based on reaction modes are covered. A few books (4,9,31) dealing with general aspects of selenium and tellurium and reviews on selenium and tellurium methods (6,64) should be mentioned.

PHYSICAL AND CHEMICAL PROPERTIES OF SELENIUM AND TELLURIUM

Selenium and tellurium belong to Group VIA of the periodic table, the sulfur subgroup, the members of which are indiscriminately referred to as the chalcogens or chalcogenins.

The stable form of selenium is the grey, semimetallic form, which consists of infinite helical chains of selenium atoms. Two red modifications can be obtained from evaporating carbon disulfide solutions of selenium. Both consist of cyclic selenium molecules. Only one crystalline form of tellurium is known. It is the silvery, semimetallic form, which is isostructural with grey selenium. The melting points of selenium (grey) and tellurium are 220.5°C and 450°C, respectively. The boiling points are 685°C and 1390°C, respectively.

Selenium and tellurium have six electrons in their ns and np valence shells, that is, those which, when completed, bring the configuration to that of the next noble gas. Therefore, their chemistry shows a markedly nonmetallic character. The properties of selenium and tellurium lie between those of the nonmetals, oxygen and sulfur, and those of the more definitely metallic polonium. Although there is much similarity in the chemistry of the Group VIA elements, the display of metallic behavior becomes appreciably greater in passing down the group.

Volatile hydrogen selenide, H_2Se, and hydrogen telluride, H_2Te, are produced by strong reducing agents or acidification of the alkali or alkaline earth-metal selenides and tellurides. Most metals react directly with selenium and tellurium to form the corresponding selenides and tellurides. The compounds are principally ionic, containing Se^{2-} and Te^{2-} ions. The principal oxidation states of selenium and tellurium, however, are 4^+ and 6^+. Selenium dioxide, SeO_2, (subl. pt. ~320°C mp 340°C) dissolves in water to form the moderately strong oxidizing agent selenious acid, H_2SeO_3 ($K_1 = 2.7 \times 10^{-3}$, $K_2 = 2.5 \times 10^{-7}$). Tellurium dioxide, TeO_2, is nonvolatile and very slightly soluble in water (15), but dissolves in base to give tellurite. Like aluminum, tellurium has amphoteric character (15). The stabilization of tellurium(IV) in solution is effected by complexation, usually by the halide ions. Citric acid and tartaric acid are complexing agents for tellurium. Ethylenediaminetetraacetic acid (EDTA) forms complexes with neither selenium nor tellurium (15). Selenium trioxide has never been prepared pure, but selenic acid, H_2SeO_4 ($K_1 > 1$, $K_2 = 8.9 \times 10^{-3}$), is formed by the oxidation of selenious acid and is similar to sulfuric acid. Telluric acid, H_6TeO_6, is a weak dibasic acid ($K_1 \sim 10^{-7}$) which may be dehydrated to TeO_3.

Selenium forms a brown-yellow liquid selenium monochloride, Se_2Cl_2. Selenium(IV) chloride, $SeCl_4$, is a colorless solid completely dissociated in the vapor to selenium(II) chloride and chlorine but much more stable than sulfur(IV) chloride. There seems to be no tellurium monochloride, Te_2Cl_2. Tellurium, on the other hand, does form a solid dichloride, $TeCl_2$, in which the bonds are presumably more stable. Tellurium(IV) chloride, $TeCl_4$, is, as expected from the lower electronegativity of tellurium and therefore greater bond polarity, much more stable than either sulfur(IV) chloride or selenium(IV) chloride.

Selenium monobromide, Se_2Br_2, is a dark-red liquid, but no corresponding Te_2Br_2 is known. Selenium(II) bromide, $SeBr_2$, is only known in the vapor state, as a decomposition product of selenium(IV) bromide. Tellurium(II) bromide, $TeBr_2$, is very unstable and rapidly disproportionates to tellurium and tellurium(IV) bromide.

Selenium(IV) bromide, $SeBr_4$, is a rather unstable, low-melting solid. In the vapor state, it is completely dissociated to selenium(II) bromide and bromine. It reacts with hydrogen bromide or bromides to form hydrogen hexabromoselenate(IV), H_2SeBr_6, or its salts. Tellurium(IV) bromide, $TeBr_4$, has many more polar bonds, is much more stable, and forms $HTeBr_5$ and H_2TeBr_6 complexes and their salts. The chlorides and bromides of selenium(IV) and tellurium(IV) are soluble in some organic solvents and have been used for their separation.

Tellurium(IV) iodide, TeI_4, in which the bonds are not very polar but are between atoms of nearly the same radius, is a black solid. It is sparingly soluble in water and is hydrolyzed rapidly only when the water is heated. Tellurium(IV) iodide reacts with potassium iodide to form the stable potassium hexaiodotellurate(IV).

Both selenium and tellurium form yellow precipitates with sulfide having various compositions. They react with sulfur trioxide to form a green solid, $SeSO_3$, and a red compound, $TeSO_3$, respectively. Selenic acid dissolves sulfur, selenium, and tellurium to form blue, green, and red solutions, respectively. Many organic thiocompounds such as dithiol, thiocarbamates, and bismuthiol II form colored complexes and precipitates with selenium and tellurium. Selenium reacts with thioamino acids such as methionine, cysteine, and cystine, but tellurium does not.

Aromatic ortho diamines react with selenium(IV) to form usually yellow complexes. The diamino group in the ortho position attached to a benzene ring (not pyridine ring or cyclohexane ring) seems to be a functional group which is reactive to selenium(IV) but not to tellurium.

Oxidation of selenium and tellurium to the quadrivalent state is easily effected by nitric acid. The sexivalent acids are quantitatively produced by strong oxidizing agents such as ceric and permanganate ions. The 6^+ state can easily be returned to the 4^+ state by boiling with dilute hydrochloric acid. A number of agents reduce the quadrivalent forms to the elements. Sols are produced in this reduction that are the basis for some of the colorimetric methods.

The selenocyanate ion, $SeCN^-$, may be formed by heating selenium with potassium cyanide. Many salts of selenocyanate are known as well as complexes similar to those formed by thiocyanate. Little is known about the formation of tellurocyanates.

I. SEPARATIONS

A. PRECIPITATION METHOD

Precipitation separations of selenium and tellurium usually involve reduction to their elemental states. The most electropositive metals interfere, some of them by reduction to the element and some by the formation of selenides and tellurides. For example,

$$2Au^+ + Te^{4+} + 3Sn^{2+} \longrightarrow Au_2Te\downarrow + 3Sn^{4+}$$

$$2Cu^{2+} + Te^{4+} + 4Sn^{2+} \longrightarrow Cu_2Te\downarrow + 4Sn^{4+}$$

The best-known precipitants for selenium and tellurium are sulfur dioxide, hydrazine, hydroxylamine, hypophosphorus acid, and stannous chloride. Of these, sulfur dioxide is perhaps the oldest and best established from the standpoint of purity of the element precipitated. From concentrated hydrochloric acid solutions, that is exceeding $8N$, selenium is precipitated in a highly pure form and free from tellurium (58). In $3-5N$ hydrochloric acid, both tellurium and selenium are quantitatively precipitated by sulfur dioxide (57). In the sulfur dioxide precipitation of tellurium, hydrazine is an effective promoter. Coprecipitation, especially of copper, cadmium, bismuth, antimony, tin, and molybdenum, increases with decreasing acidity (8). Nitric acid should be removed before sulfur dioxide reduction. If bromine is present, it is reduced in the first moment of sulfur dioxide addition preliminary to precipitation of the element.

In the separation of selenium, it is important not to permit the temperature to rise above $30°C$ because of the coprecipitation that comes about as the amorphous red selenium is transformed into the grey form (58). In warm solution, volatile selenium monochloride is always formed as an intermediate and can be lost unless certain precautions are observed (59). Hence, even at room temperature, a large excess of reducing agent should be used to reduce the possibility of formation of the monochloride.

Stannous chloride has the advantages of being more rapid in its action of being effective in the presence of considerable amounts of nitric acid and of being readily available and easily handled (23,79,80). The precipitate from stannous chloride reduction is more subject to coprecipitation. However, Noakes (70) recommends a preliminary separation with stannous chloride instead of evaporation to destroy nitric acid, when nitric acid can interfere. Crossley (23) eliminates coprecipitation of copper by complexing it with thiourea before reduction with stannous chloride.

Hydroxlamine quantitatively and selectively precipitates selenium in the presence of tellurium from a hot solution of 17% hydrochloric acid ($d = 1.085$) or in the presence of citric or tartaric acid (60).

Selective reduction by iodide has been used to separate selenium from tellurium preliminary to the colorimetric determination of the latter (46).

To facilitate quantitative separation of the minimum amounts of precipitate encountered in trace determinations of tellurium and selenium, the following mechanisms have been used:

1.   Collection on asbestos in a microfilter or on a sintered-glass or platinum filter crucible.

2.   Use of a coprecipitating collector to increase the precipitate bulk. Tellurium is collected by added selenium during reduction by hypophosphite (82). After filtration, the precipitate is dissolved in bromine—hydrobromic acid and the selenium removed by evaporation. Ferric and aluminum hydroxides coprecipitate tellurous and selenious acids. Ferric hydroxide has long been used in preliminary separations in the determinations of tellurium and selenium in various copper analyses (26,79,80). The suitability of coprecipitation on aluminum hydroxide for separations of trace concentrations is indicated by the data of Aaremae and Assarsson (1) in which good recoveries are obtained with as little as 2 ppm tellurium. They found the recoveries to be low when ferric hydroxide was used. Coprecipitation of traces of selenium by ferric hydroxide is pH dependent (see Fig. 1).

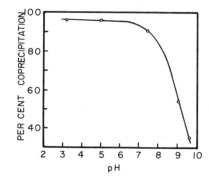

Fig. 1. Effect of pH upon coprecipitation of traces of selenium on ferric hydroxide.

3.   Extraction—analytic floatation separates selenium from the aqueous phase (46). Chloroform quantitatively extracts and floats selenium at the solvent interface from aqueous solution of tellurium(IV) and other interferences.

Tellurium may be easily precipitated at pH 5 in the presence of ethylenediaminetetraacetic acid (EDTA). The precipitation of tellurous acid in the presence of EDTA is highly selective, free from contamination of polyvalent metal ions and selenium (15). Selenium may be selectively precipitated with 2,3-diaminonaphthalene in the presence of EDTA.

Selective complexation has been used to separate tellurium and bismuth. Iodobismuthite is selectively formed in 0.1M iodide solution and may be extracted into a suitable solvent, for example, 3:1 amyl alcohol—ethyl acetate, leaving quadrivalent tellurium in water (46). From 0.6M iodide, tellurium(IV) can be extracted by 1:2 ethyl ether—amyl alcohol (33).

B. VOLATILIZATION METHOD

Selenides and tellurides can be volatilized in a stream of chlorine when heated. The quadrivalent salts can be distilled in a current of HCl gas (26,27,35). The volatility of the tellurium halides is greatly decreased in the presence of sulfuric or phosphoric acid or by dilution with water. The volatilization of selenium with hydrobromic acid—bromine solution or sulfuric acid or perchloric acid is well known. However, arsenic, germanium, mercury and antimony are also distilled with selenium. Cheng (17) separated selenium and mercury from tellurium by vaporization.

C. ION EXCHANGE METHOD

Resolution of selenium and tellurium mixtures by ion exchange is possible using Amberlite IR-120 in a solution of 0.3N hydrochloric acid (2). Also, tellurium may be separated on an anion exchange resin, Dowex 1, from antimony and tin (81). Ion exchange separation is possible with tellurium(IV) and iodide being adsorbed on an anion exchange resin, while tellurium(VI) passes through. Iodide is then eluted by a weakly acid solution and tellurium(IV), at higher acidity (39).

D. EXTRACTION METHOD

Both selenium and tellurium form yellow complexes with thiocarbamates that can be extracted into organic solvents such as chloroform, carbon tetrachloride, etc. Selenium may be selectively extracted as piazselenol into benzene or toluene at pH 7 in the presence of EDTA.

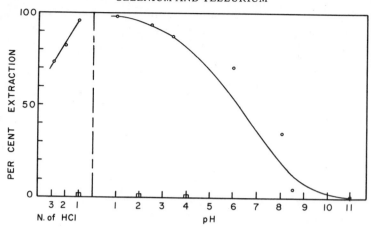

Fig. 2. Extraction of carrier-free tellurium with dithizone in carbon tetrachloride (64): (○) extraction with $1.8 \times 10^{-3} M$ dithizone in carbon tetrachloride; (□) extraction with only carbon tetrachloride ($V_0 = 2$ ml; $V_w = 10$ ml).

Tellurium may be extracted into amyl alcohol as the iodide. In approximately $4N$ hydrochloric acid solution in the presence of bromine, tellurium, along with gold, iron, molybdenum, and gallium, are extracted with ethyl acetate. From the organic phase, iron, gallium, molybdenum, and tellurium are back extracted with $0.3-0.4N$ hydrochloric acid. In an approximately $1.5N$ hydrochloric acid solution containing thiocyanate, iron, gallium, and molybdenum can be separated from tellurium solutions by extraction with ethyl acetate. With bis(2-ethylhexyl) orthophosphate and n-heptane or toluene, tellurium can be separated from molybdenum, niobium, rare earths, thorium, uranium, neptunium, plutonium, and zirconium.

Tellurium(IV) may be extracted as a yellow tellurium–dithizone complex into carbon tetrachloride (77). This extraction should be done in a dilute hydrochloric acid solution or a pH 1, as shown in Fig. 2. Only small amounts of tellurium can be quantitatively extracted with carbon tetrachloride solutions, of dithizone   (Fig. 3).   Selenium(IV)   can   also   be   extracted   using dithizone (74,75,83).

Byrne and Gorenc (11) studied the extraction of many metal ions by iodide with toluene and found selenium to be a special case, since on addition of selenium(IV) solution, a selenium precipitate is produced by reduction with hydriodic acid. The amorphous form is soluble in toluene, and hence "extraction" occurs. When addition of $12N$ to as low as $0.2N$ hydrochloric acid and potassium iodide molarities of $1.0M$ and $0.05M$ are used, the percentage of selenium found in the organic phase is always better than 99%. The behavior is observed for a selenium concentration of 20 ppm in the aqueous phase, namely;

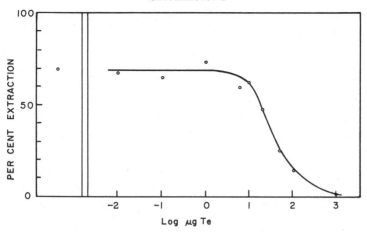

Fig. 3. Extraction of 10^{-5} to 1 mg tellurium with $6 \times 10^{-4}M$ dithizone in carbon tetrachloride showing the effect of tellurium concentration (64) (V_0 = 2 ml; V_w = 10 ml).

2.5×10^{-4}M. When larger quantities than 50 ppm selenium are present, it partially or completely precipitates from the 10 ml organic phase. The solution is not stable and tends to precipitate in the presence of foreign ions (for example, from wet-ashed material) or on standing.

Zmbova and Bzenic (97) described the use of a mixture of ethyl acetate and methyl isobutyl ketone for separation of tellurium and iodine. Iodine is extracted from $0.1M$ hydrochloric acid solution in the presence of hydrogen peroxide; tellurium(IV) is extracted from $5N$ hydrochloric acid, and tellurium(VI) remains in the aqueous solution. An enhancement of the solvent extraction seems to occur when this mixture of organic solvents is employed. It appears to be a synergetic effect; one of the extractants neutralizes the charge on the ion, thus forming a neutral complex, while the other solvent displaces a water molecule coordinated in the complex thus making it less polar in the aqueous phase.

Hikime et al. (34) found that tellurium(IV) complex with thiourea is extracted by tributyl phosphate (TBP) from a dilute hydrochloric acid solution in the presence of potassium thiocyanate and that the yellow color develops more intensely in the TBP phase than in the aquesous solution. When potassium thiocyanate or thiourea alone is present in the solution, no tellurium is extracted by TBP. However, when a large excess of thiocyanate ions is present and the hydrochloric acid concentration is higher than $0.1N$, small amounts of tellurium can be extracted by TBP. The colored tellurium thiourea complex in TBP may be used as a basis for the photometric determination of tellurium.

Iodine is extracted with TBP from $0.2N$ hydrochloric acid in the presence of hydrogen peroxide, and tellurium(IV) and (VI) remain in the aqueous phase. When the hydrochloric acid concentration is increased to 4M, tellurium(IV) can be extracted with TBP, while tellurium(VI) remains in the aqueous phase (38).

II. METHODS OF DETERMINATION

Standard Solutions of Tellurium and Selenium

Tellurium Standard Solution, 100 ppm (from Tellurium Metal). Solutions are easily prepared by dissolving pure tellurium in a nitric–hydrochloroc acid mixture. The final solution contains small amounts of nitric acid, which will not interfere in the stannous chloride reduction method or the iodotellurite method, but will interfere in the hypophosphorous acid reduction method because of the elevated temperature used. Accurately weigh a 1.0-g sample of pure tellurium and transfer to a 400-ml beaker. Add 50 ml concentrated hydrochloric acid, cover with a watch glass, and warm the solution. Add small portions of concentrated nitric acid to bring about the reaction. After the tellurium is dissolved, add 100 ml water and boil gently for 5 min to remove oxides of nitrogen. Add 30 ml hydrochloric acid and dilute to 1 liter. The resulting solution is 1000 ppm in tellurium and about 1N in hydrochloric acid. Prepare the solution for making up calibrating standards by diluting the stock solution tenfold, namely, to 100 ppm in tellurium and 1.0N in hydrochloric acid.

Tellurium Standard Solution, 100 ppm. (from Potassium Hexabromotellurate(IV)). Potassium hexabromotellurate(IV) is crystalline and dissolves easily in dilute hydrochloric acid. It is a very convenient form of tellurium(IV) for use as a reference standard. It is especially useful when a nitric acid-free standard solution is desired. Dissolve 1.074 g K_2TeBr_6 in 100 ml 4N hydrochloric acid and dilute to 2 liters. The resulting solution is 100 ppm in tellurium and 0.2N in hydrochloric acid.

Preparation of Potassium Hexabromotellurate (IV) (9). Dissolve 0.5 g tellurium dioxide in 5 ml 40% hydrobromic acid. To this solution, add two mole equivalents of potassium bromide in saturated aqueous solution (0.75 g potassium bromide in 1.4 ml water). Evaporate on a steam bath with stirring until the orange hexabromotellurate crystals settle rapidly. Recrystallize from boiling dilute (5–10%) hydrobromic acid. Dry in a vacuum desiccator first over soda lime, then over sulfuric acid. Yield is 1.60 g (75%).

Selenium Standard Solution, 200 ppm (from Selenium Dioxide). Selenium dioxide is readily available and is easily purified by sublimation.[1] It dissolves

[1] Resublimation of Selenium Dioxide (4): Pulverize the selenium dioxide and place it in an evaporating dish. Moisten with nitric acid. Into a funnel of somewhat larger diameter than the evaporating dish, insert a plug of glass wool. Nest a smaller funnel into it and invert

easily in water or dilute acid. It is deliquescent and must be dried over P_2O_5 before being weighed. Dissolve 0.281 g dried SeO_2 in water and dilute to 1 liter.[2]

Selenium Standard Solution, 200 ppm (from Calcium Selenite). Calcium selenite is a nonhygroscopic standard for selenium. Dissolve 0.424 g calcium selenite in 250 ml 0.2N hydrochloric acid and dilute to 1 liter.[2]

A. DETERMINATION OF TELLURIUM AND SELENIUM BY SOL METHODS

The sol methods differ from one another mainly according to the reductant used. Stannous chloride is a suitable reagent for both tellurium and selenium, whereas hypophosphorous acid is especially appropriate for tellurium and hydrazine has special advantages for selenium sol formation. Sols formed by stannous chloride probably contain substantial amounts of coprecipitated tin oxides, in this respect resembling the stannous-reduced gold sol, commonly called "purple of Cassius." They differ from "pure sols" such as those produced by hypophosphorous acid or hydrazine in physical characteristics and behavior. For example, stannous-reduced sols are more stable in moderately concentrated hydrochloric acid than sols formed with hypophosphorous acid. No systematic study of the stannous-reduced sols has been made. However, these sols are the basis for the oldest tellurium and selenium colorimetric sol methods (91,96). The successful applications of these methods reported in the literature testify to their value.

The hypophosphorous acid–reduced tellurium sols (32) are presumably sols of the pure elements, that is, there is a minimum of coprecipitation with these sols. Fundamental surveys of the effects of various reaction conditions upon the sol properties are available for these sols, which may be helpful in developing a method for new applications. The most striking effect observed in these studies is the dependence of position of the absorption band upon concentration of reductant; for example, tellurium sols vary from deep blue through purples and reds to amber with increasing concentration of reductant. From the practical analytic viewpoint, this means that conditions can be chosen to present an absorbance maximum to fit an available window in a medium complicated by various absorption bands (47). The studies have also proved that excellent

---

the pair over the dioxide. Heat with an open flame to sublime. CAUTION: Selenium fumes are poisonous. The sublimation temperature is 317°C. Keep the purified product in a desiccator with P_2O_5 or sulfuric acid.

[2] Particles of red selenium may appear in selenious acid solutions after long standing as a result of reduction by organic impurities.

reproducibility can be obtained with respect to particle and optical characteristics under appropriate conditions.

Adsorption of sol particles on glassware introduces an error that is significant in higher sol concentrations and is more noticeable in blue than in red sols. This adsorption can be practically eliminated by treating the glass with a suitable chlorosilane solution, for example, Desicote (Beckman Instrument Co., Pasadena, Calif.). The film is simply and readily applied and should be frequently renewed. It is recommended that the absorption cells be similarly treated.

1. Stannous Chloride Method for Tellurium and/or Selenium

In the formation of sols for analytic purposes, the stannous reduction is made at room termperature in $2-3N$ hydrogen ion concentration. These sols are relatively stable, resisting agglomeration by moderate concentrations of hydrogen ion and oxidation by dilute nitric acid. Sols stable for short periods of time can be prepared without a protective colloid and have somewhat larger absorptivity than sols prepared in the presence of a protective colloid. Addition of gum arabic, however, stabilizes the sols for several hours. The procedure given here is adapted from methods for tellurium and selenium in ores by Volkov (91) and Zemel (96) and by Crossley for tellurium in copper (23).

Reagents

Stannous Chloride, 10%. Dissolve 50 g stannous chloride in 50 ml $10N$ hydrochloric acid and dilute to 500 ml.

Gum arabic 4%. Dissolve 4 g gum arabic powder (*U.S.P.*) in 100 ml hot water. Centrifuge to remove large particulate matter. Prepare fresh daily.

Procedure

The solution to be reduced should contain $0.1-1$ mg tellurium(IV) or $0.2-2$ mg selenium(IV) in $3N$ hydrochloric acid and in a 25-ml volume (if nitric acid is present for sample solution, the hydrochloric acid concentration should be reduced accordingly). Add 2 ml 10% stannous chloride reagent with swirling. After the color is developed, add 3 ml 4% gum arabic. Dilute to 50 ml and read the absorbance using a light-blue filter and 1-cm cells. Stannous chloride absorbs strongly in the ultraviolet region at wavelengths below 290 nm. Carry a blank through the procedure and set the 100% transmittance against it. Prepare a calibration graph using standard solutions of tellurium or selenium.

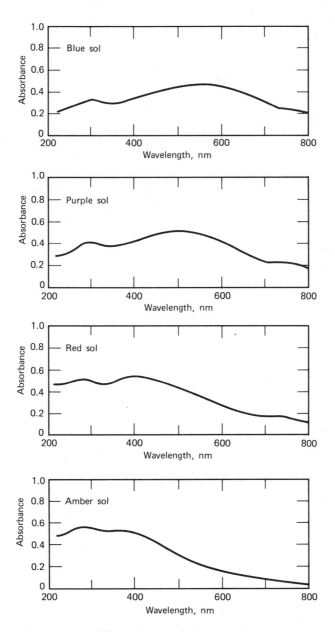

Fig. 4. Spectra of tellurium sols.

2. Hypophosphorous Acid Method for tellurium (45,47)

In hydrosols of pure tellurium, two absorption bands are found — one in the visible and one in the ultraviolet region (Fig. 4). The visible band shifts markedly from the red to the ultraviolet region with decreasing particle size, corresponding to shifts in the sol color from deep blue for particles of about 600 nm in length through purple and reddish hues to an amber color for particles of approximately 300 nm in length. There is a small but significant increase in absorptivity with decrease in particle size. The maximum of the ultraviolet band with respect to wavelength is almost independent of the particle size; the absorptivity, however, varies considerably with particle size, being about one third larger for red sols than for blue sols. Because of the variability of the optical properties with changing particle dimensions, frequent calibration checks are recommended until a method is well established in a laboratory and whenever changes are made that might affect particle formation. When the visible band is used, the choice of wavelength for reading absorbances must be chosen with regard to certain precautions for each set of conditions for sol formation. For this reason, the ultraviolet band is more convenient. Nevertheless, the visible band makes available a wide range of wavelengths for systems in which other absorbing substances may exist. For each of the absorption bands, the reproducibility of sols improves with decrease in particle size.

Sol properties are principally controlled by the concentration of reductant, as shown in Table I. The hydrogen ion concentration may vary between 0.01 and 0.3N. In more acidic medium, there is undue agglomeration. Within the prescribed acidity limits, the effect on the absorptivity is small but significant, and best precision is obtained if the acidity can be controlled between 0.1 and 0.2N [H^+].

TABLE I. Effect of Hypophosphorous Acid on Tellurium Sols

$M(H_3PO_2)$	Sol Color	Approximate wavelength of absorbance maximum (nm)
0.06	Blue	580
0.12	Purple	500
0.25	Red	400
0.4	Amber	350

For stabilizing the sols, 0.3 to 0.5% gum arabic is effective. Gelatin is an effective protective colloid at 0.01%. However, it can be used only for colorimetry in the visible because of its strong absorptivity in the ultraviolet. The reduction is highly temperature dependent. To minimize variation from this source, the reductions are carried out at the boiling point.

In the visible region, Beer's law is followed at all wavelengths. Estimates of standard error (in ppm Te) are 0.3 for blue sols, 0.2 for purple sols, and 0.1 for red sols. If the visible band is used for spectrophotometry, the variation of the band within a concentration series for any set of conditions should be taken into consideration. These changes are shown graphically in Fig. 5.

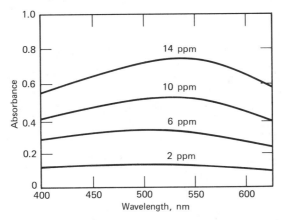

Fig. 5. Effect of tellurium concentration on visible band of purple sols.

In summary, as tellurium concentration increases, (a) the band becomes narrower, (b) the band shifts toward longer wavelengths, and (c) the absorptivity at the maximum increases. Hence, it is recommended that the wavelength for the standard working be chosen by one of the two following procedures: (1) Find the wavelength of maximum absorbance for the 10 ppm sol and use it for reading the other sols. (2) Make calibration graphs at three or four wavelengths in the vicinity of the maxima to determine the optimum wavelength for readings.

In the ultraviolet region, Beer's law is followed by blue sols in the wavelength region of 240–290 nm. The absorbance-versus-concentration plot for red sols is linear; however, the intercept on the concentration axis is slightly negative, that is, slightly positive on the absorbance axis.

The absorptivity for red sols at 290 nm is 53 and at 250 nm is 47. In this spectral region, the standard error is uniformly 0.15 ppm tellurium for red sols and 0.3 ppm tellurium for purple and blue sols.

Reagents

Hypophosphorous Acid, 3M. Dilute 32 ml purified 50% (9.5M) hypophosphorous acid with water to 100 ml. If the purified 50% acid is not available, use 60 ml 30% (5M) hypophosphorous acid.

Gum Arabic, 4%. See preceding method.

Procedure[3]

To a solution containing 0.1–0.7 mg tellurium(IV) and 1–8 mEq hydro-chloric acid in a 125-ml Erlenmeyer flask, add 3 ml 4% gum arabic and sufficient water to make the volume 35 ml. Heat to boiling. While rapidly swirling the mixture, add rapidly from a pipet 5 ml 3*M* hypophosphorous acid. Allow to digest near the boiling point for 15 min and cool in a bath of tap water for 15 min. Transfer to a 50-ml volumetric flask and dilute to volume. Read the absorbance at a wavelength in the region of 240–290 nm.[4]

3. Hydrazine Method for Selenium (32)

The absorption band for selenium sols is a very broad one, beginning at 600 nm and increasing steadily to a maximum at 250–260 nm (Fig. 6). In contrast to tellurium sols, the optical characteristics of selenium sols are

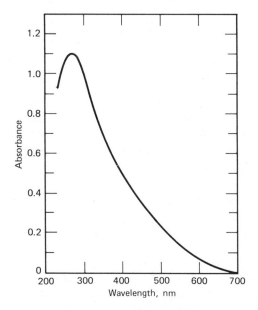

Fig. 6. Absorption spectrum of red amorphous selenium sol. Reduction of 20 ppm Se by hydrazine.

[3] This procedure yields red sols. To produce sols of other colors, vary the hypophos-phorous acid as indicated in the preceding discussion.

[4] If the 240–290 nm region is not available, longer wavelengths to approximately 420 nm may be used.

relatively independent of reductant concentration. Exceptional behavior is found only at excessive hydrazine concentrations or high alkalinities, in which cases the sol particles are of a different form and the sol presents a greyish color with a small absorption maximum at about 360 nm. Because hydrazine absorbs slighly at wavelengths below 260 nm, lower hydrazine concentrations are preferred and absorbance is measured at 260 nm

The reduction of selenium(IV) by hydrazine is acid catalyzed, the reaction being very slow at pH 11 or greater, taking place in 1 min at pH 9 and in a few seconds at lower pH values. Maximum uniformity of particle characteristics within individual sols is found in sols formed in the pH range of 8–11. Selenium sols are relatively stable in weakly basic medium, even in the absence of a protective colloid. Hydrogen ion is an effective agglomerant for selenium sols; hence, in acidic medium, the sols require protection and are sensitive to salt concentration. Unbuffered solutions of selenious acid assume approximately pH 9 upon addition of the hydrazine reagent. If there is appreciable acidity in the solution when it is ready for color development, it should be neutralized to pH 8–9 before adding the hydrazine.

The optimum hydrazine concentration range is $0.9-1.5M$. With a decrease in concentration below this range, there is a decrease in the rate of sol formation and in particle uniformity.

The selenium sols follow Beer's law. A standard error estimate based on 10 degrees of freedom is ±0.2 ppm. The molar absorptivity is 47 at 260 nm.

Reagent

Hydrazine Hydrate, 85%. *U.S.P.* reagent, $17M$.

Procedure

The solution to be analyzed should contain 0.1–1.0 mg quadrivalent selenium in 35 ml solution. The solution should be approximately neutral. To develop the color, heat the solution to the boiling point. Add rapidly from a pipet 2 ml 85% hydrazine hydrate while vigorously stirring the flask.[5] Continue heating for 2 to 5 min. Remove and cool the flask to room temperature in a water bath. Dilute to 50 ml in a volumetric flask. Measure the absorbance at 260 nm (or more generally in the region of 250–300 nm).[6]

---

[5] In the presence of interfering agglomerants, the sol may be stabilized by the addition of 3 ml 4% gum arabic (see tellurium sol methods).

[6] Longer wavelengths may be used with some sacrifice of sensitivity but little loss in precision or accuracy.

B. IODINE METHODS FOR TELLURIUM AND SELENIUM

1. Iodotellurite Method (29,46)

Formation of the iodotellurite complex is favored by increasing concentration of either iodide or hydrogen ion. In promoting the reaction, these two variables interact; that is, increase in iodide concentration enhances the effect of hydrogen ion, and vice versa. When tellurium(IV) is present in trace concentrations, the formation of a colored complex is very slight if $[H^+]$ and $[I^-]$ are less than $0.1N$. Johnson and Kwan (46) report a plateau for the intensity of the iodotellurite color as a function of $[H^+]$ and $[I^-]$ at $0.15–0.4N$ hydrochloric acid and $0.25–0.4N$ iodide. Higher absorbances are found at higher concentrations of these ions. Also, although the color develops instantly, a slow deepening of color occurs on standing (10). These effects suggest slow conversion to a higher complex, possibly $TeI_5^- + I^- \rightarrow TeI_6^{2-}$. Precipitation of tellurium(IV) iodide does not occur at the dilutions of hydrogen, tellurium, and iodide ions used in the spectrophotometric method (30).

The absorption spectrum of the iodotellurite complex shows a sharp absorbance maximum at 285 nm and a short plateau at 325–345 nm (Fig. 7).

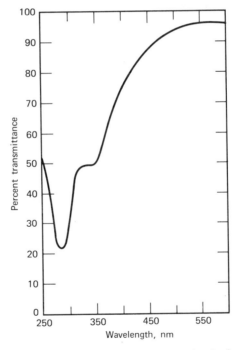

Fig. 7. Absorption spectrum of (1 ppm) iodotellurite complex in $0.4M$ potassium iodide and $0.2N$ hydrochloric acid.

The molar absorptivity of the maximum is roughly twice that of the plateau, the values being 700 and 325, respectively. Absorbances at the plateau are more reproducible than those at the maximum. Brown (10) found absorbances measured with a violet filter (maximum transmittance at 430 nm) to be highly dependent on the time of standing.[7] Results with a blue filter (maximum transmittance at 470 nm) were very unsatisfactory.

Reagents

Potassium Iodide, 2*M*. Prepare fresh daily.

Hydrochloric Acid, 2*N*.

Procedure

Place a sample containing 0.02–0.1 mg tellurium(IV) in a 50-ml volumetric flask. Dilute to about 30 ml. Add 5 ml 2*N* hydrochloric acid. Add with stirring 10 ml 2*M* potassium iodide. Dilute to the mark and mix. Read the absorbance at 335 nm in 1-cm cells within 20 min after color development.[8] As a reference solution, use a blank prepared as above, with the omission of the tellurite.

2. Iodine Method for Selenium (54)

The iodine formed in the reduction of selenious acid affords an index to the selenium present. In the triiodide form, it presents an absorption band in the ultraviolet with an absorbance maximum at 350 nm and a molar absorptivity of about 300 at this wavelength. At this wavelength, the colloidal selenium, which is also formed, contributes about 20% of the measured absorbance. When present as the starch complex, the iodine presents a maximum at 615 nm with a somewhat larger molar absorptivity, approximately 500. In this method, the system is a colloidal one involving elemental selenium either by itself or protected with starch. In the triiodide procedure of Lambert, Arthur, and Moore (54), agglomeration of elemental selenium occurs at selenium concentrations above 5 ppm (absorbance of 1.1). In the starch–iodine modification, agglomeration causes serious deviations at concentrations above 1 ppm (absorbance of 0.3). Geiersberger (29) stabilized the selenium sol by making it 0.1% in gelatin and found the $I_3^- - Se^0$ absorbances (presumably in the violet region) to follow Beer's law in the concentration range of 1–20 ppm. Also, 0.3% gum arabic should be effective in this method as in the selenium sol method.

---

[7] Brown (10) obtained satisfactory results with the above procedure using a violet filter only by standardizing the time interval between the reagent addition and absorbance reading. He used a 5-min interval.

[8] Exposure to sunlight should be minimized. The flasks may be of low actinic glass (amber or red) or may be painted to exclude light.

The selenious acid–iodide reaction is acid catalyzed. To attain completeness of reaction rapidly, Lambert et al. (54) use $2N$ sulfuric acid, and Geiersberger (29) uses $1N$ sulfuric. Decrease in hydrogen ion concentration below these levels is expected to decrease the reaction rate but to increase the stability of the resulting colloid. In the following procedure, the iodide concentration is maintained low, approximately $1N$, to control interference from air oxidation of iodide and from iodotellurite interference in the presence of tellurium(IV) up to concentrations of 2000 ppm. The air oxidation side reaction is further controlled by removal of oxygen by bubbling carbon dioxide through the solution and protecting the solution from light. Adsorption of elemental selenium on glassware can be avoided by "silicone" treatment (see sol procedures in the preceding section).

The following simple procedure is adapted from the procedures of Lambert et al. (54) and Geiersberger (29). The detailed Lambert method for trace selenium analysis in neutral waters is described under special applications. In the detailed procedure are found means to avoid most of the usual interference.

Reagents

Iodide Reagent, 5% Cadmium Iodide.[9] Dissolve 25 g CdI_2 in about 450 ml water. Dilute to 500 ml and store in a brown glass bottle.

Starch–Iodide Reagent (for the Starch–Iodine Modification), 1% Cadmium Iodide–0.25% Starch.[10] Dissolve 10 g cadmium iodide in 800–900 ml water. Boil for 15 min. Add slowly a thin paste containing 2.5 g soluble starch to the hot solution. Cool. Filter if turbid. Dilute to 1 liter and store in a dark bottle.

Sulfuric Acid, 5N.

Procedure

Transfer a sample containing $15-150\ \mu g$ selenium(IV)[11] to a 25-ml volumetric flask. (If the starch–iodine modification is used, take $10-40\ \mu g$ selenium(IV).) Adjust the solution to contain 25–50 mEq hydrogen ion. (If there is no strong acid already present, add 5–10 ml 5N sulfuric acid.) Pass carbon dioxide through the solution for 10 min. Bring the volume to 20–23 ml. Develop the color by one of the following methods:

Triiodide Method. Add 2 ml 5% cadmium iodide solution, dilute to volume, stopper, mix, and allow to stand for 5 min. Read the absorbance at 352 nm.

[9] According to Lambert et al. (54), cadmium iodide is a more stable iodide reagent than the commonly used alkali iodides.

[10] Cadmium iodide inhibits bacterial deterioration of the starch.

[11] If the selenium is present in the hexivalent form, reduction to the quadrivalent can be effected by boiling the solution for 1 hr after the addition of 1 ml saturated potassium bromide and 1 ml 1% sodium hypochlorite (see application to natural waters).

Starch—Iodine Method. Add 2 ml of the starch—iodide reagent, dilute to volume, stopper, mix, and allow to stand for 5 min. Read the absorbance at 615 nm.[12]

In both methods, carry a blank through all steps of the procedure and set the 100% transmittance reading with the blank in the reference cell. Prepare a calibration graph using the standard selenious acid solution.

C. 3,3'-DIAMINOBENZIDINE METHOD FOR SELENIUM (13,14)

Selenium(IV) forms complexes with many o-diamines (3). Particularly important is the formation of the yellow monopiazselenol with 3,3'-diamino-benzidine (DAB):

| DAB (colorless) | monopiazselenol (yellow) |

The monopiazselenol is formed when DAB is in excess; however, when selenium is in excess, dipiazselenol is formed. Both piazselenols are yellow and intensely colored. Hoste and Gillis (36,37) developed the monopiazselenol color in $0.1N$ hydrochloric acid in 50 min. In the Cheng improved method (13,14), the monopiazselenol is formed at pH 2.0–3.0 in 30 min and is then extracted at pH 5 or above by toluene. The extraction accomplishes (a) concentration of the monopiazselenol from the DAB, (b) separation from a number of interferences, and (c) solution of larger amounts of monopiazselenol, that is, the aqueous method has an upper concentration limit of 2.5 μg Se per ml imposed by the solubility limit of the monopiazselenol in water.

In aqueous solution, the monopiazselenol has a principal absorbance maximum at 340 nm and a maximum with low absorptivity at 420 nm (Fig. 8). In toluene, the absorbance maximum at 340 nm is retained and the band at 420 nm is strongly enhanced. The reagent, DAB, also has a principal absorbance maximum at 340 nm, but not at 420 nm in both water and toluene. Accordingly, absorbances of aqueous solutions are read at 340 nm and must be corrected for reagent absorbance. Absorbances of toluene solutions are read at 420 nm where the reagent absorbance is practically negligible.

The molar absorptivity values of monopiazselenol in aqueous solution and toluene are listed in Table II.

[12] Solutions of hexammine nickel ion serve as visual comparison standards. Directions for preparation of these standards are given by Lambert et al. (54).

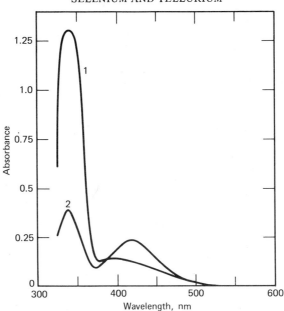

Fig. 8. Absorption spectrum of piazselenol: (1) 25 μg Se in 25 ml water; (2) 25 μg Se in 10 ml toluene.

The method is highly selective. Tellurium(IV) does not react. Vanadium(V), iron(III), copper(II), and other oxidizing substances interfere by forming colored oxidation products of the reagent. Hoste and Gillis (37) employ fluoride to mask iron and oxalate to mask copper. Cheng (15) masks all interfering ions except vanadium(V) and chromium(VI) with ethylenediaminetetraactic acid (EDTA).

TABLE II. Comparison of Molar Absorptivity Values

Aqueous solution	pH at color development	Se, μg/ml	ϵ
At 347–349 nm.	1.0	2.5	14,350
At 340 nm.	2.5	1.0	14,500
At 420 nm.	6.0	1.0	8,850
Toluene solution			
At 340 nm.	6.0	2.0	10,000
At 420 nm.	6.0	2.0	9,950

In the extraction step, colored ions are separated. Substances that reduce or complex selenium(IV) interfere, for example, stannous and iodide ions and ascorbic acid.

The following procedure developed by Cheng (13) is applicable in the presence of tellurium. For a procedure applicable in the presence of various metal ions, see the procedures for selenium in alloys and other materials given in this chapter.

Reagents

3,3′-Diaminobenzidine Hydrochloride, 0.3% in Water. Keep in a refrigerator. The reagent is commercially available (J. T. Baker Chemical Co., Phillipsburg, N.J.) or may be synthesized as described by Hoste (36). The salt is readily dissolved in water. Alternatively, a solid mixture of DAB and an indifferent salt (NH_4Cl) may be used.

Formic Acid, 2.5M.

Toluene, Reagent Grade.

Procedure

To a 100-ml beaker add an aliquot containing $1-100\,\mu g$ selenium(IV). Add 2 ml 2.5M formic acid and adjust the pH to 2–3. Dilute to approximately 50 ml with water. Add 2 ml 0.5% diaminobenzidine solution. Allow to stand for 30 to 50 min. Adjust the pH to 6–7 with dilute ammonia. Transfer to a 125-ml separatory funnel. Add 10.00 ml toluene from a pipet.[13] Extract the mono-piazselenol and determine its absorbance in the extract at 420 nm in a 1-cm cell. Use a reagent blank.

D. 1-PYRROLIDINE CARBODITHIOLATE METHOD FOR SELENIUM (61,63)

Selenium forms a light-yellowish complex with 1-pyrrolidine carbodithiolate in a weakly acidic medium. The complex is readily extractable by many organic solvents.

Procedure

To a 125-ml separatory funnel, add 10 ml buffer solution and the sample solution containing $0.01-0.25$ mg selenium(IV). Add 5 ml of the 0.15% carbodithiolate solution, mix, and allow to stand for 5 min. Extract for 1 min

---

[13] If less than 10 μg selenium is present, use only 6 ml toluene for the extraction and read the absorbance in a 5-cm cell.

with 25 ml chloroform. Transfer the organic extract to a 25-ml volumetric flask, dilute to volume with chloroform, and mix. After standing for 40–55 min, measure the absorbance at 303 nm against a reagent blank.

E. DIANTHRIMIDE METHOD FOR SELENIUM (55,56)

The organic reagent 1,1′-dianthrimide is commonly used for the photometric determination of small amounts of boron, the color of the boron–1,1′-dianthrimide complex being developed in concentrated sulfuric acid by heating. Germanium, tellurium, and selenium also react with this reagent. Langmyhr and Omang (56) reported a photometric method for selenium with 1,1′-dianthrimide, and later Langmyhr and Dahl (55) reported a second method for selenium with 2,2′-dianthrimide. The former reagent is more sensitive, but bismuth, cobalt, chromium, iron, mercury, nickel, lead, tin, and oxidizing agents interfere. The latter reagent is less sensitive but reacts only with selenium(IV).

2,2′-Dianthrimide reacts with selenium forming two species, the dimer Se_2Di_2 and the monomer $SeDi$, the dimer probably consisting of two monomers connected by hydrogen bonding. The dimer predominately exists in solutions of low total concentration and/or by decreasing the excess of 2,2′-dianthrimide. The formation constants for the monomer and the dimer complexes are 7.7×10^3 and 32×10^{12}, respectively (24).

Reagents

2,2′-Dianthrimide Solution, 0.05% (w/v). Dissolve 0.5000 g 2,2′-dianthrimide in 100 ml 96.2% sulfuric acid and dilute to 1 liter with 96.2% sulfuric acid.

Standard Selenium Solution. Selenium, 50 mg per ml in 1% sodium hydroxide solution.

Procedure

Transfer 1.0 ml 1% sodium hydroxide solution containing 1, 10, 20, 30, 40, and 50 mg selenium into 50-ml bottles. Add 5 ml 0.05% 2,2′-dianthrimide solution and 12 ml 99.9% sulfuric acid. The final sulfuric acid concentration is 95.5%. After heating the solution for 5 hr at 90°C, cool and measure the absorbance at 605 nm against a reagent blank. Beer's law is followed for 0–50 mg selenium in 18 ml.

For the 1,1′-dianthrimide procedure, 5 ml 0.5% 1,1′-dianthrimide and 20 ml 96% sulfuric acid are used. The solution is heated for 16 hr at 70°C.

F. DIPHENYLTHIOSEMICARBAZIDE METHOD FOR SELENIUM (84)

Selenium reacts with many hydrazine derivatives including 1,1′-diphenyl-hydrazine, 1-phenylthiosemicarbazide, 1,4-diphenylsemicarbazide, 1,4-diphenylthiosemicarbazide, diphenylcarbazide, and diphenylthiocarbazide to form colored products which are readily extracted by organic solvents. 1,4-Diphenylthiocarbazide has been used as an analytic reagent for copper, gold, and platinum. Selenium reacts with this reagent to form a red-brown 1:2 complex with absorbance maxima at 320 and 410 nm. The molar absorptivity is 44,000 at 320 nm and 6940 at 410 nm (85).

Because many metal ions including iron, copper, silver, mercury, and bismuth interfere, preliminary separation of selenium for its determination is necessary. Tellurium(IV) only reacts with this reagent in a very strong acid medium at $5N$.

Reagents

Standard Selenium Solution, $1 \times 10^{-4} M$.

Diphenylthiosemicarbazide Solution, $1 \times 10^{-2} M$. Dissolve an appropriate amount of 1,4-diphenylthiocarbazide in dilute 3:1 acetic acid. This solution is stable.

Procedure

Transfer various amounts of $1 \times 10^{-4} M$ selenious acid solution into 150-ml separatory funnels. Add 0.5 ml $1N$ hydrochloric acid and dilute to 9.5 ml. Add 0.5 ml of an acetic acid solution of diphenylthiosemicarbazide ($0.01M$). Mix and allow to stand for 10 to 15 min. Extract with 4 ml chloroform for 1–2 min. Measure the absorbance of the extract at 410 nm against a reagent blank.

G. METHOD FOR SELENIUM BASED ON NITRITE REACTION (71)

A photometric method for determining submicrogram concentrations of selenium is based on the oxidation of hydroxylamine hydrochloride to nitrous acid by selenious acid followed by the diazotization of sulfanilamide by the nitrite produced and subsequent coupling of the diazonium salt with N-(1-naphthyl)ethylenediamine dihydrochloride.

Reagents

Standard Selenium Solution.

Hydroxylamine Hydrochloride. Solutions of 0.1, 1, 5, 10, 12, 16, 20, 30, 40, and 50% are prepared by dissolving required amounts of hydroxylamine hydrochloride in water.

Sulfanilamide (p-Aminobenzenesulfonamide). Solutions of 0.05, 0.1, 1, 2, 3, 4, 5, 6, and 7% are prepared by dissolving required amounts of sulfanilamide in $6N$ hydrochloric acid solution. For higher concentrations of this reagent, it is necessary to effect solution by heating. When there is a visible residue after all of the sulfanilamide has dissolved, filter, while hot, through a fine-sintered glass funnel. The solution is stable for one month when kept tightly stoppered and in a refrigerator.

N-(1-Naphthyl)ethylenediamine Dihydrochloride. Solutions of 0.05, 0.1, 0.2, 0.3, 0.4, 0.5, and 0.6% are prepared by dissolving required amounts of this reagent in 1% (v/v) hydrochloric acid solution. The solutions are kept in a refrigerator in a lightproof container and tightly stoppered. They are stable for about two weeks.

Procedure

To 5 ml sample solution containing $0.01-0.20$ μg per ml selenium(IV) in lightproof reaction vessels, add 2 ml 10% hydroxylamine hydrochloride solution, 10 ml 12N hydrochloric acid solution, and 2 ml 6% sulfanilamide solution in 6N hydrochloric acid solution. Stopper the reaction vessels tightly and heat the reaction mixture in a thermostatically controlled water bath at $60 \pm 1°C$ for 90 min. Add 5 ml 0.1% N-(1-naphthyl)ethylenediamine dihydrochloride in 1% hydrochloric acid solution plus 1 ml water and continue heating 30 min in a constant-temperature water bath at $60 \pm 1°C$. Be sure that the reaction vessels are tightly stoppered. After completing the reaction, cool to room temperature and measure the absorbance at 554 nm. The amount of selenium is found from a calibration graph. Beer's law is followed for $0.01-0.20$ μg seleinium per ml.

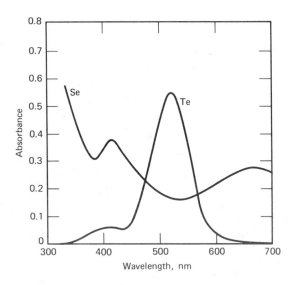

Fig. 9. Absorption spectra of Se·SO$_3$ and Te·SO$_3$ addition compounds (25).

H. M · SO₃ ADDITION COMPOUND METHODS (93)

Elemental tellurium and selenium dissolve in concentrated sulfuric acid forming colored addition compounds with the acid anhydride, namely Te · SO_3. The formation of the complexes and their stability are critically dependent upon temperature and time of heating. At lower temperatures, formation is incomplete and erratic; and at higher temperatures, fading ensues. In particular, the optimum conditions for formation of the tellurium complex involve heating at 100°C for 10 min; for selenium, heating for 15 min at 175°C is optimum. At the latter temperature, tellurium is converted to a colorless form. Consequently, selenium can be determined in the presence of tellurium, the results being approximately 5% high.

Selenium gives very erratic results in the determination of tellurium at 100°C so that a preliminary separation is required. The absorbance of the tellurium addition compound is read at the absorbance maximum, 520 nm (Fig. 9). For selenium, maxima occur at 420 nm and 680 nm, the larger absorptivity being at the former wavelength (Fig. 9). However, tests indicate that best precision is obtainable at a minimum in the absorbance curve at 350 nm. The molar absorptivity of the tellurium band is 15, and that of the selenium band (350 nm) is 4.

In most applications, the samples are first dissolved. Then, elemental selenium and tellurium are reduced and separated in finely divided form, which is suitable for reaction with sulfuric acid.

Reagents

Selenium, Pure, Red Powdered Selenium.

Tellurium, Pure, Powdered Tellurium.

Sulfuric acid, 98%, Reagent Grade.

Procedure

To a sample containing 2–10 mg selenium or 0.5–4 mg tellurium in finely divided elemental form, add 35 ml hot concentrated sulfuric acid (100°C for tellurium analysis and 175°C for selenium). Maintain at the specified temperature 10 min for tellurium and 15 min for selenium. Cool to room temperature and dilute to 50 ml with concentrated sulfuric acid. Read the absorbance for tellurium at 520 nm and for selenium at 350 nm in 1-cm cells.

I. THIOUREA METHOD FOR TELLURIUM

Tellurium reacts with thiourea in acid media to form a yellow complex which is sparingly soluble in water but dissolves readily in nonpolar solvents. A spectrophotometric study of complex formation gave a Te:Tu ratio of 1:4.

The intense yellow color produced in the reaction of tellurium(IV) with thiourea is useful for the colorimetric determination of tellurium in moderately concentrated strong acid solutions. Nielsch and Giefer have studied the spectrophotometry of the thiourea complex in nitric (68), sulfuric, and phosphoric acids (69). The recommended concentration ranges for acid and thiourea are given in Table III. Copper does not interfere. This feature, coupled with the high nitric acid concentration permitted, makes the method particularly useful in the analysis of copper (67).

TABLE III. Recommended Concentrations for the Thiourea Method (Final Concentrations)

Acid	Acid concentration	Thiourea, %
Nitric	0.8–1.6 N	10–12
Sulfuric	0.8–2.0 N	9–11
Phosphoric	0.3–1.5 M	10–12

The absorptivity of the complex increases rapidly in the violet to a maximum at 310 nm, attaining a molar absorptivity of 14 (Fig. 10). Beer's law is followed at the maximum and also at longer wavelengths. When measured spectrophotometrically at the absorbance maximum, a range of 1–40 ppm of tellurium is possible if both 5- and 1-cm cells are used. The colored product is stable for several hours.

Selenium(IV) and selenium(VI) interfere. Ovsepyan et al. (72,73) found that selenium reacts with thiourea to form a complex which is a selenium(II) derivative. In the reaction between selenious acid and thiourea, it has been found that thiourea reduces selenium(IV) to its elemental state which then dissolves in an excess of thiourea. A number of cations interfere: bismuth(III), antimony(III), tin(II), palladium(II), mercury(I), mercury(II), osmium(VIII), rhodium(III), and platinum(IV).

Tellurium(IV) reacts with thiourea in a few minutes at room temperature. Tellurium(VI) reacts only in hot solution. Jilek and Vrestal (42,43) developed a differential analysis for the two tellurium valence states using the tellurium thiourea reaction at room temperature and at the boiling point.

The following procedure is based on the studies of Nielsch and Giefer (68,69).

Procedure

The sample to be analyzed should contain 0.05–2 mg tellurium(IV) in 5–10 ml solution in a 50-ml volumetric flask. Add sufficient concentrated acid to bring the total acidity within the limits indicated in Table III, namely,

2.5–5 ml concentrated nitric acid, 1–3 ml concentrated sulfuric acid, or 1–5 ml 85% phosphoric acid.[14] Add 5 g thiourea.[15] Dilute with water to 50 ml. Prepare a reagent blank for comparison. Made absorbance readings at 310 nm on a spectrophotometer or with a violet filter in a photometer.

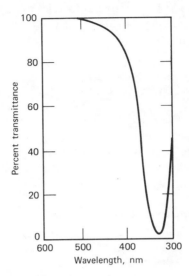

Fig. 10. Absorption spectrum of the tellurium–thiourea complex in phosphoric acid.

J. THIOUREA METHOD FOR TELLURIUM BY EXTRACTION WITH TRIBUTYL PHOSPHATE (34)

Tellurium(IV) forms a yellow complex with thiourea in a dilute hydrochloric acid solution and the complex is extracted into TBP phase when an excess of potassium thiocyanate is present. The tellurium–thiourea complex in TBP has strong absorptivity in the ultraviolet region, showing an absorbance maximum between 330 and 560 nm. The tellurium complex in TBP is more sensitive than that in a dilute hydrochloric acid solution.

The absorbance at 400 nm remains constant in the pH range 0.6–1.7. A pH 1.0 is recommended for the color development and TBP extraction. The yellow complex is very stable in TBP solution. Beer's law is followed for 1–30 ppm tellurium and for 1–80 ppm tellurium at 380 nm and 400 nm, respectively. Iron, bismuth, copper, vanadium(V), nickel, and selenium interfere. Preextraction of these interfering ions may be made prior to the extraction of tellurium.

[14] If the acid concentration is too high, thiourea is decomposed with evolution of gas.
[15] Thiourea must be added after the acid.

Procedure

To 15 ml of a solution containing up to 500 mg tellurium and approximately 0.1N hydrochloric acid, add 2 ml 50% potassium thiocyanate solution and 2 ml 8% thiourea solution. Adjust pH to 1 with ammonia solution and dilute the solution to 20 ml with 0.1N hydrochloric acid solution. After standing for 20 min, add 20 ml previously purified n-tributyl phosphate and shake for 2 min. Centrifuge the organic phase for 2 min and measure the absorbance at 400 nm.

K. HYPOPHOSPHOROUS ACID METHOD FOR TELLURIUM BY INDUCED PRECIPITATION OF GOLD (52,53)

An extremely sensitive method for the quantitative determination of tellurium is based on the induced precipitation of gold from a 6N hydrochloric acid solution containing gold chloride, copper chloride, and hypophosphorous acid. The amount of gold reduced is proportional to the amount of tellurium present. As little as 1 ng tellurium gives a measurable reaction with 1 mg gold in 50 ml solution.

The rate of the reduction of gold chloride from a 6N hydrochloric acid solution by hypophosphorous acid is governed not only by the amount of tellurium present but also by the concentrations of gold, copper, and hypophosphorous acid, as well as temperature. The reduced gold is filtered, dissolved, and determined photometrically according to the modification of the rhodamine B method (75).

Arsenic, selenium, mercury, and antimony also induce the precipitation of elemental gold, but much larger amounts of these elements are required. Bromides and iodides interfere by decreasing the sensitivity of reduction reaction. All these interferences may be eliminated by evaporating the sample solution with hydrobromic and sulfuric acids.

Procedure

To 50 ml 6N hydrochloric acid at 15°C are added, successively, accompanied by thorough mixing, 1 mg gold(III) chloride(1 g gold in 100 ml 6N hydrochloric acid), 6 mg copper as copper(II) chloride (3 g $CuCl_2$ in 100 ml 6N hydrochloric acid), 1 ml solution containing the sample dissolved in concentrated sulfuric acid, and finally 1 ml 50% hypophosphorous acid. At a fixed time (5 min) after addition of the hypophosphorous acid, the solution is filtered through a Millipore filter (0.45-μ pore size). The gold so collected is dissolved by hydrobromic acid containing a small amount of elemental bromine, and the amount of gold is determined photometrically by any sensitive method for gold such as the rhodamine B method (76) or the thio-Michler's ketone method (21). Alternatively, the amount of reduced gold in the sample solution can be determined turbidimetrically.

L. BISMUTHIOL II METHOD FOR TELLURIUM (20)

Bismuthiol II (5-mercapto-3-phenyl-1,3,4-thiadizole-2-thione) has been proposed as a reagent for determining tellurium independently by Jankovsky and Ksir (41) and Cheng (15). It was also studied by Yoshida et al. (95) and Navrtil and Sorfa (66).

In an acid medium (pH < 5), bismuthiol II forms with selenium(IV) and tellurium yellow precipitates extractable into various organic solvents. Jankovsky and Ksir found that the extraction of colored tellurium complex into benzene was quantitative when the pH of the aqueous phase is lower than 4.3, whereas the extraction of the similar selenium complex starts at pH 3.5. They did not study the kinetics of the reaction and extraction process. Cheng (16) found the optimum condition for the formation of the tellurium—bismuthiol II complex to be at pH 2.0–2.5, but this complex was extractable optimally into chloroform at pH 5–7. It was believed that the formation of nonextractable protonated colored chelate occurs in the first step. Due to the hydrolysis of tellurium at pH around 5 (15), the reaction between tellurium and bismuthiol II is slow. In alkaline medium, the tellurium—bismuthiol II complex may gradually decompose in favor of the formation of tellurite. Therefore, an acidic medium is favorable for the formation of the tellurium—bismuthiol II complex. In order to have a maximum masking effect of EDTA, the optimum pH should be as high as possible. A pH of 2.2 was chosen for this reason. It is assumed that nitrogen is easily protonated and that the extractable species should be neutrally charged.

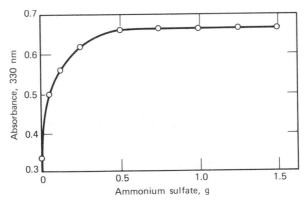

Fig. 11. Effect of ammonium sulfate on color development (0.2 micromole Te in 25 ml solution, extracted with 10 ml chloroform).

In the second step, the extractable complex in the organic phase is formed by deprotonation at increased pH according to the scheme.

$$[H_8 Te(Bis)_4]^{8+} + 8OH^- \rightleftharpoons [Te(Bis)_4]^0 + 8H_2O$$

Cheng found the composition of the complex with a ratio of 1:4 by the use of Job curves. This was confirmed by Yosida et al. (95). The structure of the tellurium–bismuthiol II complex proposed by Cheng was not confirmed by Yoshida et al. and Navratil and Sorfa, although they did not propose another structure. The infrared and Raman spectra of the bismuthiol II complexes of tellurium and other metals have been studied by Wang and Cheng (92), but the Raman spectra of the complexes did not offer much evidence for structural determination. This structural problem may be investigated by photoelectron spectroscopy (19) for definite evidence of bondings between the metal and sulfur and nitrogen.

Cheng and Goydish (20) further improved the sensitivity and speed of the bismuthiol II method for tellurium by addition of ammonium sulfate (Fig. 11).

Reagents

Bismuthiol II (Potassium Salt of 5-Mercapto-3-phenyl-1,3,4-thiadizole-2-thione), 1% Aqueous Solution.

Standard Tellurium Solution, 100 ppm. Dissolve 0.1 g pure tellurium in 5 ml concentrated sulfuric acid by heating. Heat to drive off most of the acid. Cool carefully, add 100 ml water, dissolve the tellurous acid precipitate in the minimum amount of $6M$ potassium hydroxide, and then dilute to 1 liter in a volumetric flask. More dilute solutions may be prepared from the stock solution.

Complexing Solution. Dissolve approximately 530 g ammonium sulfate, 7.5 g disodium EDTA, and 42 g citric acid in water and dilute to 1 liter.

Procedure

Transfer 0, 1.0, 2.0, 3.0, 4.0, and 5.0 ml of the standard tellurium solution (10 ppm) to a series of 100-ml beakers. Add 5 ml of the complexing solution and dilute to approximately 25 ml. Add 2 ml 1% bismuthiol II solution, adjust to pH 2.5, and allow to stand for 2 min. Adjust to pH 6.5 to 7.0 with $6M$ ammonium hydroxide solution. Transfer to 125-ml separatory funnels. Shake vigorously with 10 ml chloroform for 1 min. Filter the organic extract through glass wool and measure its absorbance at 330 nm against a reagent blank.

M. DITHIOCARBAMATE METHOD FOR TELLURIUM (18)

Both selenium and tellurium react with sulfide and compounds containing a mercapto group. It is expected that dithiocarbamates form precipitates and yellow-colored complexes with selenium and tellurium. In this respect, selenium and tellurium behave similarly as metal ions. Bode (7,8) proposed the use of dithiocarbamate for the photometric determination of selenium and tellurium. He also used EDTA and cyanide to mask interfering metal ions.

Since DAB has been universally accepted as a superior reagent for selenium, the dithiocarbamate methods have lost their significance. Bismuthiol II has been shown to be a good reagent for tellurium so that the dithiocarbamates are not attractive as reagents for traces of tellurium because of their disadvantage in photodecomposition (Fig. 13). Nevertheless, if carried out carefully, they offer alternate methods for selenium and tellurium. Other dithiocarbamates rather than the sodium salt of diethyldithiocarbamate have been tested for better photostability without success. An ultraviolet spectrophotometric method for the determination of tellurium using ammonium pyrrolidinedithiocarbamate has been proposed (62).

Bode measured the spectra of the tellurium–diethyldithiocarbamate complex between 300 and 500 nm. Cheng (18) measured the spectra of diethyldithiocarbamate complexes of tellurium and thallium in chloroform between 245 and 460 nm, with a weak absorptivity at approximately 430 nm and a strong absorptivity at 260 nm (Fig. 12). The selection of 310 nm is purely arbitrary. Absorbance measurements at the 420-nm maximum are not sensitive, and at the 260-nm maximum, serious photodecomposition may occur.

The tellurium–diethyldithiocarbamate complex decomposes rather rapidly in light (Fig. 13). For 200 μg tellurium in 25 ml chloroform, the absorbance decreased 75% in 2 hr, while in the dark it decreased only 5%.

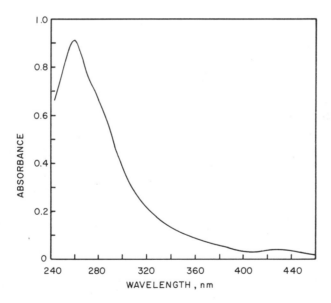

Fig. 12. Absorption spectrum of tellurium–diethyldithiocarbamate complex (25 μg Te(IV) per 25 ml CHCl$_3$).

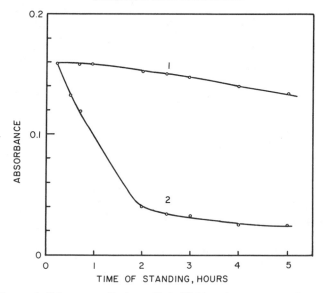

Fig. 13. Effect of light on tellurium–diethyldithiocarbamate complex. Absorbance measured at 428 nm (200 μg Te per 25 ml CHCl$_3$): (1) extract in dark; (2) extract under fluorescent light.

Reagents

Standard Tellurium Solution, 10 μg tellurium per ml.

DTAP Solution, 0.2M. Dissolve diethylenetriaminepentaacetic acid in water with the help of aqueous solution of ammonia.

Potassium Cyanide, 1%.

Ammonium Acetate Solution, 1M.

Sodium Diethyldithiocarbamate Solution, 0.25%.

Procedure

Transfer 0, 1.0, 2.0, 3.0, 4.0, 5.0, and 6.0 ml standard tellurium solution (10 μg Te per ml) to 100-ml beakers and dilute to approximately 50 ml. Add 1 ml 0.2M DTPA solution and 2 ml 1M ammonium acetate. Add 1 ml 1% potassium cyanide and adjust the pH to 7–8 with acetic acid or ammonium hydroxide. Transfer the solution into a 125-ml red separatory funnel (or regular separatory funnel covered with black electric tape). Add 2 ml 0.25% sodium dithiocarbamate solution and 10 ml carbon tetrachloride. Shake 2 min each time, extracting three times with a total 25-ml volume of carbon tetrachloride. If the extract is turbid, centrifuge for several minutes or filter through glass wool. Measure the absorbance at 310 nm against a reagent blank.

N. TUNGSTEN BLUE METHOD FOR TELLURIUM (28)

Reduction of heteropoly acids to a blue-colored compound can be used as a basis for photometric determination of tellurium. Synthesized tellurite-12-tungstic heteropoly acid proved to be tetrabasic and can be represented by the formula $H_4[TeO_4(H_2WO_4)_{12}] \cdot 13H_2O$. Yellow solutions of the tellurite-12-tungstic heteropoly acid are reduced with formation of telluritetungsten blue. By using a carbon steel carbon steel wire as a reducing agent, the blue color is developed. The blue color is only stable in the presence of a reducing agent.

The heteropoly anion $([TeO_4(H_2WO_4)_{12}]^{4-}$ is only stable in $<1.1N$ hydrochloric acid solution. Because of the acidity dependence of the blue color intensity, it is necessary to control the hydrochloric acid concentration strictly in the preparation of a calibration graph and in the color development of sample solution. It is important that the color development of standard solutions and the sample solution is done at the same acidity. If the ratio of $WO_3:TeO_3$ is <12, low results are obtained. Excess of paratungstate in solution does not interfere.

The number of ml of hydrochloric acid (x) required for adjusting the acidity of the solution to the desired value may be calculated according to the formula

$$(xC - A)/(x + B) = C_1$$

where C is the concentration of the acid used; B is the total volume of tellurite and paratungstate solutions; A is the number of hydrogen ions replacing sodium in sodium tellurite and paratungstate; and C_1 is the required acidity. In the following procedure, $C_1 = 0.04N$.

Reagents

Standard Tellurium Solution. A stock solution may be prepared as follows: To a solution containing 1.1 g sodium tellurite pentahydrate, add 12.5 g sodium paratungstate, $Na_{10}W_{12}O_{14} \cdot 28H_2O$. Take an aliquot of the sodium tellurite and sodium paratungstate solution, add the calculated amount of hydrochloric acid, and dilute to 100 ml with water so that the final solution is $0.04N$ in hydrochloric acid. From the stock solution, more dilute solutions may be prepared by appropriate dilutions with $0.04N$ hydrochloric acid.

Procedure

To 15 ml of the dilute tellurite and paratungstate solution add 3 mg finely cut steel wire. After standing for 1 hr, measure the absorbance immediately in a photometer with a red filter (no specified wavelength) against a reagent blank.

For the analysis of samples with an unknown tellurite content, sodium paratungstate can be taken in an excess of the molar ratio of 1:1. In order to obtain reliable results, the addition of various amounts of sodium paratungstate to equal aliquots of sample solution is recommended, while the acidity of the solution is kept constant.

O. OTHER REAGENTS FOR SELENIUM AND TELLURIUM

Thioglycollic acid is recommended as a selenium reagent (50). A catalytic reaction which is based on the reaction between 2,3-diaminophenazine and glyoxal in the presence of hypophosphorous acid has been suggested for determining submicrogram concentrations of selenium(IV) that acts as a catalyst (88). A sensitive method for the determination of trace amounts of selenium based on its catalytic effect in the reduction of methylene blue by sodium sulfite is described (93).

A photometric method based on the reduction of the iron(III) complex of 1,10-phenanthroline by tellurim(IV) to give an intensely colored ferroin has been suggested for determining tellurium(IV) (40).

Berg and Teitelbaum (5) and Feigl (25) have shown that selenious acid may be detected by pyrrole in the presence of selenic, tellurous, and telluric acids. The reaction depends on the oxidation of pyrrole to "pyrrole blue" by selenious acid, which is reduced to its elemental state. Suzuki (85) has developed a quantitative method to determine selenium based on this reaction.

Toluene 3,4-dithiol gives red precipitates with selenium and tellurium in an acidic medium. These precipitates are readily extractable into organic solvents (18). Any aromatic compounds containing two mercapto groups in the position should give such a reaction with selenium or tellurium.

Sawicki (78) reported that the colorless 4-dimethylamine-1,2-phenylene-diamine and 4-methylthio-1,2-phenylenediamine react with selenious acid to give stable bright-red and blue-purple colors, respectively, in appropriate media. No quantitative procedure has been developed, and the selectivity of the reactions has not been studied.

III. APPLICATIONS

A. SELENIUM IN NATURAL WATERS – IODINE METHOD (54)

Selenium in natural waters occurs in quadrivalent and sexivalent inorganic forms and as organic-bound selenium. The following method by Lamberg, Arthur, and Moore (54) is applicable to natural waters with a selenium content in the range of 0.2–6.0 ppm. The method includes processes for the decomposition of organic matter, distillation of selenium from the bulk of the interferences, removal of iodine, reduction of selenium(VI), and masking of most of the remaining interferences.

1. Removal of Iodine

Pipet a 10.0-ml sample of the water into a separatory funnel. Add 2.0 ml 12N sulfuric acid and two drops saturated bromine water. Extract with 5-ml portions of carbon tetrachloride until the extracts are clear.[16]

2. Reduction of Selenate

Transfer the extracted sample with a minimum amount of wash water into a 25 x 200 mm digestion tube. If iodine-free waters are being analyzed, start with 10 ml water and 2.0 ml 12N sulfuric acid. Add 1.0 ml saturated potassium bromide solution and 1.0 ml 1% sodium hypochlorite solution.[17] Lightly stopper with a glass stopper. Heat on a steam bath for 1 hr.[18]

3. Distillation as the Volatile Bromide

The bromide is distilled through an uncooled tube attached by a ground-glass connection to the digestion tube and leading into a 25 x 200 mm receiver, which is immersed in an ice bath and contains 2 ml 12N sulfuric acid. The solution in the digestion tube at this point should contain 10 ml sample, 2 ml 12N sulfuric acid, 1.0 ml saturated potassium bromide solution, and 1.0 ml 1% sodium hypochlorite solution. Add boiling chips to the tube and distill until only the sulfuric acid remains.[19]

4. Masking of Interferences and Development of Color

The following treatment may be applied to the solutions resulting from any of the preceding treatments or to 10 ml raw water sample plus 2 ml 12N sulfuric acid. The sample is contained in a 25-ml volumetric flask. Add 1 ml saturated sodium nitrite solution and pass in carbon dioxide at a moderate rate for 10 min. Add 1 ml saturated urea solution and rinse the wall of the test tube by carefully increasing the gas flow. Add 1 ml saturated tartartic acid solution, 1 ml 85% phosphoric acid, 1 ml 12N sulfuric acid, and 2 ml 90% formic acid. Develop the color by one of the following procedures.

[16] If there is no bromine color in the first extraction, add two more drops bromine water. Continue extraction and bromine addition until bromine is shown to be in excess.

[17] Stabilized sodium hypochlorite solution, 1% commercially available as Zonite, is recommended by Lambert, Arthur, and Moore (54).

[18] In addition to precluding formation of free selenium, the data of Lambert, Arthur and Moore (54) indicate a catalytic effect of bromine in selenate reduction. Although selenate reduction is only 88% complete at the end of 1 hr at 100°C, longer digestion is not recommended because of the possible loss of volatile selenium bromides.

[19] Most of the organic selenicals are decomposed without digestion to SO_3 fumes.

Reagents

See the colorimetric iodine method for selenium.

Procedures

Triiodide Method. Add 1.0 ml 5% cadmium iodide solution, dilute to volume, stopper, and mix. Allow to stand for 5 min or more and read the absorbance at 352 nm.

Starch–Iodine Method. Purge with carbon dioxide for 10 min. Add 1.0 ml of the starch–cadmium iodide reagent, dilute to volume, stopper, and mix. Allow to stand for at least 5 min and read the absorbance at 615 nm.

B. SELENIUM IN COPPER, STAINLESS STEEL, AND ARSENIC – PIAZSELENOL METHOD

The piazselenol method requires removal of excess acid from the dissolution step, but it does not require a preliminary separation of selenium. The procedure below has been applied successfully by Cheng to the determination of selenium in high-purity arsenic (13) or copper (14) and in stainless steel (14). Ethylene-diaminetetraacetic acid is used to mask iron(III) and copper(II).

Reagents

See the piazselenol method (Section IIC)

Procedure

Weigh a sample containing 10–100 μg selenium. Dissolve the sample as follows:

Arsenic. For 10- to 20-g samples of arsenic, add gradually approximately 20 ml 1:1 hydrochloric acid–nitric acid mixture. WARNING: The reaction may be violent if acid is added too rapidly. The reaction should be carried out in a good hood.

Copper. Dissolve a 1-g sample in 10 ml 1:1 nitric acid. Add more nitric acid if necessary to complete the solution.

Stainless Steel. To a 1-g sample add 100 ml hydrochloric acid and then 5 ml nitric acid. Heat gently in a covered beaker until solution is complete. When sample is dissolved, evaporate the solution to 3–5 ml. Dilute to 40–50 ml with water. Add ethylenediaminetetraacetic acid to chelate iron and copper; namely, for 1 g stainless steel, add 25 ml 0.1M disodium salt of EDTA; for 1 g copper, add 6.5 g disodium salt of EDTA. Adjust to pH 2.5–3.0 with ammonia and hydrochloric acid. Add 2 ml 2.5M formic acid and continue with the piazselenol procedure as given in the general methods Section IIC.

C. TELLURIUM IN LEAD

In the method of Brown (10), the lead alloy is quickly dissolved in acetic acid–hydrogen peroxide. Tellurium is separated by stannous chloride reduction from lead, which is held in solution as the chloro complex, and from copper, which complexed with thiourea. The iodotellurite method is then used.

Reagents

Acetic Acid–Hydrogen Peroxide Solution. Dilute 10 ml glacial acetic acid with 20 ml water. To this solution add 200 ml 30% hydrogen peroxide. Prepare fresh.

Sodium Chloride. Saturated aqueous solution.

Stannous Chloride. Dissolve 12 g stannous chloride in 50 ml concentrated hydrochloric acid.

Procedure

Weigh an amount of lead alloy equivalent to 0.02–0.1 mg tellurium into a 50-ml beaker. Cover with a watch glass and add dropwise 2 ml of the acetic acid–hydrogen peroxide reagent. Allow the reaction to proceed for 1 min, then swirl and warm gently. Add 10 ml saturated sodium chloride solution and 2 ml concentrated hydrochloric acid. Heat to boiling and place on a steam bath until the sample is dissolved. Rinse down the cover glass and sides with hot water. Add 0.01 g thiourea and leave for 5 min. Add 4 ml stannous chloride reagent and digest on a steam bath for 15 to 20 min. Filter on asbestos in a micro suction filter.[20] Dissolve the tellurium by adding in small portions a total of 3 ml hot concentrated nitric acid.[21] Wash the filter wih two 3-ml portions of 1N nitric acid. Boil the filtrate for 5 min to expel oxides of nitrogen. Cool, neutralize to phenolphthalein with 1:1 ammonia, transfer to a 50-ml volumetric flask, and determine the tellurium by the iodotellurite method, as described in the general methods section.

D. TELLURIUM IN COPPER – STANNOUS CHLORIDE METHOD

The sample is dissolved in 50% nitric acid and then diluted tenfold. Copper(II) ion is masked with thiourea. The tellurium sol is formed by stannous

[20] An appropriate microfilter is described by Brown (10). Johnson and Kwan (45) recommend for this purpose the Kirk calcium-type microfilter (Microchemical Specialties, Berkeley, Calif.).

[21] A film of tellurium may adhere to the glassware. This should be washed off with a little nitric acid. Adsorption of this film may be avoided by treatment of the glass with a solution of chlorosilanes, for example, Desicote (Beckman Instrument Co., Pasadena, Calif.).

chloride reduction and measured colorimetrically. The following procedure is based on the method of Crossley (23).

Reagents

Nitric Acid, 50%. 1:1 Dilution of concentrated nitric acid with water.

Stannous Chloride. Dissolve 30 g stannous chloride in 350 ml concentrated hydrochloric acid. Dilute to 500 ml with water.

Thiourea. Dissolve 100 g thiourea in 1 liter water and add 60 ml concentrated nitric acid.

Procedure

Dissolve 1 g copper in 20 ml 50% nitric acid in a covered flask. When solution is complete, bring the solution to boiling. Cool and dilute to 200 ml. Pipet a 10-ml aliquot into a 200-ml Erlenmeyer flask. Add 15 ml of the thiourea solution and 50 ml water. Add, while swirling the solution, 15 ml stannous chloride solution. Transfer to a 100-ml volumetric flask and dilute to volume. Measure the adsorbance against a reagent blank, using a violet filter. Prepare the calibration graph using copper samples of known tellurium content (or various weights of a given analyzed tellurium—copper alloy).

E. SELENIUM IN PHARMACEUTICALS (51)

Burn the sample (0.2 g containing 1—12 mg selenium in oxygen in the Schöniger flask and absorb the decomposition products in 25 ml $0.5N$ nitric acid. Mix the solution with 25 ml water, boil for 10 min, and cool. Adjust pH to 1.8—2.2 with ammonium hydroxide and add 0.2 g hydroxylamine hydrochloride and 5 ml freshly prepared 0.1% 2,3-diaminonaphthalene solution in $0.1N$ hydrochloric acid containing 0.5% hydroxylamine hydrochloride. Mix and allow to stand for 100 min. Extract with 5 ml cyclohexane and measure the absorbance at 380 nm against a reagent blank.

F. SELENIUM IN STEROIDS (90)

Place a sample containing 20—100 mg selenium in a 100-ml round-bottomed flask with a 24/40 ground joint. After dissolving the sample with 50 ml ethanol, add about 300 mg Raney nickel (fresh) prepared according to the method of Drake (Drake, N.J., Org. Synthesis Vol. 21, p.15). Fit the flask with a condenser and gently reflux the solution for 30 min. Cool the flask and decant the ethanol from the Raney nickel, which is then washed by decantation with 30 ml ethanol. Fit the flask with an air condenser and add dropwise 4 ml nitric acid (sp.

gr. = 1.42). CAUTION. Use a hood. Warm the flask for complete dissolution of the Raney nickel. After cooling, add 10 ml water and boil the solution to remove oxides of nitrogen. After again cooling, add 20 ml water and 10 mg o-phenyldiamine dihydrochloride. After standing for 15 min, adjust the solution to pH 2.5 ± 0.1 with ammonium hydroxide, transfer to a 125-ml separatory funnel, and extract twice with 10-ml portions of chloroform. Dry the chloroform extract with sodium sulfate, transfer to a 25-ml volumetric flask, and dilute to volume with chloroform. Measure the absorbance at 330 nm against a chloroform blank.

Perform a blank determination simultaneously on the Raney nickel. Subtract the absorbance of the blank from that of the sample and calculate the amount of selenium using 791 as the $E_{1\,cm}^{1\%}$ value of the benzoselenodiazol:

$$\mu g\ Se = \frac{0.432A \times 1.1}{791 \times 4}$$

where A is the absorbance, 0.432 is a conversion factor for benzoselenodiazol to selenium, and 1.1 is a correction factor for the incomplete reaction between the o-phenyldiamine and selenium.

G. SELENIUM IN SEAWATER, SILICATES, AND BIOMATERIALS – DAB METHOD (12)

1. Seawater

Filter the water sample through a membrane filter (pore size 0.5 μ) and add to 5 liters of the sample, with vigorous stirring, 2 ml ferric chloride solution (30 mg Fe per ml). Wait for 30 min. If the pH is higher than 6.5, adjust pH to 5 to 6 with 2N hydrochloric acid. After about 2 hr, when the precipitate has begun to settle, add an additional 2 ml ferric chloride solution. After this addition, the pH of the solution is normally reduced to 3–4, which is too low for complete coprecipitation of selenium. Adjust carefully the pH to 4–6 with 2N ammonium hydroxide.

After two days, when the precipitate has settled, carefully siphon off the supernatant liquid, and separate the precipitate by centrifugation. Wash the precipitate four times with 30-ml portions of 0.5% ammonium nitrate solution, centrifuging and decanting after each washing. At once, dissolve the precipitate in 1 ml concentrated nitric acid, warming on the water bath to facilitate solution. Dilute with water until its acidity is approximately 0.2N, and pass it through the ion exchange column (1.5 cm diameter and 10 cm length of the hydrogen form of Zeo-Karb 225 resin, 8% crosslinked, 52–100 mesh, cr equivalent). Desorb selenium from the column with 270 ml 0.2N nitric acid.

Combine these washings with the original percolate, add 1 ml $2N$ sodium hydroxide, and evaporate to dryness on a water bath.

Dissolve the residue in 5–10 ml water and add two to three drops bromothymol blue indicator and one to two drops $2N$ ammonium hydroxide to neutralize any free acid (until the indicator turns blue color). Add 2 ml $2.5M$ formic acid and 5 ml EDTA solution and dilute to approximately 23 ml. Adjust the pH to 2–3 with nitric acid or ammonium hydroxide. Add 2 ml 0.5% 3,3'-diaminobenzidine solution and, after 30 min, add $7N$ ammonium hydroxide dropwise until the indicator changes to greenish blue (pH 6–7). Transfer the solution to a 50-ml separatory funnel, add 5 ml toluene, and shake the funnel for 3 min. Measure the absorbance of the organic extract at 420 nm against a blank.

2. Silicates

Wash marine sediments and other samples containing appreciable amounts of chloride with water until free from chloride. Weigh 1–2 g of the dried sample (ground to pass an 80-mesh sieve) into a platinum dish and add approximately 0.2 microcuries carrier-free ^{75}Se as selenite. Add 20 ml 40% hydrofluoric acid and 20 ml concentrated nitric acid and heat the covered dish on a water bath overnight. Remove the lid from the dish and evaporate almost to dryness. Add an additional 20 ml 40% hydrofluoric acid and 20 ml concentrated nitric acid and, after swirling, repeat the evaporation. Evaporate to dryness three times with concentrated nitric acid to volatilize hydrogen fluoride.

Add 25 ml $4N$ hydrochloric acid to the residue and boil gently for 2 min to facilitate dissolution and to reduce any selenium(VI) formed to selenium(IV). Cool and centrifuge to remove any undissolved substances; wash the latter twice with approximately 10 ml water and combine the washings with the supernatant liquid. Digest the residue again with 2 ml concentrated nitric acid and 2 ml hydrofluoric acid and evaporate to dryness. Evaporate to dryness twice with 5-ml aliquots of nitric acid. Add 2–3 ml $5N$ hydrochloric acid and warm on a water bath. Centrifuge and discard any residue which may remain after washing it twice with small portions of $4N$ hydrochloric acid. Combine the solution and washings with the main hydrochloric acid solution.

Transfer the combined hydrochloric acid solutions to a 5-liter conical flask and dilute to approximately 2.5 liters with water. Stir the solution and adjust the pH to approximately 3.5–4.0 with dilute sodium hydroxide, but avoid exceeding pH 4.0. Dilute to approximately 5 liters and add 60 g sodium chloride to help the coagulation of the ferric hydroxide precipitate. Add, with shaking, 2 ml 25% ferric chloride solution (3% Fe) and adjust the pH to 4.5–5.0 with dilute ammonium hydroxide. After 2 hr, add an additional 2 ml ferric chloride solution and readjust the pH to 4.5–5.0.

Allow the solution to settle for two days. Siphon off the supernatant liquid, separate the precipitate by centrifugation, and wash it with 0.5% ammonium nitrate solution as described in the procedure for seawater. Dissolve the precipitate by warming with 1 ml concentrated nitric acid; add 0.1 ml concentrated hydrochloric acid if it is difficult to dissolve the precipitate. Should there be any undissolved substances remaining, remove them by centrifugation and redigest them with 0.2 ml concentrated nitric acid. Combine the nitric acid extracts, add sufficient water to reduce the acidity to 0.2N, and pass the solution through the ion exchange column filled with Zeo-Karb 225 hydrogen-form resin. Elute with 350 ml 0.2N nitric acid. Combine the percolate and eluate, add 1 ml 2N sodium hydroxide, and evaporate to dryness on a water bath. Dissolve the residue in water and dilute to 25 ml. Count the ^{75}Se activity of a suitable aliquot of this solution with a gamma-ray counter. Compute the chemical yield of the separation process by comparing the activity of this solution with that of the corresponding volume of the ^{75}Se solution initially added.

Transfer the aliquot used for counting to a 50-ml beaker and mix with the remainder of the selenium concentrate. Evaporate to approximately 15 ml on a water bath and follow the photometric procedure as previously given in the procedure for seawater (Section IIIG1).

3. Marine Organisms

Weigh 1–2 g of the washed, finely chopped sample into a 25-ml Kjeldahl flask and add 10 ml of the digestion mixture (dissolve 0.4 g ammonium molybdate, $(NH_4)_2MoO_4$, in 15 ml water; add with cooling 15 ml concentrated sulfuric acid and 20 ml 70% perchloric acid) and approximately 0.2 microcuries carrier-free ^{75}Se solution. Heat the flask carefully with a small flame until the organic matter has been completely decomposed and the digest becomes colorless. Pour into approximately 20 ml water and centrifuge to remove the white molybdic acid precipitate. Wash the precipitate twice with 5-ml aliquots of 2N sulfuric acid and then discard it since it is free from selenium. Combine the solution and washings and dilute to 1 liter with water. Add 15 g sodium chloride and 1 ml ferric chloride (20 mg Fe). Adjust carefully the pH of the solution with 4N sodium hydroxide to 4–6. After about 2 hr, add an additional 1 ml ferric chloride solution and readjust the pH to 4–6. Allow the precipitate to settle and separate it by centrifugation. Wash it three times with 25-ml portions of 0.5% ammonium nitrate solution. Dissolve the precipitate by warming on a water bath with 1 ml concentrated nitric acid and dilute to approximately 20 ml with water. A small amount of yellowish residue molybdophosphoric may remain and should be separated by centrifugation and discarded after washing with a small volume of 1N nitric acid. Combine the washings with the main bulk of the solution, dilute with water to approximately 0.2N acidity, and pass the solution through the ion exchange column. Continue with the procedure steps as given for seawater (Section IIIG1).

Before the photometric measurement, the chemical yield of the analytic separation should be assessed by gamma-counting a suitable aliquot of the solution. The determined amount of selenium can then be corrected for losses of selenium occurring during the analysis.

H. SELENIUM IN IRON AND STEEL – 4-METHYL-O-PHENYLENEDIAMINE METHOD (49)

Cheng (13) and Clark (22) have reported methods with DAB involving masking iron with EDTA, and with 2,3-diaminonaphthalene involving reduction of iron(III) to iron(II) with hydroxylamine hydrochloride, respectively. But for selenium contents of 0.01% or less, it is preferrable to separate iron (85). Kammori and Ono (48) reported a method of extracting iron with methyl isobutyl ketone, but it is unsatisfactory because of partial coextraction of selenium with the iron. Tanaka et al. (89) have reported the extraction of iron caprate with organic solvents and suggested the use of capric acid in the separation of iron from selenium.

Reagents

Standard Selenium Solution (1 mg Se per ml). Dissolve 2.2 g anhydrous sodium selenite in 1 liter 0.1M hydrochloric acid solution. Standardize the solution. Dilute selenium solutions may be prepared by appropriate dilutions.

4-Methyl-o-phenylenediamine Hydrochloride Solution. The reagent is purified according to Tanaka and Kawashima (87). Prepare freshly 1% solution before use.

Capric acid–Chloroform Solution, 1M. Dissolve approximately 43 g n-capric acid in 250 ml chloroform.

8-Quinolinol Solution 2%. Dissolve 2 g 8-quinolinol in 5 ml glacial acetic acid and dilute to 100 ml with water.

Procedure

Transfer 0.1–0.3 g sample (for 0.02–0.005% Se) to a 200-ml conical beaker. Dissolve in 20 ml 8M nitric acid for iron, or in 3–4 ml of a hydrochloric acid, nitric acid, and water mixture (2:1:3) for steel. Add another 20 ml 8M nitric acid to the steel sample. When dissolution is complete, add 2–3 ml 70%

perchloric acid and heat gently on a hot plate. Evaporate to fumes and allow to fume gently for 10–20 min. Cool and add a few drops of 3% hydrogen peroxide to the cooled solution in order to reduce chromium(VI) to chromium(III). Transfer the solution to a 100-ml beaker. (For a sample with higher than 0.02% selenium content, dilute the prepared sample solution to 100 ml with $0.1N$ hydrochloric acid solution and take a suitable aliquot for the subsequent steps.)

Add 3 ml $1M$ acetic acid and adjust the pH to 4.7–5 with $6M$ ammonia solution. Transfer the solution to a 150-ml separatory funnel. Extract iron twice with approximately 10 ml capric acid in chloroform. Add 3 ml 2% 8-quinolinol solution to the aqueous solution and extract metal quinolates with chloroform. Wash all the organic phase once with 5 ml of an acetate buffer solution (pH 4.7–5). Combine the aqueous solutions and then wash the solution successfully with chloroform and toluene. Adjust the pH of the aqueous solution to 1.5–2.0 with $6M$ hydrochloric acid. Add two to three drops 3% hydrogen peroxide, 1 ml $0.25M$ disodium salt of EDTA solution, and 1 ml 4-methyl-o-phenylenediamine solution. After standing for 2 hr at room temperature, extract the 5-methyl-piazselenol formed with 10.00 ml toluene, the funnel being shaken for 10 min on a shaker. After separating the two phases, wash the toluene phase once with $0.1N$ hydrochloric acid. Measure the absorbance of the toluene extract at 337 nm with toluene as reference.

I. SELENIUM IN LEAD (63)

Procedure

Add 15 ml perchloric acid–nitric acid mixture (5:1) to 10 g of the sample in a 300-ml conical flask and heat gently to dissolve the sample. Evaporate to copious fumes. Cool, add 50 ml water, boil gently, and then add 50 ml hydrochloric acid. Continue to heat for 1 min with occasional swirling to prevent bumping until the precipitate coagulates well. Cool to 10–15°C in an ice bath. Filter into a 250-ml conical flask and wash the 300-ml flask and filter paper with a fine stream of cold $6N$ hydrochloric acid. Discard the precipitate.

Add 2 ml arsenic(III) solution (0.125 g As_2O_3 in 100 ml $0.5N$ NaOH), followed by 10 ml 50% hypophosphorous acid solution, swirl, cover, and boil gently for 5 min. Cool to 70–80°C, filter through a retentive filter paper, and wash three or four times with $6N$ hydrochloric acid. Transfer the paper and precipitate to the 250-ml flask. Gather any traces of the metal from the glass funnel by wiping with a small piece of filter paper and put into the 250-ml flask. Add 10 ml nitric acid and 4 ml perchloric acid, cover, and heat to destroy the paper. When the decomposition is complete, evaporate to 2 ml and proceed with procedure given for the preparation of calibration graph for selenium with DAB (Section IIC). Carry a reagent blank through the same procedure.

J. SELENIUM IN COPPER (63)

Procedure

Dissolve 5 g of the sample in 25 ml of the perchloric–nitric acid mixture in a beaker and evaporate to copious fumes. Cool, add 50 ml water, and warm to dissolve salts. Add 2 ml concentrated hydrochloric acid and 2 ml arsenic(III) solution (0.125 g As_2O_3 in 100 ml $0.5N$ NaOH and 15 ml 50% hypophosphorous acid) and continue with the procedure as given for the determination of selenium in lead (Section III I).

K. TELLURIUM IN LEAD (63)

Follow the procedure given for the determination of selenium in lead (Section III I). When the filter paper has been completely decomposed by the final nitric acid–perchloric acid treatment, evaporate to copious fumes of perchloric acid. Add 2 ml sulfuric acid and evaporate to 1 ml. Cool and proceed with the procedure described for the preparation of calibration graph for the determination of tellurium with dithiocarbamate (Section IIM). Carry a reagent blank through this same procedure.

L. TELLURIUM IN COPPER (63)

Follow the procedure as given for the determination of selenium in copper (Section IIIJ) but wash the flask and precipitate of arsenic plus tellurium with successive 10-ml portions of $6N$ hydrochloric acid until a 10-ml water wash is not blue on being made strongly ammoniacal – when no more than about 1 mg copper is left in the flask plus paper. When the filter paper has been completely decomposed, continue with the procedure as described for the determination of tellurium in lead (Section IIIK).

REFERENCES

1. Aaremae, A., and G. O. Assarsson, *Anal. Chem.*, **27**, 1155 (1955).
2. Aoki, F., *Bull. Chem. Soc. Japan*, **26**, 480 (1953).
3. Ariyoshi, H., M. Kiniwa, and K. Toei, *Talanta*, **5**, 112 (1960).
4. Bagnall, K. S., *The Chemistry of Selenium, Tellurium and Polonium*, Elsevier, Amsterdam, 1966.
5. Berg, R., and M. Teitelbaum, *Mikrochemie*, **15**, 32 (1934).
6. Bock, R., D. Jacob, M. Fariwar, and K. Frankenfeld, *Z. Anal. Chem.*, **200**, 81 (1964).
7. Bode, H., *ibid.*, **144**, 176 (1955).

8. Bode, H., *ibid.*, **153**, 335 (1956).

9. Brasted, R. C., *Comprehensive Inorganic Chemistry*, Vol. VIII, Van Nostrand, New York, 1961.

10. Brown, E. G., *Analyst*, **79**, 50 (1954).

11. Byrne, A. R., and D. Gorenc, *Anal. Chim. Acta*, **59**, 81 (1972).

12. Chau, Y. K., and J. P. Riley, *ibid.*, **33**, 36 (1965).

13. Cheng, K. L., *Anal. Chem.*, **28**, 1738 (1956).

14. Cheng, K. L., *Chemist Analyst*, **45**, 67 (1956).

15. Cheng, K. L., *Anal. Chem.*, **33**, 761 (1961).

16. Cheng, K. L., *Talanta*, **8**, 301 (1961).

17. Cheng, K. L., *ibid.*, **9**, 501 (1962).

18. Cheng, K. L., unpublished research.

19. Cheng, K. L., J. Carvar, and T. A. Carlson, unpublished research.

20. Cheng, K. L., and B. L. Goydish, *Talanta*, **13**, 1210 (1966).

21. Cheng, K. L., and P. F. Lott, *Proceedings of International Symposium on Microtechniques*, Wiley-Interscience, New York, 1962, pp. 317–331.

22. Clark, W. E., *Analyst*, **95**, 65 (1970).

23. Crossley, P. B., *ibid.*, **69**, 206 (1944).

24. Dahl, I., and F. J. Langmyhr, *Anal. Chim. Acta*, **35**, 24 (1966).

25. Feigl, F., *Spot Tests*, 4th ed., Elsevier, New York, 1954, pp. 317–25.

26. Furman, N. H., *Scott's Standard Methods of Chemical Analysis*, 6th ed., Vol. I, Van Nostrand, New York, 1962, p. 938.

27. Furman, N. H., *ibid.*, pp. 927–928.

28. Ganelina, E. Sh., *Zh. Anal. Khim.*, **18**, 551 (1963).

29. Geiersberger, K., and A. Durst, *Z. Anal. Chem.*, **135**, 11 (1932).

30. Geiersberger, K., and A. Durst, *ibid.*, pp. 16, 21.

31. Green, T. E., and M. Turley, *Selenium and Tellurium* in *Treatise on Analytical Chemistry*, Part V, Vol. 7, Kolthoff, I. M., and P. J. Elving, Eds. Wiley-Interscience, New York, 1961, pp. 137–206.

32. Haisty, R. W., *M.S. Thesis*, University of Illinois, Urbana, 1955.

33. Hanson, C. K., *Anal. Chem.*, **29**, 1204 (1957).

34. Hikime, S., *Bull. Chem. Soc. Japan*, **33**, 761 (1960).

35. Hillebrand, W. F., G. E. F. Lundell, H. A. Bright, and J. I. Hoffman, *Applied Inorganic Analysis*, 2nd ed., Wiley, New York, 1953, p. 333.

36. Hoste, J., *Anal. Chim. Acta*, **2**, 402 (1948).

37. Hoste, J., and J. Gillis, *ibid.*, **12**, p58 (1955).

38. Inarida, M., *J. Chem. Soc. Japan, Pure Chem. Sect.*, **79**, 721 (1958).

39. Inarida, M., *ibid.*, **80**, 399 (1959).

40. Ingamells, C. O., and E. B. Sandell, *Microchem. J.*, **3**, 3 (1959).

41. Jankovsky, J., and O. Ksir, *Talanta*, **5**, 238 (1960).

42. Jilek, A., and J. Vrestal, *Chem. Zvesti,* **6**, 497 (1952).

43. Jilek, A., and J. Vrestal, *ibid.,* **1**, 33 (1953).

44. Johnson, R. A., *Anal. Chem.,* **25**, 1013 (1953).

45. Johnson, R. A., and B. R. Anderson, *ibid.,* **27**, 20 (1955).

46. Johnson, R. A., and F. P. Kwan, *ibid.,* **23**, 651 (1951).

47. Johnson, R. A., and J. P. Kwan, *ibid.,* **25**, 1017 (1953).

48, Kammeri, O., and A. Ono, *Bunseki Kagaku,* **15**, 290 (1966).

49. Kawashima, T., and A. Ueno, A. *Anal. Chim. Acta,* **58**, 219 (1972).

50. Kirkbright, J. F., *ibid.,* **35**, 116 (1966).

51. Klein, H. R., R. H. King, and W. J. Mader, *J. Pharm. Sci.,* **58**, 1524 (1969).

52. Lakin, H. W., and C. E. Thompson, *Science,* **141**, 42 (1963).

53. Lakin, H. W., and C. E. Thompson, Geological Survey Res., E128, 1962.

54. Lambert, J. L., P. Arthur, and T. E. Moore, *Anal. Chem.,* **23**, 1101 (1951).

55. Langmyhr, F. J., and I. Dahl, *Anal. Chim. Acta,* **29**, 377 (1963).

56. Langmyhr, F. J., and S. H. Omang, *ibid.,* **23**, 565 (1960).

57. Lenher, V., and A. W. Homberger, *J. Amer. Chem. Soc.,* **30**, 387 (1908).

58. Lenher, V., and C. H. Kao, *ibid.,* **47**, 769 (1925).

59. Lenher, V., and C. H. Kao, *ibid.,* p. 772.

60. Lenher, V., and C. H. Kao, *ibid.,* p. 2454.

61. Looyenga, R. W., and D. F. Boltz, *Anal. Lett.,* **2**, 491 (1969).

62. Looyenga, R. W., and D. J. Boltz, *ibid., Mikrochim. Acta,* **1971**, 507.

63. Luke, C. L., *Anal. Chem.,* **31**, 572 (1959).

64. Mabuchi, H., *Bull. Chem. Soc. Japan,* **29**, 842 (1956).

65. Murashova, V. I., and S. G. Sushkova, *Zh. Anal. Khim.,* **24**, 729 (1969).

66. Navatil, O., and J. Sorfa, *Collect. Czech. Chem. Commun.,* **34**, 975 (1969).

67. Nielsch, W., and G. Boltz, *Z. Metallkunde,* **45**, 380 (1954).

68. Nielsch, W., and L. Giefer, *Z. Anal. Chem.,* **145**, 347 (1955).

69. Nielsch, W., and L. Giefer, *ibid.,* **155**, 401 (1957).

70. Noekes, F. D. L., *Analyst,* **76**, 542 (1951).

71. Osburn, R. L., A. D. Shendrikar, and P. W. West, *Anal. Chem.,* **43**, 594 (1971).

72. Ovsepyran, E. N., G. N. Shaposhnikova, and N. G. Galfayan, *Zh. Neorg. Khim.,* **12**, 2411 (1967).

73. Ovsepyan, E. N., G. N. Shaposhnikova, and V. M. Tataryan, *Izv. Akad. Nauk SSSR, Otd. Khim. Nauk,* **18**, 225 (1965).

74. Ramakrishna, R. S. and H. M. N. H. Irving, *J. Chem. Soc. D,* **1969**(23), 1356.

75. Ramakrishna, R. S., and H. M. N. H. Irving, *Anal. Chim. Acta,* **49**, 9 (1970).

76. Sandell, E. B., *Colorimetric Determination of Traces of Metals,* 3rd ed., Interscience, New York, 1959, p. 502.

77. Sasaki, Y., *Bunseki Kagaku*, **4**, 637 (1955).

78. Sawicki, E., *Anal. Chem.*, **29**, 1376 (1957).

79. Schoeller, W. R., *Analyst*, **64**, 318 (1939).

80. Schoeller, W. R., and A. P. Powell, *Analysis of Minerals and Ores of Rarer Elements*, Hafner, New York, 1955, p. 230.

81. Smith, G. W., and S. A. Reynolds, *Anal. Chim. Acta*, **12**, 150 (1955).

82. Southern, H. K., Report BR-606, May 31, 1945 in *Analytical Chemistry of the Manhattan Project*, Rodden, C. J., Ed., McGraw-Hill, New York, 1950, p. 318.

83. Stary, J., J. Marek, K. Kratzer, and F. Sebesta, *Anal. Chim. Acta*, **57**, 393 (1971).

84. Sushkova, S. G., and V. N. Murashva, *Zh. Anal. Chim.*, **25**, 1475 (1966).

85. Suzuki, M. J., *Chem. Soc. Japan, Ind. Chem. Sect.*, **56**, 323 (1953).

86. Tanaka, K., N. Takagi, and H. Tsuzimura, *Bunseki Kagaku*, **18**, 319 (1969).

87. Tanaka, M., and T. Kawashima, *Talanta*, **12**, 211 (1965).

88. Tanaka, M., T. Kawashima, and H. Miwa, *Chem. Soc. Japan*, **37**, 1085 (1964).

89. Tanaka, M., N. Nakasuka, S. Gote, in *Solvent Extraction Chemistry*, D. Dyrssen, J. O. Lilienzine, and J. Rydberg, Eds. North Holland, Amsterdam, 1967, p. 154.

90. Throop, L. J., *Anal. Chem.*, **32**, 1807 (1960).

91. Volkov, S. T., *Zavod. Lab.*, **5**, 1429 (1939).

92. Wang, J.-C., and K. L. Cheng, *Microchem. J.*, **15**, 607 (1970).

93. West, P. W., and T. V. Ramakrishna, *Anal. Chem.*, **40**, 966 (1969).

94. Wiberley, S. E., L. G. Bassett, A. M. Burrill, and H.Lyng, *ibid.*, **25**, 1586 (1953).

95. Yoshida, H., M. Taza, and S. Hikime, *Talanta*, **13**, 185 (1966).

96. Zemel, V. S., *Zavod. Lab.*, **5**, 1433 (1936).

97. Zmbora, B., and J. Bzenic, *Anal. Chim. Acta*, **59**, 315 (1972).

CHAPTER

11

SILICON

D. F. BOLTZ

Department of Chemistry, Wayne State University, Detroit, Michigan

LOUIS A. TRUDELL

*Division of Mathematics and Science, Macomb County Community College,
Warren, Michigan*

G. VICTOR POTTER[1]

[1] G. Victor Potter, the author of the chapter on silicon in the first edition, was with the Sylvania Electric Products, Inc., Towanda, Pennsylvania, prior to his retirement.

Molybdoheteropoly chemistry is the basis of most of the spectrophotometric methods developed for the determination of traces of silicon. Thus, the formation of 12-molybdosilicic acid is the initial process involved in the seven methods to be presented in this chapter. The most extensively used method is the heteropoly blue method based on the reduction of the molybdosilicic acid to give a characteristic blue species which we shall designate "the heteropoly blue of silicon," although "molybdenum blue" is often used to designate this blue product.

The presence of silica in natural waters and silicates in glass requires that special attention be given to avoid contamination and to compensate for traces of silicic acid derived from glassware (44) and reagents. The distilled water should be redistilled using a block tin condenser and a polyethylene storage bottle. Alternately, distilled water can be passed through a column filled with a strongly basic anion exchange resin, for example, Dowex 2, and stored in a polyethylene bottle. The use of glassware should be avoided as much as possible with basic solutions. Polyethylene or Teflon beakers and platinum ware should be used whenever possible. It is recommended that all volumetric glassware and polyethylene be rinsed in sequence with 1:2 ammonium hydroxide, 1:10 sulfuric acid, and redistilled water just prior to use. The ammonium hydroxide used in procedures for the determination of silicon must be "silica free." Inasmuch as reagent-grade ammonium hydroxide is bottled in glass, it is necessary to prepare the silica-free ammonium hydroxide by saturating redistilled water with ammonia from a cylinder of compressed ammonia. Plastic tubing and polyethylene bottles are used in this laboratory preparation of silica-free ammonia solution.

The properties of silicic acid should be taken into account when applying any spectrophotometric method for the determination of silicon which is based on molybdoheteropoly chemistry. Monosilicic acid, H_4SiO_4, is a very weak acid with pK_1 and pK_2 values of 9.8 and 12.2, respectively (53). Monosilicic acid polymerizes to form disilicic acid, $H_6Si_2O_7$, and higher polysilicic acids. Alexander (2), in studying the polymerization of monosilicic acid, found that below pH 3 the polymerization increased with the square root of time and produced a relatively uniform distribution of molecular weight for the polymers formed. Fluoride ions catalyze the polymerization at this low pH. Above pH 3, for example, pH 4–5, a heterogeneous distribution of polymers is obtained; the rate of polymerization is presumably more dependent on the number of available silanol groups and the polymerization catalyzed by hydroxyl ions. At the trisilicic acid stage, it is possible to have a ring polymer with six silanol groups, whereas a chain polymer could have eight silanol groups. The polymerization of monosilicic acid is essentially instantaneous at pH 6.

The rate of polymerization increases with an increase in temperature, and the concomitant crosslinking and/or cyclization results in species that do not form molybdoheteropoly complexes. Thus, the degree of polymerization determines the rate of formation and the amount of molybdosilicic acid formed. The dilution of a polymerized silicic acid solution favors depolymerization. Alexander (1) observed the rate of formation of molybdosilicic acid from monosilicic, disilicic, and low molecular weight polysilicic acids to be very reproducible and to vary inversely with the degree of polymerization. High molecular weight polymers, "colloidal silica," do not react with molybdic acid to form molybdosilicic acid.

The conversion of polymerized silicic acid to monosilicic acid can be effected by heating a sodium carbonate or sodium hydroxide solution for 30–60 min in a platinum dish. Silica can be solubilized to give the monosilicic acid by a sodium carbonate fusion in a platinum crucible and dissolution of the melt in water.

I. SEPARATIONS

The main interfering substances in spectrophotometric methods for the determination of silicon are phosphorus, arsenic, germanium, iron, barium, bismuth, lead, cerium, antimony, aluminum, thorium, uranium, zirconium, and fluoride. Phosphate, arsenate, and germanate ions are the most serious potential interfering species because they also form molybdoheteropoly acids. Often citric, tartaric, or oxalic acid is added following the formation of molybdosilicic acid to complex the excess molybdate and thereby promote the decomposition of those less stable molybdoheteropoly complexes.

A. PRECIPITATION FOLLOWING DEHYDRATION METHOD

Silicic acid is partially decomposed in highly acidic solution to give a silica precipitate. However, in order to dehydrate all of the silicic acid and any hydrated silica, it is necessary to employ a very effective dehydration process. Although evaporation to dryness of hydrochloric acid solutions followed by a baking period, or boiling and evaporation almost to dryness of sulfuric acid solutions, will dehydrate the silica, the use of perchloric acid is recommended. Upon evaporation, a perchloric acid solution ultimately contains 72% perchloric acid and, at this composition, corresponding to $HClO_4 \cdot 2H_2O$, is a very efficient dehydrating agent and will completely dehydrate silicic acid and hydrated silica within 5 min at its boiling point (about 200°C).

Another advantage of using the perchloric acid treatment is that perchlorate salts will readily dissolve in water. Because dehydrated silica will hydrolyze rather readily to form silicic acid, it is necessary to remove the precipitated silica by filtration as rapidly as possible. A warning is given to remind the reader that hot, concentrated perchloric acid will react with organic matter so violently that an explosion often results. Hence, if organic material is present in the sample, a preliminary oxidation using nitric acid is suggested. The silica residue on filter paper should be rinsed thoroughly with water and finally once with dilute ammonia solution prior to ignition of the filter paper.

Perchloric acid has been used successfully for separating amounts of silicon as low as 0.1 mg from known synthetic samples, high-purity nickel, sulfide phosphorus, and NBS alloys with consistently good results (50). Details of the perchloric acid dehydration procedure are given in Section III of this chapter.

The removal of phosphate by precipitation of calcium phosphate from a borate-buffered solution at pH 10 before the colorimetric determination of silica in boiler water has been studied by Schwartz (54). However, he later determined silica in the presence of phosphate by destroying the molybdophosphoric acid by adding oxalic acid (55). Citric acid (34) and tartaric acid (12,13) have also been used to circumvent the interference of phosphate and arsenate.

B. DISTILLATION OR VOLATILIZATION METHOD

Silicon tetrafluoride can be distilled from a sulfuric acid or perchloric acid solution. Holt (28) has isolated microquantities of silicon by the distillation of silicon tetrafluoride from perchloric acid solution using a special all-platinum uniformly heated distillation apparatus. An absorbing solution containing boric acid, sulfuric acid, and ammonium molybdate was used to collect the distillate prior to colorimetric determination by the heteropoly blue method.

Silicon tetrafluoride has also been volatilized using the Conway micro-diffusion technique and a polypropylene cell. Ethylene glycol was the absorbent, and the heteropoly blue spectrophotometric method was used to quantitate the 10–35 μg silicon isolated (3). Arsenic and germanium can be removed from silicic acid by volatilization of arsenic(III) chloride and germanium(III) chloride upon evaporation of hydrochloric acid solutions. The corresponding bromides of arsenic and germanium can also be volatilized. It is advisable to convert to monosilicic acid after removal of impurities by volatilization.

Another volatilization method utilizes a dry hydrogen chloride gas sweep to remove certain elements from silica (20). This method involves placing the filter paper containing the contaminated silica in a platinum combustion boat and igniting the filter paper. The platinum boat is then transferred to a tube in the combustion furnace and heated to about 700°C with a slow stream of dry hydrogen chloride gas passing through the tube not exceeding 35–40 ml per min. A cover should be arranged over the boat to avoid possible contamination from the tube. Because organic matter is sometimes present in commercial hydrogen chloride gas, it is advisable to use scrubbing bottles containing carbon tetrachloride in the system as well as a scrubbing bottle containing sulfuric acid to remove any moisture. Of the elements that are likely to be found as contaminants when the silica is dehydrated by perchloric acid dehydration, iron, antimony, tin, germanium, and tungsten are removed by this treatment. Arsenic, selenium, mercury, and chromium can also be volatilized by this method.

A rapid method for separating silicon tetrafluoride from a sample solution to an absorbing solution by a current of dry air has been developed by Hozdic (29). The use of dry air, a special concentrated sulfuric acid, and the avoidance of premature condensation are essential factors in the separation of micro amounts of silicon.

C. EXTRACTION METHOD

One of the most satisfactory methods of partitioning phosphate and arsenate from silicate involves a selective extraction of the corresponding molybdoheteropoly acids. Clabaugh and Jackson (16) devised a method based on the following sequential operation: (a) formation of molybdophosphoric acid, molybdoarsenic acid, and molybdosilicic acid at pH 1.8; (b) extraction of molybdophosphoric acid with diethyl ether from an approximately $1.2M$ hydrochloric acid; (c) extraction of molybdosilicic acid with butanol from a $2.2M$ hydrochloric acid solution; and (d) extraction of molybdoarsenic acid with methyl isobutyl ketone from hydrochloric acid readjusted to $1.2M$. This separation method was used advantageously by Hurford and Boltz (30) in the sequential indirect ultraviolet spectrophotometric and atomic absorption spectrometric determination of phosphorus and silicon. Similar sequential extraction methods for involving molybdosilicic acid have been devised (21,32). Fresenius and Schneider (21) used a 1:1 benzene–isobutyl alcohol mixture to extract molybdophosphoric acid prior to determination of silicon.

Antimony, bismuth, iron, molybdenum, niobium, tantalum, tin, titanium, tungsten, vanadium, and some copper may be removed by chloroform extraction of their cupferron precipitates (22). To the cool solution, about $3.5–4.0N$ with sulfuric acid, add a 5% solution of cupferron slowly with constant stirring until no further precipitation takes place and crystals of the reagent appear. Then add an excess of the reagent and allow to stand for at least 5 min. Extract with chloroform until the chloroform layer is colorless. Transfer the aqueous layer to a platinum dish and heat gently to evaporate remaining traces of chloroform.

Many cations can be extracted from acidic solution as 1-pyrrolidine-carbodithioate complexes into chloroform. Hence, at pH 1 or about $0.1N$ hydrochloric acid, copper(II), nickel(II), cobalt(II), lead(II), arsenic(III), antimony(III), and bismuth(III) are extracted quantitatively; while at pH 2, iron(III) and vanadium(V) are also extracted completely and tin(II) and zinc(II) are extracted to the extent of about 95% (37). Methyl isobutyl ketone has also been used as an extractant (43). Platinum and palladium chelates of the 1-pyrrolidine-carbodithioic acid have also been extracted with chloroform (51). To an aliquot of the sample solution in polyethylene beaker, add hydrochloric acid or sodium hydroxide until the pH is 1–2. Transfer to a polyethylene separatory funnel, add 5 ml of a 0.2% solution of ammonium pyrrolidinecarbodithioate, and mix thoroughly. Add 10 ml chloroform, shake vigorously for at least 1 min, and remove chloroform extract. Transfer the aqueous phase to Teflon beaker, heat to boiling for 5 min, add slowly 1 ml concentrated nitric acid and continue heating to destroy excess reagent.

D. ION EXCHANGE METHOD

This method of separation is applicable to the removal of numerous interferences from solutions containing silicic acid. Cations are removed from nitric acid solution (pH = 1) using strongly acidic ion exchange resin in the H or Na form. Pietri and Wenzel (49) have removed plutonium from silicic acid, and Nemodruk, Palei, and Bezrogova (46) have removed iron(III), aluminum(III), and uranyl ions from silicic acid solution and demonstrated complete recovery of the α-silicic acid in the effluent from the cation exchange column. Polymeric forms of silicic acid are retained by ion exchange columns.

The following procedure has been recommended by Nemodruk, Palei, and Bezrogova (46) to ensure complete conversion to α-silicic acid prior to using cation exchanger columns. Transfer a 10-ml aliquot of an acidic sample solution containing 10–500 μg silicon to a Teflon beaker. Neutralize with a 20% sodium hydroxide solution until pH is about 3 and add additional 3 ml of the sodium hydroxide solution. Heat the solution on a boiling water bath for 30 min. Cool and add this basic solution dropwise with vigorous stirring to 2 ml concentrated nitric acid in a Teflon beaker. Neutralize the excess acid with a $2N$ sodium hydroxide solution until the pH is about 1. Dilute the solution to about 20 ml and transfer to a cation exchange column in the H form. Wash the column with 20 ml distilled water and combine effluent and washings before diluting to 50 ml.

Anions of strong acids, such as PO_4^{3-} and AsO_4^{3-}, can be removed by weakly basic anion exchange resins. Sussman and Portnoy (60) removed chromate ($pK_1 = 0.8$) with a weakly basic anion exchange resin in the Cl form after removal of heavy metal cations, such as iron(II) and copper(II), with a strong cationic exchanger. Andersson (9) and Luke (36) separated phosphate from silicic acid on a weakly basic anion exchange resin. The following procedure is essentially that of Sussman and Portnoy. Fill an ion exchange column to a depth of at least 15 cm with weakly basic anion resin, 20–50 mesh, which has been thoroughly soaked in water. After back washing, convert the exchanger to the chloride by passing 20 ml 5% hydrochloric acid through the tube at 10 ml per min followed by distilled water at the same rate until the effluent acidity to methyl orange has dropped below 2 mEq per liter.

Pass the filtered sample solution through the exchanger column at the rate of about 10 ml per min, rejecting the first 75–100 ml of effluent which will be diluted by the distilled water occupying the voids in the exchanger column. Collect the next 10–25 ml for determination of silicon by one of the colorimetric methods given in Section II.

When necessary to regenerate the exchange column, pass 1500 ml 1:1 ammonium hydroxide through at a rate of 10 ml per min, followed by 500 ml distilled water at the same rate. Convert to the chloride salt with hydrochloric acid as described above.

E. ELECTRODEPOSITION METHOD

The following elements can be deposited on the platinum cathode in electrodeposition from mineral acid solutions: silver, gold, mercury, copper, cadmium, lead, antimony, bismuth, rhenium, platinum, rhodium, and palladium. Zinc, cobalt, and nickel can be removed by electrodeposition from basic solution.

Electrolysis with a mercury cathode in dilute sulfuric acid solution will remove large quantities of arsenic, antimony, bismuth, cadmium, chromium, cobalt, copper, gallium, germanium, indium, iron, lead, mercury, molybdenum, nickel, rhenium, selenium, silver, tellurium, tin, zinc, gold, and the platinum metals, except ruthenium. Manganese is partially removed. Elements remaining in solution include aluminum, beryllium, magnesium, phosphorus, titanium, vanadium, zirconium, the alkaline earth metals, and the rare earth metals.

Prepare the mercury cathode cell with sufficient mercury and transfer the solution, about 1–2N with sulfuric acid, to it and electrolyze with a current of about 0.15 ampere with constant stirring. When a test drop gives no test for the elements to be removed, drain off the solution and wash the mercury with distilled water while the current is being partially continued. Filter the solution, adjust the pH to 4–5, make up to volume, and take suitable aliquots for the determination of silicon.

Minster (39–41) used the mercury cathode separation method followed by the heteropoly blue method for the determination of silicon. However, Luke (35) observed fading of the heteropoly blue following electrolysis.

II. METHODS OF DETERMINATION

A. MOLYBDOSILICIC ACID METHOD (18,33)

When excess molybdate solution is added to a solution containing monomeric silicic acid, a yellow 12-molybdosilicic acid forms. According to Strickland (59), solutions acidified with less than 1.5 mole acid per mole molybdate resulted in an α-12-molybdosilicic acid being formed while if two or more moles of acid were added per mole of molybdate a β-12-molybdosilicic acid was formed. The β-acid has a higher absorptivity than the α-acid and changes spontaneously into the α-acid, a transition accelerated by heating of the solution. Recently, Hargis (25) has made a spectrophotometric kinetic study of the formation of β-12-molybdosilicic acid and found that, whereas the reaction rate was first order in silicate and molybdate concentrations and independent of acid concentration at low acidities, at higher acidities the reaction rate was first order in silicate, sixth order in molybdate, and inverse-seventh order in acid concentration. He postulated a mechanism in which a fast initial reaction product is a silicate molybdate intermediate which then undergoes a much slower reaction with additional molybdate to produce the β-molybdosilicic acid. Excess molybdate was observed to slow the conversion from β-acid to α-acid.

Hargis (26) has also elucidated the reaction stoichiometry for the α-acid, indicating that six molybdate species combine with one silicate liberating hydrogen ions, thus indicating the probability that the molybdate species is dimeric. A spectrophotometric method based on the initial rate method has been developed on the basis of these fundamental studies on the properties of molybdosilicic acid. The molar absorptivity for α-12-molybdosilicic acid at 350 nm is $(5.88 \pm 0.24) \times 10^3$ liter $\cdot$ mole$^{-1} \cdot$ cm^{-1}.

In general, a pH of 1.4–1.6 (11,61) is recommended for the formation of molybdosilicic acid.

Special Solutions

Stock Silicate Solution. Dissolve 5.50 g sodium metasilicate nonahydrate, $Na_2SiO_3 \cdot 9H_2O$, in redistilled water and dilute to 1 liter. Standardize 100-ml aliquots of this stock solution gravimetrically.

Standard Silicate Solution. Use a microburet to transfer sufficient stock silicate solution to a 1-liter volumetric flask so that on dilution to the mark with redistilled water the final solution contains 0.020 mg silicon per ml. Store in a paraffin-lined bottle to avoid contamination.

Molybdate Reagent. Dissolve 50.0 g ammonium molybdate, $(NH_4)_6Mo_7O_{24} \cdot 4H_2O$, in about 400 ml redistilled water in a polyethylene beaker. When dissolution is complete, add "silica free" ammonia to adjust to pH 7.5–8. Filter if necessary, transfer to a 500-ml polyethylene graduated cylinder, and dilute to 500 ml with redistilled water. Store in a polyethylene bottle.

Tartaric Acid Solution, 10%. Dissolve 50.0 g tartaric acid in redistilled water and dilute to 500 ml. Store in a polyethylene bottle.

Hydrochloric acid solution, 7N. Add 58 ml concentrated hydrochloric acid to 40 ml silica-free water in a 100-ml volumetric flask and dilute to mark.

Procedure

Transfer a 50-ml aliquot of sample solution containing 0.05–0.4 mg silicon (0.1–0.8 mg SiO_2) to a 100-ml polyethylene beaker. Add 0.10 g "silica free" sodium carbonate and heat on a steam bath for 1 hr to depolymerize the silica. Cool the solution to room temperature and add 1 ml 7N hydrochloric acid solution. Transfer the solution immediately to a 50-ml Nessler tube and adjust the solution to mark with silica-free water. Add immediately 2.0 ml of the molybdate reagent and mix thoroughly. After 10 min, add 1.5 ml of the tartaric acid solution and mix thoroughly. Measure the absorbance 10 min after addition of tartaric acid at 350 nm, using 1.000-cm cells and, in the reference cell, a reagent blank solution prepared in the same manner as the sample (Fig. 1).

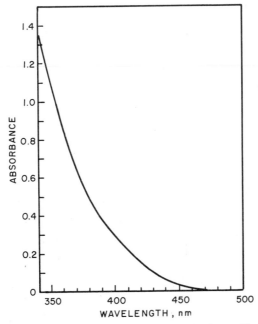

Fig. 1.   Absorption spectrum of yellow molybdosilicic acid (8 ppm Si, 1-cm cells, reagent blank solution used in reference cell).

B. EXTRACTION–MOLYBDOSILICIC ACID METHOD (19)

The advantage of this method involving the isolation of the molybdosilicic acid by extraction is the possibility of improving the precision by concentrating the absorption species in a smaller volume of organic solvent and thereby operate in the optimum concentration range.

Special Solution

Molybdate Reagent. Dissolve 8.0 g ammonium molybdate, $(NH_4)_6Mo_7O_{24} \cdot 4H_2O$, in redistilled water and dilute to 100 ml.

Sulfuric Acid, 10N. Dissolve very carefully 282 ml concentrated sulfuric acid (95%; sp. gr. 1.83) in redistilled water and, when cool, dilute to 1 liter.

Boric Acid Solution, 5%. Dissolve 5.0 g boric acid, H_3BO_3, in about 90 ml redistilled water and dilute to 100 ml.

Procedure

Transfer a 25-ml aliquot of sample solution containing not more than 200 μg silicon to a Teflon beaker. Add 2 ml 1:3 hydrofluoric acid and heat on a water bath at 60–70°C for 30 min. Cool, add 20 ml of a 5% boric acid solution, and allow solution to stand for 15 min.

Add 5 ml of the molybdate reagent, adjust to pH 2 with silica-free ammonia solution, and heat at about 100°C for 30 min. Cool to room temperature, transfer to 100-ml volumetric flask, dilute to the mark, and mix.

Transfer a 25-ml aliquot to a polyethylene separatory funnel and add 10 ml of the 10N sulfuric acid. Add 10.0 ml n-amyl alcohol and shake vigorously for 2 min. After removing the aqueous phase, add 10 ml 2N sulfuric acid solution and wash the organic phase. Measure the absorbance of the organic extract in 1.0- or 2.0-cm cells at 350 nm using an extract from a reagent blank solution in the reference cell.

C. HETEROPOLY BLUE METHOD (11)

This method is the most extensively used method for the spectrophotometric determination of traces of silicon. Numerous studies have been made to delineate optimum solution parameters for determining silicon in a variety of materials. The heteropoly blue of silicon is produced by the reduction of four or more equivalents of molybdenum(VI) in the 12-molybdosilicic acid to the molybdenum(V) state to give a charge transfer-type absorber. The method is applicable to the determination of 10–150 μg silicon, assuming that the final volume of the solutions is 100 ml, that 1.0-cm absorption cells are used, and that the absorbance is measured in the near-infrared region, namely, 815–820 nm.

The first precaution is to maintain proper pH for the formation of 12-molybdosilicic acid. A pH of 1–2 is optimum, although pH 0.8–3.8 has been reported as satisfactory (11,33,38,61). Once formed, a solution of 12-molybdosilicic acid can be acidified to a 2–3 molarity without decomposition. At low acidity, pH 1–4, many reductants, such as tin(II) chloride, will reduce the excess molybdate to form a "molybdenum blue" which interferes seriously with the measurement of the absorbance of the "heteropoly blue" of silicon. Tartaric acid is often added to destroy any molybdophosphoric acid which might form; the formation constant for citromolybdate complex is evidently much larger than that for the molybdophosphoric acid. Complexation of the excess molybdate also inhibits formation of the undesirable "molybdenum blue." Tartaric acid also forms stable complexes with aluminum(III), iron(III), and zinc(II) ions. Barium, bismuth, lead, and antimony should be absent, as these ions tend to precipitate and cause turbidity.

Many different reductants have been used with a wide variation in acidity. The reduction products often have characteristic absorbance maximum at different wavelengths and with different molar absorptivities (56). Therefore, each heteropoly blue method should be checked to ascertain optimum wavelength for measurement of the absorbance with the spectrophotometer being used.

Special Solutions

Molybdate Reagent. Dissolve 18.8 g ammonium molybdate, $(NH_4)_6 Mo_7 O_{24} \cdot 4H_2O$, in silica-free water containing 23 ml concentrated sulfuric acid and diluting to 250 ml.

Reductant. Prepare solution A by dissolving 25 g sodium hydrogen sulfite, $NaHSO_3$, in 200 ml silica-free water. Prepare solution B by dissolving 2 g anhydrous sodium sulfite, $Na_2 SO_3$, in 25 ml silica-free water and add 0.4 g 1-amino-2-naphthol-4-sulfonic acid. Mix solutions A and B and dilute to 250 ml.

Tartaric Acid, 10%. Dissolve 10 g d-tartaric acid, $H_2 C_4 H_4 O_6$, in about 75 ml silica-free water and dilute to 100 ml.

Standard Silicon Solution (1 ml = 10.0 μg Si). Dissolve 25 g sodium metasilicate, $Na_2 SiO_3 \cdot 9H_2O$, in water in a stainless steel or polyethylene beaker, dilute to 500 ml, and store in a polyethylene bottle. Standardize the stock solution by gravimetric determination in the following manner. To 50-ml aliquots add 8–10 ml perchloric acid and evaporate to copious fumes of perchloric acid (57). With cover glasses over the beakers, continue the boiling for 15 to 20 min so that the acid refluxes down the sides of the beakers. Allow to cool somewhat and then dilute to 25–30 ml with 1:99 hydrochloric acid. Filter, wash with 1:99 hydrochloric acid, and ignite in platinum, first at a low temperature to burn off the carbon of the filter paper and finally at 1050–1100°C to constant weight. Add one or two drops sulfuric acid and a few milliliters hydrochloric acid, evaporate until all the sulfuric acid has been expelled from the crucible, and again ignite and weigh. Silica is the difference between the first and second weights. Calculate the required volume of stock solution required to make 1000 ml solution containing 0.01 mg silicon per ml, dilute to 1000 ml with water, and store in a polyethylene bottle.

Procedure

Transfer a 50- or 75-ml aliquot of a sample solution neutral to litmus and containing not more than 200 μg silicon to a 100-ml volumetric flask. Dilute to about 90 ml add 1.0 ml of the molybdate reagent, and allow to stand for 5 min.

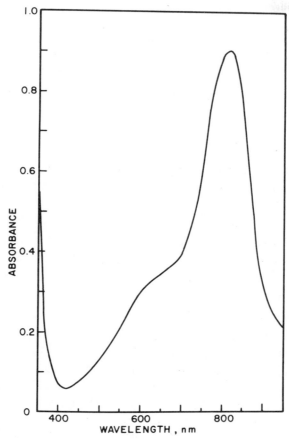

Fig. 2.   Absorption spectrum of heteropoly blue of silicon (1.2 ppm Si, 1-cm cells, water in reference cell).

Add 4 ml of the 10% tartaric acid and mix. Add 1.0 ml of the reductant, dilute to the mark, and mix. After 20 min, measure the absorbance at 815 nm, using 1.0-cm cells and a reagent blank solution in the reference cell.

Preparation of Calibration Graph. Transfer 0, 1.0, 2.0, 3.0, 4.0, 5.0, 6.0, and 7.0 ml of the standard silicon solution (1 ml = 10.0 μg Si) to 100-ml volumetric flasks, dilute to about 90 ml, and add 1 ml of the molybdate reagent. After 5 min, add 4.0 ml of the 10% tartaric acid solution and mix. Add 1 ml of the reductant, dilute to the mark, and mix. After 20 min, measure the absorbance at 815 nm in 1.000-cm cells using the reagent blank solution (0 ml standard silicon solution) in the reference cell (Fig. 2). Plot absorbance values versus μg silicon per ml solution.

D. EXTRACTION–HETEROPOLY BLUE METHOD

The heteropoly blue method can be modified by extracting the 12-molybdosilicic acid from aqueous solution into an immiscible organic solvent and thereby circumventing the deleterious effect of high concentrations of certain diverse ions. Several procedures have been developed based on the extraction of molybdosilicic acid prior to reduction to produce the heteropoly blue of silicon (16,47,48). The procedure of Trudell and Boltz (62) using 1,2-propanediol carbonate and chloroform as extractant and tin(II) chloride as reductant will be presented.

This method is applicable to the determination of 0.5–6 μg silicon. Less than 25 ppm phosphate, arsenate, and fluoride do not interfere.

Special Solutions

Acidic Molybdate Solution. Dissolve 25.0 g ammonium molybdate, $(NH_4)_6Mo_7O_{24} \cdot 4H_2O$, in silica-free water. Add 10.0 ml concentrated sulfuric acid (sp. gr. 1.84; 96% by wt. H_2SO_4), and dilute to 500 ml.

Standard Silicate Solution. Dissolve 5.50 g sodium metasilicate, $Na_2SiO_3 \cdot 9H_2O$ in water and dilute to 1 liter. Standardize this solution gravimetrically, then use a microburet to transfer sufficient silicate solution to a 1000-ml volumetric flask so that, on dilution to the mark, the final silicate solution contains 0.002 mg silicon per ml.

Sulfuric Acid Rinse Solution. Dissolve 226 ml concentrated sulfuric acid in water and dilute to 1 liter.

Tin(II) Chloride Solution. Dissolve 2.38 g $SnCl_2 \cdot 2H_2O$ in 170 ml concentrated hydrochloric acid. Dilute to 1 liter with water and add tin pellets to stabilize the solution. This solution should be prepared fresh each month.

Procedure

Transfer a sample containing 0.5–6.0 μg silicon in the form of soluble silicate to a 50-ml volumetric flask. Dilute the sample to approximately 40 ml with distilled water. Add 5.0 ml of the acidic ammonium molybdate solution, then dilute to the mark with distilled water. The pH of this solution should be about 1.4. After letting the solution stand 20 min or longer to allow for complete formation of the molybdosilicic acid, transfer the solution to a 125-ml separatory funnel. Rinse the flask twice with a total of 20 ml of the sulfuric acid rinse solution and add the rinsings to the separatory funnel. Transfer 10 ml propylene carbonate to the separatory funnel and gently shake the stoppered funnel to dissolve the organic solvent. Pipet 0.8 ml chloroform into the separatory funnel, followed by 15 ml of the tin(II) chloride solution. Shake the separatory funnel vigorously on the mechanical shaker for 1 min. Allow about 4 min for the aqueous and organic layers to separate, then transfer the chloroform extract of the heteropoly blue of the molybdosilicic acid to a 5-ml volumetric flask. Repeat the extraction step on the aqueous solution in the separatory funnel with 0.4 ml chloroform.

Combine the chloroform extracts in the volumetric flask and dilute to the mark with propylene carbonate. Measure the absorbance of the organic extract from the sample solution at 775 nm against chloroform in the reference cell. The absorbance of the organic extract from a reagent blank solution should also be measured at 775 nm against chloroform in the reference cell. The latter extract is prepared by treating a mixture of 45 ml distilled water and 5.0 ml acidic ammonium molybdate solution according to the procedure described for the solution containing soluble silicate. Subtract the absorbance of the blank from the measured absorbance.

E. INDIRECT ULTRAVIOLET SPECTROPHOTOMETRIC METHOD

A very sensitive, indirect method that is based on the absorptivity of molybdate ions resulting from the decomposition of molybdosilicic acid was developed by Trudell and Boltz (61) and modified by Hurford and Boltz (30). The essential steps involved in the procedure include (a) the formation of molybdosilicic acid, (b) the extraction of the yellow heteropoly acid with an immiscible solvent mixture, (c) the decomposition of the extracted molybdosilicic acid with a basic buffer solution, and (d) the measurement of the ultraviolet absorptivity of the molybdate ion. Twelve molybdate ions are measured in the final buffered solution for every one silicon atom that is present in the original solution taken for analysis. The absorbance measurements are made at 230 nm. Conformity to Beer's law is observed for 0.02–0.5 ppm silicon, and the optimum concentration range is 0.06–0.5 ppm silicon.

Ions that do not interfere at a concentration of 500 ppm include aluminum, ammonium, bromide, calcium, chloride, chromium(III), cobalt(II), manganese(II), nickel(II), oxalate, perchlorate, potassium, silver, sodium, sulfate, and zinc. Arsenate, arsenite, iron(III), and phosphate cause serious interference and must be absent or removed from the sample. Acetate, lead(II), nitrate, and vanadate can be tolerated in concentrations up to 250 ppm, whereas dichromate, nitrite, and tungstate should not exceed 50, 20, and 100 ppm, respectively. Up to 10 ppm copper(II) and fluoride can be tolerated.

Special Solutions

Standard Silicate Solution. Dissolve 5.50 g sodium metasilicate, $Na_2SiO_3 \cdot 9H_2O$, in water and dilute to 1 liter. Standardize this solution gravimetrically and use a microburet to transfer sufficient silicate solution to a 1-liter volumetric flask so that, on dilution to the mark, the final silicate solution contains 0.0100 mg silicon per ml.

Molybdate Solution. Dissolve 50 g ammonium molybdate, $(NH_4)Mo_7O_{24} \cdot 4H_2O$, in water and dilute to 500 ml.

Buffer Solution. Dissolve 53.5 g ammonium chloride and 70 ml concentrated ammonium hydroxide in water and dilute to 1 liter. This solution is $1M$ in ammonia and ammonium chloride.

Extractant. Mix five volumes diethyl ether and one volume 1-pentanol. Prepare all of the aqueous solutions with reagent-grade chemicals and redistilled water and store them in paraffin-lined bottles to avoid silica contamination. The extractant mixture should be prepared fresh daily with reagent-grade solvents.

Procedure

Transfer a sample containing $2-50 \mu g$ silicon in the form of soluble silicate to a 125-ml separatory funnel. Add 1.0 ml 1:1 hydrochloric acid and dilute the acidified sample solution to about 45 ml with redistilled water. Add 5.0 ml of the ammonium molybdate solution, mix, and allow to stand for 20 min to assure complete formation of the molybdosilicic acid. The pH of the solution should be about 1.4. Add 30 ml 1:1 hydrochloric acid and 20 ml 5:1 ethyl ether—1-pentanol extractant mixture. Shake the separatory funnel vigorously on a mechanical shaker for 1 min. Rinse the stopper with 1–2 ml fresh extractant mixture and collect the rinsings in the separatory funnel. Allow the layers to separate, then withdraw and discard the lower aqueous layer. Wash the organic phase twice, each time with 25 ml 1:10 hydrochloric acid. Each washing consists of shaking the organic phase with the wash solution for 15 sec, allowing the two phases to separate, and withdrawing and discarding the lower aqueous phase. After the second wash step is completed, rinse the funnel tip with distilled water to assure the complete removal of excess molybdate ions.

Transfer 40 ml of the ammonium chloride–ammonium hydroxide buffer solution to the separatory funnel and shake vigorously for 20 sec. Withdraw the lower aqueous phase into a 100-ml volumetric flask. Rinse the funnel stopper and the interior walls of the funnel twice, each time with several milliliters distilled water, and transfer the rinsings to the volumetric flask.

Dilute the solution in the 100-ml volumetric flask to the mark with distilled water and mix well. The pH of the final aqueous solution should be about 9. Measure the absorbance at 230 nm against a reagent blank in matched 1.000-cm silica cells. The reagent blank is prepared according to the procedure used for the silicate samples, but distilled water is substituted for an aliquot of a silicate sample. Refer the absorbance values to a standard calibration graph obtained by using standard silicate solutions.

F. INDIRECT 2-AMINO-4-CHLOROBENZENETHIOL METHOD

The molybdate ions resulting from the decomposition of molybdosilicic acid form an insoluble complex with 2-amino-4-chlorobenzenethiol hydrochloride at pH 2 which can be extracted by chloroform. The amount of the molybdenum complex is equivalent to a specific amount of silicon in the original aliquot of silicate sample. In chloroform solution, the complex is green and shows maximum absorbance at 715 nm. This indirect method was developed by Trudell and Boltz (64) and is applicable to the determination of 0.05–0.20 ppm silicon equivalent as molybdenum complex in chloroform solution. The optimum concentration range is 0.075–0.175 ppm silicon equivalent as the molybdenum complex in chloroform solution (3.0–7.0 μg silicon in the original sample aliquot).

Silicon may be determined without interference in the presence of 100 ppm of the following ions: acetate, bromide, calcium, chloride, chromium(III), cobalt(II), copper(II), magnesium, manganese(II), molybdate, nickel(II), nitrate, oxalate, perchlorate, permanganate, potassium, silver, sodium, sulfate, and tungstate. Up to 50 ppm aluminum and zinc can be tolerated, whereas only 25 ppm arsenite and lead(II) are permitted. Arsenate and phosphate interfere at concentrations exceeding 10 ppm. Dichromate, fluoride, iron(III), nitrite, and vanadate must be absent or removed from the sample.

Special Solutions

Standard Silicate Solution. Dissolve 5.50 g sodium metasilicate, $Na_2SiO_3 \cdot 9H_2O$, in water and dilute to 1 liter. Standardize this solution gravimetrically and use a microburet to transfer sufficient silicate solution to a 1-liter volumetric flask so that, on dilution to the mark, the final silicate solution contains 2.0 μg silicon per ml.

Molybdate Solution. Dissolve 25.0 g ammonium molybdate, $(NH_4)_6Mo_7O_{24}$ · $4H_2O$, in water and dilute to 500 ml.

Ammonia Solution. Prepare a 0.25M ammonia solution by diluting 17.5 ml concentrated ammonium hydroxide to 1 liter with water.

Buffer Solution. Dissolve 106.4 g anhydrous sodium sulfate and 20.68 g sodium bisulfate, $NaHSO_4$ · H_2O, in water and dilute to 1 liter. This solution is 0.75M in sodium sulfate and 0.15M in sodium bisulfate.

Reagent Solution. Dissolve 0.25 g 2-amino-4-chlorobenzenethiol hydrochloride (Eastman Kodak No. 3279) in 5.0 ml 95% ethanol. Prepare this reagent solution fresh every two days.

Extractant Mixture. Mix five volumes diethyl ether with one volume 1-pentanol. All aqueous solutions should be prepared with reagent-grade chemicals and redistilled water and stored in paraffin-lined bottles to avoid silica contamination. Prepare the extractant mixture daily with reagent-grade solvents.

Procedure

Transfer a sample containing 2–8 μg silicon in the form of soluble silicate to a 125-ml separatory funnel. Add 1.0 ml 1:1 hydrochloric acid and dilute the acidified sample solution to about 45 ml with redistilled water. Add 5.0 ml of the ammonium molybdate solution, mix, and allow the solution to stand for 20 min to assure complete formation of the molybdosilicic acid. The pH of this solution should be about 1.4. Add 30 ml 1:1 hydrochloric acid and 20 ml extractant mixture. Shake the separatory funnel vigorously on a mechanical shaker for 1 min. Rinse the stopper with 1–2 ml fresh extractant mixture and collect the rinsings in the separatory funnel. Allow the layers to separate, then withdraw the lower aqueous phase and discard it. Wash the organic phase twice, each time with 25 ml 1:10 hydrochloric acid. Each washing consists of shaking the organic phase with the wash solution for 15 sec, allowing the two phases to separate, and withdrawing and discarding the lower aqueous phase. After the second wash step is completed, rinse the funnel tip with distilled water.

Add 20 ml of the 0.25M ammonia solution to the funnel and shake vigorously for 30 sec. After the layers have separated, carefully transfer the lower aqueous layer containing the decomposition products of molybdosilicic acid to a 100-ml volumetric flask. Rinse the funnel and funnel stopper several times with a total of 20–30 ml distilled water and add the rinsings to the volumetric flask. Acidify the flask contents with 4.0 ml 1:10 hydrochloric acid solution. Add 25 ml of the sulfate–bisulfate buffer solution, dilute to the mark with distilled water, and mix well. The pH of this solution should be between 1.9 and 2.3.

Transfer a 25.00-ml aliquot of the buffered molybdate solution to a clean 125-ml separatory funnel that has a cotton plug inserted into its tip. Numerous samples may be simultaneously treated up to this point of the procedure, and the aliquots of the buffered molybdate solutions may be allowed to stand in their respective separatory funnels until they are used to complete the procedure that follows.

Add 0.25 ml of the ethanolic reagent solution, swirl the funnel to mix its contents, and allow the solution to stand 15—60 min. Add 10.00 ml chloroform, start a mechanical timer, and shake the funnel vigorously on a mechanical shaker for 1 min. Allow the phases to separate for several seconds, then rinse the 1.000-cm silica cells with small portions of the chloroform extract that have been drained through the cotton plug. Finally, fill the cell with the green chloroform extract, and exactly 3 min after the chloroform extraction step was begun, place the cell in the spectrophotometer and quickly measure the absorbance of the extract at 715 nm with chloroform in the reference cell. Correct for the absorbance of a reagent blank extract. Refer the corrected absorbance to a standard calibration graph.

So that the analyst may use his time most efficiently, the following sequence of steps is recommended for the completion of the determination of a series of sample solutions:

1.   Add the ethanolic reagent solution to three of the separatory funnels containing 25.00-ml aliquots of the buffered molybdate solution.

2.   After 15 min, add the ethanolic reagent solution to three other aliquots of the buffered molybdate solutions.

3.   Individually complete the extraction and measurement of each of the three solutions prepared in step 1. This will require a total of 12—15 min.

4.   Add the ethanolic reagent solution to three more aliquots of buffered molybdate solutions.

5.   Individually complete the extraction and measurement of each of the three solutions prepared in step 2.

Continue the sequence of steps until all the molybdate aliquots have been taken completely through the recommended procedure.

G. INDIRECT PEROXYMOLYBDIC ACID METHOD

Molybdate ions from the decomposition of molybdosilicic acid will react with hydrogen peroxide in strongly acidic solution to form peroxymolybdic acid, which absorbs in the ultraviolet region and shows maximum absorbance at 330 nm. The absorbance of the peroxymolybdic acid solution at 330 nm is a function of the silicon concentration in the aliquot of the original silicate sample. This indirect peroxymolybdic acid method for the determination of silicon, developed by Trudell and Boltz (63), is relatively simple, rapid, and sensitive. It suffers from fewer and less serious interferences than do the other indirect spectrophotometric methods reported in this chapter. This method also has the advantage of being applicable over an unusually wide concentration range, with conformity to Beer's law observed for 0.1 to 3.2 ppm silicon. The optimum concentration range is 0.25–2.5 ppm silicon.

Ions that do not cause interference at a concentration of 500 ppm are acetate, aluminum, ammonium, bromide, calcium, chloride, chromium(III), cobalt(II), copper(II), iron(III), lead(II), manganese(II), molybdate, nickel(II), nitrate, oxalate, perchlorate, permanganate, potassium, sodium, sulfate, tungstate, and zinc. Silver does not interfere if the final solution of peroxymolybdic acid is filtered before it is measured spectrophotometrically. Arsenite can be tolerated up to a concentration of 250 ppm. Fluoride, dichromate, and phosphate must not exceed 50 ppm, whereas arsenate, nitrite, and vanadate cannot be tolerated in excess of 25 ppm.

Special Solutions

Standard Silicate Solution. Dissolve 5.50 g sodium metasilicate, $Na_2SiO_3 \cdot 9H_2O$, in water and dilute to 1 liter. Standardize this solution gravimetrically and use a microburet to transfer sufficient silicate solution to a 1-liter volumetric flask so that, on dilution to the mark, the final silicate solution contains 2.0 μg silicon per ml.

Molybdate Solution. Dissolve 50 g sodium molybdate, $Na_2MoO_4 \cdot 2H_2O$, in water and dilute to 500 ml.

Ammonia Solution. Dilute 35 ml concentrated ammonium hydroxide to 1 liter with water.

Extractant. Mix five volumes diethyl ether and one volume 1-pentanol.

All the aqueous solutions should be prepared with reagent-grade chemicals and redistilled water and stored in paraffin-lined bottles to avoid silica contamination. The extractant mixture should be prepared daily with reagent-grade solvents.

Procedure

Transfer a sample containing 5–170 μg silicon in the form of soluble silicate to a 125-ml separatory funnel. Add 1.0 ml 1:1 hydrochloric acid and dilute the acidified sample solution to about 45 ml with redistilled water. Add 5.0 ml of the sodium molybdate solution, mix, and allow the solution to stand for 20 min to assure complete formation of the molybdosilicic acid. The pH of this solution should be about 1.4. Add 30 ml 1:1 hydrochloric acid and 20 ml 5:1 ethyl ether—1-pentanol extractant mixture. Shake the separatory funnel vigorously on a mechanical shaker for 1 min. Rinse the stopper with 1—2 ml fresh extractant mixture and collect the rinsings in the separatory funnel. Allow sufficient time for the layers to separate, then withdraw the lower aqueous layer and discard it. Wash the organic phase twice, each time with 25 ml 1:10 hydrochloric acid. Each washing consists of shaking the organic phase with the wash solution for 15 sec, allowing the two phases to separate, and withdrawing and discarding the lower aqueous phase. After the second wash step is completed, rinse the funnel tip with distilled water.

Add 25 ml of the dilute ammonia solution to the funnel and shake vigorously for 20 sec. After the layers have separated, carefully transfer the lower aqueous phase containing the molybdate from the decomposition of the extracted molybdosilicic acid to a 50-ml volumetric flask. Rinse the funnel stopper with 2—3 ml distilled water, collect the rinsings in the funnel, and transfer the rinsings to the flask. Rinse the funnel once more with 3—5 ml distilled water and transfer the rinsings to the volumetric flask.

Acidify the flask contents with 14 ml 1:1 perchloric acid. Add 2.0 ml 3% hydrogen peroxide, dilute the resulting peroxymolybdic acid solution in a 50-ml volumetric flask to the mark with distilled water, and mix well. Measure the absorbance at 330 nm against a reagent blank in matched 1.000-cm silica cells. Distilled water is substituted for the aliquot of silicate sample in the preparation of the reagent blank solution according to the recommended procedure. Refer the absorbance values to a standard calibration graph obtained by using standard silicate solutions.

III. APPLICATIONS

A. SILICA IN WATER (4,5,42,58)

The molybdosilicic acid method is applicable to the determination of 4—25 ppm silica (2—12 ppm Si) in water when 1.0-cm absorption cells are used. The heteropoly blue method can be used for the 0.3—2 ppm silica range when 1.0-cm cells are used and absorbance is measured at 815 nm or for the 1—5 ppm silica range when the absorbance is measured at 650 nm. In the case of seawater, it has been found that the salinity decreases the absorbance, and therefore equivalent amounts of salt must be added to the standards used in preparing the calibration graph (45,52).

Special Solutions

Molybdate Reagent. Dissolve 10.0 g ammonium molybdate tetrahydrate, $(NH_4)_6Mo_7O_{24} \cdot 4H_2O$, in distilled water. Filter and dilute the filtrate to

100 ml and adjust the pH to 7.5—8 with ammonia. Store in a polyethylene bottle.

Stock Silicate Solution. Dissolve 4.75 g sodium metasilicate nonahydrate, $Na_2SiO_3 \cdot 9H_2O$, in carbon dioxide-free water and dilute to 1 liter. Use 100-ml aliquots and standardize gravimetrically using perchloric acid dehydration.

Standard Silicate Solution. Use a microburet to transfer sufficient volume of stock silicate solution to a 500-ml volumetric flask so that 5 mg silica will have been transferred to the flask. Dilute to volume with redistilled water. This solution contains 10 μg SiO_2 per ml. Store in a polyethylene bottle.

Oxalic Acid Solution. Dissolve 10 g oxalic acid dihydrate, $H_2C_2O_4 \cdot 2H_2O$, in redistilled water and dilute to 100 ml.

Sulfuric Acid, 1N. Add very carefully 28 ml concentrated sulfuric acid to about 800 ml redistilled water, stir, cool, and dilute to 1 liter.

Reductant. Dissolve 0.5 g 1-amino-2-naphthol-4-sulfonic acid and 1 g anhydrous sodium sulfite, Na_2SO_3, in 50 ml redistilled water. Dissolve 30 g sodium bisulfate, $NaHSO_3$, in 150 ml distilled water, add reductant solution, and mix. Filter, and store the filtrate in a polyethylene bottle.

Procedure A1, Molybdosilicic Acid Method

Transfer a 50-ml aliquot to a 100-ml platinum dish and add 0.20 g silicate-free sodium bicarbonate. Digest on a steam bath for 1 hr to convert silica to active state. Cool, add very slowly 2.4 ml of the 1N sulfuric acid with stirring, and transfer to a 100-ml polyethylene beaker. Adjust volume to exactly 50 ml. Add immediately 1.0 ml 1:1 hydrochloric acid and 2.0 ml of the molybdate reagent. Mix thoroughly and allow to stand for 5—10 min. Add 1.5 ml of the oxalic acid solution and again mix thoroughly. Measure the absorbance of solutions, including a reagent blank solution, against water in the reference cell at 350 nm, using 1.0-cm cells.

Procedure A2, Heteropoly Blue Method

If polymerized silicic acid is presumed to be present in the filtered water sample, transfer a 50.0-ml aliquot to a 100-ml platinum dish. Add 0.20 g pure sodium carbonate (silica free) and digest on a steam bath for 1 hr. Cool and add slowly, with constant stirring, 2.4 ml of a 1N sulfuric acid solution.

Transfer a 50.0-ml aliquot of filtered water sample or the entire prepared sample solution to a polyethylene beaker and add rapidly in sequence 1.0 ml 1:1 hydrochloric acid and 2.0 ml of the molybdate reagent. Mix very thoroughly and allow to stand for 10 min. Add 1.5 ml of the oxalic acid solution and mix well. After 5 min, add 2.0 ml of the reductant and mix. After 10 min, measure the absorbance at 650 or 815 nm, using 1.0-cm cells with water in the reference cell.

If the water sample had any turbidity or color, a background absorbance correction can be determined by following the same procedure but adding equivalent volumes of silica-free water in place of molybdate and reductant solutions and measuring the absorbance against water. Thus, the absorbance of the reagent blank solution and this background absorbance would be subtracted from the measured absorbance before using the calibration graph.

Preparation of Calibration Graph. Prepare a series of standard silica solutions by transferring 0, 2.0, 5.0, 10.0, 15.0, and 20.0 ml of the standard silica solution (1 ml = 10.0 μg SiO_2) to 100-ml volumetric flasks. Dilute to the mark with silica-free water. Transfer 50.0-ml aliquots of these solutions to polyethylene beakers and follow the same procedure as given for determination of silica in water samples. Plot corrected absorbance values against μg silica in 50 ml of sample solution taken for analysis.

B. SILICON IN FERROUS MATERIALS

Method B1

The method requires the use of a prescribed volume of acid and controlled evaporation during the process of solution in order to ensure that all of the silicon in solution is reactive, that none is rendered insoluble during solution of the sample, and that, on dilution, the solution will have a certain acidity (24). Since excessive evaporation leads to low results, it is essential that the solution volume (originally 50 ml for a 0.25-g sample) should not fall below the following figures:

Silicon	Minimum volume	
	3.5% $H_2 SO_4$	$HCl-HNO_3$
0.5%	15 ml	24 ml
2.0%	35 ml	45 ml

For samples containing 2–4% silicon, an extra 0.25 g high-purity iron should be added and 100 ml of the dissolving acid used.

Under certain conditions, the $HCl-HNO_3$ dissolving acid gives low results and even forms the hydrogel of silicic acid, presumably by hydrolysis of silicon and particularly in the presence of graphite. In such cases, $H_2SO_4-HNO_3$ gives satisfactory results, and the method finally adopted used three solvent acids in the following way: (1) H_2SO_4, 3.5% by volume for all unalloyed steels; (2) $HCl-HNO_3$ for alloyed steels; (3) $H_2SO_4-HNO_3$ for all cast irons.

It has been shown (10,23) that iron(II) sulfate in the presence of oxalic acid will satisfactorily reduce the molybdosilicic acid complex preferentially in presence of similar complexes of phosphorus, vanadium, tungsten, etc., and an excess of molybdate. Presumably, this depends on the lowering of the oxidation potential of the iron(II)–iron(III) system by the oxalic acid. The maximum color intensity is developed in 30 sec and remains stable for at least 24 hr.

It has been found that the final color varies with the amount of iron present, possibly because an increase in the concentration of the ferric ion decreases the reducing power of the system. If, however, the correct amount of oxalic acid is present, it is possible to have a wide variation in iron content over which constant readings can be obtained.

It is recommended that temperatures of the reduced solutions should not be allowed to depart more than $\pm 2°C$ from that at which the calibration curve was made while taking transmittance readings.

Reagents

Dissolving Acid 1. Add 35 ml sulfuric acid (sp. gr. 1.84) to 800 ml water, cool, and dilute to 1 liter.

Dissolving Acid 2. To 500 ml water add 125 ml hydrochloric acid (sp. gr. 1.18) and 45.5 ml nitric acid (sp. gr. 1.42), cool, and dilute to 1 liter.

Dissolving Acid 3. To 500 ml water add 36 ml sulfuric acid (sp. gr. 1.84) and 45.5 ml nitric acid (sp. gr. 1.42), cool, and dilute to 1 liter.

Potassium Permanganate, 0.225 g per Determination.

Hydrogen Peroxide, 5%.

Ammonium Molybdate, 2.5% Solution.

Oxalic Acid, 5% Solution.

Iron(II) Ammonium Sulfate. Dissolve 6 g $Fe(NH_4)_2(SO_4)_2 \cdot 6H_2O$ in water, add 1 ml dissolving acid 1, and dilute to 100 ml.

Procedure

Dissolving acid 1 is used for all plain carbon and low-alloy steels. Dissolving acid 2 is used for all high-alloy material; it can be used for plain carbon steel, but occasionally an insoluble residue is found which may necessitate an extra filtering operation. Acid 2 is not generally suitable for cast irons. Dissolving acid 3 should be used for all plain and alloy irons; it is also suitable for many types of steel.

Range 0–2.0% Silicon. Transfer 0.25 g sample to a 250-ml volumetric flask and dissolve by gentle heating in 50 ml of suitable acid. When solution is complete, treat in the following manner: (a) If acid 1 or 3 has been used, add 0.225 g potassium permanganate and boil for 2 min. Discharge the manganese dioxide precipitate by the dropwise addition of hydrogen peroxide. Cool, dilute to the mark, and mix. (b) If acid 2 has been used, boil off the nitrous fumes for 2 min, dilute to the mark, and mix.

Range 2.0–4.0% Silicon. Transfer 0.25 g sample and 0.25 g silicon-free iron to a 500-ml volumetric flask and dissolve in 100 ml of a suitable acid. If acid 1 or 3 has been used, add 0.45 g potassium permanganate, treat with hydrogen peroxide, cool, dilute to the mark, and mix. If acid 2 has been used, omit the permanganate, boil, cool, dilute to the mark, and mix.

If the solutions obtained by the above procedures are found to be cloudy due to the presence of undecomposed carbides, filter off 50–60 ml into clean, dry beakers using a dry filter paper, discarding the first few milliliters of the filtrate. Transfer suitable aliquots to 50-ml volumetric flasks, add 10 ml ammonium molybdate solution, and allow to stand for 10 min. Then add to 10 ml oxalic acid solution, mix, immediately add 5 ml iron(II) ammonium sulfate solution, dilute to the mark, and mix.

Prepare a reference solution by treating silicon-free iron in the same manner as the sample and measure absorbance at approximately 815 nm. To correct for background color, obtain the transmittance readings of a similar size aliquot of the sample solution, using an aliquot of the silicon-free iron solution in the reference cell.

Obtain the mg silicon from a previously constructed calibration graph and calculate the percentage silicon as follows:

$$\text{silicon } \% = \frac{A - B}{C \times 10}$$

where A = mg silicon in sample or aliquot used, B = mg silicon in background color correction, and C = g sample taken or represented in the aliquot.

Method B2 (35)

This method is not applicable to alloys where solution of silicon is incomplete, but almost all ferrous, ferromagnetic, nickel, and copper alloys can be successfully analyzed. Confirmatory data on a wide variety of samples indicate that the silicon is readily soluble and that good results can be obtained. The method consists of solution of the sample in a mixture of hydrochloric and nitric acids, destruction of the nitric acid by heating with formic acid, removal of all interfering metals by a carbamate–chloroform extraction, and, finally, determination of silicon by the usual photometric molybdenum blue method using essentially the technique described by Minster (41,42). The time of analysis is about 2 hr. The method is applicable in the range of 0.01—2% silicon if the rather poor precision and accuracy in the upper ranges, which result from the limitations of colorimetric techniques, can be tolerated.

The colored metal ions, manganese, chromium(III), iron(III), cobalt, molybdenum(V), and molybdenum(VI), exhibit little or no absorptivity at 765 nm. On the other hand, vanadium(IV), vanadium(V), copper(II), and, to a lesser extent, nickel show absorptivity in this region and will therefore interfere in the silicon determination unless removed or compensated for. In addition, iron(III), vanadium(V), molybdenum(VI), and copper(II) interfere by oxidizing the tin(II) chloride which is added to reduce the molybdosilicic complex. Under the conditions of the method, not much more than about 2 or 3 mg iron(III) can be present at the time of the addition of tin(II) chloride; otherwise, low results for silicon are obtained. The iron interference can be virtually eliminated by reducing the metal to the iron(II) state with the aid of a silver reductor before the addition of the ammonium molybdate. The iron(II) partially reduces any molybdosilicate complex to molybdenum blue, but does not appear to reduce any of the ammonium molybdate itself.

Unfortunately, the same cannot be said for molybdenum(V) and copper(I). When molybdenum(VI) and copper(II) are separately reduced to molybdenum(V) and copper(I) in a silver reductor and treated with ammonium molybdate, the solutions turn dark blue. Aluminum, chromium(III), and lead are not extracted by carbamate under the conditions recommended. Extraction of manganese is very incomplete. Iron(III), vanadium(V), nickel, cobalt, molybdenum(VI), copper(II), zinc, and tin(IV) are all extracted, although the completeness varies with the metal and is probably not quantitative in any instance. Fortunately, however, the amounts remaining cause little or no error in the silicon determination. Of the commonly encountered elements, only vanadium may cause trouble. The unextracted vanadium usually has an absorbance equivalent to about 1 μg silicon.

A higher ratio of acid to silicon is required for complete solution of high-silicon samples than for low-silicon samples. Because of the need for high initial acidity, it is best, whenever possible, to dissolve the sample in the 10 ml hydrochloric acid—nitric acid mixture rather than to dissolve it in 4 ml 1:1 nitric acid and then add 6 ml 2:1 hydrochloric acid. After complete solution has been accomplished, there is not harm in lowering the acidity by dilution of the sample.

The acidity of the solution at the time of the carbamate extraction must be high enough to permit efficient extraction by the chloroform and yet not so high as to prevent the formation of the heteropoly complex in the subsequent color development. If the acidity is too low at the time of extraction, the carbamates may tend to remain in the aqueous layer.

Once an analysis is started, it should be carried through to completion the same day in order to keep the blanks from the glassware as low as possible. Carbamate solutions dissolve glassware fairly rapidly and should always be stored in plastic containers. If reagents and distilled water of sufficient purity have been used, the reagent blank should not amount to more than about 3 μg silicon.

The Vaughn method of eliminating interference from phosphorous used by Minster (40) has proved to be very satisfactory.

Apparatus and Reagents

Standard Silicon Solution (1 ml = 0.01 mg Si). Transfer 0.1000 g powdered silicon dioxide to a platinum crucible and heat to constant weight at 1100°C to remove water. Treat the dry residue with hydrofluoric and sulfuric acids, evaporate to dryness, reignite, and weigh. The difference between the first and second weights represents the silicon dioxide in the sample taken.

Transfer an accurately weighed portion of the undried powdered silicon dioxide which is equivalent to 0.107 g silicon dioxide to a platinum dish. Add 10 ml redistilled water and about 0.5 g potassium hydroxide. Warm until complete solution occurs. (If the particular sample available is not soluble in potassium hydroxide, it may be necessary to resort to a fusion with 1 g anhydrous sodium carbonate, followed by solution of the cooled melt in water.)

Quickly pour the solution into 350 ml redistilled water in a 400-ml beaker and wash out the dish into the beaker. Transfer the solution to a 500-ml volumetric flask, dilute to the mark with redistilled water, and transfer to a clean, dry plastic container. Then quickly pipet 50 ml of this solution to a 500-ml volumetric flask, dilute to the mark with redistilled water, and store in a clean, dry plastic container.

Sulfuric Acid Solution, 1:3. Pour 100 ml sulfuric acid into 300 ml redistilled water. Cool and transfer to a plastic container.

Ammonium Molybdate Solution. Dissolve 10 g of the larger crystals of ammonium molybdate, $(NH_4)_6Mo_7O_{24} \cdot 4H_2O$, in 200 ml redistilled water and store in a plastic container.

Tin(II) Chloride Solution. Transfer 1 g of the larger crystals of tin(II) chloride, $SnCl_2 \cdot 2H_2O$, to a 100-ml volumetric flask and add 2 ml hydrochloric acid. Warm gently until a clear solution is obtained, then cool, dilute to the mark, and mix. Prepare fresh each day.

Hydrochloric Acid–Nitric Acid Mixture. Mix 40 ml hydrochloric acid, 40 ml redistilled water, plus 20 ml nitric acid. Prepare fresh each day as required.

Carbamate Solution. Dissolve 5 g sodium diethyldithiocarbamate in 100 ml redistilled water. Filter the solution if cloudy and store in a plastic container. Prepare fresh each day as required.

Procedure

Preparation of Calibration Graph. Transfer 0, 1, 2, 3, 4, 5, and 6 ml standard solution (0.01 mg Si per ml) from a good-quality buret to 100-ml volumetric flasks and add 1 ml 1:3 sulfuric acd. Dilute to 55 ml with redistilled water and mix. Add 10 ml ammonium molybdate solution, mix, and allow to stand 5 min. Add 30 ml 1:3 sulfuric acid, stopper the flask, and invert once or twice to mix the solution. (The tin(II) chloride that is added subsequently must not be allowed to come in contact with nonacidified molybdate solution on the walls of the flasks; if this occurs, some reduction of molybdate will result.) Add 1 ml tin(II) chloride solution and swirl to mix. Immediately dilute to the mark and mix. Transfer a portion of the solutions to absorption cells and read the % transmittance at approximately 765 nm, 5 min after the addition of the tin(II) chloride, using the blank in the reference cell. Prepare a calibration graph by plotting the absorbance values against mg silicon per 100 ml solution.

Analysis of Sample. Transfer up to 0.2 g of the milled sample to a 125-ml conical flask. The sample taken should not contain more than about 0.5 mg silicon, or, if the expected silicon content is over 1%, not more than about 0.3 mg silicon. Carry a reagent blank through the entire analysis. Add 10 ml hydrochloric acid–nitric acid mixture, cover, and warm gently to dissolve the sample. (In the event that nickel metal is being analyzed, dissolve the sample in 4 ml 1:1 nitric acid by heating gently and then add 6 ml 2:1 hydrochloric acid.) Avoid as much as possible loss of acid by volatilization. Ignore insoluble carbides or other material such as tungsten.

When solution is complete, remove the cover and add about 1 ml formic acid. When decomposition of the nitric acid starts, remove the sample to a cooler place and allow the reaction to proceed. Add further 1-ml portions of formic acid from time to time, warming occasionally, if necessary, to maintain the decomposition. About 3 or 4 ml formic acid will be required to destroy the nitric acid. Finally, when the gassing subsides, return the flask to the low-temperature hot plate, rinse down the walls with about 0.25 ml formic acid, and heat to gentle boiling to expel brown fumes completely. Avoid the addition of unnecessary excess formic acid and avoid loss of too much hydrochloric acid by boiling. Cool the sample at room temperature.

Filter into a 100-ml volumetric flask and dilute to the mark with redistilled water. Transfer a suitable aliquot, containing 0.01–0.06 mg silicon, to a 150-ml Squibb-type separatory funnel. Add 1 ml 1:3 sulfuric acid and dilute to 25 ml. Add a 5-, 10-, or 25-ml portion of the carbamate solution, respectively, to the 5-, 10-, or 25-ml aliquot of the sample solution. Stopper and shake once or twice to mix the sample thoroughly. Add 35 ml chloroform, stopper, and shake vigorously for 30 sec. Remove the stopper and allow the layers to separate. When separation is complete, swirl the solution to loosen the floating drop of chloroform and then drain off and discard as much as possible of the lower layer. (Avoid too rapid a drain-off; otherwise, some of the aqueous layer may be withdrawn into the stem of the funnel.) Repeat the extraction two or more times with 10- to 15-ml portions of chloroform to remove traces of extractable carbamates. In the wash extractions, 10-sec shaking is adequate.

When the chloroform layer remains colorless after the second or subsequent wash extraction, drain off the chloroform as completely as possible and transfer the aqueous solution back to the 125-ml conical flask with a minimum amount of washing. Heat just to gentle boiling to expel most of the chloroform. If large drops of chloroform persist, they can be removed by touching with a glass stirring rod. Small drops of chloroform remaining in solution will do no harm. Cool to room temperature and transfer to a 100-ml volumetric flask with a minimum of washing. Add 10 ml ammonium molybdate solution and proceed as directed for preparation of the calibration graph. With the aid of the calibration graph, determine the amount of silicon present in the sample.

C. SILICON IN STEEL

Molybdosilicic Acid Method

This method has been developed using a series of National Bureau of Standards samples ranging from 0.009 to 0.44% silicon and has been used for the determination of silicon in several thousand samples with results that correlated well with gravimetric determinations (27).

For most accurate results, the acid concentration should be maintained within 10% of the amount specified. Since the intensity of the color is affected by temperature fluctuations, it is advisable to maintain the temperature of solutions within a few degrees of the temperature at which the calibration curve is made. Maximum color intensity will form within 5 or 6 min after addition of the ammonium molybdate. Heating will accelerate the color formation. After the addition of the fluoride, however, time becomes an important factor. No interference from elements commonly present in low-alloy steels has been observed. Excessive carbon may be removed by filtration. The amount of fluoride added must be measured from a buret or pipet. The amount of persulfate may be varied within wide limits without influencing the accuracy of the determination.

The method is satisfactory for steels containing as high as 3.0% silicon. With low silicon contents, both the sample and blank will remain constant for 15 min or longer; but for higher silicon contents, the sample will begin fading in a few minutes. For these samples, the transmittance values must be obtained within a few minutes. The blank may be stabilized by the addition of a small amount of citrate, but the citrate should not be added to the sample as it will cause fading of the color.

Reagents

Nitric acid, 3N. Dilute 200 ml of the concentrated acid to 1 liter.

Ammonium Persulfate, 12%. Dissolve 120 g ammonium persulfate in water and dilute to 1 liter.

Ammonium Molybdate, 8%. Dissolve 80 g of the large crystals in water and dilute to 1 liter. Filter if necessary.

Sodium Fluoride, 2.4%. Dissolve 24 g in water and dilute to 1 liter.

Hydrochloric Acid, 3N. Dilute 250 ml of the concentrated acid to 1 liter.

Nitric acid, 0.6N. Dilute 50 ml 3N acid to 250 ml.

Procedure

Transfer 0.5 g sample to a 250-ml Erlenmeyer flask, add 50 ml 3N nitric acid, and heat to dissolve. Add 5 ml 12% ammonium persulfate, boil until clear (about 1 min), remove, and cool to room temperature. Dilute to 250 ml and mix. Pipet two 25-ml aliquots into two dry beakers or flasks. To one flask, add 5 ml 8% ammonium molybdate solution, mix, and allow to stand 6 min. To the blank, add 10 ml 2.4% sodium fluoride solution. At the end of 6 min, add 10 ml 2.4% sodium fluoride to the sample and 5 ml of the 8% ammonium molybdate solution to the blank. Obtain the transmittance reading at 400 nm, using the blank in the reference cell. From a previously constructed calibration graph, obtain the mg silicon and calculate the % silicon in the sample.

If the sample does not go into solution with 3N nitric acid, use 25 ml 3N nitric acid and 25 ml 3N hydrochloric acid and proceed as above. When the silicon concentration is in excess of that shown on the calibration graph, dilute 100 ml of the original sample with 100 ml 0.6N nitric acid, and develop color on a 25-ml aliquot as before.

D. SILICON IN COPPER AND COPPER ALLOYS

The molybdosilicic acid method (7), the extraction molybdosilicic acid method (19), the heteropoly blue method (15), and the extraction heteropoly blue method (47) have been utilized in the determination of traces of silicon in pure copper and copper-base alloys. The molybdosilicic acid method suitable for the determination of alloys containing 0.02–0.4% silicon and not more than 0.05% phosphorous will be presented in this section.

Special Solutions

Standard Silicon Solution. Fuse 0.0856 g pure anhydrous silica, SiO_2, with 1.0 g anhydrous sodium carbonate in a platinum crucible. Cool and dissolve the melt in silica-free water. Transfer to a 1-liter volumetric flask and dilute to the mark with silica-free water. Each ml of this solution contains 0.040 mg silicon. Store in a polyethylene bottle.

Molybdate Solution. Dissolve 25.0 g ammonium molybdate, $(NH_4)_6Mo_7O_{24}$ · $4H_2O$, in silicon-free water and dilute to 250 ml. Store in a polyethylene bottle.

Boric Acid Solution. Dissolve 30 g boric acid, H_3BO_3, in about 500 ml hot, distilled water. Cool and store in a polyethylene bottle.

Citric Acid Solution. Dissolve 5.0 g citric acid in silica-free water and dilute to 100 ml. Prepare fresh each day.

Urea Solution. Dissolve 10 g urea in water and dilute to 100 ml. Prepare fresh each day.

Pure Copper Turnings, or wire containing less than $10^{-3}\%$ silicon.

Procedure

Weigh 1.00 g sample and transfer to a platinum crucible. Add about 0.3 ml hydrofluoric acid (to dissolve any silicicides) and 11.0 ml 1:2 nitric acid. Place a platinum cover on the crucible and allow to digest for 5 min. (Heat the crucible on an asbestos-topped hot plate at about 60°C if the sample is not dissolved completely.) Transfer the cool solution to a 100-ml volumetric flask using a polyethylene funnel which is immersed in the 25 ml boric acid added previously to the volumetric flask. Dilute to the mark with silica-free water and mix thoroughly.

Transfer a 50-ml aliquot to a 100-ml volumetric flask. Add 5 ml of the urea solution and swirl the flask to allow nitrogen produced by reduction of residual nitrous acid to escape. Add 5.0 ml of the molybdate reagent and dilute to approximately 90 ml, mix thoroughly, and allow the solution to stand for 10 min. Add 5.0 ml of the citric acid, dilute to the mark with silica-free water, and mix. Measure the absorbance at either 350 or 400 nm, using 1.000-cm cells and reagent blank solution prepared from pure copper in the reference cell.

Preparation of Calibration Graph. Transfer 1.00-g samples of the pure copper (less than $10^{-3}\%$ silicon) to four platinum crucibles. The copper taken should be within 90% of the amount present in the alloy taken for analysis. Add 0.3 ml hydrofluoric acid to each crucible and then 11.0 ml 1:2 nitric acid. Place a platinum cover on the crucible and allow to digest for 15 min. Heat gently to effect complete dissolution if necessary. Transfer the cool solution to a 100-ml volumetric flask using long-stem polyethylene funnels dipping into 25 ml of the boric acid solution added previously to each flask. Dilute each flask to the mark with silica-free water and mix thoroughly.

By means of a dry pipet, transfer 50-ml aliquots from each flask to a 100-ml volumetric flask, giving eight flasks, each containing 50 ml solution. Add 0, 1.0, 2.0, 5.0, 10.0, 15.0, 20.0, and 25.0 ml of the standard silicon solution to the eight flasks. Add 5 ml of the urea solution to each flask and swirl for several minutes. Add 5.0 ml of the molybdate solution, dilute to the mark with silica-free water, and mix. After 10 min, measure the absorbance at 350 nm or 400 nm, using 1.00-cm cells and the reference blank solution (no standard Si solution added) in the reference cell. Plot absorbance against mg silicon per 100 ml solution.

E. SILICON IN ALUMINUM ALLOYS

The molybdosilicic acid method is extensively used to determine silicon in aluminum and aluminum alloys, especially when the silicon concentration is in the 0.1–1.0% range (6). Careful control of pH and precise measurement of volume of reagents used are important. If the silicon is not completely solubilized after dissolution of the sample, a turbidity will be observed. The stabilization may usually be effected by heating for a longer time in the sodium hydroxide solution. High manganese alloys may give a suspension of hydrous manganese(IV) oxide following acidification. In this case, the dropwise addition of a saturated solution of sodium sulfite to the hot solution reduces the manganese(IV). The slight excess of sulfite is destroyed by adding dropwise sufficient $0.1N$ potassium permanganate solution to give a slight purplish hue. The permanganate color is discharged by the dropwise addition of a 1% oxalic acid solution.

Special Solutions

Standard Silicate Solution. Transfer 0.1070 g anhydrous silicon dioxide to a platinum crucible containing 1.0 g anhydrous sodium carbonate. Mix flux and silica and fuse in the covered platinum crucible. After cooling, dissolve the melt in a polyethylene beaker, cover, and heat on a steam bath for 30 min. Cool, transfer to a 1-liter volumetric flask, dilute to the mark with silica-free water, and mix thoroughly. Store in a polyethylene bottle. Each ml of this solution contains 50.0 μg silicon.

Ammonium Molybdate Solution. Dissolve 50 g ammonium molybdate, $(NH_4)_6 Mo_7 O_{24} \cdot 4H_2 O$, in silica-free water and dilute to 500 ml.

Sodium Hydroxide. Dissolve 150 g sodium hydroxide pellets in about 300 ml silica-free water in a large nickel crucible. Cool and transfer to a 500-ml polyethylene graduated cylinder. Dilute to 500 ml and transfer to a polyethylene bottle for storage.

Aluminum Nitrate Solution. Transfer a 1.0-g sample of pure aluminum (NBS melting point aluminum, Sample No. 446) to a 250-ml nickel beaker, add 100 ml of the sodium hydroxide reagent, cover, and allow to stand. If necessary, heat gently to effect complete dissolution. Cool somewhat and transfer to a beaker containing 125 ml 1:1 nitric acid. Heat the covered beaker until the solution becomes clear. Cool to room temperature, transfer to a 250-ml volumetric flask, and dilute to mark.

Procedure

Transfer 0.1 g sample to a nickel crucible, add 10.0 ml of the sodium hydroxide solution, and cover. After the reaction subsides, rinse the cover and wash down the sides of the crucible with a minimum amount of water. Boil gently until dissolution is complete.

Cool and transfer the contents of the crucible to a 150-ml beaker containing 12.5 ml of the 1:1 nitric acid. Cover and warm until a clear solution is obtained. Cool and transfer to a 100-ml volumetric flask. Dilute to the mark and mix well.

Transfer a 50- or 75-ml aliquot to a 150-ml beaker, add 1.0 ml 1:1 nitric acid, and dilute to about 80 ml. Use a pH meter and adjust the pH of the sample solution to within 1.1–1.3 by using 1:1 nitric acid or the sodium hydroxide solution. Add 10 ml of the molybdate reagent, transfer the solution to a 100-ml volumetric flask, dilute to the mark, and mix thoroughly. After 5 min, measure the absorbance in 1.0-cm cells at 400 nm, using a reagent blank solution in the reference cell.

The reagent blank solution is prepared by the following procedure: Transfer 10.0 ml of the sodium hydroxide solution to a 100-ml volumetric flask containing 11.5 ml 1:1 nitric acid. Dilute to about 50 ml, cool, dilute to the mark, and mix. Transfer a 50-ml aliquot to a 100-ml beaker, add 1 ml 1:1 nitric acid, dilute to about 80 ml, and mix. Adjust to pH 1.1–1.3 as previously described for sample solution. Add 10.0 ml of the molybdate reagent, transfer to a 100-ml volumetric flask, dilute to the mark, and mix.

In case the sample solution has a yellow hue, a background absorbance correction must be applied. Use another aliquot of sample solution and the same volume of 1:1 nitric acid and dilute to volume in a 100-ml volumetric flask. Measure the absorbance of this solution against the solution obtained by taking the 50 ml solution remaining in the volumetric flask in the procedure for preparation of a reagent blank solution.

Preparation of Calibration Graph. Use a buret to transfer 12.5 ml of the aluminum nitrate solution to eight 100-ml beakers. Add 1.0 ml 1:1 nitric acid to each beaker. Add 0, 2.0, 5.0, 8.0, 10.0, 15.0, 20.0, and 25.0 ml of the standard silicate solution (1 ml = 50 μg Si) to beakers. Dilute to about 80 ml and mix. Use a pH meter and adjust the pH of each solution to 1.1–1.3 using 1:1 nitric acid or the sodium hydroxide solution. Add 10 ml of the molybdate solution, transfer each solution to a 100-ml volumetric flask, dilute to volume, and mix. After 5 min and no longer than 20 min, measure the absorbance of the seven standard solutions at 400 nm against the reagent blank solution, the solution containing no standard silicate solution.

F. SILICON IN NICKEL (18)

This procedure is applicable to the determination of the low silicon content (less than 0.01%) in electronic nickel.

Special Solutions

Molybdate Solution. Dissolve 10 g ammonium molybdate, $(NH_4)_6 Mo_7 O_{24} \cdot 4H_2 O$, in distilled water and dilute to 200 ml. Store in a polyethylene bottle.

Tin(II) Chloride Solution. Transfer 1.0 g tin(II) chloride, $SnCl_2 \cdot 2H_2 O$, to a 100-ml volumetric flask and add 2 ml hydrochloric acid. Mix until dissolution is complete and dilute to the mark with distilled water with constant mixing. This solution should be prepared fresh each day.

Sulfamic Acid Solution. Dissolve 10 g sulfamic acid, $HSO_3 NH_2$, in distilled water and dilute to 100 ml. This solution should be prepared fresh each day.

Procedure

Transfer a suitable-size sample containing about 0.02–0.06 mg silicon to a 150-ml beaker, add 5 ml 1:1 nitric acid, cover, and warm gently to dissolve, avoiding as much as possible the loss of acid by evaporation. Or, if the silicon is so high as to make it necessary to weigh an extremely small sample, take 0.2–0.5 g of the sample, dissolve as above, and heat gently to expel most of the brown fumes. Make up to volume and take suitable aliquots. To the sample solution or aliquot containing a total of 5 ml 1:1 nitric acid, add 5 ml 10% sulfuric acid and proceed as in preparation of the calibration curve above.

To correct for background absorbance, transfer a similar portion or aliquot of the sample solution to a 150-ml beaker and proceed as with the analysis of the sample, omitting the addition of ammonium molybdate and tin(II) chloride solutions. Measure the absorbance of this solution at 705 nm, using the blank in the reference cell, and calculate the % silicon as follows:

$$\text{silicon } \% = \frac{A - B}{C \times 10}$$

where A = mg silicon in sample or aliquot used, B = mg silicon in background absorbance correction, and C = g sample represented.

Preparation of Calibration Graph. Transfer 0, 0.5, 1.0, 2.0, 4.0, and 6.0 ml of the standard silicon solution (1 ml = 0.01 mg Si) to 150-ml beakers, add 5 ml 1:1 nitric acid and 5 ml 10% sulfamic acid, and dilute to approximately 45 ml. Carefully neutralize with approximately $6N$ silica-free ammonium hydroxide, added dropwise until congo red paper just turns red, and then add 1 ml 1:3 sulfuric acid. Transfer the solutions to 100-ml volumetric flasks and dilute to about 55 ml. Add 10 ml ammonium molybdate solution, mix, and allow to stand 5 min. Add 30 ml 1:3 sulfuric acid, stopper the flask, and invert once or twice to mix the solution. (The tin(II) chloride that is added subsequently must not be allowed to come in contact with non-acidified molybdate solution on the walls of the flasks; if this occurs, some reduction of molybdate will result.) Add 1 ml tin(II) chloride solution and swirl to mix. Immediately dilute to the mark and mix. Transfer a portion of the solutions to absorption cells and read the absorbance at approximately 765 nm, 5 min after the addition of the tin(II) chloride, using the blank in the reference cell.

G. SILICON IN BERYLLIUM, ZIRCONIUM, AND TITANIUM ALLOYS (14,17)

Reagents

Ammonium Molybdate. Dissolve 25 g of the larger crystals in about 200 ml redistilled water, add 20 ml concentrated sulfuric acid, and after cooling dilute to 250 ml.

Tartaric Acid. A 20% solution to which a few crystals of thymol are added as preservative.

Ammonium Hydroxide. Silica-free or low-silica ammonium hydroxide may be prepared by bubbling ammonia through redistilled water contained in a plastic bottle immersed in an ice bath.

Boric Acid, Saturated Solution.

Zirconium Fluoride Solution. Dissolve 28 g zirconyl chloride octahydrate in redistilled water, adjust the volume to 384 ml, transfer to a plastic bottle, and add 16 ml 48% hydrofluoric acid.

Reducing Solution. This solution is prepared by dissolving 11 g sodium bisulfite, 0.8 g sodium hydroxide, and 0.2 g 1-amino-2-naphthol-4-sulfonic acid in redistilled water and diluting to 100 ml.

Beryllium

Silicon may be determined in the presence of beryllium without difficulty. Samples of the oxide or other compounds may be dissolved in either sulfuric or hydrochloric acid, care being taken to avoid excess acid.

Procedure

Oxide. Transfer 1 g sample to a plastic beaker and dissolve in a minimum quantity of hydrochloric acid. Dilute with 20 ml water, add 0.5 ml 24% hydrofluoric acid, and warm at 60–65°C for 0.5 hr. Cool the solution and determine silicon in the following manner. Add 50 ml saturated boric acid solution and mix. Add 4 ml 10% ammonium molybdate and adjust the pH to 1.2–1.3 using the special ammonia solution. After 10 min, add 4 ml 20% tartaric acid and mix. Then add 1 ml of the reducing solution, transfer to a 100-ml volumetric flask, dilute to the mark, and mix. After 20 min, measure the absorbance at approximately 815 nm, using a blank of the reagents in the reference cell. Obtain the mg silicon from a previously constructed calibration graph and calculate the % silicon.

Metal. Transfer 0.2 g sample to a plastic beaker and dissolve in an alkaline solution containing 3 g sodium hydroxide so that the total volume is about 20 ml. Acidify the solution with dilute sulfuric acid, add 0.5 ml 24% hydrofluoric acid, and warm at 60–65°C for 0.5 hr. Cool, add 40 ml saturated boric acid solution, and proceed as for oxide. In this case, however, the solution for the calibration graph must contain the same amount of sodium hydroxide as the sample.

Zirconium

The determination of silicon in the presence of zirconium presents no difficulties other than solution of the sample. As it is necessary to use hydrofluoric acid to dissolve the metal and oxide, caution must be exercised in order to avoid loss by volatilization. However, there is no danger of this if solutions are kept dilute and low temperatures maintained (60–65°C).

Procedure

Metal. Transfer 1 g sample to a plastic beaker and add 15–20 ml redistilled water and 10 drops 1:1 sulfuric acid. Cover the beaker with a plastic cover and dissolve the metal by slow addition of about 3 ml 48% hydrofluoric acid. After most of the metal has dissolved, heat the solution to 60–65°C until solution is complete. After solution is complete, cool, add 40 ml saturated boric acid solution, and proceed as above for beryllium. Since premature reduction often occurs after addition of the ammonium molybdate, it should be destroyed by the addition of a minimum quantity of 0.06% potassium permanganate.

Oxide. Transfer 1 g sample to a Teflon beaker, add 20 ml water, 10 drops 1:1 sulfuric acid, and 4 ml 48% hydrofluoric acid. Cover with a plastic cover and heat at 60–65°C until solution is complete, maintaining the volume at about 20 ml by the addition of water if necessary. When solution is complete, add 40 ml saturated boric acid solution and proceed as above. Refractory oxide will not dissolve when this procedure is used.

Tetrafluoride. Transfer 1 g sample to a Teflon beaker, add 20 ml water and 0.5 ml 1:1 sulfuric acid. Heat at 60–65°C until solution is complete, maintaining the volume at 20 ml by the addition of water if necessary. Some samples may require the addition of hydrofluoric acid for complete solution. After solution is complete, proceed as above.

Other compounds such as zirconyl chloride are treated like zirconium tetrafluoride, except that they are treated with 0.5 ml 48% hydrofluoric acid for 0.5 hr at 60–65°C.

Titanium Alloys.

Containing 0.01–0.14% Silicon (17).

Reducing Solution. Dissolve 15 g sodium bisulfite, 0.5 g anhydrous sodium sulfite, and 0.25 g 1-amino-2-naphthol-4-sulfonic acid in redistilled water and dilute to 100 ml. Store in a plastic container.

Procedure

Preparation of Calibration Graph. Transfer 0, 1, 2, 3, 5, and 7 ml silicon standard (1 ml = 0.01 mg Si) to plastic beakers. Add 20 ml redistilled water and 0.5 ml 48% hydrofluoric acid. Add 40 ml of a saturated solution of boric acid, mix, and transfer to 100-ml volumetric flasks. Add 10 ml 5% ammonium molybdate, mix, and allow to stand for 10 min. Add 5 ml 20% tartaric acid solution and mix. Add 1 ml reducing solution, dilute to the mark, and mix. After 20 min, obtain transmittance readings at 700 nm, using the blank in the reference cell, and prepare a suitable calibration graph. The transmittance readings are made at 700 nm rather than at the point of maximum absorbance in order to obtain analyses over a wider range of concentrations.

Analysis of Sample. Transfer a sample containing 0.01–0.14 mg silicon to a plastic bottle and add 20 ml water and 0.5 ml 48% hydrofluoric acid. Cover and allow to dissolve by standing overnight. Add 40 ml of a saturated solution of boric acid and mix. Add 3% potassium permanganate solution until the solution in the bottle is pink, and then add five drops in excess. Rinse down the wall of the bottle, place it in a 1-liter beaker containing 800 ml boiling water, and heat for 90 min, swirling occasionally during the heating. Remove the bottle from the beaker and cool to room temperature. Filter, wash, and dilute the filtrate to about 75 ml in a 100-ml volumetric flask. Add 10 ml 5% ammonium molybdate solution and proceed as in the preparation of the calibration curve. Obtain the mg silicon from the calibration graph and calculate the % silicon.

H. SILICON IN BIOLOGIC MATERIALS

Reducing Solution. Dissolve 0.2 g 1-amino-2-naphthol-4-sulfonic acid, 2.4 g sodium sulfite, $Na_2SO_3 \cdot 7H_2O$, and 12 g sodium metabisulfate in water and dilute to 100 ml.

Tissue (31)

Procedure

Transfer 0.05–0.5 g sample to a platinum crucible, add six times its weight of sodium carbonate, burn off, and fuse to a clear melt. Dissolve the melt in water by warming, cool, add 5 ml 0.44% potassium dihydrogen phosphate, KH_2PO_4, neutralize to congo red with 10N sulfuric acid, and add 0.2 ml in excess. Transfer to a 250-ml volumetric flask and dilute to the mark.

Transfer suitable aliquots, containing 0.01–0.06 mg silicon, to 25-ml volumetric flasks and dilute to about 15 ml. Add 2 ml 5% ammonium molybdate in 1N sulfuric acid and allow to stand 10 min. Then add 5 ml 10N sulfuric acid and 0.5 ml of the aminonaphtholsulfonic acid reductant, dilute to the mark, and mix. After 10 min, measure absorbance at approximately 815 nm, obtain values from a previously constructed calibration graph, and calculate the % silicon. Compensate and correct for a blank.

Urine (31)

Some urines are difficult to analyze colorimetrically for silicon because of very high phosphate and, frequently, protein. By removing most of the phosphate with calcium, the ammonia with nitrite, and protein with trichloro-acetic acid, these difficulties are overcome.

Procedure

By Removal of Phosphate. Treat 0.3–3.0 ml urine with 10 ml 0.5% $CaCl_2$ and saturated $Ca(OH)_2$ until pink to phenolphthalein and dilute to about 15 ml. Add about 50 mg carbon, shake the mixture, warm to about 70°C for 15 min, then cool and filter into a 25-ml volumetric flask. Add 2 ml 5% ammonium molybdate in $1N$ sulfuric acid and mix. After 10 min, add 5 ml $10N$ sulfuric acid, mix, add 0.5 ml aminonaphtholsulfonic acid reductant, dilute to the mark, and mix. Measure absorbance at approximately 815 nm and calculate silicon in ppm.

By Removal of Ammonia (and Urea). Treat 2 ml sample with 1 ml 5% $NaNO_2$, 2 ml 20% trichloroacetic acid, 50 mg carbon, and 1 ml caprylic alcohol and then filter into a 25-ml volumetric flask. Dilute to about 15 ml, add 2 ml 7% sodium molybdate in $1N$ sulfuric acid, mix, and allow to stand for 10 min. Then add 5 ml $10N$ sulfuric acid, mix, add 0.5 ml aminonaphtholsulfonic acid reductant, dilute to the mark, and obtain values for silicon as above.

Blood (31)

Procedure

Mix 2 ml whole blood, serum, or plasma with 20 ml 5% trichloroacetic acid and filter after 5 min. Transfer 15 ml of the filtrate to a 25-ml volumetric flask and heat for 15 min on a boiling-water bath. Carefully shake the contents of the flask while hot to expel carbon dioxide. Cool and add 2.0 ml 5% ammonium molybdate in $1N$ sulfuric acid. Mix and after 10 min add 5 ml $10N$ sulfuric acid. After mixing, add 0.5 ml reductant. Measure the absorbance at approximately 815 nm.

I. SILICON IN FOODS AND AGRICULTURAL PRODUCTS

Many of these materials, particularly grains, are high enough in silica content for the determination to be made by gravimetric methods. However, for foods such as dairy products, meat, and eggs, colorimetric methods are adaptable.

Procedure

Treat 0.1–0.5 g dried sample with nitric and sulfuric acids or nitric and perchloric acids in a platinum dish to oxidize organic matter. Wash down the sides of the dish with a little water, evaporate to dryness, and ignite gently. Fuse the residue with 1–2 g anhydrous sodium carbonate until a clear melt is obtained. Cool, take up melt in water, filter, neutralize, adjust pH to 4.5–5.0

with 6N sulfuric acid, and take suitable aliquots for determination of silicon by one of the methods previously described.

Interference from phosphate and iron can be avoided by using the oxalic acid–iron(II) ammonium sulfate method as follows: To suitable aliquots adjusted to 0.25N with sulfuric acid, add 10 ml 5% oxalic acid and 5 ml 6% iron(II) ammonium sulfate containing 1.25 ml 1N sulfuric acid per 100 ml, mixing after the addition of each reagent. Dilute to 100 ml and measure absorbance at approximately 815 nm. Compensate and correct for a blank and determine silicon from a previously constructed calibration graph.

J. SILICON IN PLANTS (65)

Special Solutions

Molybdate Reagent. Dissolve 20 g ammonium molybdate tetrahydrate, $(NH_4)_6Mo_7O_{24} \cdot 4H_2O$ in distilled water containing 60 ml concentrated hydrochloric acid and dilute to 1 liter.

Metol Solution. Dissolve 10.0 g metol, p-methylaminophenol sulfate, in water containing 6.0 g sodium sulfite and dilute to 500 ml. Store at room temperature in a foil-covered polyethylene bottle. Prepare a fresh solution if this reagent becomes discolored.

Oxalic Acid, 10%. Dissolve 100 g oxalic acid in distilled water and dilute to 1 liter.

Reductant. Transfer 100 ml of the metol solution to a 500-ml polyethylene bottle and add 60 ml of the oxalate acid solution, 120 ml 9N sulfuric acid, and 20 ml distilled water. Cool to room temperature.

Standard Silicate Solution. Dilute sufficient volume of stock silicate solution to obtain standard silicate solution containing 0.005 mg silicon per ml.

Procedure

Dry plant tissue at 70°C. Pulverize to pass a 20-mesh screen and mix thoroughly. Weigh accurately a 20- to 25-mg sample and transfer to a platinum crucible. Have an empty crucible available to use for preparing the reagent blank. Add six drops concentrated sulfuric acid to each crucible and heat on a hot plate until the sulfuric acid causes fuming. Ignite very carefully the charred material over a burner, increasing the heat gradually until a white ash is obtained. Add about 0.4 g anhydrous sodium carbonate to each crucible and fuse for 20 min. Swirl the molten melt to ensure that all of the ash is removed from the sides of the crucible and incorporated in the melt. Cool the crucible and place in a 100-ml polyethylene beaker containing 30 ml redistilled water and 3 ml 1:3

460   SILICON

hydrochloric acid. Heat beaker and contents on a boiling water bath to speed dissolution. Remove the crucible and wash thoroughly with water, adding the rinsings to the original solution. Evaporate to about 30 ml and filter if any hydrous oxide precipitate has formed. Adjust pH to about 7, transfer to a paraffin-lined 50-ml volumetric flasks with recalibrated 50-ml graduation mark, and dilute to the mark with redistilled water.

Transfer a 20-ml aliquot of the solution containing not more than 50 μg silicon to a recalibrated 50-ml paraffin-lined volumetric flask and dilute to 20 ml if necessary. Use the same volume of reagent blank solution in the development of the reference blank solution. Use a microburet to add 3.0 ml of the molybdate solution to each flask, mix thoroughly, and allow to stand for exactly 10 min before adding rapidly 15 ml of the reductant. Dilute each solution to 50 ml, mix, and allow to stand for 3 hr. Measure absorbance in 1.000-cm cells at 810 nm.

REFERENCES

1. Alexander, G. B., *J. Amer. Chem. Soc.*, **75**, 5655 (1953).
2. Alexander, G. B., *ibid.*, **76**, 2094 (1954).
3. Alon, A., B. Bernas, and M. Frankel, *Anal. Chim. Acta,* **31**, 279 (1964).
4. American Public Health Association, *Standard Methods for the Examination of Water and Wastewater,* 13th ed., Washington, D.C., 1971, pp. 303–309.
5. American Society for Testing and Materials, *1973 Annual Book of ASTM Standards,* Part 23, ASTM Designation D 859–68, Philadelphia, pp. 403–404.
6. American Society for Testing and Materials, *1974 Annual Book of ASTM Standards,* Part 12, ASTM Designation E 34–72, Philadelphia, pp. 113–115.
7. American Society for Testing and Materials, *ibid.*, ASTM Designation E 62–72, pp. 325–328.
8. American Society for Testing and Materials, *ibid.*, ASTM Designation E 107–64, pp. 388–391.
9. Andersson, L. H., *Arkiv Kemi,* **19**, 243 (1962).
10. Andrew, T. R., and C. H. R. Gentry, *Analyst,* **81**, 339 (1956).
11. Boltz, D. F., and M. G. Mellon, *Ind. Eng. Chem., Anal. Ed.,* **19**, 873 (1947).
12. Boyle, A. J., and V. Hughey, *ibid.*, **15**, 618 (1943).
13. Bunting, W. E., *ibid.*, **16**, 612 (1944).
14. Carlson, A. B., and C. V. Banks, *Anal. Chem.,* **24**, 472 (1952).
15. Case, O. P., *Ind. Eng. Chem., Anal. Ed.,* **16**, 309 (1944).
16. Clabaugh, W. S., and A. Jackson, *J. Nat. Bur. Stand.,* **62**, 201 (1959).

17. Codell, M., C. Clemency, and G. Norwitz, *Anal. Chem.*, **25**, 1432 (1953).

18. De Sesa, M. A., and L. B. Rogers, *ibid.*, **26**, 1278 (1954).

19. Donaldson, E. M., and W. R. Inman, *Can. Dept. Mines Tech. Surv.*, Mines Branch Tech. Bulletin T B 77, 1965, 17 pp.

20. Fowler, R. M., *Ind. Eng. Chem., Anal. Ed.*, **4**, 382 (1932).

21. Fresenius, W., and W. Schneider, *Z. Anal. Chem.*, **214**, 341 (1965).

22. Furman, N. H., W. B. Mason, and J. S. Pekola, *Anal. Chem.*, **21**, 1325 (1949).

23. Gentry, C. H. R., and L. G. Sherrington, *J. Soc. Chem. Ind.*(London), **65**, 90 (1946).

24. Glasgow Absorptiometric Panel, *Metallurgia*, **50**, (No. 299), 145 (1954).

25. Hargis, L. G., *Anal. Chem.*, **42**, 1494 (1970).

26. Hargis, L. G. *ibid.*, p. 1497.

27. Hill, U. T., *ibid.*, **21**, 589 (1949).

28. Holt, B. D., *ibid.*, **32**, 124 (1960).

29. Hozdic, C., *ibid.*, **38**, 1626 (1966).

30. Hurford, T. R., and D. F. Boltz, *ibid.*, **40**, 379 (1968).

31. King, E. J., B. D. Stacy, P. F. Holt, D. M. Yates, and D. Pickles, *Analyst*, **80**, 441 (1955).

32. Kirkbright, G. F., A. M. Smith, and T. S. West, *Analyst*, **92**, 11 (1967).

33. Knudson, W. H., C. Juday, and V. W. Meloche, *Ind. Eng. Chem., Anal. Ed.*, **12**, 270 (1940).

34. Lindsay, F. K., and R. G. Bielenberg, *ibid.*, **12**, 460 (1940).

35. Luke, C. L., *Anal. Chem.*, **24**, 148 (1953).

36. Luke, C. L., *ibid.*, **36**, 2036 (1964).

37. Malissa, H., and S. Goniscek, *Z. Anal. Chem.*, **169**, 401 (1959).

38. Milton, R. F., *Analyst.*, **76**, 431 (1951).

39. Minster, J. T., *Analyst.*, **71**, 74 (1946).

40. Minster, J. T., *ibid.*, p. 428.

41. Minster, J. T., *ibid.*, **73**, 507 (1948).

42. Morrison, I. R., and A. L. Wilson, *Analyst*, **88**, 446 (1963).

43. Mulford, C. E., *At. Abs. Newsletter*, **5**, 88 (1966).

44. Mullin, J. B., and J. P. Riley, *Analyst*, **80**, 73 (1955).

45. Mullin, J. B., and J. P. Riley, *Anal. Chim. Acta*, **12**, 162 (1955).

46. Nemodruk, A.A., P. N. Palei, and E. V. Bezrogova, *Zh. Anal. Khim.*, **25**, 319 (1970).

47. Pakaln, P., *Anal. Chim. Acta*, **40**, 328 (1968).

48. Pakalns, P., and W. W. Flynn, *ibid.*, **38**, 403 (1967).

49. Pietri, C. E., and A. W. Wenzel, *Anal. Chem.*, **35**, 209 (1963).

50. Potter, G. V., *ibid.*, **22**, 927 (1950).

51. Pyle, J. T., and W. D. Jacobs, *Anal. Chem.*, **36**, 1796 (1964).

52. Robinson, R. J., and H. J. Spoor, *Ind. Eng. Chem., Anal. Ed.*, **8**, 455 (1936).

53. Roller, P. S., and G. Ervin, *J. Amer. Chem. Soc.*, **62**, 461 (1940).

54. Schwartz, M. C., *Ind. Eng. Chem., Anal. Ed.*, **6**, 364 (1934).

55. Schwartz, M. C., *ibid.*, **14**, 893 (1942).

56. Shakhova, Z. F., E. N. Dorokhova, and N. K. Chuyan, *Zh. Anal. Khim.*, **21**, 707 (1966).

57. Smith, G. F., *Perchloric Acid*, 45th ed., G. Frederick Smith Chemical Co., Columbus, Ohio, 1940, p. 18.

58. Straub, F. G., and H. Grabowski, *Ind. Eng. Chem., Anal. Ed.*, **16**, 574 (1944).

59. Strickland, J. D. H., *J. Amer. Chem. Soc.*, **74**, 862, 868, 872 (1952).

60. Sussman, S., and I. L. Portnoy, *Anal. Chem.*, **24**, 1644 (1952).

61. Trudell, L. A., and D. F. Boltz, *ibid.*, **35**, 2122 (1963).

62. Trudell, L. A., and D. F. Boltz, *Anal. Chim. Acta*, **52**, 343 (1970).

63. Trudell, L. A., and D. F. Boltz, *Anal. Lett.*, **3**, 465 (1970).

64. Trudell, L. A., and D. F. Boltz, *Talanta*, **19**, 37 (1972).

65. Volk, R. J., and R. L. Weintraub, *Anal. Chem.*, **30**, 1011 (1958).

SULFUR

GORDON D. PATTERSON, Jr.

E. I. du Pont de Nemours & Co., Inc., Experimental Station, Wilmington, Delaware

JAMES M. PAPPENHAGEN

Department of Chemistry, Kenyon College, Gambier, Ohio

Sulfur, the 13th most abundant element, occurs widely in inorganic and organic compounds as well as in free state. Extensive use of both sulfur and its compounds in commercial products and processes has led to the frequent use of colorimetric methods for their quantitative determination. Although many sulfur-containing compounds exhibit color, relatively few types have found widespread use in quantitative/colorimetric analysis. These have primarily been barium sulfate, several metal sulfides, methylene blue, copper diethyldithiocarbamate, and ferric thiocyanate (94,117,118,214, 238,239).

I. SEPARATIONS

The two most important methods of separation of sulfur compounds for colorimetric measurement are (a) precipitation, and (b) volatilization followed by sorption. Usually, volatilization involves sweeping an acidified liquid sample with an inert gas to carry over a gaseous sulfur-containing compound to an absorption vessel, where the desired constituent is collected for later treatment and measurement.

A. EVOLUTION AND SORPTION OF HYDROGEN SULFIDE

Sulfides and sulfates in liquid samples can be converted to hydrogen sulfide, and the gaseous hydrogen sulfide can then be swept out by a stream of inert carrier gas. Hydrogen sulfide is removed from such systems, as well as from samples originally in the gaseous state, by absorption in liquid reagents which react quantitatively with the hydrogen sulfide to form nonvolatile products.

Various reagents have been used for this purpose. Field and Oldach (75) used a caustic soda solution. Sands et al. (203) made a detailed study of the determination of low concentrations of hydrogen sulfide in gas by the methylene blue method, finding that 2% zinc acetate solution was the preferred absorbent.

The difficulties of converting sulfur compounds in solid samples to hydrogen sulfide depend on the chemical nature of both the matrix and the desired constituent. Acidification alone is usually adequate if the sulfur is present as sulfides, although solubility is sometimes a problem. Pomeroy's work (191) on the estimation of sulfides in wastewaters led to a procedure adopted by the American Public Health Association (10) (see later under methylene blue procedures).

Jaboulay (109) found that sulfur in mild steels is present entirely as FeS and that treatment with dilute hydrochloric acid serves to convert all sulfur to hydrogen sulfide. Some other steels contain sulfur as both FeS and Fe_2S_3 and these require concentrated hydrochloric acid for complete conversion. Still other steels, especially those quenched from above $1200°C$, contain much Fe_2S_3 which reacts only partly with hydrochloric acid so that low results are obtained. Hence, the use of a stream of carbon dioxide may be recommended to remove the last traces of hydrogen sulfide from the solution, in contrast to the usual assumption that the large volume of hydrogen evolved takes care of this.

Oxidative and reductive methods for pyritic sulfur in solid fuels were critically reviewed by Radmacher and Mohrhauer (196). The usual reduction method was to treat with zinc and hydrochloric acid containing some stannous chloride, Reduction was incomplete in some cases but was always improved by finer grinding of the fuel sample. A more powerful reducing agent, consisting of chromium powder plus hydrochloric acid, completely reduced pyritic sulfur without attacking organic sulfur.

Sulfides and reducible sulfur in alkalies can be converted to hydrogen sulfide by acidification or by stannite and nascent hydrogen reduction, followed by passage into zinc acetate solution. Using the conditions described (46), 0.2–100 ppm could be measured on a 3-g sample. Budd and Bewick (46) confirmed the use of stannous chloride as well as aluminum metal for the quantitative reduction of thiosulfate to sulfide. They also point out that complete removal of oxygen from the apparatus is necessary before releasing trace quantities of hydrogen sulfide. This method surmounts the difficulty of incomplete reduction with aluminum alone by reducing in two steps, first an alkaline reduction with sodium stannite, and second the acid reduction with aluminum, the evolved hydrogen sulfide being absorbed in ammoniacal cadmium chloride solution.

Johnson and Nishita (113) have briefly reviewed the techniques for reducing sulfates to sulfides. From 0.5 to 300 μg sulfur, as sulfate or as total sulfur, may be determined with a precision of about 2% at the 95% confidence level. Sulfur-free nitrogen is bubbled into a boiling flask containing the sample in a reducing mixture of hydroiodic acid, formic acid, and red phosphorus. The nitrogen, with the evolved hydrogen sulfide, bubbles through pyrogallol–sodium phosphate reagent and then delivers the hydrogen sulfide to a zinc acetate–sodium acetate solution.

B. PRECIPITATION METHODS

Precipitation of trace sulfides from aqueous solutions can be quantitatively accomplished by treatment with zinc acetate followed by the addition of enough base to form a large carrier precipitate. Pomeroy (189,190) recommended this procedure to overcome difficulties due to interferences (in the methylene blue method) by iodides, sulfites, and thiosulfates. After the precipitate settles, most of the supernatant liquid is discarded before proceeding with the analysis. Colorimetric measurement of the amount of sulfide present before and after precipitation and redissolution gave checking results. This pretreatement with zinc acetate has additional merit as a preservative method for sulfide samples, since zinc sulfide has been shown to be relatively resistant to oxidation.

Not infrequently, two types of separation are used, an example being the procedure of Polson and Strickland (188). Microgram amounts of sulfur in nitrate or chloride solutions can be collected through the use of barium chromate carrier precipitation in solutions at a pH of about 1. The sulfate in the precipitate is then converted to hydrogen sulfide by heating in a mixture of hydrogen and hydrogen chloride vapor.

Iwasaki et al. (108) determined traces of sulfate in aqueous solutions by adding an excess of a weighed amount of barium chromate, precipitating and filtering the barium sulfate, then determining the chromium in the filtrate with diphenyl carbazide. A similar approach was recommended by Broekhuysen and Bechet (43).

Some procedures call for precipitation of benzidine sulfate followed by colorimetric determination based on reaction of the benzidine portion of the molecule. The precipitant may be benzidine hydrochloride, and special conditions are frequently used, such as the addition of nonaqueous solvents, low temperatures, and centrifuging. Solubility losses are least when the medium is slightly acidic (pH 3), the temperature low (close to 0°C), and the sample large enough to contain 100 mg or more sulfate. Washing should be with 80% alcohol or saturated aqueous benzidine sulfate solution.

Sulfate and sulfide precipitations can be used in a number of ways in the colorimetric determination of sulfur. Those based on turbidimetry will be discussed below, since physical separation by filtration is not involved.

The precipitation of calcium phosphate by the addition of calcium chloride, sodium hydroxide, and sodium carbonate before the spectrophotometric determination of sulfate has been reported (181). The calcium carbonate formed serves as a collector of the calcium phosphate.

C. EXTRACTION METHODS

Free sulfur is readily removed from samples by extraction with various solvents. All sorts of samples are amenable to this operation, from hydrated iron oxide to rubber. Sulfur in hydrocarbons is extracted with acetone according to the procedure of Bartlett and Skoog (30). Lowen (151) recommended several solvents for removing dithiocarbamate insecticides from food crop residues.

So many restrictions are imposed by the nature of the sample and by the possible need for not altering the chemical nature of the desired constituent that details of extraction problems will be discussed along with the procedure given later.

D. SORPTION METHODS

Since many gaseous sulfur compounds are quantitatively absorbed on silica gel, this phenomenon may be employed in determining trace quantities. Fogo and Popowsky (77) studied such separations from air and gaseous fuels in connections with their work in the conversion of sulfur compounds to hydrogen sulfide for subsequent determination by the methylene blue method. The desired compounds are sorbed on the silica gel at $25°C$ and then desorbed at $500°C$ and hydrogenated over a quartz catalyst. Under the conditions used, any hydrogen sulfide and carbon oxysulfide present in the sample are not sorbed. The apparatus consists of a sampling train having a calcium chloride drying tube, one or two sorption tubes containing purified 8- to 14-mesh silica gel, and a suitable gas meter. Thiols may be excluded selectively ahead of the train by scrubbing with an alkaline solution of cadmium sulfate.

Stratmann (231) developed a method for sulfur dioxide in air involving separation by adsorption on silica gel, followed by reduction to hydrogen sulfide and measurement by the color given with ammonium molybdate.

Sorption separations in liquids of course play an important part in methods involving evolution of hydrogen sulfide or sulfur dioxide. Sampling of sulfur dioxide at less than 0.1 ppm was found to be inefficient using the bubbler system and the West-Gaeke method (246). Similar methods involving solid sorbents which change color are discussed under miscellaneous methods, Section II.

E. ION EXCHANGE METHODS

In the determination of sulfate by the displacement—barium chloranilate method, interfering cations are removed by using cation exchange resins. Carlson, Rosell, and Vallejos (52) used a cation exchanger in ammonium form in the analysis of water. Schafer (208) used cation exchange resins in either the hydrogen or sodium form and the batch technique to remove cations before, determining sulfate in coal ash, precipitator dusts, and boiler deposits. A weakly basic anion exchange resin, in the chloride form, has been used to isolate thiocyanate from phenolic compounds before elution and colorimetric determination of the thiocyanate (255).

II. METHODS OF DETERMINATION

Sulfur methods usually depend on reactions of the sulfate radical or hydrogen sulfide, although methods involving the element itself or such forms as sulfur oxides, thio, dithio, and thiocyanate compounds are useful when circumstances permit. Much use is made of the many insoluble sulfates and sulfides amenable to turbidimetric measurement. The extreme sensitivity of the methylene blue method for hydrogen sulfide has led to its application to other types of samples where prior conversion to hydrogen sulfide is possible. Gases and vapors containing reactive sulfur groups form useful colors with impregnated papers or granular gels.

Separations, although avoided when possible, still are used frequently, owing to the many insoluble sulfur compounds and the readily volatile gases containing sulfur which are known. Fortunately, it often is possible to carry out the separation and color formation reactions simultaneously.

A. BARIUM SULFATE TURBIDIMETRIC METHOD

In spite of the well-known uncertainties inherent in turbidimetric methods, barium sulfate suspensions have been studied widely and found useful for quantitative estimations of sulfur. Sulfur already present as sulfate may be determined directly, whereas sulfides and organic sulfur compounds may be converted to sulfate by wet or dry oxidation. The conditions for oxidation vary with the nature of the sample. Many hydrocarbons may be oxidized in a flame, the basis of the widely used lamp method. More drastic conditions are necessary for difficultly combustible samples, and then the Parr bomb, Carius tube, and furnace-type oxidations are resorted to. Wet oxidations generally involve nitric acid with an accelerator and sometimes perchloric acid.

Other turbidimetric methods are discussed separately in Section B.

Canals et al. (50) found nephelometric sulfate determinations critically dependent on the crystalline form and size distribution of the light-scattering particles. In general, for an amorphous substance consisting of particles less than 1 micron in diameter, the light-scattering intensity is proportional to the mass of scattering particles, provided no disturbing foreign ions are present. Unfortunately, this rarely is exactly the case in practice, although studies on the effects of sodium, potassium, and lithium chlorides have shown that none of these affects the essentially spherical shape and size of the particles. The light-scattering intensity varied linearly with the mass of barium sulfate present, and the absorbance was unchanged by the presence of sodium or potassium chloride, although it was increased by lithium chloride. Also it was found that the light-scattering intensity depended on the nature and relative amounts of the precipitant and such factors as temperature and aging time before measurement.

Tananaev and Rudnev (236) studied the merits of ethanol in decreasing the barium sulfate solubility. The light absorbance increases with increased ethanol content up to 30%, after which it decreases sharply. Excess barium chloride has striking effects on the maxima which first diminish, then increase sharply as the barium chloride concentration is raised. Excess barium chloride is believed to affect not only the dispersion of the barium sulfate but also its crystalline form and bulk, due to coprecipitation. Since both the crystalline form and the bulk of the precipitate affect the light absorption, the measured light absorbance depends in a complicated manner on the circumstances of precipitation, primarily the alcohol content and the concentration of excess barium chloride.

Very often, the sample must be oxidized to convert the sulfur compounds to sulfate. The lamp method for determining sulfur compounds in hydrocarbons has had much study and is widely used, generally for concentrations above 20 ppm (16,19,157). Early difficulties have been surmounted largely by ingenious improvements in the burner and absorber designs, but the method is still limited to samples having sufficient flammability, low solids content, and sulfur compounds sufficiently volatile to burn completely.

Lane's (142) critical investigation of the variables in the method as proposed by Zahn (258) resulted in several modifications to the standard ASTM method. Such details of technique as position of the wick tip, careful rinsing of the lamp chimney, and the use of Pyrex wool wicks were found to be critical for some types of samples and ranges of concentration.

The nature of both the sample matric and the desired constituent influence the utility of this method. It should not be used for samples containing elemental sulfur, since incomplete oxidation may occur. Special conditions are necessary when the sample contains volatile organic sulfur compounds. Nonluminous flames are essential, since they indicate complete combustion. Aromatic hydrocarbons normally give smoky flames, but some improvement can be obtained by mixing the sample with an equal volume of isopropanol. This method of assuring a nonluminous flame also aids in the complete combustion of volatile organic sulfur compounds.

Hinden and Grosse (98) suggested an alternative modification of the lamp method for materials boiling up to $500°C$. The lamp is threaded with three to five cotton wicks trimmed flush with the top of the burner. The sample of 5−10 ml is placed in an upper reservoir, the lamp is weighed, and the sample then run into the lower reservoir and given at least 10 min to diffuse to the top of the wick. The lamp is then lighted and combustion carried out, after which the lamp is reweighed to determine the amount of sample consumed. In work with petroleum cuts containing up to 35% aromatic compounds, they found that no fractionation occurred during burning. They solved the problem of maintaining the burner warm enough to prevent solidification of sample material in several ways: by wrapping a Nichrome heating coil around the tube from the reservoir to the tip; or suspending a similar coil from the chimney into which the burner fits; or using a metallic burner tube for heat conduction; or warming the incoming air. An alternative design featuring a secondary air supply directly to the burner (a 1-mm metal tube inside, and concentric with, the burner tube, with the wick wrapped around it) allowed burning even methylnaphthalene without smoking.

Turbidimetric methods involving the formation of barium sulfate have been used for sulfur dioxide in air. Oxidation of the sulfur dioxide may be accomplished by hydrogen peroxide (49), and other oxidants have also been suggested. The Schòniger oxygen flask method has been found to be applicable to the determination of sulfur in fuel oil (147).

Sands et al. (204) studied various combustion methods for determining sulfur compounds in synthesis gas. Their recommended procedure measures as little as 0.0001 grain of sulfur (as $BaSO_4$) per 250 ml solution. They found that other methods for small amounts of sulfate had no advantages over turbidimetric methods.

General Procedure

The measurement of the transmittance of a dilute suspension of barium sulfate produced under carefully controlled conditions is the basis of this turbidimetric method for the determination of sulfate. The addition of an electrolyte solution containing sodium chloride and hydrochloric acid gives solutions of similar ionic strength and pH, and the addition of rather uniformly sized solid particles of barium chloride to give a controlled rate of dissolution of the precipitant tends to give a suspension of uniformly sized barium sulfate particles. The addition of less polar solvents such as glycerol and ethanol decreases the solubility of the suspended particles and thus stabilizes the suspension.

Reagents

Standard Sulfate Solution. Dissolve 0.1479 g anhydrous sodium sulfate, Na_2SO_4, in distilled water and dilute to 1 liter. Each ml of this solution contains 0.100 mg sulfate.

Stabilizer Solution. Mix 100 ml glycerol and 200 ml ethanol.

Electrolyte Solution. Dissolve 240 g sodium chloride in about 900 ml distilled water, add 20 ml concentrated hydrochloric acid (sp. gr. 1.19; 37% by wt. HCl), and dilute to 1 liter.

Barium Chloride. Use analytic reagent-grade quality barium chloride, $BaCl_2 \cdot 2H_2O$, which has been finely pulverized and collect that fraction which passes through a 20-mesh sieve but is retained by a 30-mesh sieve.

Procedure

Standardization. Transfer 0, 5, 10, 15, 20, 25, 30, 35, and 40 ml of the standard sulfate solution to 100-ml volumetric flasks. Add 10 ml of the electrolyte solution and 20 ml of the stabilizer solution. Dilute to volume with distilled water and mix thoroughly. Add 0.300 g of the barium chloride crystals and mix the contents of each flask for 2 min by continuously inverting the flask. After 5 min, measure the transmittance of each suspension at 420 nm against a reagent blank solution in the reference cell. Prepare a calibration plot.

Sample. Transfer 50 ml of the sample solution to a 100-ml volumetric flask and repeat the procedure used in the preparation of the calibration plot.

B. OTHER TURBIDIMETRIC METHODS

Many metals form reasonably stable sulfide colloids which are useful for quantitative measurement. Trieber et al. (240) found the sols of bismuth, silver, palladium, and copper to be quite sensitive and stable. Cadmium, lead, and thallium have also been used. In many cases it is necessary or desirable to add protective colloids, such as gelatin, gum arabic, or carboxymethyl cellulose, or poly(N-vinylpyrrolidone). Generally, the absorbance curves are critically dependent on the environment and reaction conditions. Such factors include composition, reagent concentrations, temperature, and the pH of the system in which the precipitation takes place, as well as additives (protective colloid, etc.) and frequently the age of the sol.

The potentialities of lead sulfide sols have been repeatedly suggested because of the extreme sensitivity. Unfortunately, they coagulate rapidly, and hence considerable care is necessary to obtain quantitative results. Photometric measurement of the colloidal lead sulfide sol was described by Luke (152).

Laboratory techniques for the formation of lead sulfide on treated filter paper have been studied by Scherer and Sweet (209). Reducible sulfur in paper can be determined (209) by heating the sample in 30 ml 0.5% sodium hydroxide just below the boiling point, then cooling to room temperature, adding 30 ml 11N hydrochloric acid and two strips of aluminum foil, and stoppering immediately with a tight stopper carrying a 6-mm glass tube somewhat constricted near the bottom end and with a wad of cotton inserted below it. Above the constriction is placed a strip of lead acetate-impregnated filter paper (the paper having been soaked in 1% lead acetate solution, then dried). After 2 hr, the lead sulfide deposit is compared with that obtained similarly with known quantities of sulfur. A suitable standard is 0.5N sodium thiosulfate, diluted so that 1 ml contains 0.001 mg sulfur. Quantities in the range of 0.001–0.015 g sulfur are measurable.

Treiber et al. (240) found silver sulfide sols very stable. Alkali sulfide solutions were treated directly with 0.05N silver nitrate solutions, and it was found that protective colloids could be safely omitted (223).

Bismuth salts are very sensitive and specific to hydrogen sulfide, in contrast to the photosensitive silver solutions, which give turbidities also with sulfur dioxide. They are also superior to palladium chloride, which reacts with carbon disulfide and carbon oxysulfide. Treiber et al. (240) also pointed out that bismuth sulfide sols, unlike most of the colloidal metal sulfides, do not adsorb excess free hydrogen sulfide. Certain limitations are imposed, however, by the occasional occurrence of practically unfilterable turbidities in the bismuth nitrate—colloid buffer reagent used. Also, in the presence of higher sulfide concentrations, partial flocculation occurs when the bismuth nitrate exceeds about 0.25% and the gum arabic concentration is lower than about 0.3%.

The recommended bismuth reagent (135,240) is made up by dissolving 0.200 g $Bi(NO_3)_3 \cdot 5H_2O$ in 25 ml 3.2% mannitol solution. After completion of dissolution, 8 ml pure glycerol and then 36 ml 2.5% filtered gum arabic solution are added. The mixture is then brought to 100 ml with an acetic acid—sodium acetate buffer (6 parts by volume 0.2M sodium acetate solution and 1 part 0.2M acetate acid). After standing 12 hr, the solution is filtered and should be clear. The turbidity of bismuth sulfide sols prepared with this reagent reaches a maximum in 10 min and remains constant for 2 hr. As low as 5 μg sulfide in 100 ml solution is detectable.

Field and Oldach (75) determined hydrogen sulfide in gases by absorption in sodium hydroxide solution and formation of a bismuth sulfide sol. Sands et al. (203) confirmed their results in determining as little as 0.7 ppm sulfide sulfur with a precision of ±10%.

C. DISPLACEMENT—BARIUM CHLORANILATE METHOD

Bertolacini and Barney (33) developed an indirect method for sulfate based on the displacement of chloranilate from barium chloranilate by sulfate. Solid barium chloranilate, when added to a sulfate solution, precipitates barium sulfate and releases an equivalent amount of the hydrogen chloranilate ion, which has a purple color and exhibits absorbance maxima in the visible and ultraviolet regions. The excess barium chloranilate and barium sulfate is removed by centrifugation before absorbance measurements. A 50% by volume aqueous ethanolic solution is used to decrease the solubility of the barium chloranilate and thereby give lower blanks. The solution is buffered at pH 4 to form the hydrogen chloranilate species which gives the desired absorption spectrum.

Many cations, such as Ca^{2+}, Cu^{2+}, Zn^{2+}, Pb^{2+}, Fe^{3+}, and Al^{3+}, interfere by forming insoluble chloranilates so that the removal of cations by a strong cation exchange resin in the hydrogen form is recommended. Phosphate, oxalate, bicarbonate, chloride, and nitrate do not interfere at the 100-ppm level.

Measurement of the absorbance at 530 nm is recommended when the sample solution contains 10 to 300 μg per ml sulfate, and measurement at 320 nm is suggested when the solutions contain 0.5 to 10 μg per ml sulfate. Carlson et al. (52) reported that increased sensitivity can be achieved for absorbance measurements at 530 nm by adjustment of the pH of the solution to 1.8 after the removal of the excess barium chloranilate.

Applications of this method to the determination of sulfate are cited in Table IV.

Reagents

Standard Sulfate Solution. Dissolve 1.479 g anhydrous sodium sulfate in distilled water and dilute to 1 liter. Each ml of this sulfate solution contains 1.000 mg sulfate.

Strong Cation Exchange Resin. A strong cation exchange resin (Dowex 50 or Amberlite IR-120) in the hydrogen form and 20–50 mesh should be placed in a 15 cm x 1.5 cm column.

Barium Chloranilate. Available from commercial supply firms.

Buffer Solution. Dissolve 10.2 g potassium hydrogen phthalate in distilled water and dilute to 1 liter.

Procedure

Transfer 25 ml of a solution containing between 4 and 2500 μg sulfate per ml to the top of the ion exchange column. Collect effluent and washings in a 100-ml beaker. Adjust the pH of the solution to about 4 using dilute 1:100 hydrochloric acid or 1:100 ammonia solution, transfer to a 50-ml volumetric flask, and dilute to volume.

Transfer a 25-ml aliquot of the sample solution to a 100-ml volumetric flask (this aliquot should contain no more than 30 mg sulfate). Add 10 ml of the buffer solution and 50 ml 95% ethanol. Dilute to volume with distilled water and mix thoroughly. Add 0.300 g barium chloranilate and shake the flask for 10 min. Remove the excess barium chloranilate by centrifugation or filtration. Measure the absorbance of the supernatant solution against a reagent blank solution prepared according to this same procedure. The absorbance can be measured at either 530 nm or 320 nm depending on the initial concentration of sulfate. If the ultraviolet method is used, only 0.200 g of the barium chloranilate should be added.

A calibration plot can be constructed by preparing a series of solutions in which 0, 1, 5, 10, 15, 20, 25, 30, and 35 ml of the standard sulfate solution instead of the 25-ml aliquot of sample solution is used in the color development procedure.

If the ultraviolet method is to be used, prepare a 1:10 dilution of the standard sulfate solution and use 0, 0.5, 1, 2, 5, 7, and 10 ml of this dilute standard solution in preparing the colored solutions to be used in preparing the calibration plot.

D. BENZIDINE METHOD

Diazotization of stoichiometric amounts of benzidine obtained from benzidine sulfate precipitates is useful for obtaining a colorimetric measure of sulfate. The method is versatile and sensitive, and other color reactions of benzidine may also be used.

Klein's procedure (130) utilizing N-(1-naphthyl)ethylenediamine is said to have a maximum error of 2% on samples containing 0.05—0.150 mg sulfate. Chloride ion in large amounts and phosphate interfere. Letonoff and Reinhold (146) used the intense color caused by the reaction of sodium β-naphthoquinone-1-sulfonate with benzidine. The procedure included the presence of a sodium borate—sodium hydroxide buffer and the addition of acetone to reduce the color of the reagent itself. For serum and urine samples, uranium acetate is recommended to remove proteins and avoid errors from phosphate that also precipitates with benzidine.

E. METHYLENE BLUE METHOD

Many applications of the methylene blue procedure are reported in the literature, and the continued appearance of new modifications indicates both that it has reliability and lacks perfection (57,155,173). It depends on the formation of methylene blue from the reaction of sulfide with p-amino-N,N-dimethylaniline in the presence of ferric chloride. The reaction may be represented as

The molar absorptivity of the methylene blue solution used by Johnson and Nishita (113), corresponding to the range 5—50 μg sulfur in 100 ml was about 34,000, very close to that for the maximum sensitivity of colorimetric methods. Sands et al. (203) were able to detect as little as 0.00001 grain of sulfide sulfur per 50 ml solution.

Better sensitivity and reproducibility by using p-aminoaniline instead of p-amino-N,N-dimethylaniline have been reported in the determination of sulfides in acid concentrations below $0.1N$ (121).

Methylene blue solutions, although stable several days in the dark, fade rapidly in sunlight. Also their color is affected by acidity, and some care is necessary to maintain reproducible acid conditions in all determinations. This is especially true when dark methylene blue solutions are to be diluted.

The digestion of the sample (if necessary), the heating rate of the digestion, and nitrogen flow rates are not critical, but extremes should be avoided. Fogo and Popowsky (77) point out that at higher temperatures, the color reaction is rapid but hydrogen sulfide escapes to the space above the liquid; whereas at lower temperatures, the reaction is so slow that side reactions become important. It is fortunate that the maximum yields occur at $24°C$ and that small deviations do not cause serious trouble.

Various recipes have been proposed for the color-forming reagent. Roth (202) suspended 1 g p-amino-N,N-dimethylaniline (sulfate or hydrochloride) in 100 ml water and added 400 ml concentrated sulfuric acid slowly with cooling, diluted to 2 liters, and then added 200 ml of a solution of 25 g $Fe(NH_4)(SO_4)_2 \cdot 42H_2O$ and 5 ml sulfuric acid.

Pure, fresh p-amino-N,N-dimethylaniline and its sulfate dissolve in water with a pale-brown color. The red hue caused by ferric chloride fails to appear in the presence of much hydrochloric acid. On the other hand, most commercial preparations develop a dark color (when dry) owing to the influence of light and air, and the solutions have an orange-red color which is little influenced by hydrochloric acid.

Sands et al. (203) selected the methylene blue method as the basis of an ultrasensitive technique for hydrogen sulfide in gases. This was accomplished by studying the variables affecting the results and recommending a procedure to be followed exactly. The preferred absorbent is 2% acidic zinc acetate solution, and the sulfate of p-amino-N,N-dimethylaniline is preferred over the hydrochloride. Measurement should be made at 745 nm within 2 hr after completion of gas scrubbing. Their temperature study resulted in a recommendation that the solution should be cooled to $10°C$ before adding the diamine reagent. Gas samples of only 0.1 cubic foot are needed when the concentration is 0.05 grain sulfur (as hydrogen sulfide) per 100 cubic feet, and the result is obtained in less than 30 min.

The standard American Public Health Association method (10) for sulfide and reducible sulfur in sewage is based on Pomeroy's modification (191). It calls for a sufficient concentration of ferric chloride to cause complete color development in 1 min, followed by the addition of ammonium monohydrogen phosphate to eliminate the ferric color.

The color-developing process may consist of the liberation of hydrogen sulfide by acidification of the aqueous sample and sweeping it out with an inert gas into an acidic zinc acetate solution. Some types of samples may be colorized directly without the necessity of hydrogen sulfide volatilization and absorption. To the sulfide solution is added dilute p-amino-N,N-dimethylaniline solution and ferric chloride. The color forms fairly rapidly, and the absorption may be measured at 670 nm or with an appropriate filter photometer.

Special precautions are necessary for the evolution of hydrogen sulfide in the presence of carbonates, since too rapid an evolution of carbon dioxide may cause low results. Some investigators have not adequately emphasized the critical importance of sweeping the apparatus free of oxygen before releasing trace qua quantities of hydrogen sulfide. More rapid color development results from the use of relatively high concentrations of reagents, and acidity is also important.

Total sulfur in plant materials is effectively measured through oxidation by any of several oxidation methods, for example, the magnesium nitrate method, the Parr bomb peroxide method and the nitric acid—perchloric acid wet oxidation method. The latter method requires special attention to preparing a solution sufficiently low in nitrate and perchlorate so as not to interfere in the subsequent reduction of the sulfate.

Sulfate in plant materials (113) may be digested at 115°C in a mixture of hydriodic acid, formic acid, and red phosphorus. The hydrogen sulfide gas is swept out by washed nitrogen into a zine acetate—sodium acetate solution. To this is added aqueous p-amino-N,N-dimethylaniline sulfate and ferric ammonium sulfate. The reading is taken after 10 min but within 24 hr. Special precautions in using sulfur-free distilled water, filtered solutions, and sulfur-free ground-joint lubricant are recommended.

Nitrate if present in too large an amount causes low and erractic results. Normally, its effect can be reduced to acceptable levels by reducing the sample size. It is generally unsatisfactory to attempt to reduce the nitrate with finely divided iron in acidic solution. Heating the sample to dryness with hydrochloric acid may cause hydrolysis of organic sulfur compounds which otherwise do not interfere. Johnson and Nishita (113) found the most successful technique for eliminating nitrate interference to be the extraction of the sample with 10% barium chloride solution, followed by centrifugation and transfer of the residue to the digestion flask. Certain organic sulfur compounds containing thio, dithio, mercapto, and sulfuric acid groups were found not to interfere in this method.

Hydrogen peroxide interferes through oxidation of the iodide to iodine, just as would excess perchloric acid if the wet oxidation procedure were used. Moderate amounts of chromate do not interfere, but since it may contain sulfate impurities, barium chromate is not recommended as a barium sulfate carrier in barium chloride extraction procedures. Equilibration of the digestion-distillation apparatus with a preliminary trial run is advisable, due to the adsorption and solubility of hydrogen sulfide in the apparatus and wash solution.

The presence of more than 0.01 ppm copper in the distilled water used in the preparation of wash solutions and the zinc acetate solution was found deleterious by Johnson and Arkley (112). Under certain conditions, the copper precipitates sulfide, and low recoveries result.

The methylene blue method was applied by Pursglove and Wainwright (195) in the determination of carbonyl sulfide in synthesis gas. Their modification is especially useful when other organic sulfur compounds are present, since carbon disulfide, ethyl mercaptan, and thiophene do not interfere. It is based on hydrolysis in a dilute base (0.1N potassium hydroxide). The slow absorption and hydrolysis characteristics of carbonyl sulfide were compensated by increasing the time of contact between the gas and liquid. The conditions proposed by Sands et al. (203) for low concentrations of hydrogen sulfide in gas were used.

The use of 0.1N cadmium sulfate solution as the absorbing solution was found reliable by Oka and Matsuo (176), although other authors caution against the oxidative tendencies of the cadmium sulfide precipitate. They also found it advantageous to partially neutralize the colored solution with sodium hydroxide after completion of the color reaction.

A special procedure for estimating volatile sulfur compounds in onions and other plants having sulfur-containing essential oils was described by Currier (60). It involved extraction with cold water, reduction with zinc powder, followed by colorization with the p-amino-N,N-dimethylaniline hydrochloride and ferric ammonium sulfate.

For iron and steel, the procedure proposed by Roth (202) can be used. The usual approach is to treat the sample with hydrochloric acid and pass the gases directly into a solution in a colorimetric tube containing the p-amino-N,N-dimethylaniline in hydrochloric acid followed by addition of the ferric chloride and measurement. Standard steel samples should be used for calibration.

F. ULTRAVIOLET ABSORPTION METHODS

Aliphatic sulfides form complexes with molecular iodine which intensely absorb in the ultraviolet region of the spectrum (near 307 nm). In the absence of other ultraviolet absorbers, this method can be used to determine aliphatic sulfides with good accuracy and precision. The complex is a 1:1 combination of sulfide and iodine but is very sensitive to excesses of either. Due to the experimental difficulty of adding just enough iodine (especially when aromatic of olefinic hydrocarbons may be present to tie up varying amounts of the iodine), the recommended technique (92) is to make the iodine concentration very much larger than the sulfide concentration. This makes the total absorbance significantly affected only by changes in the aliphatic sulfide content.

It is possible to analyze petroleum stocks having considerable absorption near

310 nm, owing to the small amounts of sample necessary and to the large dilutions made before measurement. The sulfide sulfur may be in acyclic or cyclic saturated systems or in side chains attached to aromatic nuclei.

Increased sensitivity is obtained for low-sulfide samples by treating a portion with solid mercurous nitrate to remove the sulfides (92). This is followed by a wash with 10% aqueous mercuric nitrate solution to assure complete removal. This treated sample is then used as a blank to be compared with the iodine-complex absorption of an untreated one. Unfortunately, the mercurous nitrate–sulfide complex is oil soluble and hence cannot be used for petroleum samples of the gas oil boiling range.

Heatley and Page (93) determined elemental sulfur by its ultraviolet absorption in ethanolic solutions. Beer's law applies for $0-40$ μg per ml. Water in the ethanol up to 40% does not interfere. Various other sulfur compounds and proteins do interfere and must be removed.

G. PARAROSANILINE–FORMALDEHYDE METHOD FOR SULFUR DIOXIDE

Pararosaniline, formaldehyde, and sulfur dioxide react in acidic solution to give a purple color. It has been postulated that sulfur dioxide reacts with formaldehyde to give an intermediate, CH_2OSO_2, which then reacts with pararosaniline to give a purple sulfonic acid derivative (168). The absorbance is usually measured at 560 nm, and the method is applicable to the determination of $0-25$ μg SO_2. The purity of the pararosaniline (182), pH, and temperature (139) must be carefully controlled for reliable results.

Sulfur dioxide is absorbed in a $0.05-0.1M$ sodium or potassium tetrachloro-mercurate(II) solution to form a stable, nonvolatile dichlorosulfitomercurate(II) ion. This reaction is known as the West-Gaeke reaction (253). Although the dichlorosulfitomercurate(II) is quite stable, even to air oxidation, the sulfur dioxide is readily removed from this complex by formaldehyde. Nitrogen dioxide interferes seriously with this method. However, the addition of sulfamic acid prior to the development of the color eliminates this interference (183,207). Ozone also interferes (237). Huitt and Lodge (105) have found that the color formed by the pararosaniline–formaldehyde method is intensified in the presence of N,N-dimethylformamide.

This method has been used extensively for the determination of atmospheric sulfur dioxide (100,207,246,253). Tables VII and X should be consulted for further applications. The following general procedure is essentially that of West and Gaeke (253).

Reagents

Sodium Tetrachloromercurate(II) (TCM). Dissolve 54.4 g mercury (II) chloride and 23.4 g sodium chloride in distilled water and dilute to 2 liters. Use only reagent-grade chemicals.

Pararosaniline. Transfer 4.0 ml of a 1% aqueous solution of pure pararosaniline hydrochloride to a 100-ml volumetric flask and add 6.0 ml concentrated hydrochloric acid. Dilute to volume with distilled water.

Formaldehyde Solution. Transfer 5.0 ml of a 40% formaldehyde solution to a 1-liter volumetric flask and dilute to volume with distilled water.

Standard Sulfite Solution. Transfer 1.484 g sodium metabisulfite, $Na_2 S_2 O_5$, to a 1-liter volumetric flask and dilute to volume with distilled water. Transfer 5.00 ml of this stock solution to a 1-liter volumetric flask and dilute to volume with the TCM solution. Each ml of this solution is equivalent to 5.0 μg SO_2.

Procedure

Bubble a volume of sample containing not more than 50 μg SO_2 through 20.0 ml of the 0.1M TCM solution using a small coarse-fritted glass bubbler in a 25-ml volumetric flask.

Add 2.0 ml of the pararosaniline solution and 2.0 ml of the formaldehyde solution. Dilute to volume with the TCM reagent. Prepare a reagent blank reference solution by taking 20.0 ml of the TCM reagent and using this same color development procedure. After 25 min, read the absorbance at 560 nm against the reagent blank solution.

Prepare a calibration graph using 1, 2, 4, 6, 8, and 10 ml of the standard bisulfite solution and the same color development procedure.

H. IRON 1,10-PHENANTHROLINE METHOD FOR SULFUR DIOXIDE

Sulfur dioxide reduces iron(III) to iron(II), and the iron(II) reacts with 1,10-phenanthroline to form the orange tris(1,10-phenanthroline)−iron(II) complex. Fluoride is added to prevent the hydrolysis of iron(III). Nitrogen dioxide and ozone do not interfere. Hydrogen sulfide interferes. This indirect spectrophotometric method, developed by Stephens and Lindstrom (228), is applicable to the determination of up to 75 μl SO_2. The absorbance is measured at 510 nm.

Reagents

Iron(III) Solution, 0.001M. Dissolve 0.482 g ferric ammonium sulfate, $NH_4 Fe(SO_4)_2 \cdot 12H_2O$, in distilled water and dilute to 1 liter.

1,10-Phenanthroline, 0.03M. Dissolve 2.97 g 1,10-phenanthroline monohydrate in 450 ml water and heat to 80°C to hasten dissolution. Dilute to 500 ml.

Procedure

Transfer 10.0 ml of the 0.001M iron(III) solution to a bubble bottle fitted with a coarse gas-dispersion tube. Add 10.0 ml of the 0.03M 1,10-phenanthroline solution. Adjust solution to pH 5.5 ± 0.5 with a 0.1N NaOH solution. Add 1.0 ml *n*-octyl alcohol and insert the bottle in a water bath maintained at exactly 50°C. After the gas sample has passed through the solution, transfer the contents of the bubbler bottle to a 100-ml volumetric flask, rinse, and add the washings to the flask. Add 2.0 ml of a 5% ammonium bifluoride solution. Cool to room temperature and adjust to a volume with distilled water. Measure the absorbance at 510 nm. A gas absorption train suitable for use in the preparation of a calibration graph has been described by Stephens and Lindstrom (228).

I. MISCELLANEOUS METHODS

Elemental sulfur in acetone solution reacts rapidly and quantitatively with cyanide to give thiocyanate. Addition of ferric chloride produces a useful complex which can be measured at 465 nm. The method is applicable to elemental sulfur in hydrocarbons (30) and has a sensitivity of 2 ppm. Peroxides, sulfides, disulfides, and mercaptans do not interfere or can be removed, and an accuracy of 98—99% is possible.

Sulfide ions can be determined by the use of ammonium molybdate. In the presence of hydrogen sulfide, the molybdate assumes first a yellowish-green, than a bluish-green hue. Color intensity reaches a maximum in about 1 hr and is stable for several hours. Due to the various complex stages of oxidation of the molybdenum oxides, the color is very sensitive to reagent impurities, pH, and temperature conditions (240).

Hydrogen polysulfides, when treated with an excess of cyanide, produce hydrogen sulfide and thiocyanate. The thiocyanate is then determined colorimetrically as iron(III)thiocyanate complex (211). The color reaction of nitroprusside and sulfites or sulfides can result in the formation of either 1:1 or 1:2 complexes depending on the proportion of the reactants (70).

A method for dithiocarbamate residues on food crops was studied by Lowen (151). Microquantities of carbon disulfide in gases are measured by the yellow copper dithiocarbamate formed when it is sorbed in copper acetate—dimethylamine reagent. The sample containing the dithiocarbamate is treated with boiling sulfuric acid and a gentle vacuum applied. The evolved vapors pass first through a trap of 10% lead acetate solution, then to the carbon disulfide trap containing copper acetate—diethylamine solution. The colors are effectively measured with a filter photometer. Beer's law is followed up to at least 13 μg copper diethyldithiocarbamate per ml., and the sensitivity is approximately

1 microgram per ml for 1-cm cells. The distillation procedure, calling for 30–45 min of refluxing was found to recover essentially 100% of the carbon disulfide added in known amounts to synthetic samples. Ferric and zinc dimethyldithiocarbamates were also 100% recovered when known quantities were adsorbed on filter paper from chloroform solutions, followed by decomposition and distillation with 50 ml boiling $1N$ sulfuric acid. Zinc ethylenebisdithiocarbamate recovery is more difficult, but 75% recovery is reproducible when the experimental conditions are rigidly controlled. Sampling techniques include solvent stripping, scrubbing, or pulping of dithiocarbamate-treated food crops, depending on the nature of the sample.

The various colorimetric methods for the determination of sulfur in organic compounds have been discussed and compared (7,8,66,67,68).

Analytic methods for mercaptans and other sulfur compounds have been studied critically (59). Mercaptans react with p amino-N,N-dimethylaniline in the presence of acidic ferric alum to give a blue color. Sulfides, of course, interfere.

Sulfhydryl groups in wool and hair samples can be measured using 1-(4-chloromercuriphenylazo)naphthol-2. Burley's modification (48) avoids pulverizing the wool by dissolving the 1-(4-chloromercuriphenylazo)naphthol-2 in formamide and using this solution directly on the wool. The formamide swells the fibers sufficiently to allow reproducible penetration.

Thiophene can be determined in sythesis gas containing low amounts of unsaturated hydrocarbons by use of an isatin-acid reagent. Wainwright and Lambert (248) reviewed various methods and developed a technique sensitive to as little as 0.0001 grain thiophene sulfur in 55 ml test solution. Careful control of conditions is necessary. For example, the temperature of absorption of the thiophene in the isatin-acid reagent must be held constant. The absorbance decreases as the temperature during absorption is increased from $15°$ to $60°C$, although the solution temperature after completion of absorption has no effect on the color. The age of the isatin acid reagent also markedly affects the color. A 1-hr aging period is recommended. Color development is accelerated by the presence of ferric sulfate as an oxidizing agent. Carbon disulfide up to at least 12 times the thiophene concentration does not interfere. Mercaptans cause high results but can be removed satisfactorily by bubbling the sample gas through alkaline cadmium chloride solution. Unsaturated hydrocarbons do interfere to varying extents depending on the type and amount present, although ethylene below 2% does not interfere seriously, and propylene and butylene can be tolerated in larger amounts.

The procedure calls for the use of an aqueous solution of isatin and ferric sulfate mixed with sulfuric acid just 1 hr before use. The sample gas is first passed over iron oxides shavings to remove hydrogen sulfide, then bubbled through alkaline cadmium chloride solution (to remove hydrogen sulfide), then bubbled through the isatin-acid reagent and metered with a wet-test meter. The gas flow is continued until the test solution turns blue. The absorbance is then measured in a 1-cm cell in a spectrophotometer and 580 nm. The sulfur concentration is then obtained from a calibration curve prepared from known solutions of thiophene in dibutyl phthalate.

Sulfate ion can be determined colorimetrically in the range of 0–400 ppm by use of the insoluble thorium borate–Amaranth dye reagent (141). Dye molecules are released stoichiometrically by the sulfate, and the concentration is determined indirectly by measuring the dye absorbance at 521 nm. Interferences from fluoride, phosphate, and bicarbonate are eliminated by the addition of lanthanum nitrate and passage through a weak acid cation exchange resin. Good agreements were obtained for water samples in the range of 15–200 ppm sulfate when compared with gravimetric analyses.

Gaseous or vaporous sulfur compounds can be determined by sorption on treated granules which change color in proportion to the concentration. Mainly hydrogen sulfide and sulfur dioxide have been studied for this sort of application. The specially prepared color-sensitive gel is placed in a glass tube, and a measured volume of the sample gas is passed through. The resultant color change may be observed visually or measured photometrically. For hydrogen sulfide, a variety of color-producing agents have been impregnated onto various granular materials, for example, silver cyanide on silica (129,148).

The preferred visual measurement technique for all these gels is a general color comparison of the whole gel with standards. Several alternatives may be used, however. If only part of the column of gel is reacted, these include measuring the length of the colored portion or the time necessary (at constant sample flow) to obtain a desired degree of coloring.

Sulfur dioxide decolorizes solutions of fuchsin sufficiently reproducibly to have been the subject of a number of quantitative studies, among them Atkin (22) and Stang et al. (226).

The use of iodine–potassium iodide–starch solutions is a reliable colorimetric method for sulfur dioxide, and Katz (122) has described the use of stabilized starch–iodine solutions for this purpose. Using photoelectric measurement, one can cover a range of 0.01–1.00 ppm sulfur dioxide or more.

Techniques for the preparation of known concentrations of sulfur dioxide in air and methods for determining trace concentrations of sulfur dioxide in air have been reviewed (96). A single-step syringe sampling technique applicable to the colorimetric determination of sulfur dioxide and the use of Teflon permeation tubes filled with sulfur dioxide have been described (161).

III. APPLICATIONS

A. HYDROGEN SULFIDE IN GASES BY METHYLENE BLUE

The method of Sands et al. (203) is applicable to determining as little as 0.001 grain sulfide sulfur per 100 cubic feet when the sample volume taken is 1 cubic foot. Mercaptans, carbon disulfide, and thiophene do not interfere if photometric measurement is made at 745 nm. Modifications of this method are applicable to other types of samples. Carefully purified distilled water must be used throughout.

Reagents

Stock Zinc Acelate Solution 20%. Dissolve 200 g C.P. zinc acetate in 1 liter distilled water.

Dilute Zinc Acelate Solution, 2%. Dilute 100 ml of the stock solution to 1 liter with distilled water and acidity with about three drops acetic acid.

Diamine Reagent. Dissolve 0.15 g p-amino-N,N-dimethylaniline sulfate in a cooled mixture of 100 ml concentrated sulfuric acid and 50 ml distilled water.

Ferric Chloride Solution. Dissolve 2.7 g $FeCl_3 \cdot 6H_2O$ in 50 ml concentrated hydrochloric acid and dilute to 100 ml with distilled water.

Iodine Solution. $0.1N$ I, standardized solution.

Sodium Thiosulfate Solution. $0.1N$ $Na_2S_2O_3$ standardized solution.

Procedure

Bubble the gas sample into 50 ml of the 2% zinc acetate solution contained in a wide-mouthed bottle until a faint turbidity forms, measuring the volume of gas used. For concentrations below 13.5 grains per 100 cubic feet, a gas-washing rate of 1.8 cubic feet per hr is slow enough to avoid loss. Wash the contents of the bubbling bottle into a 1-liter volumetric flask with the 2% zinc acetate solution, rinsing out the bottle carefully, adjust the volume to 1000 ml with the 2% zinc acetate solution, and mix. Pipet 50 ml of this solution into a beaker or flask and cool to at least 10°C in an ice-water bath. Add 5.0 ml of the diamine reagent, stir, and add 1.0 ml of the ferric chloride solution. Stir and measure the color after standing 15–30 min.

Prepare a blank similarly with all reagents and use it to zero the instrument (frequent reblanking is desirable). Changes in temperature after the color is developed have no effect. If the color is too intense, make a suitable dilution of the liter zinc acetat–zinc sulfide solution and again develop the color.

Calibration standards are prepared as follows: pass 5–50 bubbles hydrogen sulfide gas (depending on the concentration desired), with vigorous stirring, into exactly 500 ml distilled water contained in a 600-ml beaker. Pipet 10 ml of this hydrogen sulfide solution into a 1-liter volumetric flask containing 100 ml of the stock zinc acetate solution (20%). Add about 800 ml distilled water and three drops acetic acid. Adjust to volume with distilled water. Immediately upon withdrawal of the 10-ml aliquot of the hydrogen sulfide solution, pipet 20 ml of the standard $0.1N$ iodine solution into the remaining 490 ml. Back-titrate the excess iodine with the standard $0.1N$ sodium thiosulfate solution, endpointing with starch indicator. Calculate the concentration of the 10-ml aliquot of the hydrogen sulfide solution. Speed is essential in this procedure to avoid loss of hydrogen sulfide and iodine.

B. FREE SULFUR IN AIR BY THALLOUS ACETATE PAPER

Free sulfur in the atmosphere is conveniently determined according to a procedure described by Magill et al. (154). The gases are drawn through filter paper previously soaked with 5% thallous acetate then dried. The method is sensitive to as low as 0.05 ppm. Interferences from dust or dark oils may be troublesome, especially if the atmospheric sulfur content is low. The extent of air dust interference may be determined by a blank run with untreated paper. Oils may be removed effectively without interference by washing the paper with alcohol or benzene after the initial spraying and drying of pyridine. Selenium interferes. A series of 15 determinations on known concentrations in air of 0.5–1.0 ppm averaged 101% recovery, with an average deviation of 8.7% (maximum of 25%). Concentrations as low as 0.05 ppm in 1 cubic foot of air are detectable to within 90% accuracy.

Reagents

Impregnated Filter Paper. Soak appropriate paper (Chemical Warfare Service Type 6 or Whatman No. 1, 7 cm in diameter) in 0.5% aqueous thallous acetate. Suspend the papers on horizontal glass rods in an oven at 80–105°C until thoroughly dry. Storage in covered glass containers for more than several weeks is possible, although a gradual decrease in sensitivity may be expected.

Standard Sulfur Solutions. Select an eyedropper which drops 40 or more drops reagent-grade pyridine per ml. Determine the exact number of drops per ml, then prepare solutions of sulfur in reagent-grade pyridine such that each drop will contain 1.0, 1.2, 1.4, 2, 4, 6, 8, 10, 15, 20, 25, 30, 40, 50, 60, 80, 100, 140, 160, 180, and 200 mg sulfur per drop.

Procedure

Clamp a portion of a sheet of the impregnated paper in the holder of a sampling device (16) that permits exposure of an area equal to the portion of paper wet by one drop from the calibrated eyedropper. Draw the air through the paper with a vacuum pump, measuring the volume (with a wet-test meter or orifice-type gas meter) after passage through the paper. Remove the paper and spray it with pyridine from an atomizer of the DeVilbiss No. 15 type, containing no rubber parts in contact with the pyridine. Allow the paper to dry and then treat as described below for the standard papers. Analysis is made by direct visual comparison with known standard papers.

Prepare standard papers with one drop standard solution, covering the concentration range expected. Prepare these by allowing the spots to dry, then spray with pyridine and nearly dry them. Place the papers in a jar of hydrogen sulfide for about 30 sec and then remove the thallous sulfide by washing in a beaker of 0.5N nitric acid. Wash out the acid by transferring the papers, suspended on a glass rod, to a container of distilled water. Remove the papers and place on a white porcelain surface to examine visually the polysulfide spots.

C. FREE SULFUR IN GASOLINE BY REACTION WITH MERCURY

Free sulfur in gasoline can be determined turbidimetrically according to Uhrig and Levin (242). A 100-ml sample is shaken with 2 ml oleic acid and 3 ml mercury for 5 min, and the resulting suspension is compared with gasoline standards containing 0.1–1.0 mg sulfur per 100 ml. Removal of hydrogen sulfide if present (with cadmium chloride solution) is necessary, but mercaptans to not interfere. Peroxides if present must be removed with ferrous sulfate solution.

Reagents

Mercury, C.P., and Benzene, C.P. Test the benzene for sulfur by shaking 100 ml with Hg in a 4-oz bottle for 5 min. If the mercury remains clean, sulfur is absent.

Oleic Acid (U.S.P.). Test for free sulfur by shaking a 1:1 oleic acid–benzene mixture with mercury. The mercury remains clean if sulfur is absent.

Gasoline (S Free). Remove sulfur from unleaded cracked gasoline by shaking with mercury until a spot check with a portion shows it to be sulfur free.

Standard Sulfur Solution. Dissolve 1.00 g sulfur crystals in 1000 ml benzene. Flowers of sulfur are too difficult to dissolve.

Standard Mercuric Sulfide Suspensions Place 100 ml mercury-treated gasoline in each of 10 narrow-mouthed, 4-oz bottles. Add 2.0 ml oleic acid to each. Add sufficient standard sulfur solution to give a range of 0.1–1.0 mg sulfur per 100 ml gasoline. Stopper and shake each with 3 ml mercury for 5 min. These standards store satisfactorily for about a week.

Procedure

Place 100 ml of the sample in a 4-ox bottle and add 2.0 ml oleic acid and 3 ml mercury. Stopper and shake for 5 min and compare the turbidity immediately with freshly shaken standards. Quantitative dilution of the sample with sulfur-free gasoline will be necessary if the 100-ml sample contains more than 1.0 mg free sulfur. Removal of hydrogen sulfide may be accomplished by shaking the sample with acidic cadmium chloride solution and filtering through paper. Aged samples may contain peroxides, and these can be removed by shaking the sample with aqueous saturated ferrous sulfate solution.

D. FREE SULFUR IN HYDROCARBONS BY THE FERRIC THIOCYANATE COMPLEX

The method of Bartlett and Skoog(30) is based on the reaction of elemental sulfur (dissolved in acetone) with cyanide to form thiocyanate, which then is determined colorimetrically by the formation of the ferric thiocyanate complex. The reaction between elemental sulfur and cyanide occurs rapidly in various solvents. Although considerable increase in the sensitivity of the method is possible by using absolutely anhydrous acetone, several disadvantages in technique would be imposed also. Hence, acetone containing 5% water was chosen, since it was shown that small variations in water content at this level do not appreciably influence the absorbance.

The ferric chloride reagent should contain about 50 mg iron per 100 ml. Higher concentrations decrease the absorbance, possibly due to the increased acidity resulting from hydrolysis of the iron. The cyanide concentration is not critical to the cyanate conversion, since rapid quantitative conversion occurs over a wide range. However, increased cyanide concentration causes increased absorption. This is probably due to decreased acidity of the solution with resultant effects on the complex formation equilibria. Iron precipitates in the presence of cyanide concentrations above 200 mg per 25 ml. Color stability is good in the dark for 3 hr, but exposure to sunlight should be avoided. The interference of mercaptans is due to their reaction with elemental sulfur in basic solutions. These, as well as sulfides and hydrogen sulfide, can be removed by shaking with mercuric chloride or silver nitrate. Checks with several organic peroxides and thiophane, thiophene, toluene, and amylene showed no serious effect on the accuracy.

Reagents

Aqueous Acetone Solvent. Dilute 50 ml water to 1 liter with technical-grade acetone.

Sodium Cyanide Solution. Dissolve 0.1 g NaCN in 100 ml aqueous acetone solvent. Let stand a few hours to clear.

Ferric Chloride Solution. Dissolve 0.4 g $FeCl_3 \cdot 6H_2O$ in 100 ml acetone solvent. Let the precipitate settle for 24 hr, then decant the supernatant liquid into a dry bottle. This solution stores satisfactorily for several weeks.

Mercuric Chloride Solution. Dissolve about 20 g mercuric chloride and 20 g KCl in 1 liter distilled water.

Standard Sulfur Solution. Dissolve 50 mg powdered roll sulfur in exactly 1 liter petroleum ether (bp 77–110°C, density 0.73 at 25°C). This gives a concentration of 50 ppm sulfur (w/v).

Procedure

a. Samples Containing No Mercaplans, Sulfides, or Disulfides. Place a 5-ml solution containing 5–50 ppm sulfur (dilute with petroleum ether if necessary) in a 25-ml volumetric flask. Mix in 15 ml of the sodium cyanide solution and let stand for 2 min. Dilute to 25 ml with the aqueous acetone solvent. To a 5-ml aliquot mix in exactly 5 ml of the ferric chloride solution. Measure the absorbance within 10 min at 465 nm in a 1-cm Corex cell against a blank formed by mixing 5 ml of the acetone solvent with 5 ml of the ferric chloride solution. If the sample itself is colored, make a blank correction by treating a 5-ml aliquot to the same reagents, leaving out the cyanide. Obtain the sulfur concentration by comparison with a calibration curve prepared by similar treatment of a suitable series of the standard sulfur solutions.

b. Procedure for Samples Containing Sulfides, Disulfides, or Mercaptans. To about 20 ml of the sample add 50 ml of the mercuric chloride solution and shake thoroughly. Use 5-ml aliquots directly if the amount of precipitate is small and settles readily. If it is large, filter and take the 5-ml aliquot from the filtrate and follow procedure a.

E. SULFATE BY BENZIDINE PRECIPITATION AND DIAZOTIZATION

Klein's procedure (130) calls for precipitation of sulfate as benzidine sulfate, purification, and dissolution in 0.2N hydrochloric acid, followed by diazotization and coupling (after destruction of excess nitrous acid) with N-(1-naphthyl)ethylenediamine dihydrochloride. An intense, stable purple color results, and Beer's law is followed. Traces of chloride and phosphate do not interfere in the precipitation. Phosphate, if present, should be removed. Chloride should not be present in more than a 30 to 1 ratio to sulfur. Klein obtained an accuracy of 2% for samples containing 0.05–0.150 mg sulfate.

Reagents

Sodium Nitrate. Prepare 0.1% fresh daily.

Ammonium Sulfamate. Make 0.5% in distilled water.

N-(1-Naphthyl)ethylenediamine Dihydrochloride. Make 0.1% in distilled water and store refrigerated in a dark bottle.

Benzidine Hydrochloride Reagent. Dissolve 4.0 g pure benzidine hydrochloride in a small volume of distilled water and dilute to 250 ml with 0.2N hydrochloric acid.

Standard Sulfate Solutions. Dissolve 0.5437 g reagent-grade potassium sulfate in 1 liter distilled water (1 ml contains 0.3 mg sulfate). Dilute 1 to 10 to obtain the working standard of 0.03 mg sulfate per ml.

Standard Benzidine Hydrochloride. Dissolve 0.4014 g pure benzidine hydrochloride in a small volume of 0.2N hydrochloric acid and then dilute to 100 ml with the same acid (1 ml is equivalent to 1.5 mg sulfate). Dilute 1 to 100 with 0.2N hydrochloric acid to obtain the working standard (1 ml equivalent to 0.015 mg sulfate).

Acetone–Ethanol. Prepare a 1:1 mixture with 95% ethanol and reagent-grade acetone.

Procedure

To a sample containing 0.015–0.15 mg sulfate in a 15-ml centrifuge tube (having a narrow conical tip), add 1 ml glacial acetic acid and 1 ml benzidine hydrochloride solution. Mix, add 2 ml 1:1 acetone–ethanol mixture, and cool to 0°C for 30 min for complete precipitation. Centrifuge at 2500 rpm for 10 min, remove the liquid completely, and wash twice with the acetone–ethanol mixture. Dissolve the precipitate with 2 ml 0.2N hydrochloric acid, cool to 0°C, and add 1 ml fresh 0.1% sodium nitrite solution. Mix and let stand for 3 min, add 1 ml 0.5% ammonium sulfamate solution, and let stand for 2 min more. Then add and mix with 1 ml 0.1% aqueous solution of N-(1-naphthyl)ethylenediamine dihydrochloride, let stand for 20 min, and then dilute with water to 50 ml in a volumetric flask. Measure the absorbance with a green filter or at an appropriate wavelength in a spectrophotometer.

F. TRACE ORGANIC SULFUR BY BARIUM SULFATE TURBIDITY

Holeton and Linch's (101) procedure for traces of sulfur in organic compounds was designed to circumvent the necessary limitations on sample size in combustion methods for sulfur, especially in the range of 1–200 ppm sulfur. The authors reviewed various methods of oxidizing sulfur-containing compounds, pointing out the limitations of each for low concentration ranges. Their method overcomes many of the difficulties by spraying the sample (dissolved in an appropriate solvent, if necessary) as a fine mist in a stream of excess air which is carried into a combustion tube. The gaseous oxidation products are chilled and the sulfur oxides are converted to sodium sulfate by absorption in dilute aqueous sodium perborate. Carbon dioxide is boiled off in the presence of nitric acid and then excess ammonium hydroxide and a protective colloid are added. Crystalline barium chloride is then added to an aliquot and the turbidity measured, using a barium-free aliquot as reference.

The authors point out the necessity of having sufficient oxygen present (as excess air) for complete combustion, expecially for organic compounds deficient in hydrogen or oxygen. Often, oxidation is still incomplete unless an oxygenated solvent (e.g., alcohol) is used as a diluent. This also aids in the combustion of viscous, low vapor pressure, and high flash point liquids. Absorbing oxidants other than sodium perborate can be used, such as hydrogen peroxide, potassium iodate, and sodium hypobromite.

The protective colloid was found essential, and the precipitate was found to be more stable when formed in alkaline medium. The presence of more than 1 or 2% halides in the sample prevents satisfactory results by this procedure. Accuracy of 5–10% of the amount of sulfur present and a precision of ±2 ppm in the range of 2–100 ppm are attainable.

Reagents

Sodium Perborate Solution. Dissolve 1 g sodium perborate in 100 ml distilled water.

Dilute Nitric Acid. Dilute 20 ml concentrated C.P. nitric acid to 100 ml with distilled water.

Dilute Ammonium Hydrozide. Dilute 30 ml concentrated C.P. ammonium hydroxide to 100 ml with distilled water.

Gum Arabic. Dissolve 10 g *U.S.P.* gum arabic in 100 ml distilled water by heating on a steam bath. Filter off any sediment.

Procedure

Dissolve the sample (20–25 g for samples containing less than 5 ppm sulfur), if solid, in an appropriate solvent. Alcohols, ethylene glycol, ethers, dioxane, and dimethylformamide are all suitable. Introduce the liquid into a combustion tube through an atomizer (the setup given by the authors (101) is recommended). Absorb the sulfur oxides in scrubbers (101) containing the sodium perborate solution, sweeping the apparatus with air after combustion is complete. Boil the absorption solutions in a beaker with additional sodium perborate, and add 2 ml of the dilute nitric acid (1 ml more, if not acid to Congo red indicator). Boil down to a volume less than 40 ml, cool, and add 1 ml of the dilute ammonium hydroxide solution (1 ml more, if not basic to brilliant yellow indicator). Add 5 ml of the gum arabic solution, filter into a 50-ml volumetric flask, and dilute to volume using distilled water rinsings of the beaker. Mix and transfer 25 ml back to the beaker, add 0.5 g C.P. reagent crystalline barium chloride, and swirl to dissolve. Add 0.5 ml aliquots of acid if not acidic to Congo red indicator). Measure the absorbance of the barium sulfate suspension in a suitable instrument with a 1-cm cell, using the barium-free aliquot as the blank. Obtain the total grams of sulfur from calibration curve (using pure sodium sulfate) for both. Subtract that for the blank solution from that for the sample solution. Blank values obtained for 2-hr scrubbing periods should not run higher than 1.5 ppm.

G. SULFUR IN PLANTS AND WOOL BY BARIUM SULFATE TURBIDITY

Procedures for sulfur in plants and in wool proposed by Steinbergs (227) call for digestion with a 10:1 ratio of concentrated nitric acid and hydrochloric acids accelerated with selenium dioxide, followed by the addition of hydrochloric acid and sodium chloride and three evaporations to dryness (necessary to ensure reduction of nitrates). The sulfate is precipitated in acidic media in the presence of glycerol or gum arabic.

Using 0.1–0.3 g oven-dried sample, it was found possible to estimate 0.01–1.50% sulfur with an error not exceeding 3–10%.

Reagents

Gum Arabic Solution. Add to a solution of 10 g gum arabic in 100 ml water a few drops of hydrochloric acid and a little barium sulfate. Stir and filter after standing a few minutes, neutralize with sodium carbonate powder, and centrifuge. Decant the supernatant liquid and add 10 ml 6N hydrochloric acid followed by 95% ethanol until all arabic acid is precipitated. Centrifuge, wash the precipitate with ethanol, and dissolve it in 200 ml hot water.

Barium Chloride Solution. Dissolve 20 g barium chloride in 100 ml 2% hydrochloric acid.

Selenium Dioxide Solution. Dissolve 4.1 g H_2SeO_3 in 50 ml distilled water.

Dilute Hydrochloric Acid. Prepare 2.9N hydrochloric acid in distilled water.

Glycerol Solution. Prepare a 3% solution in distilled water.

Procedure

Digest 0.3 g sample with 0.1 ml of the selenium dioxide solution, 10 ml concentrated nitric acid, 1 ml concentrated hydrochloric acid, and 508 mg sodium chloride for 30–45 min. Evaporate carefully to dryness, add to the hot residue 2 ml of the dilute hydrochloric acid, and again evaporate to dryness. Repeat the last treatment twice more if necessary to ensure reduction of nitrates, then add selenium dioxide solution to dehydrate the SiO_2. Finally, add 25 ml distilled water and 1.0 ml of the dilute hydrochloric acid. Boil, filter, and add 10 ml of the glycerol solution to the filtrate. Precipitate the sulfate with the barium chloride solution and measure the turbidity by comparison with potassium sulfate standards similarly prepared.

Alternatively (107), one may dissolve the plant ash containing 0.2–1.2 mg sulfur in dilute hydrochloric acid, add 3 ml of the dilute hydrochloric, acid and 2.4 ml 20% sodium chloride solution, and dilute to 40 ml with distilled water. Add 5 ml of the gum arabic solution and dilute to 50 ml in a volumetric flask with water. Heat the mixture in a beaker containing the evaporated residue of 1 ml of the barium chloride solution. Stir and return to the 50-ml volumetric flask, adding water if necessary. Compare the light absorbance with standards similarly prepared.

H. SULFUR IN PLANTS AND SOILS BY METHYLENE BLUE

The method of Johnson and Nishita (113) is applicable to the micro-determination of sulfate and total sulfur in a variety of biologic samples. Two procedures are given for sulfate (in the absence and presence of nitrate) and one, for total sulfur. They obtained Beer's law linearity in the range of 1–50 μg sulfur, using a blue-sensitive phototube at 670 nm and 5 nm effective bandwidth. Since the absorption maximum is quite sharp, nonlinear calibration curves may be expected if instruments of lower spectral purity are used. Mutual interactions of methylene blue molecules occurred when more than 50 μg sulfur were in the sample. The resulting nonlinear deviations can be eliminated, if desired, by moderate dilution of the concentrated methylene blue solution, using a diluent of the same concentrations of acid, p-aminodimethylaniline, and ferric ammonium sulfate.

Reagents

Reducing Mixture. Place 15 g reagent-grade red phosphorus, 100 ml hydriodic acid (sp. gr. 1.7, methoxyl grade), and 75 ml 90% formic acid (A.C.S. reagent grade) in a 250-ml, boiling flask. Provide a nitrogen stream at the bottom to remove the evolved hydrogen sulfide and to prevent bumping. Heat slowly to

115°C and hold at 115–117°C for 1 to 1.5 hr, during which time about 30% of the reagent evaporates. This reagent is stable two to three weeks and can be used still longer with reduced sensitivity and precision. Regeneration of used reducing mixture can be accomplished by adding, to each 100 ml, 1 g red phosphorus, 10 ml hydriodic acid, and 25 ml formic acid and digesting 1.5–2 hr at 115–117°C.

Nitrogen Purification Solution. Add 5–10 g mercuric chloride to 100 ml 2% potassium permanganate solution.

Pyrogallol–Sodium Phosphate Solution. Dissolve 10 g sodium dihydrogen phosphate and 10 g pyrogallol in 100 ml sulfur-free distilled water with the aid of a stream of nitrogen bubbling through the solution. Prepare fresh daily.

Sulfur-Free Distilled Water. If necessary, distill from alkaline permanganate in an all-glass still.

Zinc Acetate–Sodium Acetate Solution. Dissolve 50 g $Zn(OAc)_2 \cdot 2H_2O$ and 12.5 g $NaOAc \cdot 3H_2O$ in sulfur-free distilled water. Dilute to 1 liter and filter.

Aminodimethylaniline Solution. Dissolve 2 g p-aminodimethylaniline sulfate (E.K. No. 1333) in 1500 ml distilled H_2O. Add 400 ml concentrated sulfuric acid, cool, and dilute to 2 liters.

Ferric Ammonium Sulfate Solution. Dissolve 25 g $FeNH_4(SO_4)_2 \cdot 12H_2O$ in 5 ml concentrated sulfric acid and 195 ml distilled water.

Standard Sulfate Solution. Dissolve 5.434 g reagent-grade potassium sulfate in 1 liter. This solution contains 1 mg sulfur per ml.

Sulfur-Free, Ground-Joint Lubricant. Mix about 5 g silicone stopcock lubricant with 10 ml of a 1:1 mixture of hydriodic acid and hypophosphorous acid. Boil and stir for about 45 min, then pour off the acids and wash the lubricant thoroughly with sulfur-free water. Lubricate all joints with small amounts of this lubricant.

Procedure

Procedure When Nitrate Is Absent. Place 10 ml of the pyrogallol–sodium phosphate reagent in the gas-washing vessel. Place 10 ml of the zinc acetate–sodium acetate solution with 70 ml sulfur-free distilled water in a 100-ml volumetric flask. This serves as the receiver, and the gas is bubbled into it through a glass tube connected to the top of the gas-washing column by a piece of sulfur-free rubber tubing (boiled in dilute alkali and washed). Transfer an aliquot of the sample smaller than 2 ml and containing less than 300 μg sulfur to the digestion flask. Add 4 ml of recently agitated reducing mixture, Quickly attach the flask to the condenser and begin the nitrogen flow into the bottom of the digestion flask, using nitrogen washed by bubbling through the mercuric chloride solution in 2% potassium permanganate. Adjust the nitrogen flow rate to 100–200 ml per min. Begin heating and boil gently for 1 hr. Remove the receiving (volumetric) flask, leaving the connecting tube in it, so as not to lose adhering zinc sulfide. Quickly add 10 ml of the p-aminodimethylaniline solution to the receiving flask. Stopper, mix, add 2 ml of the ferric ammonium sulfate solution, stopper, and mix again. Dilute to volume and mix thoroughly. Read the absorbance after 10 min but before 24 hr at 670 nm or with a suitable filter photometer.

Procedure When Nitrate Is Present. When the sample contains more than 6 mg nitrate and this cannot be avoided by reducing the sample size, the following procedure is recommended. Place the sample in a 15-ml conical centrifuge tube and wet it with 95% ethanol. Add five drops 10% barium chloride solution. Place the tube 0.5 in. in a boiling-water bath for 10–15 min. This ensures wetting sample particles adhering to the sides of the tube. Add 5 ml 10% barium chloride solution, mix, wash down the sides with another 5 ml barium chloride solution, and centrifuge. Discard the aqueous phase and transfer the solid to the digestion flask quantitatively, washing with sulfur-free distilled water. Reduce the volume to less than 2 ml and proceed in the usual manner.

Procedure for Total Sulfur by Wet-Ashing Technique. Place 100 mg (or less, depending on the estimated sulfur content) of the plant material in a 30-ml micro-Kjeldahl flask, add 2 ml concentrated nitric acid, and digest on the steam bath for 30 min. Add 1 ml 60% perchloric acid and heat slowly over a microburner, continuing the digestion for at least 1 hr after fumes of perchloric acid appear. Cool and add 3 ml $6N$ hydrochloric acid to aid complete removal of nitrate. Reheat until perchloric acid fumes reappear. Cool and dilute to 50 ml in a volumetric flask. Aliquots of this solution, usually not larger than 2 ml, are used for the sulfate determination by the usual procedure.

I. METHYLENE BLUE DETERMINATION OF HYDROGEN SULFIDE IN GASES

In the method of Fogo and Popowsky (77), hydrogen sulfide is precipitated as zinc sulfide and converted to methylene blue using p-aminodimethylaniline sulfate and ferric chloride. Measurement is made at 670 nm and is sensitive to 3 μg. The technique is designed for occasional as well as frequent use.

Reagents (Deviations of 5% in Concentrations are Allowable)

Zinc Acetate Solution. Make a 1% solution of C.P. salt in distilled water.

Sodium Hydroxide Solution. Make a 12% solution of C.P. sodium hydroxide in distilled water.

Ferric Chloride Solution. Make a 0.023M solution in 1.2M hydrochloric acid.

p-**Aminodimethylaniline Solution**. Dissolve 0.5 g E.K. white label grade in 500 ml 5.5M hydrochloric acid.

Procedure

Obtain two gas samples simultaneously at the source by connecting one arm of a glass cross to the source by Tygon tubing with an attached screw clamp. Charge two receivers with 130 ml 1% zinc acetate solution and 5 ml 12% sodium hydroxide solution and mix. Attach two arms of the glass cross, through tubing fitted with screw clamps, to the inlet arms of the two receivers and test meters to the outlet ends. Fit the fourth arm of the glass cross with a short length of bleeder tubing and screw clamp. With all four screw clamps open, turn on the gas sufficiently to flow at a rate considerably exceeding the sampling rate. Slowly close the bleeder tube until gas passes through the gas absorption bottles at about 170 liters per hr (6 cubic feet per hr). Collect an amount that will contain between 35 and 350 μg hydrogen sulfide. Sampling rates for samples smaller than 50 liters should be lower.

Close the gas-receiving bottles and adjust their temperature and that of the amine reagent to 24 ± 3°C. Raise the top of the bottle and pipet in 25 ml of the amine reagent. Close the bottle and swirl to dissolve all the precipitate quantitatively. Again raise the top, add 5 ml ferric chloride solution, close, and mix. Use fast-delivery pipets rather than slower, more accurate ones. Allow the closed bottle to stand 10 min, transfer quantitatively to a 250-ml volumetric flask, rinse, and dilute to the mark with distilled H_2O. Measure after 20 min but before 20 hr, keeping the colored solution out of direct sunlight. Make a blank solution by mixing the same amounts of all reagents in a 250-ml volumetric flask. Use the blank solution as the reference liquid for the colorimetric measurement at 670 nm. Very dark solutions may be measured satisfactorily at 750 nm.

Calibration standards are prepared with carefully measured quantities of sodium sulfide or hydrogen sulfide. Lumps of sodium sulfide (washed to remove sodium sulfite) dissolved in oxygen-free distilled water to give a concentration of 20 μg sulfur per ml are convenient. Standardize the sodium sulfide solution iodometrically. Avoid contact with oxygen during storage.

J. SULFIDES IN WATER BY METHYLENE BLUE (173)

A pretreatment procedure has been proposed by Pomeroy (189,190) to circumvent certain kinds of interference, especially when very low concentrations of hydrogen sulfide are to be measured. The suggested alternative is to add zinc acetate to the sample solution together with enough sodium hydroxide or sodium carbonate to provide large carrier precipitate. Soluble potentially interfering compounds remain dissolved unless they precipitate with the zinc. The precipitate is settled, the supernatant liquid decanted, water added (with further decanting if necessary), and the colorimetric procedure followed in the usual way.

Precipitated zinc sulfide is quite resistant to air oxidation, and hence sulfide samples may be stored in this state.

Reagents

Zinc Acetate Solution. Dissolve 220 g $Zn(OAc)_2 \cdot 2H_2O$ in 1 liter water to make a $2N$ solution.

Procedure

Add an amount of $2N$ zinc acetate solution dependent on the nature of the sample (generally 1 ml in a 250 ml bottle is adequate to ensure an excess and the formation of a good precipitate). Add an equal volume of $1N$ sodium carbonate solution (a sodium hydroxide solution seems to give a less compact precipitate but should be used where there are low sulfide concentrations, due to the possible loss of hydrogen sulfide when much carbonate is present). The amount of alkali may be reduced or eliminated for samples of high alkalinity. Mix the sample until good flocculation is attained, and proceed to develop the methylene blue color as in other methods (see Section I or K).

K. SULFIDES IN ALKALIES BY METHYLENE BLUE

The method of Budd and Bewick (46) is applicable to small quantities of sulfide and reducible sulfur in alkalies. The hydrogen sulfide evolved upon acidification (or by stannite-nascent hydrogen reduction, if necessary) is absorbed by zinc acetate solution and converted to methylene blue. The range is 0.2–100 ppm using a 3-g sample. Losses due to oxidation of sulfide in alkali carbonates are minimized by pelleting the sample.

Reagents

Boiled Distilled Water. Boil distilled water for at least 30 min, cool, and store in a closed container. Prepare fresh daily.

Stannous Chloride Solution. Dissolve 10 g $SnCl_2 \cdot 2H_2O$ in 10 ml hydrochloric acid and dilute to 100 ml with water, adding a few pieces of mossy tin. Store no longer than three days at room temperature.

Aluminum Strips. Cut $\frac{1}{4}$ x $1\frac{1}{4}$-in. strips of reagent-grade $\frac{1}{8}$-in. aluminum foil. Immediately prior to use, dip in boiling 1:5 hydrochloric acid for 30 sec, then rinse with water.

Zinc Acetate Stock Solution. Dissolve 208 g $Zn(OAc)_2 \cdot 2H_2O$ in distilled water and dilute to 1 liter.

Zinc Acetate, 2% Solution. Add several drops acetic acid to 100 ml of the zinc acetate stock solution and dilute to 1 liter with distilled water.

Stock Amine Solution. Dissolve 27.2 g p-aminodimethylaniline sulfate in 75 ml cold 1:1 sulfuric acid and dilute to 100 ml with more 1:1 acid. This may be stored for several months.

Dilute Amine Solution. Dilute 25 ml stock amine solution with cold 1:1 sulfuric acid to 1 liter. This solution will store for about a month at room temperature.

Ferric Chloride Solution. Dissolve 100 g $FeCl_3 \cdot 6H_2O$ in distilled water and dilute to 100 ml.

Iodine Solution. $0.1N$, standardized accurately.

Sodium Thiosulfate Solution. $0.1N$, standardized accurately.

Standard Sodium Sulfide Solution. Prepare an approximately $0.01N$ sodium sulfide solution. Weigh about 1.2 g large-crystal $Na_2S \cdot 9H_2O$ and rinse the crystal surfaces with water, discarding the washings. Dissolve the washed crystals in boiled and cooled distilled water, diluting to 1 liter. Stopper quickly and mix. Pipet 100 ml of the sulfide solution into a beaker containing 250 ml water, 20.00 ml $0.1N$ standardized iodine solution, and 25 ml of approximately $0.1N$ hydrochloric acid. Titrate the excess iodine with $0.1N$ standardized thiosulfate to the starch endpoint. Calculate the ml $0.1N$ iodine that reacted with the sulfide. One ml $0.100N$ iodine is equivalent to 0.0120 g $Na_2S \cdot 9H_2O$ or 0.00160 g sulfur. Calculate the volume of sulfide solution containing 1.000 mg sulfur, dilute this volume to 500 ml with boiled and cooled distilled water, and stopper quickly. Both sulfide solutions are unstable and should not be stored.

Special Hydrochloric Acid. Place one or two strips aluminum foil in a small beaker of concentrated hydrochloric acid before the acid is used for sulfide determination. After a violent reaction for several seconds, the acid is poured off, ready for use.

Procedure

Details of apparatus and appropriate volumes of acid and absorption solution to use are given by Budd and Bewick (46). The reaction flask contains dilute hydrochloric and methyl orange indicator, while the 2% zinc acetate solution is in the absorption flask. The latter is placed in a water bath held at $23-25°C$.

Flush the whole apparatus with a rapid flow of nitrogen (or carbon dioxide) for 10 min, then reduce the flow rate to three to five bubbles per sec and maintain this rate during the remainder of the test.

For samples not containing carbonates, the procedure calls for quickly adding to the reaction flask approximately 3 g sample from a weighing bottle. Restopper immediately; the accurate weight of sample used is determined from the before-and-after weights of the weighing bottle. Sufficient acid is added through a dropping funnel to make the pH in the reaction flask just acid to methyl orange. After complete neutralization of the sample, maintain the inert gas flow of three to five bubbles per sec to sweep all liberated hydrogen sulfide into the absorption flask. Carbonate samples are best pelleted by pressing 1.5 g at 2000 psi to 0.5-in. diameter. The use of these solid pellets results in slow evolution of carbon dioxide upon acidification. Otherwise, the sample is in alkaline solution long enough to oxidize sulfide and produce low results.

To develop the color, add quickly to the solution in the absorption flask 1.5 ml of the dilute amine solution and swirl. Add five drops of the ferric chloride solution and swirl. Keep the gas inlet tube in the absorption flask during this treatment to ensure reaction of any zinc sulfide which may cling to it. Transfer the colored solution to a 50-ml mixing cylinder, rinsing with boiled distilled water, and bring the volume to the mark and mix. After 15 min, read the absorbance in a 1-cm cell at 670 nm or with an appropriate filter photometer. Subtract the absorbance of the blank from that of the sample and refer to a similarly prepared standard curve for the amount of sulfur present.

Prepare standards by running through the same procedure as the samples. This is necessary since the hydrogen sulfide volatilization is incomplete though reproducible. Standards may, of course, be directly colorized if the hydrogen sulfide volatilization and absorption steps are not used on the sample or if total reducible sulfur is being determined.

Add to the reaction flask an accurately measured volume of standard sulfide solution, one drop methyl orange indicator, enough boiled cooled water to bring the volume to 30 ml, and 0.1 g reagent-grade sodium bicarbonate. Quickly assemble the apparatus and proceed in the same manner as for the unknown sample. Subtract the absorbance obtained for a blank run in the same manner and obtain the micrograms of sulfide sulfur recovered by comparison with the absorbances of standard sulfide solutions directly colorized. Since sulfides are unstable, daily checks on the standard solutions are necessary. The main calibration curve is conveniently obtained by taking 0, 1, 3, 5, 7.5, 10, 12.5, 15, 17.5, and 20 ml of the working standard sulfide solution (containing 2.00 μg sulfide sulfur per ml). Dilute each to 3–5 ml with boiled distilled water (at 23–25°C) in a mixing cylinder. Immediately add 1.5 ml of the dilute amine solution to each, swirl, and add five drops of the ferric chloride solution, swirl, dilute to 50 ml, and mix. Read all absorbances after 15 min but before 2 hr.

Solutions too dark to be measured should be aliquoted and diluted, then compared with standards similarly diluted. The presence of the reagents, especially excess ferric chloride, affects the absorbance of the methylene blue, although not the wavelength of maximum absorption. Blanks are carried through the same procedure, but the amount of hydrochloric acid added to the reaction flask should be equal to that used with the sample rather than just sufficient to acidify methyl orange. The temperature of all absorption solutions should be held at $23-25°C$ when adding the amine and ferric chloride reagents, the color being more intense for lower temperatures and less intense for higher temperatures.

Procedure for Total Reducible Sulfur

A similar procedure may be used to determine sulfide plus reducible sulfur compounds (sulfite and thiosulfate). Weigh a 3-g sample to 0.01 g and place in the reaction flask with 25 ml cool, boiled distilled water and 2 ml of the stannous chloride reagent, swirling to avoid loss of hydrogen sulfide due to concentration of the acid in the reagent. Boil gently for 1 min, then cool the flask with tap water. Connect the flask to the absorption flask (held at $23-25°C$ in a water bath). Flush with nitrogen, then bleed it through at three to five bubbles per sec as in the regular procedure. Quickly add two of the treated aluminum strips and 10 ml hydrochloric acid to the flask. Heat to boiling unless the presence of caustic generates enough heat. Sweep the nitrogen through for 20 min and carry out the color development as in the regular procedure. Run a blank in the same way, deduct its absorbance, and read the μg total reducible sulfur from the direct colorization standardization curve.

Sulfate sulfur is not reduced in this procedure. Thiosulfite is readily recovered in amounts of 20 μg, whereas sulfite recovery is low at concentrations above 5 ppm. Concentrations of both up to 100 ppm do not interfere in the sulfide sulfur determination.

L. ALIPHATIC SULFIDES IN HYDROCARBONS BY ULTRAVIOLET ABSORPTION

This method of Hastings and Johnson (91,92) permits a sensitivity of 1 ppm under ideal conditions. The complex formed with iodine absorbs strongly near 308 nm, and good precision is obtained by adding the iodine reagent in an excess far exceeding the stoichiometric 1:1 ratio. Potential interferences from petroleum samples containing ultraviolet-absorbing compounds are usually negligble due to the minute amounts of sample required and also to the fact that such interferences are largest below 300 nm.

The procedure given here is that of Drushel and Miller (69) and was designed especially for following separations of aliphatic sulfides from crude petroleum by chromatography or thermal diffusion. They suggest a somewhat higher iodine concentration than that used by Hastings and Johnson (91,92) in order to further enhance the mass action effect in the formation of the sulfide—iodine complex Also, carbon tetrachloride is used as the solvent instead of isooctane.

Nitrogen- and oxygen-containing compounds also form complexes with iodine but are considered likely to be present in small enough concentrations in petroleum to cause negligible interference. However, Drushel and Miller's calculations show that until the aromatic content of a sample and its interference are better known, the method should not be applied to such samples containing less than 0.05 wt-% aliphatic sulfur. Use of photomultiplier detection is worthwhile when backgrounds are necessarily higher than are usually encountered in routine work.

Temperature effects were estimated as an absorptivity decrease of about 2 3% per degree centrigrade increase. When it is not possible to use thermospacers with the spectrophotometer, the 7-thiatridecane—iodine complex is recommended as the model compound to use in calculating temperature corrections.

Calibration solutions may be prepared from known pure aliphatic sulfides. If the identity of the sulfides in the sample is unknown, blends should be prepared by mixing equimolar amounts of all the readily available sulfides in the pure state. They should be diluted to a point where the absorbance for the iodine complex for a 1-cm cell falls between 0.1 and 1.0.

Reagent

Iodine Reagent. Dissolve 10.00 g reagent-grade I_2 in 1 liter technical carbon tetrachloride.

Procedure

Dissolve a sample containing about 0.03—0.07 mg sulfide sulfur in 25 ml carbon tetrachloride. Dilute 1 ml of the iodine reagent to 10 ml with the sample solution. Measure the absorbance at 310 nm immediately in a 1-cm, fused-quartz cell using a spectrophotometer. Use thermospacers, or note the temperature. For the reference cell, use 1 ml carbon tetrachloride diluted to 10 ml with the same sample solution. Obtain a blank value by diluting 1 ml of the iodine reagent to 10 ml with carbon tetrachloride using plain solvent as the reference. The percent aliphatic sulfide sulfur can be obtained by referring to a calibration curve or using the formula recommended by Drushel and Miller (69). In either case, the absorbance of the iodine (blank) solution is subtracted from that for the sample—iodine blend. An absorptivity of 400 liters per g -cm was used by Drushel and Miller for their experimental conditions.

Tables I—XI list and group various methods listed in the bibliography according to sample or material, technique or treatment, method or reagents used, wavelength, and comments.

TABLE I. Determination of Sulfur as Alkylbenzenesulfonate and Anionic Detergents

Sample or material	Method or reagent(s)	Wavelength, nm	Comments	Reference
Sewage, water	methylene blue, alkaline borate; $CHCl_3$ extraction	650	bcd	(1)
Surface water, sewage	methylene blue, amine–$CHCl_3$ and amine–hexane extractions; $CHCl_3$ extractions	650	cde	(73)
Water	methylene blue, H_2SO_4, NaH_2PO_4,; $CHCl_3$ extraction	652	cd	(97, 179)
Sewage, sewage effluents, river waters	methylene blue, alkaline phosphate; $CHCl_3$ extraction	650	cd	(150)
Surface waters, sewage	methyl green, glycine–HCl; benzene extraction	615	cde	(165)
Fresh and saline waters	methylene blue, alkaline borate; $CHCl_3$ extraction	650	automated, cde	(225)
Sewage, effluents (sulfate and sulfonate type)	methylene blue, alkaline phosphate, $CHCl_3$ extraction (hydrolyzed and nonhydrolyzed samples)	650	hydrolysis of alkyl sulfates with HCl	(233)
General	azure A, H_2SO_4, $CHCl_3$ extraction	630	cd	(247)
General	rosaniline hydrochloride; $CHCl_3$–EtOAc extraction	540		(119)
General	direct measurement in KH_2PO_4 solution	225	c	(249)

[a] Discusses structures.
[b] Discusses theoretical principles.
[c] Discusses interferences.
[d] Compares results with those of other methods.
[e] Discusses statistical treatment of data.

TABLE II. Determination of Miscellaneous Sulfur Functional Groups and Compounds

Sample or material	Sample treatment	Method or reagent(s)	Wavelength nm	Comments	Reference
Sulfonyl halides, general Organic materials	combustion in Pt-lined oxygen bomb	Alkaline pyridine turbidimetric, $BaCl_2$	395,550 400	bc	(12) (26)
Thiophene in benzene		isatin, ferric sulfate		Block comparator	(14, 17, 28)
Piperazine—CS_2 compounds and alkyl derivations	acid hydrolysis, extraction with $CHCl_3$	CS_2 in $CHCl_3$	319	e	(40)
Biological materials	oxidation in oxygen bomb	$BaCl_2$, then K_2CrO_4, determination excess chromate	370		(43)
Reducible sulfur in alkalies	stannous chloride, Al strips	methylene blue, $Zn(OAc)_2$, PADA, $FeCl_3$	670	bc	(46)
H_2SO_4 in air		benzidine sulfate, diazotization, and coupling with m-MeC_6H_4OH	482	c	(51)
Isomeric toluenesulfonic acids in sulfuric acid		multicomponent spectrophotometric analysis	see ref. (54)	e	(54)
Thiourea in sewage and industrial effluents		Pentacyano-ammonia-ferroate, HOAc	610	c	(65)
Sulfuric esters	Acid hydrolysis	$BaCl_2$, then K_2CrO_4, determination excess	375	c	(72)
Hydrocarbons	vaporized in H_2 stream, pyrolyzed, Pt catalyst	methylene blue, $Zn(OAc)_2$, PADA, $FeNH_4(SO_4)_2$	667	bde	(74)
Rocks	ignition with vanadium pentoxide in stream of N_2, reduction by Cu	p-rosaniline—formaldehyde, Na_2HgCl_4	550	d	(86)

TABLE II. cont'd

Sample or material	Sample treatment	Method or reagent(s)	Wavelength nm	Comments	Reference
Aliphatic sulfides paper, cellulose fibers		intermolecular complexes of aliphatic sulfides and I_2 (procedures for sulfide, sulfite, sulfate, elemental sulfur total sulfur)	308 see ref. (103)	b c	(91, 92, 104)
Thiourea	digestion with HNO_3 and NaCl in sealed tube	nitrous acid, determination SCN^- with iron(III) perchlorate	455	c e	(106)
Organic sulfur		precipitation of SO_4^{2-} with 4-amino-4-chlorodiphenyl (CAD), determination excess CAD	254	c d	(114)
Total sulfur oxides in atmosphere	absorption by H_2O_2 or lead peroxide	displacement—barium chloranilate	530		(116)
Ester sulfate	hydrolyze with formic acid	benzidine sulfate, diazotization, coupling with alkaline thymol	485	c	(124)
Phenyl and tolylsulfonylurea	extraction with organic solvents	Nitration, reduction to amines, diazotization, coupling with N(1-naphthyl)ethylenediamine	usually 547	c	(125)
Organic sulfur	oxidation with Cr(VI)—H_3PO_4, reduction with Sn(II)—H_3PO_4	methylene blue, $Zn(OAc)_2$, PADA, $FeCl_3$	655		(126)
Naphthas	lamp combustion	displacement—barium chloranilate	330 or 530	d e	(131)
Total sulfur in rubber products	oxidation with HNO_3—Br_2, $HClO_4$	lead nitrate, lead sulfate, HCl, determination lead chloride complex	270	c d e	(136)

Substance	Procedure	Reagent	Wavelength	Methods	Ref.
Uranium trioxide, sodium ziconium fluoride, hydrofluoric acid	general procedure fusion with vanadium pentoxide, reduction over Cu	p-rosaniline—formaldehyde, Na_2HgCl_4	560	c d	(143)
Mononitrothiophene, dinitrothiophene in benzene		determination dinitrothiophene: ethyl alc–NaOH, determination mononitrothiophene: nitration to dinitrothiophene	540	c e	(145)
Organic compounds	combustion of sample in Pt spiral in flask containing ammoniacal H_2O_2 and filled with O_2	displacement—barium chloranilate	332 or 530	d e	(153)
Sulfadiazine, sulfamerazine, sulfathiazole		sulfadiazine: 2-thiobarbituric acid sulfamerazine, sulfathiazole–HCl	533 243 and 280	c	156
CS_2, potassium ethylxanthate		CS_2 ethanolic KOH, determination potassium ethylxanthate	301	e	160
Carbonyl sulfide in synthesis gas	hydrolysis in KOH	methylene blue, $Zn(OAc)_2$, PADA, $FeCl_3$	Klett No. 66 filter	c d	195
Sulfide endgroups, number-average molecular weight in high polymers		intermolecular complexes of sulfides and I_2	308	d e	200
General	oxidation in microbomb or combustion in flask filled with O_2, reduction with $HI—HCO_2H$	methylene blue, $Zn(OAc)_2$, PADA, $FeNH_4(SO_4)_2$	Zeiss 366 filter		202
Sulfuric acid aerosol	decomposition in stream of N_2, reaction with Cu	p-rosaniline—formaldehyde, K_2HgCl_4	548 or 575	c d e	207
$S_2O_3^{2-}$, spent liquor from sulfite cellulose cooking		AgSCN, determination dissolved SCN^- as iron(III)–thiocyanate complex	460	b c	212
$S_2O_3^{2-}$, presence of polythionates and hydrogen sulfite		p-benzoquinone, HOAc, determination quinone thiosulfuric acid	400	c	213

503

TABLE II. cont'd

Sample or material	Sample treatment	Method or reagent(s)	Wavelength nm	Comments	Reference
Naphthylaminemono-sulfonic acids	oxidation with ammonium vanadate in H_2SO_4	determination of colored oxidation products	various	e	215
4-Biphenylsulfonic acid and 4,4'-biphenyl-disfulfonic acid	extraction with methylene chloride, separation by continuous electrophoresis	determination mono—di ratio, then determination total concentration in unfractionated mixture	about 255–266		(221)
Reducible sulfer in pulp board paper	reduction with stannous chloride, aluminium squares	methylene blue, $Zn(OAc)_2$, PADA, $FeCl_3$	Corning filters 4784 and 2408	d	(15, 56, 224)
Organic substances	oxidation with HNO_3 or aqua regia (Carius tubes)	displacement—barium chloranilate	327.5 or 530	c e	(229)
Gases, solid materials	reduction by H_2 at a platinum contact (various reduction processes)	ammonium molybdate, H_2SO_4, determination molybdenum blue	570–590		(230)
Thiourea in ammonium thiocyanate		simultaneous spectrophotometric technique	270 and 245		(235)
Thiosulfate, trithionate, polythionates above trithionate		$NaCN$, $CaCl_2$, $Fe(NO_3)_3$—HNO_3, determination iron(III)—thiocyanate complex	460	c	(243, 244)
Thiophene in synthesis gas	pass gas over iron oxide, bubble through $CdCl_2$	isatin-acid, ferric sulfate, determination indophenine	580	c	(248)
Sulfuric acid in concentration range 85–99% sulfuric acid		quinalizarin, determination ratio of absorbances at two wavelengths	535 and 630	c	(259)
Spent sulfite liquor		nitrosolignin method	Corning filters 5113 and 3389		(83)

[a]Discusses structures.
[b]Discusses theoretical principles.
[c]Discusses interferences.
[d]Compares results with those of other methods.
[e]Discusses statistical treatment of data.

TABLE III. Determination of Sulfur in Metals and Fuels

Sample or material	Sample treatment	Method or reagent(s)	Wavelength nm	Comments	Reference
A. In Alloys and Metals					
Steel, iron, cobalt	various procedures; reduce by HI and hypophosphorous acid	methylene blue; Zn(OAc)$_2$, PADA, FeCl$_3$	Spekker filter 608	d	(241)
Selenium	ignition in O$_2$	p-rosaniline–formaldehyde, Na$_2$HgCl$_4$	557	e	(2)
Blister and refined copper	combustion in stream of O$_2$	p-rosanaline–formaldehyde, Na$_2$HgCl$_4$	560	c e	(27)
Metals	combustion in O$_2$ stream in an induction furnace	p-rosanaline–formaldehyde, Na$_2$HgCl$_4$	560	d e	(47)
Iron alloys	dissolution in HCl	methylene blue, Zn(OAc)$_2$, PADA, FeCl$_3$	670	c d e	(137)
Metals, alloys	dissolution in HCl, HNO$_3$, add HCO$_2$H, reduction with hydroiodic–hypophos-phorous acid mixture	colloidal lead sulfide, lead citrate	370	c	(152)
Iron, steel	combustion with O$_2$	fuchsin, NaHCO$_3$ saturated with CO$_2$	530		(166)
Copper	combustion in stream of air	p-rosaniline–formaldehyde, Na$_2$HgCl$_4$	560		(194)
Selenium	NaCl, HNO$_3$, HBr, reduction with HI, hypophosphorous acid in HOAc	methylene blue, Zn(OAc)$_2$, PADA, NH$_4$Fe(SO$_4$)$_2$	667	e	(220)

505

TABLE III. cont'd

Sample or material	Sample treatment	Method or reagent(s)	Wavelength nm	Comments	Reference
B. In Fuels and Related Materials					
Organic sulfur in coals and related products	oxidation with HNO_3	4-amino-4′-chlorodiphenyl (CAD), determine excess CAD	254	d	(4)
Petroleum	oxidation in high-temperature induction furnace	displacement—barium chloranilate	530	c d e	(33)
Aliphatic sulfides in crude petroleum oils and their chromatographic fractions		intermolecular complexes of aliphatic sulfides and I_2	310	b c	(69)
Coal, pyritic sulfur		indirect; iron determined with 1,10-phenanthroline	510	d e	(71)
Petroleum fractions	oxyhydrogen combustion	turbidimetric; $BaCl_2$	410	e	(102)
Mercaptan odorant in liquified petroleum gas		gas-detection tube, measure colored length		c	(187)
Petroleum products	combustion in H_2 stream with Ni catalyst	methylene blue; $Zn(OAc)_2$, PADA, $FeCl_3$	670	c d	(210)
Oils, organic solids	combustion in stream of O_2 in two furnaces	p-rosaniline—fomaldehyde; Na_2HgCl_4	560	c e	(219)

[a] Discusses structures.
[b] Discusses theoretical principles.
[c] Discusses interferences.
[d] Compares results with those of other methods.
[e] Discusses statistical treatment of data.

TABLE IV. Determination of Sulfate and Sulfur Trioxide in Various Samples

Sample or material	Sample treatment	Method or reagent(s)	Wavelength nm	Comments	References
General		displacement–barium chloranilate	332	b	(3, 29)
General		barium complex–nitchromazo, $BaCl_2$	640	c e	(31)
General		displacement–barium chloranilate	530, 332	c d e	(33)
General		displacement–barium chloranilate	530	c d e	(34)
Plant material		benzidine sulfate in HCl	250	c e	(37)
General		displacement–barium chloranilate	530	c e	(52)
Phosphoric acid		turbidimetric; $BaCl_2$–gelatin	420	b e	(61)
Natural waters		$BaCl_2$; then K_2CrO_4; detn. of excess chromate	375	e	(72)
Technical phosphoric acid and phosphates		turbidimetric; $BaCl_2$			(80)
Natural waters		displacement–barium chloranilate	520	automated d e	(81)
Natural waters		ferric perchlorate, aluminium perchlorate, mercuric perchlorate, perchloric acid; detn. of $FeSO_4^+$ complex	325–360	c d e	(82)
General	reduction with $NaH_2PO_2 - H_2O$, HOAc, HI, in stream of N_2	methylene blue; $Zn(OAc)_2$, PADA, $FeNH_4(SO_4)_2$	667	b c e	(87, 88)
Plant ash		turbidimetric; $BaCl_2$, gum arabic	525		(107)
Natural waters		>20 ppm: displacement–barium chromate	370	c	(108)
		<20 ppm: barium chromate; then diphenyl carbazide, detn. Cr(VI)	545		

TABLE IV. cont'd

Sample or Material	Sample treatment	Method or reagent(s)	Wavelength nm	Comments	References
General		detn. of SO_4^{2-} with 4-amino-4'-chlorodiphenyl (CAD); detn. of excess CAD	254	c d	(114)
General		barium molybdate; detn. released molybdate with thioglycolic acid	365	c d	(115)
General		benzidine sulfate; diazotization, coupling with alkaline thymol	485	c	(124)
General	reduction with Sn(II)–H_3PO_4	methylene blue; $Zn(OAc)_2$, PADA, $FeCl_3$	655	c	(127)
General		benzidine sulfate; diazotization, coupling with N-(1-naphthyl)-ethylenediamine	green filter	c	(130)
General, water		thorium borate–amaranth dye; detn. of released dye	521	c d	(141)
Water		SO_4^{2-} to H_2SO_4; dehydration action on saccharose, residue dissolved in water	420	c	(175)
Water		displacement–barium chloranilate	530		(193)
Water		turbidimetric; $BaCl_2$, buffer	420	c e	(13, 201)
Coal ash, related materials		displacement–barium chloranilate	310 or 530	c	(208)
General		turbidimetric; $BaCl_2$ (no additives)	480	d e	(256)
Flue gases		turbidimetric, $BaCl_2$, HCl–gum acasia, isopropanol	Neutral density filter	e	(76, 167)
Flue gases		displacement–barium chloranilate		Automated c	(144)

[a]Discusses structures.
[b]Discusses theoretical principles.
[c]Discusses interferences.
[d]Compares results with those of other methods.
[e]Discusses statistical treatment of data.

508

TABLE V. Determination of Hydrogen Sulfide

Sample or material	Sample treatment	Method or reagent(s)	Wavelength nm	Comments	References
Biological materials, enzymic reactions		methylene blue; Zn(OAc)$_2$, PADA FeNH$_4$(SO$_4$)$_2$	675		(42)
Natural waters		methylene blue, PADA, FeCl$_3$ (mixed reagent)	670	c e	(58)
Soluble sulfide, enzymic desulfhydration of cysteine		colloidal lead sulfide; Pb(OAc)$_2$, HOAc, gum arabic	500	c	(64)
Gases		colloidal bismuth sulfide; bismuth nitrate, HOAc	350	c d	(75)
Gases		methylene blue; Zn(OAc)$_2$, PADA FeCl$_3$	670		(77)
Seawater, natural water		methylene blue; PADA, FeCl$_3$	660	automated,e	(85)
Air		methylene blue; Cd(OH)$_2$, PADA, FeCl$_3$	670		(110)
Acetylene		gas—detection tube; measure discolored length			(128)
General		crystal violet—tetraiodomercurate(II)	590	c e	(140)
Natural gas		methylene blue; Zn(OAc)$_2$, PADA, FeCl$_3$	745	c e	(18, 158)
Air		lead acetate on motion picture film		automated	(174)
Flue gases		UV detn. as SO$_2$ following oxdn. in furnace		automated	(198)
Gas		methylene blue; Zn(OAc)$_2$, PADA, FeCl$_3$	Klett No.66 filter	c	(192, 203)
Gaseous sulfide in the environment		methylene blue; Cd(OH)$_2$ PADA, FeCl$_3$	749 (or both 675 and 749)		(260)
Air		sodium nitroprusside			(163)
Air		starch—iodine			(170)

[a] Discusses structures.
[b] Discusses theoretical principles.
[c] Discusses interferences.
[d] Compares results with those of other methods.
[e] Discusses statistical treatment of data.

509

TABLE VI. Determination of Sulfur in Food, Grain, Plants, Soils

Sample or material	Sample treatment	Method or reagent(s)	Wavelength, nm	Comments	References
Wheat	oxdn. in O_2 – filled flask	barium complex–nitchromazo, $BaCl_2$	640	c e	(31)
Foodstuffs	oxygen flask, Pt spiral, NaOH, H_2O_2	turbidimetric; $BaCl_2$ – Tween 80	neutral density filter or Ilford 601 filter	comparison with (34) barium chloranilate method, c e	(34)
Soils, total sulfur	combustion with vanadium pentoxide and cupric oxide in a stream of N_2; redn. with copper	p-rosaniline–formaldehyde; $Na_2 HgCl_4$	550	c d	(39)
Plant materials, total sulfur	oxdn. with HNO_3 and $HClO_4$	$BaCl_2$; then $(NH_4)_2 CrO_4$; detn. excess $(NH_4)_2 CrO_4$	400	c d	(41)
Soils and plants, total sulfur	combustion with $KNO_3 - HNO_3$ (use fuming HNO_3 first for plants)	turbidimetric, $BaCl_2$, gum acacia	490	d e	(55)
Onions, volatile sulfur	extn. with H_2O; acid hydrolysis; redn. with Zn powder	methylene blue; PADA, $FeNH_4 (SO_4)_2$	670	c d e	(60)
Soils, plants	see ref. (113)	colloidal bismuth sulfide; bismuth nitrate, HOAc, gelatin	400	d e	(63)
Plant materials	wet digestion, ammonium metavanadate, $K_2 Cr_2 O_7$, acid mixture	turbidimetric; $BaCl_2$ – Tween 80	410	e	(79)

Plant materials, soils, irrigation waters	SO_4^{2-} direct; redn. with HI, HCO_2H, red P in stream of N_2; total S: oxdn. to SO_4^{2-}	methylene blue; $Zn(OAc)_2$, PADA, $FeNH_4(SO_4)_2$	670	c d e	(113)
Acid clay	redn. with Sn(II)–H_3PO_4	methylene blue; $Zn(OAc)_2$, PADA, $FeCl_3$	655	c	(127)
Dithiocarbamate residue on food crops	decomposition–absorption	copper acetate–diethylamine; detn. CS_2 as copper diethyldithiocarbamate	blue filter		(151)
Plants	wet oxdn. with HNO_3, $HClO_4$	$Ba(NO_3)_2$; then K_2CrO_4; detn. excess K_2CrO_4	366	c d e	(162)
Soil, total sulfur	Combustion in induction furnace in stream of O_2 with iron powder, molybdenum, and chromium trioxides; redn. with HI,HCO_2H, and red P in stream of N_2	methylene blue; $Zn(OAc)_2$, PADA, $FeNH_4(SO_4)_2$	673	d e	(217)
Plants, lime-sulfur deposits	redn. with HCO_2H, HI, red P in stream of N_2	methylene blue; $Zn(OAc)_2$, PADA, $FeCl_3$	670		(222)
Plant material	digestion with HNO_3, HCl	turbidimetric; $BaCl_2$	Hilger H508 filter	d	(227)

[a] Discusses structures.
[b] Discusses theoretical principles.
[c] Discusses interferences.
[d] Compares results with those of other methods.
[e] Discusses statistical treatment of data.

511

TABLE VII. Determination of Sulfur Dioxide

Sample or material	Sample treatment	Method or reagent(s)	Wavelength nm	Comments	References
Gases (in presence of SO_3)		basic fuchsin H_2SO_4–formaldehyde	580	d	(5, 22)
General		ferric ion, 1,10-phenanthroline; detn. tris(1,10-phenanthroline)–iron(II) complex	510	c	(23)
General		acetate buffer, ferric ion, ferrozine; detn. iron(II) complex	562	c	(24)
Malt, beer		p-rosaniline–formaldehyde; Na_2HgCl_4	560	c d e	(32)
General		p-nitroaniline–formaldehyde; Na_2HgCl_4	387	c	(35)
General		direct, UV	198	c d	(36)
General		lead acetate; detn. excess plumbous ion	208	c	(38)
Atmosphere		p-rosaniline–formaldehyde		automated c	(95)
Air		increase in transmittance of starch–iodine solutions		automated	(122)
General		p-aminoazobenzene–formaldehyde; Na_2HgCl_4	505	a b c d	(132)
Air		gas-detection tube; measure colored length		d	(133)
Soft drinks (free SO_2)		p-rosaniline–formaldehyde; Na_2HgCl_4	560	d	(149)
General		p-rosaniline–formaldehyde; Na_2HgCl_4	550		(171)
Atmosphere		gas-detection tube; use std. color cards, measure colored length, or relate color or length to sample volume		c	(184, 185)

Application	Method / Reagent	Wavelength (nm)	Notes	Ref.
Atmosphere	basic fuchsin H_2SO_4–formaldehyde	570	d	(186)
Flue Gases	UV detn. of SO_2		automated	(198)
Atmosphere	p-rosaniline–formaldehyde; K_2HgCl_4	548 or 575	b c e	(206)
General, process stream and source analysis	H_2SO_4	276	c e	(216)
Air	basic fuchsin H_2SO_4–formaldehyde	green filter (525)	c	(226)
Atmosphere	Ferric ion, 1,10-phenanthroline; detn. tris(1,10-phenanthroline)–iron(II) complex	510	c e	(228)
General	modified atomic absorption spectrophotometer, flow-through cell	215	c d	(234)
Atmosphere	basic fuchsin H_2SO_4–formaldehyde	580	d	(245)
Atmosphere	p-rosaniline–formaldehyde; Na_2HgCl_4		automated, d	(250)
Atmosphere	acetate buffer, ferric ion, ferrozine, NH_4F; detn. Fe(II) complex	560	c e	(252)
General atmosphere	p-rosaniline–formaldehyde; Na_2HgCl_4	560	b c	(20, 253)
General atmosphere	p-rosaniline–formaldehyde; Na_2HgCl_4	560	c e	(254)

[a] Discusses structures.
[b] Discusses theoretical principles.
[c] Discusses interferences.
[d] Compares results with those of other methods.
[e] Discusses statistical treatment of data.

TABLE VIII. Determination of Mercaptans, Thiols, Sulfhydryl Groups

Sample or material	Sample treatment	Method or reagent(s)	Wavelength nm	Comments	References
Sulfhydryl, general		N-ethylmaleimide, phosphate buffer	300	c	(6)
Sulfhydryl, heme proteins		4,4'-dipyridinesulfide	307, 324	b	(9)
Thiols, hexane solutions		pyridine, benzidrine hydrochloride	525	oxdn. with Br_2, addn. of CN^- to give CNBr	(25)
Thiols, general		N-ethylmaleimide, alc. KOH	515		(44)
Sulfhydryl, wool		1-(4-chloromercuriphenylazo)-naphthol-2 formamide	blue green filter	c e	(48)
Sulfhydryl, general		p-chloromercuribenzoate; detn. excess mercurial with dithizone—CCl_4	625	d	(78)
Sulfhydryl groups (thiols, proteins, biological materials)		heterocyclic disulfides	various wavelengths	follow formation of product (thione) or disappearance of reagent	(84)

Mercaptans, air	mercuric acetate, PADA, FeCl$_3$	500	c e	(164)
Mercaptan odorant in liquefied petroleum gas	Gas-detection tube, measure colored length		c	(187)
Sulfhydryl groups, general	N-ethylmaleimide, phosphate buffer	300		(199)
Thiols, general	NaNO$_2$, H$_2$SO$_4$, ammonium sulfamate; HgCl$_2$, sulfanilimide; N-1-naphthylethylenediamine; detn. azo dye	yellow-green filter	b c	(205)
Sulfhydryl, tissue (total, protein bound, non-protein)	5,5'-dithiobis-(2-nitrobenzoic acid); detn. 2-nitro-5-mercaptobenzoic acid	412	c	(218)

[a]Discusses structures.
[b]Discusses theoretical principles.
[c]Discusses interferences.
[d]Compares results with those of other methods.
[e]Discusses statistical treatment of data.

TABLE IX. Determination of Sulfides

Sample or material	Sample treatment	Method or reagent(s)	Wavelength nm	Comments	References
Alkalies		methylene blue; $Zn(OAc)_2$, PADA, $FeCl_3$	670	b c	(46)
Water samples		sodium nitroprusside, alkaline soln.	536	automated c d e	(53)
General		demasking; potassium bis(7-iodo-5-sulfoxino)–palladium(II), $FeCl_3$	650	c	(89)
Air		methylene blue; $Cd(OH)_2$, PADA, $FeCl_3$	670		(111)
General		methylene blue–iodine complex; detn. liberated methylene blue	665	c d	(120)
General		bismuth sulfide sol; $Bi(NO_3)_3$ in aq. soln. of mannitol, glycerol, gum arabic, Na OAc, HOAc	see Refs. 75 & 240		(135)
General		iron(III) ammonium sulfate, nitrilo-triacetic acid, NH_3, Na_2SO_3	635	b c e	(197)
Sulfate pulp black liquor		p-phenylenediamine, $FeCl_3$	600	c d	(232)

[a]Discusses structures.
[b]Discusses theoretical principles.
[c]Discusses interferences.
[d]Compares results with those of other methods.
[e]Discusses statistical treatment of data.

TABLE X. Determination of Sulfite and Elemental Sulfur

Sample or material	Sample treatment	Method or reagent(s)	Wavelength nm	Comments	References
A. Sulfite					
General		p-rosaniline–formaldehyde; with and without Na_2HgCl_4	562	c	(11)
General		direct, UV	198	c d	(36)
General		mercuric thiocyanate; detn. released SCN^- with ferric ion	variable	c	(99)
General		mercury(II), H_2SO_4; detn. mercury-(II)–sulfite complex	237 or 250	c	(177)
General		mercuric nitrate, KBr; detn. excess mercuric reagent with diphenyl-carbazone (benzene extn.)	562	b c	(178)
B. Elemental Sulfur					
Hydrocarbons		NaCN in acetone, $FeCl_3$ in acetone; detn. SCN^- (add $HgCl_2$ to samples containing sulfides, disulfides, mercaptans)	465	c	(30)

517

TABLE X. cont'd

Sample or material	Sample treatment	Method or reagent(s)	Wavelength nm	Comments	References
Soil	extn. with acetone	colloidal sulfur	420	c	(90)
General		ethanolic solutions of sulfur	264–274	c	(93)
Petroleum fractions	oxyhydrogen combustion	turbidimetric, $BaCl_2$	410	e	(102)
Paper, cellulose fibers	redn. with N-(4,4'-dimethoxy-benzohydrilidine)benzyl-amine	methylene blue; $Zn(OAc)_2$, PADA, $FeCl_3$	670		(103)
General		n-hexane solutions of S	276	e	(159)
General		N-(4,4'-dimethoxybenzohydrilidine)-benzylamine, extn. with benzene	590	b	(180)
Gasoline		mercury, oleic acid; detn. mercury sulfide	comparison	c	(242)
Sulfur sols		NaCN, $CuCl_2$, $Fe(NO_3)_3 - HNO_3$; detn. iron thiocyanate complex	460	c	(244)

[a]Discusses structures.
[b]Discusses theoretical principles.
[c]Discusses interferences.
[d]Compares results with those of other methods.
[e]Discusses statistical treatment of data.

TABLE XI. Determination of Thiocyanate

Sample or material	Sample treatment	Method or reagent(s)	Wavelength nm	Comments	References
Plasma, serum	Aeration; Br_2 water, arsenous acid	Benzidine in pyridine	532	c	(45)
General		Dithiocyanatodipyridine copper(II) complex, extn. with $CHCl_3$	407	c e	(62)
General		Mercury-EDTA	235	c	(134)
Industrial wastes		Dipyridine-copper(II)-thiocyanate complex; extn. with $CHCl_3$	410	c e	(138)
General		Potassium perrhenate, HCl, $SnCl_2$, acetone; detn. rhenium-SCN^- complex	390	b c e	(169)
General		Mercury(II), methylthymolblue in CH_3OH	620	c e	(172)
General		1,10-Phenanthroline, Fe(II) soln., phosphate buffer, nitrobenzene extn., detn. tris(1,10-phenanthroline)-iron(II) thiocyanate	515	c	(257)

REFERENCES

1. Abbott, D. C., *Analyst,* **87**, 286 (1962).

2. Acs, L., and S. Barabas, *Anal. Chem.,* **36**, 1825 (1964).

3. Agterdenbos, J., and N. Martinius, *Talanta,* **11**, 875 (1964).

4. Ahmed, M. N., and G. J. Lawson, *Talanta,* **1**, 142 (1958).

5. Alekseeva, M. V., and K. A. Bushtueva, *Gigi Sanit.*, No. 4, 13 (1954).

6. Alexander, N. M., *Anal. Chem.,* **30**, 1292 (1958).

7. Alicino, J. F., *Microchem. J.,* **2**, 83 (1958).

8. Alicino, J. F., A. I. Cohen, and M. E. Everhard, *Organic Analysis: Sulfur, Treatise on Analytical Chemistry,* Part II, Vol. 12, Kolthof, I. M., and P. J. Elving, Eds., Interscience, New York, 1965, p. 57B.

9. Ampulski, R. S., V. E. Ayers, and S. A. Morell, *Anal. Biochem.,* **32**, 163 (1969).

10. American Public Health Association, *Standard Methods for the Examination of Water and Wastewater,* 13th ed., New York, 1971, p. 555.

11. Arikawa, Y., T. Ozawa, and I. Iwasaki, *Bull. Chem. Soc. Japan,* **41**, 1454 (1968).

12. Ashworth, M. R. F., and G. Bohnstedt, *Anal. Chim. Acta,* **36**, 196 (1966).

13. ASTM Designation D-516-68 (1974), American Society for Testing and Materials, Philadelphia.

14. *Ibid.,* D-931-50. (1972)

15. *Ibid.,* D-984-74 (1974).

16. *Ibid.,* D-1266-70 (1975).

17. *Ibid.,* D-1685-66 (1971).

18. *Ibid.,* D-2725-70 (1975).

19. *Ibid.,* D-2784-70 (1975).

20. *Ibid.,* D-2914-70T (1970).

21. *Ibid.,* D-2785-70 (1975).

22. Atkins, S., *Anal. Chem.,* **22**, 947 (1950).

23. Attari, A., T. P. Igielski, and B. Jaselskis, *Anal. Chem.,* **42**, 1282 (1970).

24. Attari, A., and B. Jaselskis, *Anal. Chem.,* **44**, 1515 (1972).

25. Bakes, J. M., and P. G. Jeffrey, *Talanta,* **8**, 641 (1961).

26. Bailey, J. J., and D. G. Gehring, *Anal. Chem.,* **33**, 1760 (1961).

27. Barabas, S., and J. Kaminski, *Anal. Chem.,* **35**, 1702 (1963).

28. Barnett, H. A., C. E. Bole, and C. F. Glick, *Anal. Chem.,* **32** 842 (1960).

29. Barney, J. E., II, *Talanta,* **12**, 425 (1965).

30. Bartlett, J. K., and D. A. Skoog, *Anal. Chem.,* **26**, 1008 (1954).

31. Basargin, N. N., V. L. Menshikova, Z. S. Belova, and L. G. Myasishcheva, *Zh. Anal. Khim.,* **23**, 732 (1968).

32. Beetch, E. B., and L. I. Oetzel, *J. Agr. Food Chem.,* **5**, 951 (1957).

33. Bertolacini, R. J., and J. E. Barney, II, *Anal. Chem.,* **29**, 281 (1957); *ibid.,* **30**, 202, 498 (1958).

34. Beswick, G., and R. M. Johnson, *Talanta,* **17**, 709 (1970).

35. Bethge, P. O., and M. Carlson, *Talanta,* **16**, 144 (1969).

36. Bhatty, M. K., and A. Townshend, *Anal. Chim. Acta,* **55**, 263 (1971).

37. Bingley, J. B., and A. T. Dick, *J. Agr. Food Chem.,* **15**, 539 (1967).

38. Blevin, R. C., and F. A. Gunther, *Analyst,* **86**, 675 (1961).

39. Bloomfield, C., *Analyst,* **87**, 586 (1962).

40. Booth, R. E., and E. H. Jensen, *J. Amer. Pharm. Assoc., Sci. Ed.,* **45**, 535 (1956).

41. Bosman, M. S. M., *Chem. Weekbl.,* **56**, 716 (1960).

42. Bray, R. C., *Analyst,* **83**, 379 (1958).

43. Broekhuysen, J., and J. Bechet, *Anal. Chim. Acta,* **13**, 277 (1955).

44. Broekhuysen, J., *Anal. Chim. Acta,* **19**, 542 (1958).

45. Bruce, R. B., J. W. Howard, and R. F. Hanzal, *Anal. Chem.,* **27**, 1346 (1955).

46. Budd, M. S., and H. A. Bewick, *Anal. Chem.,* **24**, 1536 (1952).

47. Burke, K. E., and C. M. Davis, *Anal. Chem.,* **34**, 1747 (1962).

48. Burley, R. W., *Text. Res. J.,* **26**, 332 (1956).

49. Cadle, R. D., M. Rolston, and P. L. Magill, *Anal. Chem.,* **23**, 475 (1951).

50. Canals, E., A. Charra, and G. Riety, *Bull. Soc. Chim., Belgrade,* **12**, 1055 (1945).

51. Capkeviciene, E., *Gig. Sanit,* **35**, 59 (1970).

52. Carlson, R. M., R. A. Rosell, and W. Vallejos, *Anal. Chem.,* **39**, 688 (1967).

53. Casapieri, P., R. Scott, and E. A. Simpson, *Anal. Chim. Acta,* **45**, 547 (1969).

54. Cerfontain, H., H. G. S. Duin, and L. Vollbracht, *Anal. Chem.,* **35**, 1005 (1963).

55. Chaudhry, I. A., and A. H. Cornfield, *Analyst,* **91**, 528 (1966).

56. Chazin, J. D., *Tappi,* **53**, 1514 (1970).

57. Clayton, E. E., and J. H. Jones, *Oil Gas J.,* **61**, 82 (1963).

58. Cline, J. D., *Limnol. Oceanogr.,* **14**, 454 (1969).

59. Colombo, P., D. Corbetta, A. Pirotta, and A. Sartori, *Tappi,* **40**, 490 (1957).

60. Currier, H. B., *Food Res.,* **10**, 177 (1945).

61. Dahlgren, S. E., *Acta Chem. Scand.,* **14**, 1279 (1960).

62. Danchik, R. S., and D. F. Boltz, *Anal. Chem.,* **40**, 2215 (1968).

63. Dean, G. A., *Analyst,* **91**, 530 (1966).

64. Delwiche, E. A., *Anal. Biochem.,* **1**, 397 (1960).

65. Dickinson, D., *Analyst,* **91**, 809 (1966).

66. Dixon, J. P., *Modern Methods in Organic Microanalysis,* Van Nostrand, London, 1968, Chap. 7.

67. Dokládalová, J., *Z. Anal. Chem.,* **208**, 92 (1965).

68. Dokládalová, J., *Mikrochim. Ichnoanal. Acta,* **1965**, 344.

69. Drushel, H., V., and J. F. Miller, *Anal. Chem.,* **27**, 495 (1955).

70. Dworzak, R., and K. H. Becht, *Z. Anorg. Allg. Chem.,* **285**, 143 (1956).

71. Edwards, A. H., G. N. Daybell, and W. J. S. Pringle, *Fuel,* **37**, 47 (1958).

72. Egami, F., and N. Takahashi, *Bull. Chem. Soc. Japan,* **30**, 442 (1957).

73. Fairing, J. D. and F. D. Short, *Anal. Chem.,* **28**, 1827 (1956).

74. Farley, L. H., and R. A. Winkler, *Anal. Chem.,* **40**, 962 (1968).

75. Field, E., and C. S. Oldach, *Ind. Eng. Chem., Anal. Ed.,* **18**, 665 (1946).

76. Fielder, R. S., P. J. Jackson, and E. J. Raask, *Inst. Fuel,* **33**, 84 (1960).

77. Fogo, J. K., and M. Popowsky, *Anal. Chem.,* **21**, 732 (1949).

78. Fridovich, I., and P. Handler, *Anal. Chem.,* **29**, 1219 (1957).

79. Garrido, M. L., *Analyst,* **89**, 61 (1964).

80. Gassner, K., and H. Friedel, *Z. Anal. Chem.,* **152**, 420 (1956).

81. Gales, M. E., Jr., W. H. Kaylor, and J. E. Longbottom, *Analyst,* **93**, 97 (1968).

82. Goguel, R., *Anal. Chem.,* **41**, 1034 (1969).

83. Goldschmid, O., and L. F. Maranville, *Anal. Chem.,* **31**, 370 (1959).

84. Grossetti, D. R., and J. F. Murray, Jr., *Anal. Chim. Acta,* **46**, 139 (1969).

85. Grasshoff, K., and K. Chan, *Advan. Automat. Anal. Technicon Int. Congr.,* 1969, **2**, 147 (1970).

86. Gupta, J. G. S., *Anal. Chem.,* **35**, 1971 (1963).

87. Gustafsson, L., *Talanta,* **4**, 227 (1960).

88. Gustafsson, L., *Talanta,* **4**, 236 (1960).

89. Hanker, J. S., A. Gelberg, and B. Witten, *Anal. Chem.,* **30**, 93 (1958).

90. Hart, M. G. R., *Analyst,* **86**, 472 (1961).

91. Hastings, S. H., *Anal. Chem.,* **25**, 420 (1953).

92. Hastings, S. H., and B. H. Johnson, *Anal. Chem.,* **27**, 564 (1955).

93. Heatley, N. E., and E. J. Page, *Anal. Chem.,* **24**, 1854 (1952).

94. Heinrich, B. J., M. D. Grimes, and J. E. Puckett, *Sulfur, Treatise on Analytical Chemistry,* Kolthoff, I. M., and P. J. Elving, Eds., Part II, Vol. 7, Interscience, New York, 1961, p. 1.

95. Helwig, H. L., and C. L. Gordon, *Anal. Chem.,* **30**, 1810 (1958).

96. Herrmann, G., *Z. Chem.,* **2**, 42 (1962).

97. Hill, W. H., M. A. Shapiro, and Y. Kobayashi, *J. Amer. Water Works Assoc.,* **54**, 409 (1962).

98. Hinden, S. G., and A. V. Grosse, *Anal. Chem.,* **20**, 1050 (1948).

99. Hinze, W. L., J. Elliott, and R. E. Humphrey, *Anal. Chem.*, **44**, 1511 (1972).

100. Hochheiser, S., U.S. Public Health Service *Publ. AP-6*, Washington, D.C., 1964, 47 pp.

101. Holeton, R. E., and A. L. Linch, *Anal. Chem.*, **22**, 819 (1950).

102. Houghton, N. W., *Anal. Chem.*, **29**, 1513 (1957).

103. Huber, O., H. Kolb, and J. Weigl, *Z. Anal. Chem.*, **227**, 416 (1967).

104. Huber, O., H. Kolb, and J. Weigl, *Z. Anal. Chem.*, **227**, 420 (1967).

105. Huitt, H. A., and J. P. Lodge, Jr., *Anal. Chem.*, **36**, 1305 (1964).

106. Hutchinson, K., and D. F. Boltz, *Anal. Chem.*, **30**, 54 (1958).

107. Iljin, W. S., *Agron. Trop.* (Maracay, Venezuela), **1**, 5 (1951).

108. Iwasaki, I., S. Utsumi, K. Hagino, T. Tarutani, and T. Ozawa, *Bull. Chem. Soc. Japan*, **30**, 847 (1957).

109. Jaboulay, B. E., *Chim. Anal.*, **33**, 48 (1951).

110. Jacobs, M. B., *J. Air Pollution Control Assoc.*, **15**, 314 (1965).

111. Jacobs, M. B., M. M. Braverman, and S. Hochheiser, *Anal. Chem.*, **29**, 1349 (1957).

112. Johnson, C. M., and T. H. Arkley, *Anal. Chem.*, **26**, 1525 (1954).

113. Johnson, C. M., and H. Nishita, *Anal. Chem.*, **24**, 736 (1952).

114. Jones, A. S., and D. S. Lethan, *Analyst*, **81**, 15 (1956).

115. Kanno, S., *Japan Analyst*, **8**, 180 (1959).

116. Kanno, J., *Intn. J. Air Pollution*, **1**, 231 (1959).

117. Karchmer, J. H., Divalent Sulfur-Based Functions, in *Treatise on Analytical Chemistry*, Part 4, Vol. 13, Kolthoff, I. M., and P. J. Elving, Eds., Interscience, New York, 1966, pp. 337–517.

118. Karchmer, J. H., *The Analytical Chemistry of Sulfur and Its Compounds*, Parts I and II, Wiley-Interscience, New York, 1970, 1972.

119. Karush, F., and M. Sonenberg, *Anal. Chem.*, **22**, 175 (1950).

120. Kato, T., *Nippon Kagaku Zasshi*, **82**, 210 (1961).

121. Kato, T., S. Takei, and K. Ogasawara, *Technol Repts. Tohoku Univ.*, **19**, 85 (1954).

122. Katz, M., *Anal. Chem.*, **22**, 1040 (1950).

123. Keily, H. S., and L. B. Rogers, *Anal. Chim. Acta*, **14**, 356 (1956).

124. Kent, P. W., and M. W. Whitehouse, *Analyst*, **80**, 630 (1955).

125. Kern, W., *Anal. Chem.*, **35**, 50 (1963).

126. Kiba, T., and I. Akaza, *Bull. Chem. Soc. Japan*, **30**, 482 (1957).

127. Kiba, T., and I. Kishi, *Bull. Chem. Soc. Japan*, **30**, 44 (1957).

128. Kitagawa, T., *J. Japan Chem. Ind. Soc.*, **No. 33**, Feb. (1951).

129. Kitagawa, T., Japan Pat. 178082 (Mar. 7, 1949).

130. Klein, B., *Ind. Eng. Chem. Anal. Ed.*, **16**, 536 (1944).

131. Klipp, R. W., and J. E. Barney, II, *Anal. Chem.,* **31**, 596 (1959).

132. Kniseley, S. J., and L. J. Throop, *Anal. Chem.,* **38**, 1270 (1966).

133. Kobayashi, Y., *Bunseki Kagaku,* **9**, 229 (1960).

134. Komatsu, S., and T. Nomura, *Nippon Kagaku Zasshi,* **87**, 841 (1966).

135. Koren, H., and W. Gierlinger, *Mikrochim. Acta,* **1953**, 220.

136. Kress, K. E., *Anal. Chem.,* **27**, 1618 (1955).

137. Kriege, O. H., and A. L. Wolfe, *Talanta,* **9**, 673 (1962).

138. Kruse, J. M., and M. G. Mellon, *Anal. Chem.,* **25**, 446 (1953).

139. Lahman, E., and K. E. Prescher, *Z. Anal. Chem.,* **251**, 300 (1970).

140. Lambert, J. L., and D. J. Manzo, *Anal. Chim. Acta,* **48**, 185 (1969).

141. Lambert, J. L., S. K. Yasuda, and M. P. Grotheer, *Anal. Chem.,* **27**, 800 (1955).

142. Lane, W. H., *Anal. Chem.,* **20**, 1045 (1948).

143. Larson, R. P., L. E. Ross, and N. M. Ingber, *Anal. Chem.,* **31**, 1596 (1959).

144. Laxton, J. W., and P. J. Jackson, *J. Inst. Fuel* (London), **37**, 12 (1964).

145. Leibmann, W., and J. T. Woods, *Anal. Chem.,* **29**, 1845 (1957).

146. Letnoff, T. V., and J. G. Reinhold, *J. Biol. Chem.,* **114**, 147 (1936).

147. Levaggi, D. A., and M. Feldstein, *J. APCA,* **13**, 380, 387 (1936).

148. Littlefield, J. B., U.S. Pat. 2,174,349 (1939).

149. Lloyd, W. J. W., and B. C. Cowle, *Analyst,* **88**, 394 (1963).

150. Longwell, J., and W. D. Maniece, *Analyst,* **80**, 167 (1955).

151. Lowen, W. K., *Anal. Chem.,* **23**, 1846 (1951).

152. Luke, C. L., *Anal. Chem.,* **21**, 1369 (1949).

153. Lysyj, I., and J. E. Zarembo, *Microchem. J.,* **3**, 173 (1959).

154. Magill, P. L., M. V. Rolston, and R. W. Bremner, *Anal. Chem.,* **21**, 1411 (1949).

155. Maier, H. G., and W. Diemair, *Z. Anal. Chem.,* **227**, 187 (1967).

156. Marzys, A. E. O., *Analyst,* **86**, 460 (1961).

157. Mason, D. McA., and C. E. Hummel, *Anal. Chem.,* **30**, 1885 (1958).

158. Mason, D. M., *Hydrocarbon Process Petrol. Refiner,* **43**, 145 (1964).

159. Maurice, M. J., *Anal. Chim. Acta,* **16**, 574 (1957).

160. Maurice, M. J., and J. L. Mulder, *Mikrochim. Acta,* **1957**, 661.

161. Meador, M. C., and R. M. Bethea, *Environ. Sci. Technol.,* **4**, (10), 853 (1970).

162. Middleton, K. R., *Analyst,* **87**, 444 (1962).

163. Mokhov, L. A., and S. A. Matseeva, *Lab. Delo,* **8**, 4 (1962).

164. Moore, H., H. L. Helwig, and R. J. Graul, *Amer. Indust. Hyg. Assoc. J.,* **21**, 466 (1960).

165. Moore, W. A., and R. A. Kolbeson, *Anal. Chem.,* **28**, 161 (1956).

166. Mukaewaki, K., *Bunseki Kagaku,* **9**, 774 (1960).

167. Napier, D. H., and M. H. Stone, *J. Appl. Chem.*, **8**, 787 (1958).

168. Nauman, R. V., P. W. West, F. Tron, and G. C. Gaeke, Jr., *Anal. Chem.*, **32**, 1307 (1960).

169. Neas, R. E., and J. C. Guyon, *Anal. Chem.*, **41**, 1470 (1969).

170. Nichols, P. N. R., *Chem. Ind.* (London), **1964**, 1654.

171. Nietruch, F., and K. E. Prescher, *Z. Anal. Chem.*, **226**, 259 (1967).

172. Nomura, T., *J. Chem. Soc. Japan, Pure Chem. Sect.*, **88**, 961 (1967).

173. Nusbaum, I., *Water Sewage Works*, **112**, 113, 150 (1965).

174. Offutt, E. B., and L. V. Sorg, *Anal. Chem.*, **27**, 429 (1955).

175. Ohlweiler, O. A., and J. O. Meditsch, *Anal. Chim. Acta*, **25**, 233 (1961).

176. Oka, Y., and S. Matsuo, *J. Chem. Soc. Japan, Pure Chem. Sect.*, **74**, 618 (1953); F. Ishihawa Anniversary Volume Science Repts., Tohoku Univ., **37**, 96 (1953).

177. Okutani, T., S. Ito, and S. Utsumi, *Nippon Kagaku Zasshi*, **88**, 1296 (1967).

178. Okutani, T., and S. Utsumi, *Bull. Chem. Soc. Japan*, **40**, 1386 (1967).

179. Orsanco Detergent Subcommittee, *J. Amer. Water Works Assoc.*, **55**, 369, 375 (1963).

180. Ory, H. A., V. L. Warren, and H. B. Williams, *Analyst*, **82**, 189 (1957).

181. Ozawa, T., *Nippon Kagaku Zasshi*, **87**, 855, 859 (1966).

182. Pate, J. B., J. P. Lodge, Jr., and A. F. Wartburg, *Anal. Chem.*, **34**, 1660 (1962).

183. Pate, J. B., B. E. Ammons. G. A. Swanson, and J. P. Lodge, Jr., *Anal. Chem.*, **37**, 942 (1965).

184. Patterson, G. D., Jr., and M. G. Mellon, *Anal. Chem.*, **24**, 1586 (1952).

185. Patterson, G. D., Jr., U.S. Pat. 2,785,959 (1957).

186. Paulas, H. J., E. P. Floyd, and D. H. Byers, *Amer. Indust. Hyg. Assoc. J.*, **15**, 277 (1954).

187. Peurifoy, P. V., M. J. O'Neal, Jr., and I. Dvoretsky, *Anal. Chem.*, **36**, 1853 (1964).

188. Polson, D. S. C., and J. D. H. Strickland, *Anal. Chim. Acta*, **6**, 452 (1952).

189. Pomeroy, R., *Anal. Chem.*, **26**, 571 (1954).

190. Pomeroy, R., *Petroleum Eng.*, **15** (No. 13), 156 (1944).

191. Pomeroy, R., *Sew. Works J.*, **13**, 498 (1941).

192. Prince, C. G. T., *J. Appl. Chem.* (London), **5**, 364 (1955).

193. Prochazkova, L., *Z. Anal. Chem.*, **182**, 103 (1961).

194. Pugh, H., and W. R. Waterman, *Anal. Chim. Acta*, **55**, 97 (1971).

195. Pursglove, L. A., and H. W. Wainwright, *Anal. Chem.*, **26**, 1835 (1954).

196. Radmacher, W., and P. Mohrhauer, *Glückauf*, **89**, 503 (1953).

197. Rahim, S. A., and T. S. West, *Talanta*, **17**, 851 (1970).

198. Risk, J. B., and F. E. Murray, *Can Pulp Paper Ind.*, **17**(10), 31 (1964).

199. Roberts, E., and G. Rouser, *Anal. Chem.*, **30**, 1291 (1958).

200. Rosenthal, I., G. J. Frisone, and J. K. Coberg, *Anal. Chem.*, **32**, 1713 (1960).

201. Rossum, J. R., and P. A. Villarruz, *J. Amer. Water Works Assoc.*, **53**, 873 (1961).

202. Roth, H., *Mikrochemie ver Mikrochim. Acta*, **36**, 379 (1951).

203. Sands, A. E., M. A. Grafius, H. W. Wainwright, and M. W. Wilson, U.S. Bur. Mines, *Rept. Invest. No. 4547*, Washington, D.C., 1949.

204. Sands, A. E., H. W. Wainwright, and G. C. Egleson, *Amer. Gas Assoc. Proc.*, **32**, 564 (1950).

205. Saville, B., *Analyst*, **83**, 670 (1958).

206. Scaringelli, F. P., B. E. Saltzman, and S. A. Frey, *Anal. Chem.*, **39**, 1709 (1967).

207. Scaringelli, F. P., and K. A. Rehme, *Anal. Chem.*, **41**, 707 (1969).

208. Schafer, H. N. S., *Anal. Chem.*, **39**, 1719 (1967).

209. Scherer, P. C. Jr., and W. W. Sweet, *Ind. Eng. Chem. Anal. Ed.*, **4**, 103 (1932).

210. Schluter, E. C., E. P. Pary, and G. Matsuyama, *Anal. Chem.*, **32**, 413 (1960).

211. Schmidt, M., and G. Talsky, *Z. Anal. Chem.*, **166**, 274 (1959).

212. Schöön, N. H., *Acta Chem. Scand.*, **12**, 1730 (1958).

213. Schöön, N. H., *Acta Chem. Scand.*, **13**, 525 (1959).

214. Schroeter, L. C., *Sulfur Dioxide Applications in Foods, Beverages, and Pharmaceuticals*, Pergamon Press, Oxford, 1966, pp. 168–184.

215. Schulze, G., *Z. Anal. Chem.*, **241**, 196 (1968).

216. Scoggins, M. W., *Anal. Chem.*, **42**, 1091 (1970).

217. Searle, P. L., *Analyst*, **93**, 540 (1968).

218. Sedlak, J., and R. H. Lindsay, *Anal. Biochem.*, **25**, 192 (1968).

219. Seefield, E. W., and J. W. Robinson, *Anal. Chim. Acta*, **22**, 61 (1960).

220. Sjöberg, B. L., *Talanta*, **14**, 693 (1967).

221. Skelly, N. E., *Anal. Chem.*, **37**, 1526 (1965).

222. Skerrett, E. J., and G. J. Dickes, *Analyst*, **86**, 69 (1961).

223. Smith, A. F., D. G. Jenkins, and D. E. Cunningworth, *J. Appl. Chem.*, **11**, 317 (1961).

224. Sobolev, I., R. Bhargava, N. Geacinter, and R. Russell, *Tappi*, **39**, 628 (1956).

225. Södergren, A., *Analyst*, **91**, 113 (1966).

226. Stang, A. M., J. E. Zatek, and C. D. Robson, *Amer. Ind. Hyg. Assoc. Quart.*, **12**, 5 (1951).

227. Steinbergs, A., *Analyst*, **78**, 47 (1953).

228. Stephens, B. G., and F. Lindstrom, *Anal. Chem.*, **36**, 1308 (1964).

229. Stoffyn, P., and W. Keane, *Anal. Chem.*, **36**, 397 (1964).

230. Stratmann, H., *Mikrochim. Acta*, **1956**, 1031.

231. Stratmann, H., and M. Buck, *Int. J. Air Water Pollution*, **9**, 199 (1965).

232. Strickland, J. D. H., and J. B. Risk, *Tappi*, **40**, 91 (1957).

233. Ströhl, G. W., and D. Kurzak, *Talanta*, **16**, 135 (1969).

234. Syty, A., *Anal. Chem.*, **45**, 1744 (1973).

235. Talreja, S. T., P. M. Oza, and P. S. Rao, *Anal. Chim. Acta*, **36**, 238 (1966).

236. Tananaev, I. V., and N. A. Rudnev, *Zh. Anal. Khim.*, **5**, 82 (1950).

237. Terraglio, F. P., and R. M. Manganelli, *Anal. Chem.*, **34**, 675 (1962).

238. Thomas, M. D., and R. E. Amtower, *J. Air Pollution Control Assoc.*, **16**, 618 (1966).

239. Tolg, G., *Ultramicro Analysis*, Wiley-Interscience, New York, 1970.

240. Treiber, E., H. Koren, and W. Gierlinger, *Mikrochim. Acta*, **40**, 32 (1952).

241. Tyou, P., and L. Humblet, *Talanta*, **3**, 232 (1960).

242. Uhrig, K., and H. Leven, *Anal. Chem.*, **23**, 1334 (1951).

243. Urban, P. J., *Z. Anal. Chem.*, **179**, 415, 422 (1961).

244. Urban, P. J., *Z. Anal. Chem.*, **180**, 110, 116 (1961).

245. Urone, P. F., and W. E. Boggs, *Anal. Chem.*, **23**, 1517 (1951).

246. Urone, P. F., J. B. Evans, and C. M. Noyes, *Anal. Chem.*, **37**, 1104 (1965).

247. Van Steveninck, J., and J. C. Riemersma, *Anal. Chem.*, **38**, 1250 (1966).

248. Wainwright, H. W., and G. J. Lambert, U.S. Bur. Mines, *Rept. Invest. No. 4753*, Washington, D.C., 1950.

249. Weber, W. J., J. C. Morris, and W. Stumm, *Anal. Chem.*, **34**, 1844 (1962).

250. Welch, A. F., and J. P. Terry, *Amer. Indust. Hygiene A. J.*, **21**, 316 (1960).

251. Welcher, F. J., Ed., *Standard Methods of Chemical Analysis*, Vol. 3, Part A, Van Nostrand, Princeton, 1966, Chapts. 1 and 2.

252. Wende, R. D., R. T. Clancy, and G. B. Jackson, *Amer. Lab.*, **4**, 46 (1972).

253. West, P. W., and G. C. Gaeke, *Anal. Chem.*, **28**, 1816 (1956).

254. West, P. W., and F. Ordoveza, *Anal. Chem.*, **34**, 1324 (1962).

255. Whiston, T. G., and G. W. Cherry, *Analyst*, **87**, 819 (1962).

256. Wimberley, J. W., *Anal. Chim. Acta*, **42**, 327 (1968).

257. Yamamoto, Y., T. Tarumoto, and Y. Hanamoto, *Bull. Chem. Soc. Japan*, **42**, 268 (1969).

258. Zahn, V., *Ind. Eng. Chem., Anal. Ed.*, **9**, 543 (1937).

259. Zimmerman, E., and W. W. Brandt, *Talanta*, **1**, 374 (1958).

260. Zutshi, P. K., and T. N. Mahadevan, *Talanta*, **17**, 1014 (1970).

APPENDIX

TRANSMITTANCE-ABSORBANCE CONVERSION TABLE

% Transmittance	0.0	0.1	0.2	0.3	0.4	0.5	0.6	0.7	0.8	0.9
99	0.0043	0.0039	0.0035	0.0031	0.0026	0.0022	0.0017	0.0013	0.0009	0.0004
98	.0088	.0083	.0079	.0074	.0070	.0066	.0061	.0057	.0052	.0048
97	.0132	.0128	.0123	.0119	.0114	.0110	.0106	.0101	.0097	.0092
96	.0177	.0173	.0168	.0164	.0159	.0155	.0150	.0146	.0141	.0137
95	.0223	.0218	.0214	.0209	.0205	.0200	.0195	.0191	.0186	.0182
94	.0269	.0264	.0259	.0255	.0250	.0246	.0241	.0237	.0232	.0227
93	.0315	.0311	.0306	.0301	.0297	.0292	.0287	.0283	.0278	.0273
92	.0362	.0357	.0353	.0348	.0343	.0339	.0334	.0329	.0325	.0320
91	.0410	.0405	.0400	.0395	.0391	.0386	.0381	.0376	.0372	.0367
90	.0458	.0453	.0448	.0443	.0438	.0434	.0429	.0424	.0419	.0414
89	.0506	.0501	.0496	.0491	.0487	.0482	.0477	.0472	.0467	.0462
88	.0555	.0550	.0545	.0540	.0535	.0530	.0525	.0520	.0515	.0510
87	.0605	.0599	.0595	.0589	.0585	.0580	.0575	.0570	.0565	.0560
86	.0655	.0650	.0645	.0640	.0635	.0630	.0625	.0620	.0615	.0610
85	.0706	.0701	.0696	.0691	.0685	.0680	.0675	.0670	.0665	.0660
84	.0757	.0752	.0747	.0742	.0737	.0731	.0726	.0721	.0716	.0711
83	.0809	.0804	.0799	.0794	.0788	.0783	.0778	.0773	.0768	.0762
82	.0862	.0857	.0851	.0846	.0841	.0835	.0830	.0825	.0820	.0814
81	.0915	.0910	.0904	.0899	.0894	.0888	.0883	.0878	.0872	.0867
80	.0969	.0964	.0958	.0953	.0947	.0942	.0937	.0931	.0926	.0921
79	.1024	.1018	.1013	.1007	.1002	.0986	.0991	.0985	.0980	.0975
78	.1079	.1073	.1068	.1062	.1057	.1051	.1045	.1040	.1035	.1029
77	.1135	.1129	.1124	.1118	.1113	.1107	.1101	.1096	.1090	.1085
76	.1192	.1186	.1180	.1175	.1169	.1163	.1158	.1152	.1146	.1141
75	.1249	.1244	.1238	.1232	.1226	.1221	.1215	.1209	.1203	.1198
74	.1308	.1302	.1296	.1290	.1284	.1278	.1273	.1267	.1261	.1255
73	.1367	.1361	.1355	.1349	.1343	.1337	.1331	.1325	.1319	.1314
72	.1427	.1421	.1415	.1409	.1403	.1397	.1391	.1385	.1379	.1373
71	.1487	.1481	.1475	.1469	.1463	.1457	.1451	.1445	.1439	.1433
70	.1549	.1543	.1537	.1530	.1524	.1518	.1512	.1506	.1500	.1494
69	.1612	.1605	.1599	.1593	.1586	.1580	.1574	.1568	.1561	.1555
68	.1675	.1669	.1662	.1656	.1649	.1643	.1637	.1630	.1624	.1618
67	.1739	.1733	.1726	.1720	.1713	.1707	.1701	.1694	.1688	.1681
66	.1805	.1798	.1791	.1785	.1778	.1772	.1765	.1759	.1752	.1746
65	.1871	.1861	.1858	.1851	.1844	.1838	.1831	.1824	.1818	.1811
64	.1938	.1931	.1925	.1918	.1911	.1904	.1898	.1891	.1884	.1878
63	.2007	.2000	.1993	.1986	.1979	.1972	.1965	.1959	.1952	.1945
62	.2076	.2069	.2062	.2055	.2048	.2041	.2034	.2027	.2020	.2013
61	.2147	.2140	.2132	.2125	.2118	.2111	.2104	.2097	.2090	.2083
60	.2218	.2211	.2204	.2197	.2190	.2182	.2175	.2168	.2161	.2153
59	.2291	.2284	.2277	.2269	.2262	.2255	.2248	.2240	.2233	.2226
58	.2366	.2358	.2351	.2343	.2336	.2328	.2321	.2314	.2306	.2292
57	.2441	.2434	.2426	.2418	.2411	.2403	.2396	.2388	.2381	.2373
56	.2518	.2510	.2503	.2495	.2487	.2420	.2472	.2464	.2457	.2449
55	.2596	.2588	.2581	.2573	.2565	.2557	.2549	.2541	.2534	.2526
54	.2676	.2668	.2660	.2652	.2644	.2636	.2628	.2620	.2612	.2604

529

APPENDIX

TRANSMITTANCE-ABSORBANCE CONVERSION TABLE *(Continued)*

% Trans- mit- tance	0.0	0.1	0.2	0.3	0.4	0.5	0.6	0.7	0.8	0.9
53	.2757	.2749	.2741	.2733	.2725	.2716	.2708	.2700	.2692	.2684
52	.2840	.2832	.2823	.2815	.2807	.2798	.2790	.2782	.2774	.2765
51	.2924	.2916	.2907	.2899	.2890	.2882	.2874	.2865	.2857	.2848
50	.3010	.3002	.2993	.2984	.2976	.2967	.2958	.2950	.2941	.2933
49	.3098	.3089	.3080	.3072	.3063	.3054	.3045	.3036	.3028	.3019
48	.3188	.3179	.3170	.3161	.3152	.3143	.3134	.3125	.3116	.3107
47	.3279	.3270	.3261	.3251	.3242	.3233	.3224	.3215	.3206	.3197
46	.3372	.3363	.3354	.3344	.3335	.3325	.3316	.3307	.3298	.3288
45	.3468	.3458	.3449	.3439	.3429	.3420	.3410	.3401	.3391	.3382
44	.3565	.3556	.3546	.3536	.3526	.3516	.3507	.3497	.3487	.3478
43	.3665	.3655	.3645	.3635	.3625	.3615	.3605	.3595	.3585	.3575
42	.3768	.3757	.3747	.3737	.3726	.3716	.3706	.3696	.3686	.3675
41	.3872	.3862	.3851	.3840	.3830	.3820	.3809	.3799	.3788	.3778
40	.3979	.3969	.3958	.3947	.3936	.3925	.3915	.3904	.3893	.3883
39	.4089	.4078	.4067	.4056	.4045	.4034	.4023	.4012	.4001	.3989
38	.4202	.4191	.4179	.4168	.4157	.4154	.4134	.4123	.4112	.4101
37	.4318	.4306	.4295	.4283	.4271	.4260	.4248	.4237	.4225	.4214
36	.4437	.4425	.4413	.4401	.4389	.4377	.4365	.4353	.4342	.4330
35	.4559	.4547	.4535	.4522	.4510	.4498	.4486	.4473	.4461	.4449
34	.4685	.4672	.4660	.4647	.4634	.4622	.4609	.4597	.4584	.4572
33	.4815	.4802	.4789	.4776	.4763	.4750	.4737	.4724	.4711	.4698
32	.4949	.4935	.4921	.4908	.4895	.4881	.4868	.4854	.4841	.4828
31	.5086	.5072	.5058	.5045	.5031	.5017	.5003	.4989	.4976	.4962
30	.5229	.5214	.5200	.5186	.5171	.5157	.5143	.5129	.5114	.5100
29	.5376	.5361	.5346	.5331	.5317	.5302	.5287	.5272	.5258	.5243
28	.5528	.5513	.5498	.5482	.5467	.5452	.5436	.5421	.5406	.5391
27	.5686	.5670	.5654	.5638	.5622	.5607	.5581	.5575	.5560	.5544
26	.5850	.5834	.5817	.5800	.5784	.5766	.5751	.5735	.5719	.5702
25	.6021	.6003	.5986	.5969	.5952	.5935	.5918	.5901	.5884	.5867
24	.6198	.6180	.6162	.6144	.6126	.6108	.6091	.6073	.6055	.6038
23	.6383	.6364	.6345	.6326	.6308	.6289	.6271	.6253	.6234	.6216
22	.6576	.6556	.6536	.6517	.6498	.6478	.6459	.6440	.6421	.6402
21	.6778	.6757	.6737	.6716	.6696	.6676	.6655	.6635	.6615	.6596
20	.6990	.6968	.6946	.6925	.6904	.6882	.6861	.6840	.6819	.6799
19	.7212	.7190	.7167	.7144	.7122	.7100	.7077	.7055	.7033	.7011
18	.7447	.7423	.7399	.7375	.7352	.7328	.7305	.7282	.7258	.7235
17	.7696	.7669	.7645	.7620	.7595	.7570	.7545	.7520	.7496	.7471
16	.7959	.7932	.7905	.7878	.7852	.7825	.7799	.7773	.7747	.7721
15	.8239	.8209	.8182	.8153	.8125	.8097	.8069	.8041	.8013	.7986
14	.8539	.8508	.8477	.8447	.8416	.8386	.8356	.8327	.8297	.8268
13	.8861	.8827	.8794	.8761	.8729	.8697	.8665	.8633	.8601	.8570
12	.9208	.9172	.9136	.9101	.9066	.9031	.8996	.8962	.8928	.8894
11	.9586	.9547	.9508	.9469	.9431	.9393	.9355	.9318	.9281	.9245
10	1.0000	.9947	.9914	.9872	.9830	.9788	.9747	.9706	.9666	.9626
9	1.046	1.041	1.036	1.032	1.027	1.022	1.017	1.013	1.009	1.004
8	1.097	1.092	1.086	1.081	1.076	1.071	1.066	1.061	1.056	1.051
7	1.155	1.149	1.143	1.137	1.131	1.125	1.119	1.114	1.108	1.102
6	1.222	1.215	1.208	1.201	1.194	1.187	1.180	1.174	1.167	1.161
5	1.301	1.292	1.284	1.276	1.268	1.260	1.252	1.244	1.237	1.229
4	1.398	1.387	1.377	1.366	1.356	1.346	1.337	1.328	1.319	1.309
3	1.523	1.509	1.495	1.481	1.468	1.455	1.442	1.432	1.420	1.409
2	1.699	1.678	1.658	1.638	1.619	1.602	1.585	1.569	1.553	1.538
1	2.000	1.959	1.921	1.886	1.854	1.824	1.796	1.769	1.744	1.721

INDEX

531